APPLIED BIOREMEDIATION - ACTIVE AND PASSIVE APPROACHES

Edited by **Yogesh B. Patil** and **Prakash Rao**

Applied Bioremediation - Active and Passive Approaches

http://dx.doi.org/10.5772/50859
Edited by Yogesh B. Patil and Prakash Rao

Contributors

Ranjith Kumavath, Pratap Devarapalli, Nivia Coelho, Naofumi Shiomi, Antonio Bevilacqua, Leonardo Petruzzi, Maria Rosaria Corbo, Milena Sinigaglia, Ismael Acosta, Aare Selberg, Bogdana Koumanova, Silviya Lavrova, André LM Porto, Natália Alvarenga Da Silva, Mirna Helena R. Seleghim, Jittima Charoenpanich, Shivesh Sharma, Phalazane Mtimunye, Carlos Eduardo Hernández Luna, Tomaz Langenbach, Blanka Vrchotova, Yogesh Bhagwan Patil

Published by InTech

Janeza Trdine 9, 51000 Rijeka, Croatia

Publishing Process Manager Sandra Bakic
Typesetting InTech DTP Team
Cover InTech Design Team

Additional hard copies can be obtained from orders@intechopen.com

Applied Bioremediation - Active and Passive Approaches, Edited by Yogesh B. Patil and Prakash Rao
p. cm.
ISBN 978-953-51-1200-6

For EU product safety concerns:
IN TECH d.o.o., Prolaz Marije Krucifikse Kozulić 3, 51000 Rijeka, Croatia,
info@intechopen.com or visit our website at intechopen.com.

Meet the editors

Dr. Yogesh Patil earned his MSc and PhD in Environmental Sciences from Pune University and currently holds the position of Associate Professor and Head - Research & Publications with Symbiosis Institute of Research and Innovation (SIRI), Symbiosis International University (SIU), Pune, India. He is teaching to MBA students in the broad areas of Energy and Environmental Management. His research focuses on developing low-cost integrated strategies/technologies for urban and industrial waste management. He has successfully completed six research projects funded by national and international agencies like UGC, IFS and World Bank. He is author of over 20 journal articles, book chapters and presented over 30 conference papers. Dr. Patil is a recipient of several honours and awards for his research.

Dr. Prakash Rao has 30 years of experience in the field of energy and environmental management with interests in environment, climate change and energy sectors. He holds a PhD in Environmental Biology from the University of Bombay and has coordinated several multidisciplinary research projects ranging from natural resources to climate change and energy. He has led the Climate Change and Energy Programme at World Wide Fund for Nature-India for nearly ten years, coordinating its global research, energy policies and community action. He has published more than 20 research papers in reputed journals and published two books including one with Prof. S.K. Dash of Indian Institute of Technology, Delhi on Assessment of Climate Change in India and mitigation policies.

Contents

Preface XI

Chapter 1 **Bioremediation of Chlorobenzoic Acids 1**
Blanka Vrchotová, Martina Macková, Tomáš Macek and Kateřina Demnerová

Chapter 2 **Development of a Bioremediation Technology for the Removal of Thiocyanate from Aqueous Industrial Wastes Using Metabolically Active Microorganisms 31**
Yogesh B. Patil

Chapter 3 **Bioremediation of Olive Mill Wastewater by Yeasts – A Review of the Criteria for the Selection of Promising Strains 51**
Antonio Bevilacqua, Leonardo Petruzzi, Maria Rosaria Corbo and Milena Sinigaglia

Chapter 4 **Nutrients and Organic Matter Removal in a Vertical-Flow Constructed Wetland 67**
Silviya Lavrova and Bogdana Koumanova

Chapter 5 **Removal of Acrylamide by Microorganisms 99**
Jittima Charoenpanich

Chapter 6 **Biodegradation and Leaching of Surfactants During Surfactant-Amended Bioremediation of Oil-Polluted Soil 121**
Aare Selberg, Kätlin Juuram, Jana Budashova and Toomas Tenno

Chapter 7 **Biodegradation of the Organophosphate Pesticide Profenofos by Marine Fungi 147**
Natália Alvarenga da Silva, Willian Garcia Birolli, Mirna Helena Regali Seleghim and André Luiz Meleiro Porto

Chapter 8 **Bioremediation of Radiotoxic Elements under Natural Environmental Conditions 179**
Phalazane Johanna Mtimunye and Evans M. N. Chirwa

Chapter 9 **Removal of Hexavalent Chromium from Solutions and Contaminated Sites by Different Natural Biomasses 207**
Ismael Acosta-Rodríguez, Juan F. Cárdenas-González, María de Guadalupe Moctezuma-Zárate and Víctor M. Martínez-Juárez

Chapter 10 **Bioremediation of Waters Contaminated with Heavy Metals Using Moringa oleifera Seeds as Biosorbent 225**
Cleide S. T. Araújo, Dayene C. Carvalho, Helen C. Rezende, Ione L. S. Almeida, Luciana M. Coelho, Nívia M. M. Coelho, Thiago L. Marques and Vanessa N. Alves

Chapter 11 **Resource Recovery from Industrial Effluents Containing Precious Metal Species Using Low-Cost Biomaterials — An Approach of Passive Bioremediation and Its Newer Applications 255**
Nilisha Itankar, Viraja Bhat, Jayati Chourey, Ketaki Barve, Shilpa Kulkarni, Prakash Rao and Yogesh Patil

Chapter 12 **A Novel Bioremediation Method for Shallow Layers of Soil Polluted by Pesticides 285**
Naofumi Shiomi

Chapter 13 **Persistence and Bioaccumulation of Persistent Organic Pollutants (POPs) 305**
Tomaz Langenbach

Chapter 14 **Rhizoremediation: A Promising Rhizosphere Technology 331**
Keshav Prasad Shukla, Shivesh Sharma, Nand Kumar Singh, Vasudha Singh, Sandeep Bisht and Vivek Kumar

Chapter 15 **Purification and Partial Characterization of a Thermostable Laccase from Pycnoporus sanguineus CS-2 with Ability to Oxidize High Redox Potential Substrates and Recalcitrant Dyes 351**
Sergio M. Salcedo Martínez, Guadalupe Gutiérrez-Soto, Carlos F. Rodríguez Garza, Tania J. Villarreal Galván, Juan F. Contreras Cordero and Carlos E. Hernández Luna

Chapter 16 **Scientific Swift in Bioremediation: An Overview** **375**
Ranjith N. Kumavath and Pratap Devarapalli

Preface

With the advancement of science and technology, wide range of toxic and hazardous (T&H) chemicals are being synthesized and produced by man on grand scale for its use in agricultural, commercial, household and industrial systems. Consequently, these entities emanate large volume of wastes containing T&H chemicals ranging from heavy metals to hydrocarbons, pesticides, phenols, PAHs, cyanides, PCBs, etc. Most of these chemicals are recalcitrant and xenobiotic in nature. If managed inappropriately in terms of handling, storage, transport, treatment and disposal, T&H wastes can get mobilized and carried into the food web as a result of leaching from waste dumps, contaminated soils and waters. This might led to deleterious irreversible, incapacitating and intangible impact on the overall human health, environment and ecology. This has originated to challenging problems confronting the present day so called 'technological society'. Since waste is inevitable, it has to be managed in precise manner before it is discharged or recycled to safeguard the environment. Conventionally, numerous physical-chemical processes are being adopted for the treatment industrial effluents and contaminated soil/land systems. Although efficient these methods are beset with several problems like high capital investment, high operational cost, reduced efficiency of treatment in the presence of complex chemical matrix, high sludge production, handling, processing and disposal, requirement of special equipment's, need of human skills, risk to human health and highly energy intensive. Such processes are therefore always on the back foot as far its use in most of the Asian countries is concerned. Thus, there is a big technological breach, which needs to be bridged immediately. Furthermore, environmental regulations in most countries demand strict actions against haphazard waste disposal.

Biological treatment technologies (bioremediation), in the recent times, are gaining immense credibility in the field of waste management. It is known that microorganisms (both prokaryotes and eukaryotes) in nature have immense potential to interact, utilize, degrade and detoxify range of pollutants/substances and therefore being explored by the researchers worldwide. They offer several advantages over conventional methods in terms of cost effectiveness, efficiency, low sludge production and also provides eco-compatible means of treating industrial effluents and reclaiming land. Other than microorganisms, plants and waste biomass from different sources also play crucial role in the management of waste. Biomass of all types are known for their capability of interacting and confronting with pollutants in both active (live) and passive (dead) way; thereby offering numerous opportunities of exploring them for environmental clean-up. Biomass, whether dead or alive, differ in their intrinsic capabilities and the mechanism of pollutant removal. They can degrade and remove variety of organic pollutants from waste by utilizing it as a suitable growth substrate. Biomass interaction with inorganics (especially heavy metals) can be based on the localization site of pollutant such as extracellular, exocellular and intracellular. Biomass, especially microorganisms are capable of mobilizing, precipitating, reducing, transforming, accumulating, coordinating, exchanging and adsorbing the inorganic pollutants and form complexes depending upon the nature of pollutant. Biomass surfaces are also usually charged. Functional groups like carboxyl, hydroxyl, phosphoryl and sulphahydryl, membrane proteins, lipids and other cell wall components are responsible for adsorption of diverse contaminants. At times, biodegradation using live microorganisms/biomass are subject to toxicity of T&H pollutants, which is completely dependent on the nature and concentration of pollutant in the given system. In such cases, employing passive biomass for the removal of contaminants from water and soil environment may be of choice.

In the 21st century, the entire world is witnessing a paradigm shift in the overall waste management practices, which is rapidly changing its face and orientation. Waste is no longer considered as waste but is recognized as a valuable 'Resource'. This lost resource could potentially be recovered from the wastes using suitable strategies using bioremediation technologies. One such novel strategy could be the use of combined active-passive biomass for the development of integrated bioremediation technology, which in the editor's and co-editor's opinion is desperate need of the hour. Application of such concepts in the modern day bioremediation processes can certainly lead to overall resource conservation and sustainable economic development. In view of this, the present edited book on bioremediation will certainly add to the advancement of knowledge in the broad field of sustainable development and in particular in the area of industrial pollution management and land/soil reclamation. This will further help the profitability of business community at large. Moreover, this book will also provide the required valuable resource and stimulus to the researchers worldwide. Bioremediation technologies require interdisciplinary knowledge of science and management, involving microbiology, chemistry, hydrogeology, engineering, plant sciences, geology and ecology, economics, operations management, etc. In the past biological remediation technologies have been successfully used to treat polluted soils, oily sludge's, cyanide, heavy metals, nitriles, groundwater contaminated by petroleum hydrocarbons, solvents, pesticides and other chemicals and have been implemented at a relatively low cost.

This edited book on "Applied Bioremediation - Active and Passive Approaches" consists of diverse mix of interesting chapters that focusses on use of active as well as passive biomass for the development of bioremediation technologies for the management of industrial effluents and contaminated land. Topics in the book include - bioremediation of chlorobenzoic acids, resource recovery from industrial waste by passive bioremediation approach, bioremediation of olive mill wastewater, nutrients and organic matter removal by constructed wetlands, bioremediation of oil polluted site, hexavalent chromium removal by natural biomass, bioremediation of radiotoxic elements, rhizoremediation, biodegradation of profenofos, removal of acrylamide by microorganisms, bioremediation of thiocyanate, purification and characterization of a thermostable enzymes, persistent organics, scientific swift in bioremediation, in situ bioaugmentation process, and biosorbents for heavy metal removal from wastes.

Research scientists and experts in the area of chemistry, biotechnology, bioremediation, environmental microbiology, energy and environmental management from diverse institutions and universities world across have contributed to this book. This book on "Applied Bioremediation - Active and Passive Approaches" should prove to be useful to the graduate and postgraduate students, research scholars, professors, scientists and bioremediation professionals in the diverse areas of life sciences like biotechnology, microbiology, biochemistry, molecular biology, and soil and environmental sciences and management. We gratefully acknowledge the cooperation and support of all the contributing authors as well as the officials of InTech Publishing Group for encouragement and help throughout the preparation of this volume.

Dr Yogesh Patil, Editor
Dr Prakash Rao, Co-Editor
Symbiosis International University (SIU), Pune, India

Chapter 1

Bioremediation of Chlorobenzoic Acids

Blanka Vrchotová, Martina Macková,
Tomáš Macek and Kateřina Demnerová

Additional information is available at the end of the chapter

http://dx.doi.org/10.5772/56394

1. Introduction

Chlorobenzoic acids (CBAs) can be released into the environment from many different sources. One possible source of CBAs is usage as herbicides or insecticides in agriculture. As a herbicide was used 2,3,6-CBA [1]. CBAs may also be formed as intermediates during the degradation of same herbicides. Namely 2,6-CBA is formed as an intermediate during microbial degradation of dichlobenil [2], 2,5-CBA in the chlorambene degradation [3] or 4-CBA is a final degradation product of the insecticide DDT [4].

Another large group of substances from whose metabolism in living organisms including mammals arise CBAs, are drugs such as indomethacin, bupropion or benzafibrate. Indomethacin is an anti-inflammatory drug used to reduce fever, pain, stiffness and swelling by inhibition of the production of prostaglandins, compounds that cause these problems. From the decomposition of indomethacin arises 5-methoxy-2-methylindoleacetic acid and the same quantity of 4-CBA [5]. In the case of antidepressant bupropion just small amount (0.3%) of 3-CBA is formed next to threohydrobupropion, erythrohydrobupropion and hydroxybupropion [6, 7]. Decomposition of bezafibrate, an anti-obesity drug, leads to the formation of equal quantity of 4-CBA and 4-(2-aminoethyl)-α,α dimethyl-benzeneacetic acid. [8].

Well known is the formation of CBAs during degradation of polychlorinated biphenyls (PCBs). PCBs are degraded by bacteria by the so called upper degradation pathway when CBAs are formed as final degradation products (Figure 1.) [10-12]. CBAs are also formed during degradation of PCBs by white rot fungi [13] [14]. The result of these metabolic pathways is a mixture of CBAs with different position and number of chlorine atoms on benzene ring in dependence of chlorination of the degraded PCB congeners [9].

The accumulation of this way formed CBAs in waste water or in soil can lead to the deceleration or inhibition of degradation of substances of which the CBAs are degradation products [15,

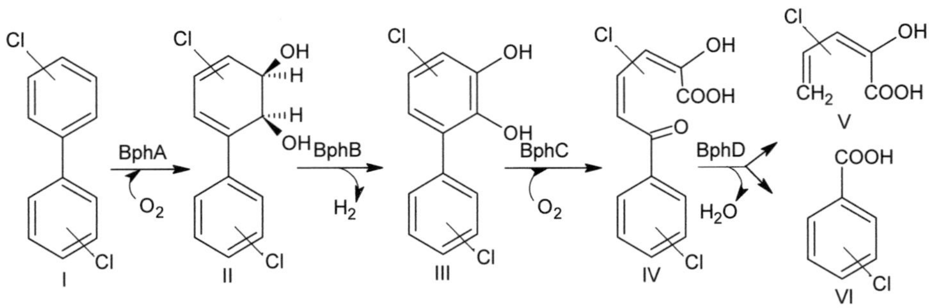

Figure 1. Degradation pathway of polychlorinated biphenyls by aerobic organisms [9]. BphA – biphenyl-2,3-dioxygenase; BphB – biphenyl-dihydrodiol-dehydrogenase; BphC – 2,3-dihydroxy-biphenyl-1,2-dioxygenase; BphD – 2-hydroxychlor-6-oxo-6-phenylhexa-2,4-dienoathydrolase; I – polychlorinated biphenyl; II – dihydrodiol; III – 2,3-dihydroxychlor biphenyl; IV – 2-hydroxychlor-6-oxo-6-phenylhexa-2,4-dienoic acid; V – chlorobenzoic acid; VI – 2-hydroxychlor-2,4-dienoate

16]. Therefore, it is important to understand bioremediation mechanisms of CBAs and to know the impact of other organisms, xenobiotics composition and the influence of other CBA isomers present in contaminated area that affect this process.

2. Bacterial degradation of chlorobenzoic acids

Bacterial degradation of CBAs can be done under aerobic or anaerobic conditions. Under anaerobic conditions is dechlorination the first step of degradation followed by degradation of the aromatic ring [17, 18].

The strategy of CBAs degradation under aerobic conditions depends on the position of chlorine atom or atoms on the aromatic ring. Crucial step is the dechlorination. Dechlorination step can be before or after degradation of aromatic ring. This depends on the structure of CBA as well as on the enzymatic equipment of the bacteria.

2.1. Anaerobic degradation of chlorobenzoic acids

Initial step of microbial anaerobic degradation of CBAs is done by process named reductive dehalogenation. During the dehalogenation are from the more chlorinated CBA by dehydrogenation process formed less chlorinated CBAs or benzoic acid and chlorine is released as hydrochloric acid [19]. Time scale of this process depends on the number and position of the chlorine atoms in the molecule of CBA and on composition of anaerobic consortium or on strain with dehalogenation activity. Dehalogenation of CBAs can be performed under methanogenic [20], denitrification [21] or sulfate reducing conditions [22]. Also phototrophic bacteria capable of reductive dehalogenation of CBAs are known, e. g. *Rhodopseudomonas palustris* DCP3 [23], strain able to degrade under anaerobic conditions 2-CBA; 3-CBA; 4-CBA and 3,4-CBA. This strain is also capable of anaerobic degration of benzoic acid.

The next step after reductive dehalogenation is the degradation of benzoic acid. Benzoic acid can be anaerobically degraded by two different pathways [24]. The first one is initiated by reduction of benzoic acid aromatic ring. Subsequent ring cleavage and degradation of intermediates occurs by reactions similar to β-oxidation of aliphatic carboxylic acids [25].

In the second pathway is the molecule of benzoic acid activated by binding to the CoA. CoA remains bound to the molecule during all degradation steps. In this pathway is benzoate-CoA degraded to the acetyl-CoA [26]. Acetyl-CoA is then degraded in tricarboxylic acid cycle to the CO_2 and water.

From previous it can be concluded that in anaerobic degradation is the molecule of CBA degraded to methane, CO_2 and water with benzoic acid as intermediate (Figure 2.) [17].

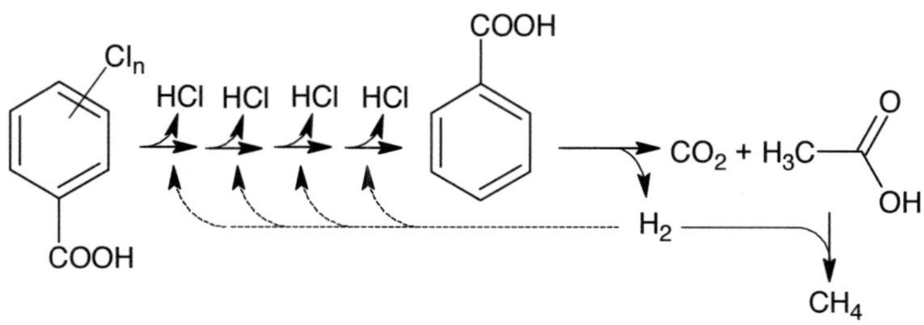

Figure 2. Anaerobic degradation of chlorobenzoic acid adapted from [17].

Anaerobic degradation of CBAs can be done by pure cultures of microorganisms as well as by consortia of anaerobic microorganisms. The first pure strain that has been observed to be able to anaerobically degrade CBA was strain *Desulfomonile tiedjei* [27]. This strain is capable of 3-CBA dehalogenation, by this process reaching enough ATP for its growth. So it does not need additional source of energy. Another strain able to anaerobic degradation of CBA is strain *Desulfomicrobium escambiense* [22] and the aforementioned strain *Rhodopseudomonas palustris* DCP3 [23].

2.2. Aerobic degradation of chlorobenzoic acids

Aerobic microorganisms evolved many different degradation strategies for CBAs degradation. Microbial aerobic degradation of CBAs depends on bacterial strain as well as on the structure and chlorination of CBA.

CBAs can be degraded via chlorocatechole (*clc* degradation genes) [28, 29] or by hydrolytic dehalogenation with hydroxybenzoic acid as an intermediate like in 4-CBA degradation (*fcb* degradation genes) [30-32] or by 4,5-dioxygenation like in 3-CBA and 3,4-CBA degradation where is 5-chloroprotocatechuic acid formed (*cba* degradation genes) [33, 34] or by 1,2-dioxygenase reaction [9, 35] or by degradation involving the formation of gentisate as an intermediate of CBA degradation [36, 37].

Strains able to aerobically degrade CBAs belong to the gram negative strains as *Rhodococcus* or *Bacillus* as well as gram positive strains like *Pseudomonas, Burkholderia* or *Achromobacter* [9].

2.2.1. Degradation of 2-chlorobenzoic acid

In case of 2-CBA degradation there are known three different ways of degradation. All of them are about dioxygenation reaction catalyzed by 2-halobenzoate-1,2-dioxygenase (EC 1.14.12.13) (Figure 3.).

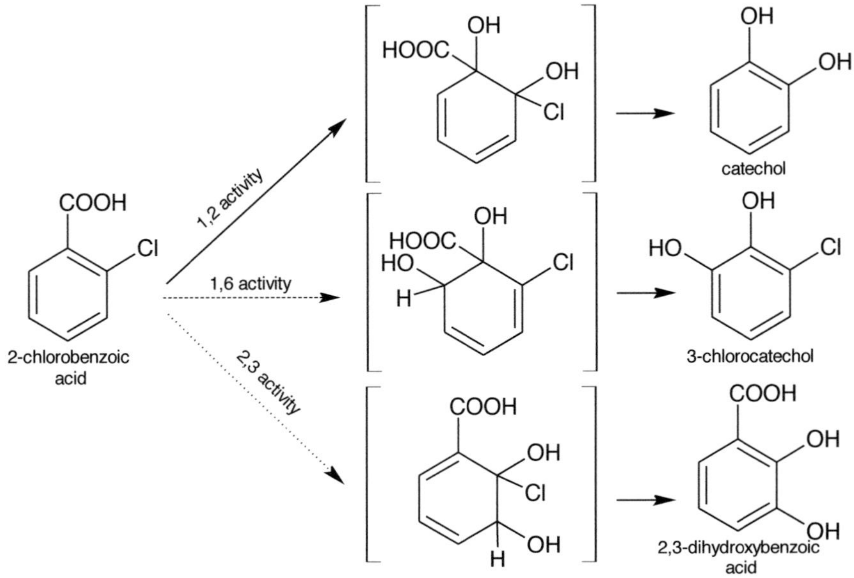

Figure 3. Aerobic degradation of 2-chlorobenzoic acid by the enzyme 2-chlorobenzoate-1,2-dioxygenase [9, 38].

Main activity of 2-halobenzoate-1,2-dioxygenase is 1,2-dioxygenase reaction. In this reaction is 2-CBA degraded to catechol. During this reaction are released carbon dioxide and chlorine. Enzyme 2-halobenzoate-1,2-dioxygenase has also 1,6-dioxygenase activity. In 1,6-dioxygenase reaction is formed 3-chlorocatechol. Last activity of 2-halobenzoate-1,2-dioxygenase is 2,3-dioxygenase reaction leading to the formation of 2,3-dihydroxybenzoic acid. Latter two reactions proceed only in a small degree [9].

Degradation of 2-CBA by 2-halobenzoate-1,2-dioxygenase was reported in two groups of microorganisms. First group degraded only 2-CBA and second next to 2-CBA also 2,3-CBA and 2,5-CBA. Both groups use for degradation 1,2-dioxygenase reaction. In case of 2-CBA this reaction leads to the formation of catechol (Figure 3.). 2,4-CBA and 2,5-CBA are degraded into 4-chlorocatechol (Figure 4.) [35].

For the 4-chlorocatechol degradation formed in 1,2-dioxygenase reaction with 2,4-CBA or 2,5-CBA it is necessary to have chlorocatechol degradation pathway. It is therefore assumed that the difference between both groups of microorganisms is given by presence of the chlorocatechol degradation pathway [9] despite the fact that two different 2-halobenzoate-1,2-dioxygenases are known.

http://dx.doi.org/10.5772/56394

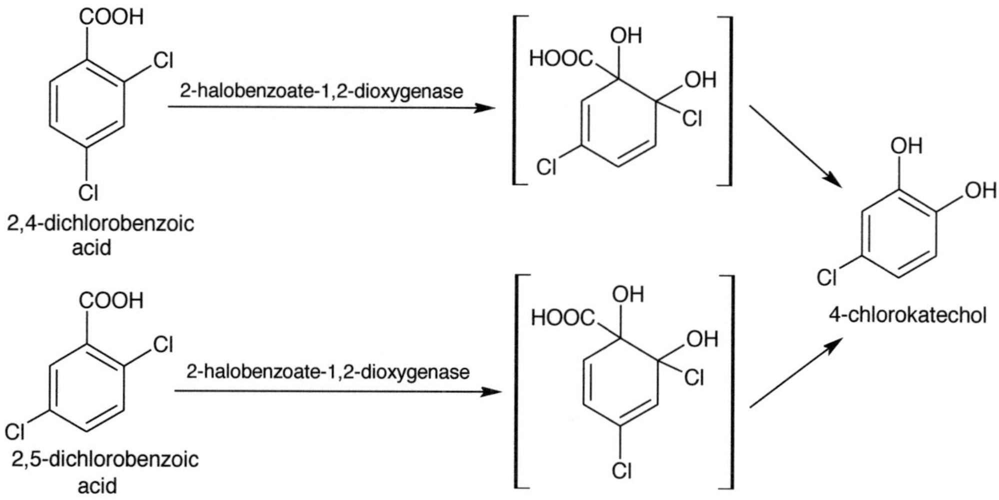

Figure 4. Degradation of 2,4-dichlorobenzoic and 2,5-dichlorobenzoic acid catalyzed by 2-halobenzoate-1,2-dioxygenase.

First 2-halobenzoate-1,2-dioxygenase is a two componential enzymatic system [39]. This 2-halobenzoate-1,2-dioxygenase has high homology with the toluate and benzoate-1,2-dioxygenase. Two component 2-halobenzoate-1,2-dioxygenase has high affinity to 2-CBA but low to 4-CBA; 2,4-CBA and 2,5-CBA[9, 40].

Second 2-halobenzoate-1,2-dioxygenase is three componential [35] and this enzyme catalyzes degradation of 2-CBA; 2,4-CBA and 2,5-CBA.

Next to 1,2-dioxygenase activity the enzyme 2-halobenzoate-1,2-dioxygenase has also 1,6-dioxygenase activity [9, 41]. Strain *Pseudomonas* sp. B-300 was in presence of glucose able to degrade 2-CBA to the 3-chlorocatechol [41].Whereas *Pseudomonas* sp. B-300 was cultivated with 2-CBA as the only carbon and energy source just catechol was identified. From that it is obvious that 1,6-dioxygenase reaction occurs only in much smaller degree than 1,2-dioxygenase reaction or in the presence of a rich source of energy.

Dioxygenase reaction in the 1,6- position is followed by 3-chlorocatechol ring cleavage forming chloromuconic acid which is dehalogenated.

Another metabolite identified in medium after cultivation of strain *Burkholderia cepacia* 2CBS with 2-CBA was 2,3-dihydroxybenzoic acid [42]. Strain *Burkholderia cepacia* 2CBS degraded most of 2-CBA by 1,2-dioxygenase reaction, 2,3-dihydroxybenzoic acid was in medium accumulated just in small quantities. Formation of 2,3-dihydroxybenzoic acid indicates that two-component 2-halobenzoate-1,2-dioxygenase of strain *Burkholderia cepacia* 2CBS has next to 1,2-dioxygenase activity also 2,3-dioxygenase activity (Figure 3.). 2,3-dihydroxybenzoic acid is the dead-end product [9].

2.2.2. Degradation of 3-chlorobenzoic acid

Degradation of 3-CBA can be done by several different degradation pathways. Benzoate-1,2-dioxygenase catalyzes the conversion of 3-CBA into 3-chlorocatechol or 4-chlorocatechol. 3-CBA can be also transformed to the protocatechuate (3,4-dihydroxybenzoic acid) or 5-chloroprotocatechuate (5-chloro-3,4-dihydroxybenzoic acid) by the enzyme 3-chlorobenzoate-4,5-dioxygenase. Another possibility is the degradation of 3-CBA via 3-hydroxybenzoic acid to the gentisic acid (2,5-dihydroxybenzoic acid).

First mentioned 3-CBA degradation is degradation in the benzoate degradation pathway. Breakdown of 3-CBA is catalyzed by the enzyme benzoate-1,2-dioxygenase (EC 1.14.12.10). Benzoate-1,2-dioxygenase is a wide spread enzyme which has been identified in many microorganisms. Substrate specificity of this enzyme is relatively narrow. Benzoate-1,2-dioxygenasa catalyzes only conversion of benzoate, 3-CBA and 3-methylbenzoate [9, 43, 44].

The mechanism of reaction catalyzed by the benzoate-1,2-dioxygenase is based on the double hydroxylation in 1,2- or 1,6-position of benzene ring. The final product of this reaction is catechol in case of benzoic acid or 4-chlorocatechol or 3-chlorocatechol in case of 3-CBA degradation (Figure 5.).

Benzoate-1,2-dioxygenase of the strain *Alcaligenes eutrophus* JMP134 [43] form from 3-CBA 3-chlorocatechol and 4-chlorocatechol in a 1:2 ratio, as well as the benzoate-1,2-dioxygenase of strain *Pseudomonas* sp. B13 [44] do.

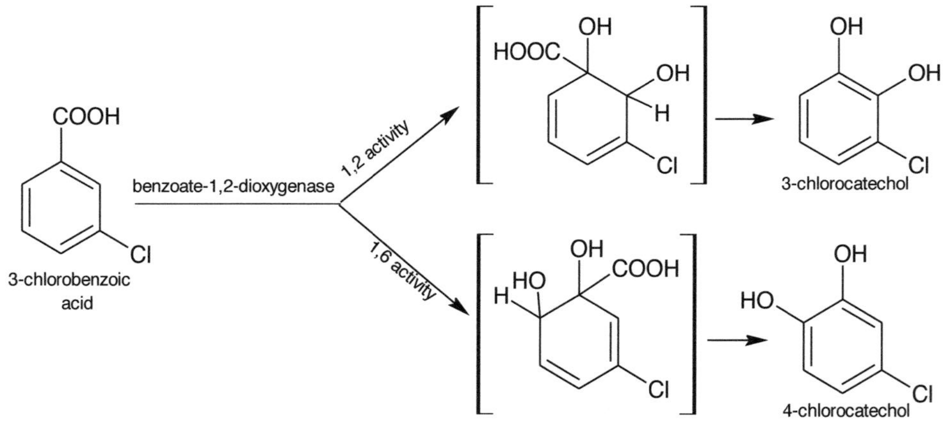

Figure 5. Degradation pathway for 3-chlorobenzoic acid by benzoate-1,2-dioxygenase [9] with indicated 1,2-dioxygenase and 1,6-dioxygenase activity.

Next to benzoate-1,2-dioxygenase 3-CBA can be degraded also by the 3-chlorobenzoate-3,4-dioxygenase [45-47]. Main activity of this enzyme is 4,5-dioxygenase reaction with formation of 5-chloroprotocatechuic acid. Besides the 4,5-dioxygenase reaction is a small amount of 3-CBA transformed by 3,4-dioxygenase reaction to the protocatechuic acid (Figure 6.).

The enzyme 3-chlorobenzoate-3,4-dioxygenase can catalyze also degradation of 3,4-CBA. 3,4-CBA is degraded to the 5-chloroprotocatechuic acid [34, 48] which means that 3,4-CBA is degraded by the 4,5-dioxygenase reaction.

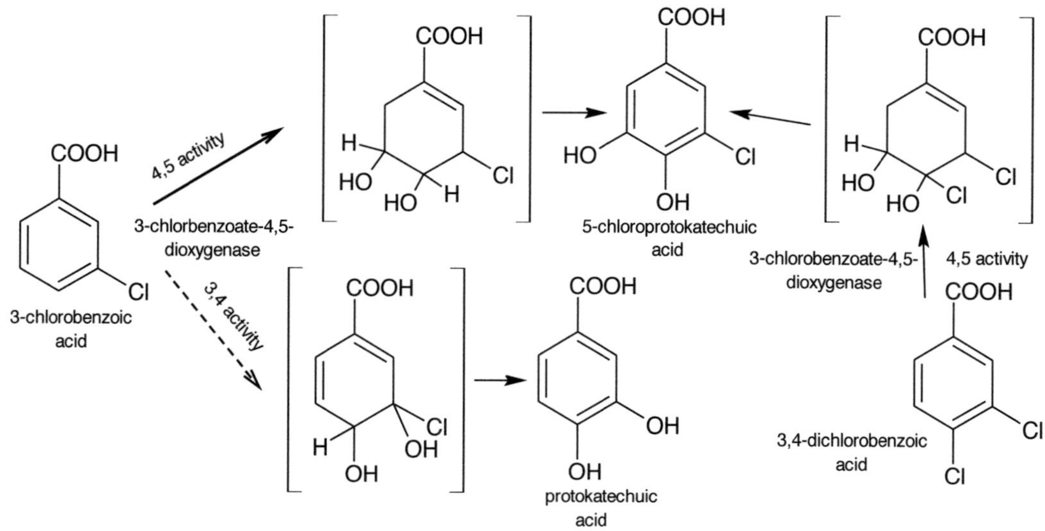

Figure 6. Degradation pathway of 3-chlorobenzoic and 3,4-dichlorobenzoic acid by 3-chlorobenzoate-3,4-dioxygenase.

3-CBA can be also degraded by monooxygenase reaction. This reaction leads to the gentisic acid with 3-hydroxybenzoic acid as an intermediate (Figure 7.) [37]. The enzyme for the conversion of 3-CBA to the 3-hydroxybenzoic acid is not yet known, the second reaction is catalyzed by 3-hydroxybenzoate-6-hydroxylase (EC 1.14.13.24) [49].

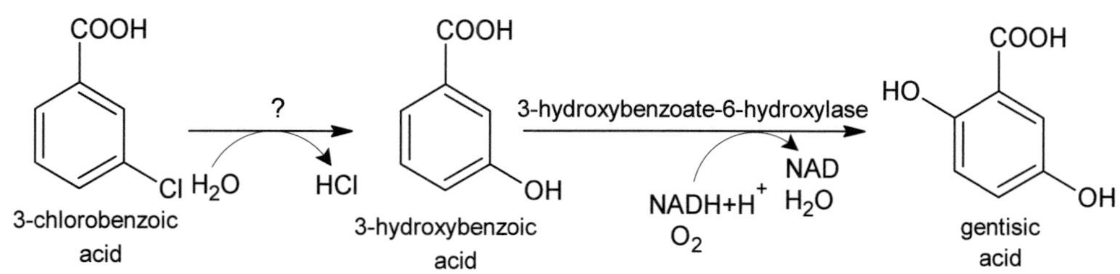

Figure 7. Degradation pathway of 3-chlorobenzoic acid with gentisic acid as an intermediate.

Monooxygenase reaction was reported in strains *Pseudomonas* sp. [50] and *Alcaligenes* sp. L6 [37]. Strain *Alcaligenes* sp. L6 has next to this pathway also pathway for 3-CBA degradation with protocatechuic acid as the final product. Strain L6 was isolated under low oxygen concentration. From this it can be assumed that 3-CBA degradation by monooxygenase reaction is held only under low oxygen concentration. When such conditions do not occur, the preferred microorganisms are those which can 3-CBA degrade by dioxygenase reaction. It may be also the reason why are described only a few strains degrading 3-CBA via gentisic acid.

2.2.3. *Degradation of 4-chlorobenzoic acid*

Microorganisms able of the degradation of 4-CBA mostly belong to the strains *Alcaligenes, Arthrobacter* and *Pseudomonas* [51]. Until now, two pathways for 4-CBA degradation were described. In first, the more common, is 4-CBA dehalogenation followed by ring cleavage. In the second is 4-CBA converted to 4-chlorocatechol which is further subjected to the ring cleavage and only then is dehalogenated [52].

In first mentioned 4-CBA degradation pathway is 4-hydroxybenzoic acid formed (Figure 8.). This pathway begins with the conversion of 4-CBA to 4-chlorobenzoate-CoA catalyzed by 4-chlorobenzoate:CoA ligase with the consumption of 1 molecule of ATP [53]. This reaction is followed by replacement of chlorine atom with hydroxyl group derived from water catalyzed by 4-chlorobenzoate:CoA dehalogenase [54]. The last step of 4-CBA dehalogenation is the hydrolysis of 4-hydroxybenzoate-CoA thioester by the enzyme 4-hydroxybenzoate:CoA thioesterase with formation of 4-hydroxybenzoate [32].

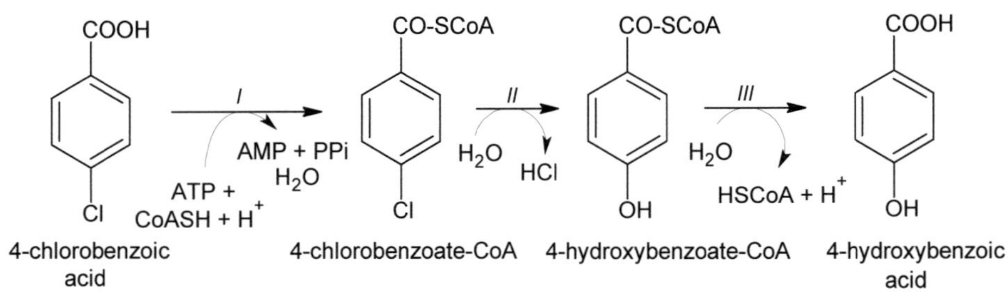

Figure 8. Degradation pathway of 4-chlorobenzoic acid with formation of 4-hydroxybenzoic acid. *I* – 4-chlorobenzoate:CoA ligase; *II* - 4-chlorobenzoate:CoA dehalogenase; *III* - 4-hydroxybenzoate:CoA thioesterase

All of the enzymes required for this conversion are organized in one operon. This operon is regulated by the presence of 4-CBA. Genes coding this operon can be located in the chromosomal DNA as well as on a plasmid [32].

The order of genes in the operon is for each bacterial strain different. *Pseudomonas* strain has dehalogenase-ligase-thioesterase and strain *Arthrobacter* has ligase-dehalogenase-thioesterase [32]. This indicates that the genes coding dehalogenation of 4-CBA have been rearranged. The low agreement of protein sequences indicates that this pathway is not the result of a recent adaptation. Therefore it is not the result of the recent release of PCBs and thus also 4-CBA, but it can be assumed that the 4-CBA dehalogenation pathway had enough time to arise from random mutations and selections. Indirectly it can be assumed that there is a natural source of 4-CBA [55].

Temporary formation of CoA thioester, one step after it was created, is special due to the energy consuming production of thioester bond. At the same time thioester formation is not required for aerobic degradation of other CBAs.

In case of anaerobic degradation of 3-CBA by strain *Rhodopseudomonas palustris* RCB100 is 3-chlorobenzoate-CoA formed. Its formation is followed by dehalogenation with release of benzoate-CoA. Benzoate-CoA is further degraded to the acetyl-CoA [56]. Same pathway is used for benzoic acid degradation. This pathway begins with formation of benzoate-CoA and CoA remains bound through the whole degradation pathway [24]. For the aerobic degradation of 4-CBA is thioester required only for the efficient dehalogenation. Thus, consumption of one molecule of ATP is acceptable due to the fact that it allows degradation of the molecule and enables its use as a source of carbon and energy.

The second possibility of 4-CBA degradation is degradation with 4-chlorocatechol as an intermediate (Figure 9.). Formation of 4-chlorocatechol is followed by ring cleavage and dehalogenation. This way of 4-CBA degradation is much less investigated than 4-CBA degradation leading to the formation of 4-hydroxybenzoate.

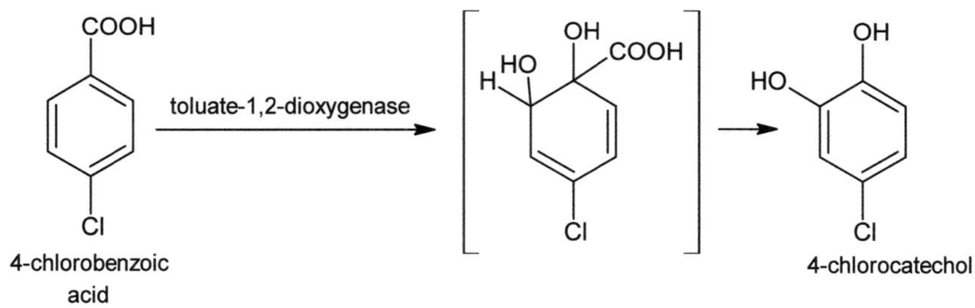

Figure 9. Degradation pathway of 4-chlorobenzoic acid leading to the 4-chlorocatechol

Enzyme of first step of 4-CBA degradation to the 4-chlorocatechol can be either the toluate-1,2-dioxygenase (EC 1.14.12.10) [31] like in case of strain *Pseudomonas aeruginosa* mt-2 or by not yet closer identified chlorobenzoate-1,2-dioxygenase described for example in the strain *Pseudomonas aeruginosa* 3mT [36].

Strain *Pseudomonas aeruginosa* 3mT was also able to degrade 3-CBA with formation of 3-chlorocatechol but formation of 4-chlorocatechol was not recorded. This activity suggest that *Pseudomonas aeruginosa* 3mT dioxygenase has only 1,2-dioxygenase activity. Not like benzoate-1,2-dioxygenase and 2-halobenzoate-1,2-dioxygenase 1,2- and 1,6-dioxygenase activity. From this can be assumed that dioxygenase from strain *Pseudomonas aeruginosa* 3mT is not identical with either of these enzymes.

2.2.4. Degradation of more chlorinated chlorobenzoic acids

Aerobic degradation of CBA with two chlorine atoms in molecule has been reported for many strains [17]. Degradation of CBA with three chlorine atoms is relatively rare, in the literature is described only a few strains and in most cases CBAs with three chlorine atoms in molecule are degraded cometabolically.

Dichlorinated CBA can be degraded by a dioxygenase reaction. For example 2,4-CBA and 2,5-CBA which are degraded by three componential 2-halobenzoate-1,2-dioxygenase with release of 4-chlorocatechol, as described in section 2.2.1. Also degradation of 3,4-CBA catalyzed by 3-chlorobenzoate-3,4-dioxygenase with formation of 5-chloroprotocatechuate has been previously described in section 2.2.2.

In case of 2,4-CBA is known one more degradation pathway similar to the 4-CBA degradation described in section 2.2.3. In this pathway is 2,4-CBA degraded via 4-hydroxybenzoate to the protocatechuic acid (Figure 10.). This pathway has been described in strains *Corynebacterium sepedonicum* KZ-4 [57] and *Alcaligenes denitrificans* NTB-1 [58].

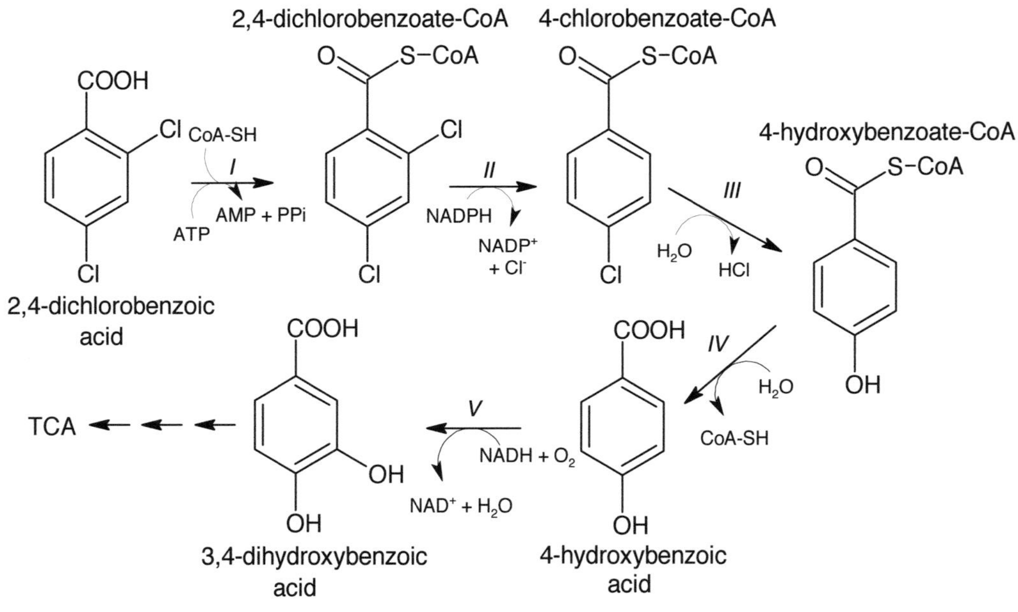

Figure 10. Degradation pathway of 2,4-dichlorobenzoic acid by strain *Corynebacterium sepedonicum* KZ-4 [57]. *I* - 2,4-dichlorobenzoate:CoA ligase; *II* – 2,4-dichlorobenzoate:CoA reductase; *III* – 4-chlorobenzoate:CoA dehalogenase; *IV* – 4-chlorobenzoate:CoA thioesterase; *V* – 4-hydroxybenzoate 3-monooxygenase; TCA – tricarboxylic acid cycle

Strain *Pseudomonas* sp. WR912 [59] can degrade 3,5-CBA by 1,2-dioxygenase reaction with release of 3,5-dichlorocatechol, which was by the same strain degraded to the succinate. Another way of 3,5-CBA degradation is the cometabolic degradation by consortia 2m.c [60]. 3,5-CBA was cometabolically converted to the 3,5-dichlorocatechol as a final intermediate by enzyme 2-halobenzoate-1,2-dioxygenase with presence of 2-CBA and 2,5-CBA. Presence of 3,5-CBA leads to the loss of 2-CBA degradation ability of the 2m.c consortium.

Same 2m.c consortium can also cometabolically degrade 2,3-CBA via 3-chlorocatechol to the 2-chloromuconic acid with presence of 2-CBA [60]. Strain *Pseudomonas aeruginosa* JB2 [61] degraded 2,3-CBA to the 4-chlorocatechol. Strain *Pseudomonas aeruginosa* JB2 can use also 2-CBA, 3-CBA, 2,5-CBA and 2,3,5-CBA as the sole carbon and energy source.

As it has been said before, CBAs with three chlorine atoms in molecule are converted in most cases cometabolically. Strain *Brevibacterium* sp. converted 2,3,6-CBA by a set of cometabolic steps to carbon dioxide and water [1]. This aerobic degradation proceeds via 2,3,6-trichloro-4-hydroxybenzoic acid or 2,3,6-trichloro-5-hydroxybenzoic acid to the 2,3,5-trichlorophenol or 2,4,5-trichlorophenol and finally to the 3,5-dichlorocatechol, which is than degraded by chlorocatechol degradation pathway (Figure 11.).

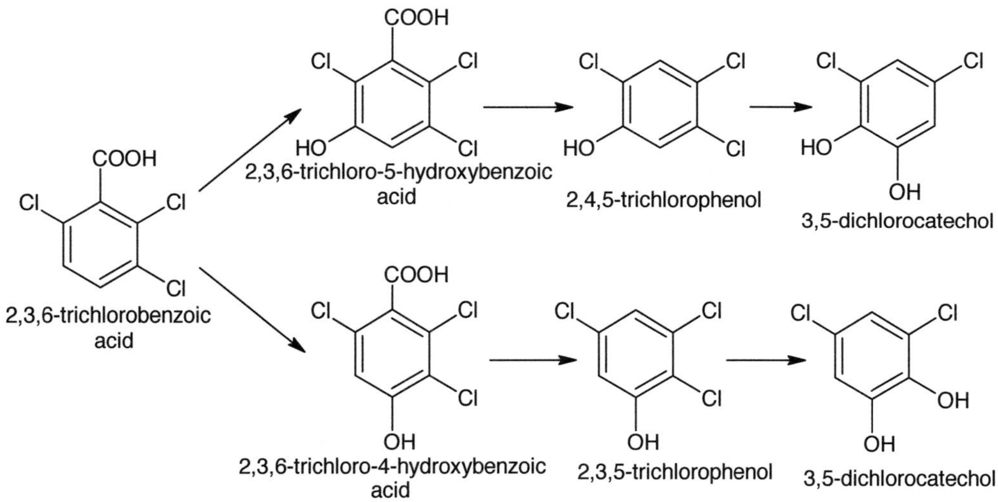

Figure 11. Cometabolic degradation of 2,3,6-trichlorobenzoic acid [1].

Strain *Pseudomonas putida* P111 degraded 2,3,5-CBA to the 3,5-dichlorocatechol (Figure 12.) but use different degradation pathway than strain *Brevibacterium* sp., 2,3,5-CBA was degraded by 1,2-dioxygenation followed by dehalogenation [62].

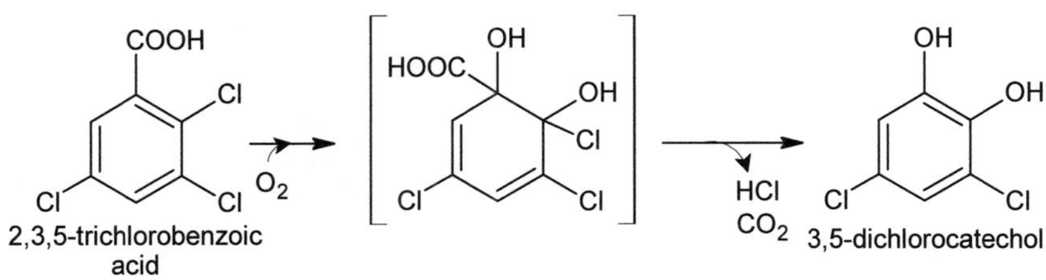

Figure 12. Degradation of 2,3,5-trichlorobenzoic acid by strain *Pseudomonas putida* P111 by 1,2-dioxygenase reaction [62].

2.2.5. Degradation of main chlorobenzoic acids degradation intermediates

In CBAs degradation are formed catechol, 3-chlorocatechol, 4-chlorocatechol, 4-hydroxybenzoic acid, gentisic acid, protocatechuic acid, chloroprotocatechuic acid and more chlorinated catechols.

Catechols can be degraded via *ortho-* or *meta-* cleavage pathway or possibly by modified *ortho-* cleavage pathway, which leads to the 3-oxoadipoic acid pathway. 3-oxoadipoic acid pathway is used for 4-hydroxybenzoic acid and protocatechuc acid degradation and leads to the tricarboxylic acid cycle.

In *ortho-* cleavage pathway is catechol or chlorocatechol degraded by catechol-1,2-dioxygenase or chlorocatechol-1,2-dioxygenase (Figure 13.). The first step of this pathway is dioxygenase reaction leading to the cleavage of the bond between the first and second position of the benzene ring. This reaction produces *cis,cis*-muconic acid or chloro-*cis,cis*-muconic acid with chlorine atom in a different position [63].

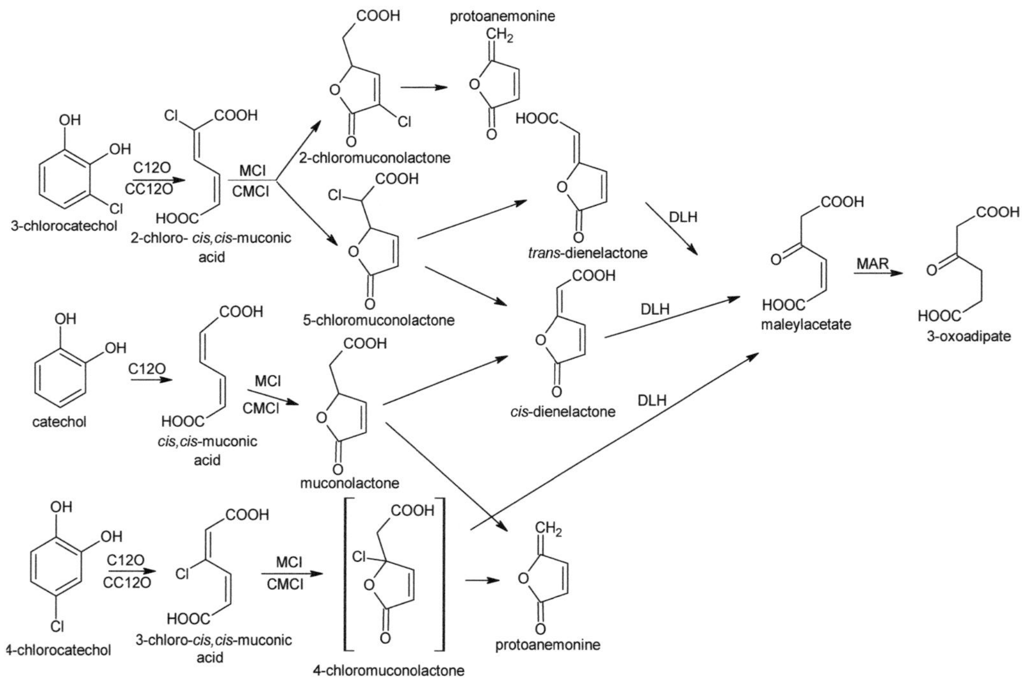

Figure 13. *Ortho-* and modified *ortho-* cleavage pathway of catechol or chlorotachols adapted from [9]. C12O – catechol-1,2-dioxygenase; CC12O – chlorocatechol-1,2-dioxygenase; MCI – muconate cycloisomerase; CMCI – chloromuconate cycloisomerase; DLH – dienelaktone hydrolase; MAR - maleylacetate reductase.

The second enzyme of this pathway is muconate cycloisomerase or chloromuconate cycloisomerase. This is a key enzyme in the degradation of chlorinated substances. The specificity of this isomerase determines whether degradation will be by *ortho-* cleavage pathway or by a modified *ortho-* cleavage pathway.

In *ortho-* cleavage pathway chloromuconate cycloisomerase catalyzes the conversion of chlorinated muconic acid to the *cis-* or *trans*-dienelactone in dependence of chlorine atom position in muconic acid. In this step is the molecule also dehalogenated [9].

Further enzyme of *ortho-* cleavage pathway is the dienelactone hydrolase. This enzyme is capable of conversion of *cis-* or *trans*-dienelactone to the maleylacetate. Maleylacetate is by maleylacetate reductase transformed into 3-oxoadipoic acid, which is further eliminated by 3-oxoadipoic acid degradation pathway (will be described later).

Majority of known strains, use for the catabolism of catechols or chlorocatechols a modified *ortho-* cleavage pathway [64]. In this pathway is chloro-*cis,cis*-muconate transformed by chloromuconate cycloisomerase to the protoanemonine (Figure 13.). Protoanemonine is a dead-end product of this pathway. Protoanemonine has also antimicrobial properties thus formation of protoanemonine from chlorinated catechols in modified *ortho-* cleavage pathway leads to the poor survival of degrading microorganisms in the soil.

Catechol or chlorocatechols can be also degraded by the enzyme katechol-2,3-dioxygenase or chlorocatechol-2,3-dioxygenase in *meta-* cleavage pathway which is initiated by this dioxygenation (Figure 14.). The result of dioxygenase reaction is the ring cleavage between second and third position of the benzene ring.

From catechol and 3-chlorocatechol is in this step 3-hydroxy-*cis,cis*-muconic acid formed. This acid is further degraded to the pyruvate and acetaldehyde [65]. The risk of this pathway is the possibility of formation of a reactive acylchloride from 3-chlorocatechol. Acylchloride irreversibly inactivate catechol-2,3-dioxygenase [66, 67].

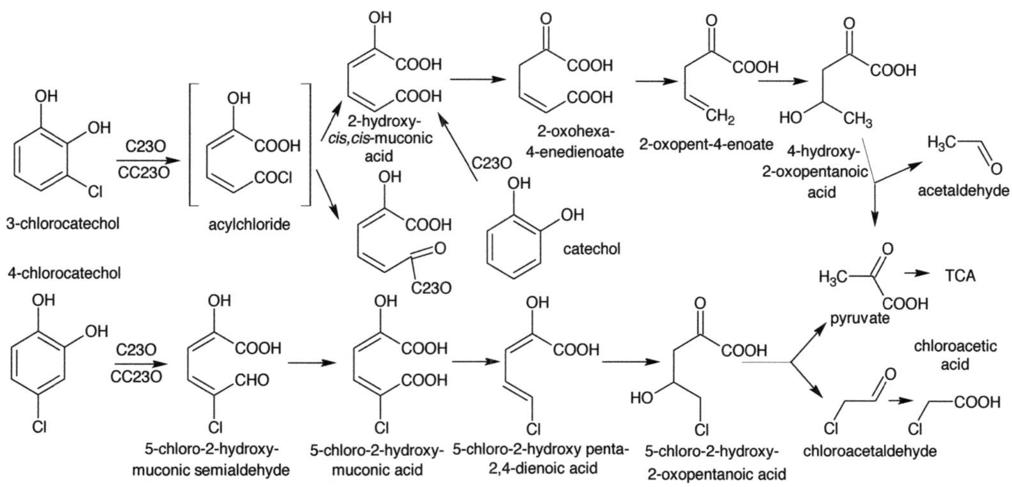

Figure 14. *Meta-* cleavage pathway of catechol or chlorotachols [9]. C23O - catechol-2,3-dioxygenase; CC23O - chlorocatechol-2,3-dioxygenase; TCA – tricarboxylic acid cycle.

The 4-chlorocatechol is by the 2,3-dioxygenase reaction converted to the 5-chloro-2-hydroxy-muconic semialdehyde [16]. This semialdehyde is in several steps transformed into the pyruvate and chloroacetic acid. Pyruvate is further degraded in the tricarboxylic acid cycle. In case of chloroacetic acid is expected further degradation due to the fact that during the course of degradation of 4-CBA via 4-chlorocatechol by strain *Pseudomonas cepacia* P166 only a temporary accumulation of chloroacetic acid was registered [68].

In CBAs degradation next to catechol and chlorocatechols can be 4-hydroxybenzoic acid formed. 4-hydroxybenzoic acid is product of 4-CBA degradation (section 2.2.3). 4-hydroxybenzoic acid is further degraded in the 3-oxoadipoic acid degradation pathway (Figure 15.) [16]. The first step of this pathway is oxidation catalyzed by 4-hydroxybenzoate-3-monooxygenase with the formation of protocatechuic acid (3,4-dihydroxybenzoic acid). Benzene ring of protocatechuic acid is in next step cleaved between third and fourth position. This way formed 3-karboxy-*cis,cis*-muconic acid is than transformed into the 3-oxoadipoic acid [51, 69].

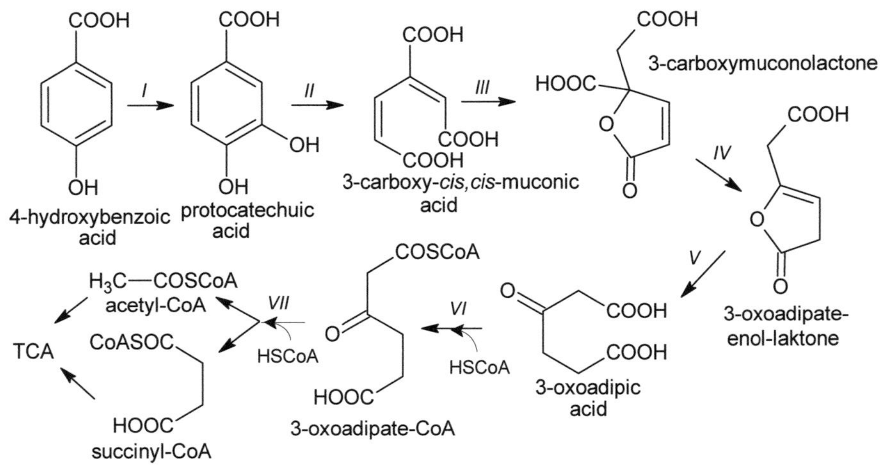

Figure 15. Degradation pathway of 3-oxoadipoic acid [9, 51, 69]. *I* – 4-hydroxybenzoate-3-monooxygenase; *II* – protokatechute-3,4-dioxygenase; *III* – 3-carboxymuconate-cykloisomerase; *IV* – 4-carboxymuconolaktone-dekarboxylase; *V* – 3-oxoadipate-enol-laktone hydrolase; *VI* - 3-oxoadipate:sukcinyl-CoA transferase; *VII* - 3-oxoadipate-CoA thiolase; TCA – tricarboxylic acid cycle.

3-oxoadipoic acid, which is also formed in the *ortho-* cleavage pathway from chlorinated catechols is then in two steps catalyzed by 3-oxoadipate:succinyl-CoA transferase and 3-oxoadipoate-CoA thiolase converted into the intermediates of the trikarboxylic acid cycle [9, 51, 69].

In 3-CBA degradation pathway initiated by monoxygenase reaction is gentisic acid formed (section 2.2.2). Gentisic acid (2,5-dihydroxybenzoic acid) is, as well as protocatechuic acid, degraded in the 3-oxoadipoic acid degradation pathway into the intermediates of the tricarboxylic acid cycle (Figure 16.).

The first step in gentisic acid degradation is the formation of maleylpyruvate by dioxygenase reaction catalyzed by gentisate-1,2-dioxygenase. Maleylpyruvate can be by isomerization transformed into the fumarylpyruvate. Maleylpyruvate and fumarylpyruvate are by respec-

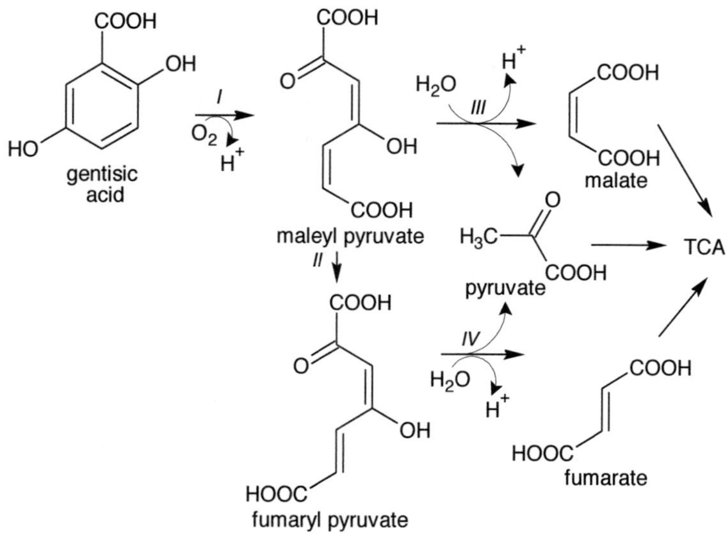

Figure 16. Gentisic acid degradation pathway [9]. *I* – gentisate-1,2-dioxygenase; *II* – maleyl pyruvate-*cis,trans*–isomerase; *III* – maleyl pyruvate-dehydrogenase; *IV* – fumaryl pyruvate-dehydrogenase; TCA – tricarboxylic acid cycle

tive hydrolase transformed into the malate or pyruvate, the tricarboxylic acid cycle intermediates [9].

2.2.6. Inhibition and activation of chlorobenzoic acids degradation

Due to the complexity of the described CBAs degradation pathways it can be expected that the CBAs degradation pathways will affect each other. This interaction can be positive as well as negative.

CBAs itself can affect microbial degradation of xenobiotics which are CBAs degradation products. For example, microbial degradation of PCBs can be by high CBAs concentration slowed down or even stopped [15, 44].

The influence of monochlorinated CBAs on the degradation of low chlorinated PCBs congeners by five different bacterial strains is already documented [15]. It was found that 3-CBA is the most potent inhibitor. The same result was obtained with monochlorinated PCBs degradation by strain *Pseudomonas stutzeri* [70], but in the case of microbial degradation of dichlorinated and trichlorinated PCBs was the inhibition effect of CBAs less significant. In the case of tests with *Pseudomonas testosteroni* B-356 and impact of monochlorinated CBAs on low chlorinated PCBs degradation, it was again found that the 3-CBA has the highest inhibition effect followed by 4-CBA and the least inhibition showed 2-CBA [16].

These findings can be explained by the formation of chlorocatechols in 3-CBA degradation with a higher probability than from other monochlorinated CBAs. Degradation of chlorocatechols in *meta-* cleavage pathway can lead to the reactive acylchloride formation. Acylchloride can inhibited 2,3-dihydroxybifenyl-1,2-dioxygenase from the upper cometabolic PCBs degradation pathway (Figure 1.). How was proved by accumulation of hydroxylated biphenyls during PCBs degradation by strain *Pseudomonas testosteroni* B-356 [16].

Acylchloride can be also irreversibly bound on catechol or chlorocatechol-2,3-dioxygenase, the first enzyme in the *meta-* cleavage pathway. Its inactivation leads to catechols or chlorocatechols accumulation and therefore to the inhibition of CBAs degradation [71].

Another potentially dangerous intermediate is protoanemonine, substance with antibiotic properties. Protoanemonine is formed in the modified *ortho-* cleavage pathway from muconic or chloromuconic acid. The main effect of protoanemonine is growth inhibition of degrading microorganisms[72].

The presence of a mixture of CBAs can also cause inhibition or activation of CBA degradation. Strain *Burkholderia cepacia* JHR22 can degrade 2-CBA, 3-CBA, 4-CBA and 3,5-CBA [44] when present individually. This strain loses 2-CBA degradation ability when 2-CBA is present in mixture with 2,3-CBA or 3,4-CBA. On the other hand addition of 2,5-CBA or 2,6-CBA had no effect on 2-CBA degradation. When was 2-CBA added with 2,4-CBA strain *Burkholderia cepacia* JHR22 in addition to the 2-CBA degradation can cometabolically degrade 2,4-CBA.

Catechols or chlorocatechols are also potential inhibitors of bacterial CBAs degradation. If these intermediates of CBAs degradation are not enough quickly degraded they can be subject to the auto-oxidation or enzymatic polymerization with formation of brown or black pigment [73, 74]. The presence of this pigment inhibits CBAs degradation by affecting the shape of bacterial cells which may consequently lead to their death as in case of strain *Pseudomonas fluorescens* [75].

3. Phytoremediation

Together with the microorganisms participate on the elimination of xenobiotics green plants. Green plants use for the xenobiotics elimination four different strategies: extraction of contaminants from soil and water (mostly heavy metals), uptake and detoxication, in some cases even degradation (organic pollutants), volatilization (organic compounds and some metaloids Se and As) and stimulation of microbial degradation in rhizosphere or by endophytic microorganisms. All this processes are called phytoremediation (Figure 17.), the use of green plants for removal or transformation of pollutants from the environment [76, 77].

Phytoextraction is an accumulation of substances from the environment to the plant biomass. Precondition for the successful application of this process is that the plant is capable to take up contaminant by roots and ideally transport it to the aboveground parts, where contaminant is deposited in relatively high concentrations (more than 1 g per 1 kg of plant biomass). The disadvantage of this process is the possibility to re-release of contaminants from plant biomass, and their introduction into the food chain in the case of animals grazing on such plants [47, 76-78].

Phytovolatilization is uptake of pollutant by plant, which is followed by pollutant conversion to the volatile form and release in to the atmosphere [78, 79]. This type of phytoremediation is

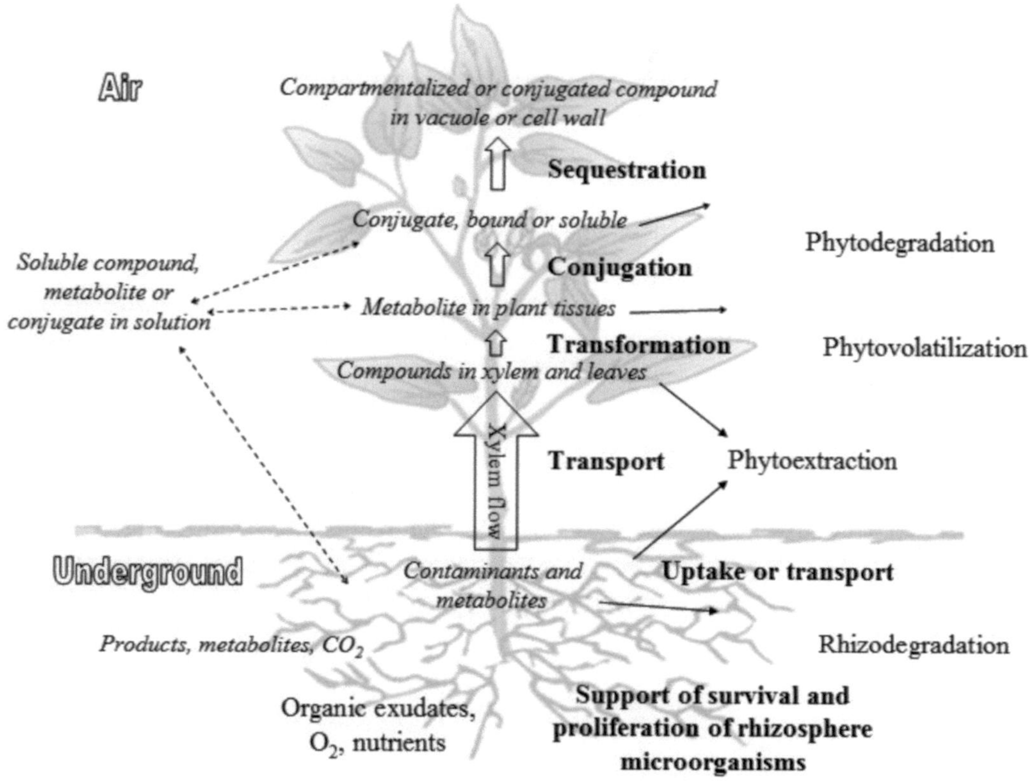

Figure 17. Schematic presentation of phytoremediation processes, adapted from [47].

controversial because it does not reduce contamination, only transfer pollutants from soil to the air [47, 76].

Rhizodegradation is stimulation of degrading microorganisms in plant rhizosphere. This stimulation can be done by support of proliferation or by creation of conditions suitable for better survival of microorganisms with degradation activity [47, 76].

Phytodegradation is a process when pollutants are taken up into the plant body, transformed and eliminated by plant metabolism. Phytodegradation can be considered also as a process in which contamination is reduced by enzymes released by the plant into the soil. Phytodegradation is most suitable for organic pollutants [47].

Plant metabolism of xenobiotics is in many aspects similar to the mammalian metabolism, therefore plants are called green liver of the planet. Uptake of xenobiotics is done by roots, from roots are then xenobiotics transported by xylem to the aboveground (harvestable) parts of plant, where xenobiotics are transformed to the non-phytotoxic metabolites. Metabolisation of xenobiotics in plant body is done in three phases. First is transformation followed by conjugation and the last phase is sequestration in plant tissues.

During transformation phase xenobiotics are subjected to enzymatic hydrolysis, oxidation or reduction. Reactive groups exposed or generated in transformation stage react with moieties such as glucuronate or glutathione in conjugation phase. Soluble conjugates are transported in the final stage to the vacuole or incorporated into the cell wall, thus preventing disturbing of the cell function by them [80]. Some water plants can transport conjugates outside the plant [47].

Using plants for decontamination has many advantages. It is a cost-saving process. After planting the area, are the costs for growing and harvesting biomass relatively low, and the obtained biomass can also be used to produce heat or electricity. It is aesthetically pleasing process positively accepted by the public. It causes minimal disruption to the environment. It is also applicable to the areas with trace or low pollution and is usable for a wide range of xenobiotics [77, 78, 81].

Phytoremediation has also drawbacks. The main disadvantage is that phytoremediation compared with physico-chemical methods is slower. It should take a few years to eliminate pollution and during the process most of the decontaminated areas can not be used for commercial purposes [78]. Therefore, the goal of the phytoremediation study is to increase the efficiency and speed of this process, by the understanding and improvement of the mechanism of phytoremediation and metabolic pathways involved in the conversion of xenobiotics by plants.

3.1. Plant metabolization of chlorobenzoic acids

About the CBAs metabolism in plants it is known, in comparison with microbial metabolism, very little. Although works have been published that deal with the CBA metabolisation in plants, it was not yet clearly demonstrated that plants degrade CBAs and not just accumulate CBAs in plant tissues.

Deavers et al. [82] investigated the metabolism of 4-CBA by cells of willow (*Salix viminalis*) under sterile conditions, as well as by whole plants. Cells were able to remove 65% of 4-CBA from the media during 360 hours with an initial concentration of 4-CBA 50 mg/l. For the whole willow plants, the concentration 50 mg/l of 4-CBA was toxic, and therefore the loss of 4-CBA was only in the range of 10-30%. When was initial concentration of 4-CBA reduced to 5 mg/l, willow plants managed during 305 hours to remove 70-90% of 4-CBA in dependence on the pH of the culture medium [82].

Other uses of the plants for metabolisation of CBAs has been described in plant tissue cultures of black nightshade (*Solanum nigrum*), tobacco (*Nicotiana tabacum*), horseradish (*Armoracia rusticana*) and alfalfa (*Medicago sativa*) [10]. In this case, the two initial concentrations were tested, 200 and 50 mg/l, and 11 different CBAs. From tested plant species the best ability to metabolize CBAs demonstrated black nightshade and horseradish. For *in vitro* grown cells of tobacco and alfalfa was the concentration of CBAs 200 mg/l toxic. So that named plant cells have not demonstrated metabolisation.

Tissue cultures of black nightshade and horseradish were within 14 days able to remove from the medium 90% of 2-CBA and about 30% of 2,3-CBA, 2,4-CBA and 2,5-CBA at an initial concentration of 200 mg/l, if the initial concentration was reduced to 50 mg/l this cell cultures removed 100% of the mentioned CBAs. Culture of black nightshade with the same efficiency metabolized 3-CBA, 4-CBA and 2,3,5-CBA. Culture of horseradish metabolized other tested CBAs with 30-60% efficiency [10].

Siciliano and Germida [83] tested the ability of 16 different kinds of grasses grow in soil contaminated with 2-CBA (816 mg/kg). Only five of them had this ability, namely it were grasses *Bromus inermis, Agropyron intermedium, Bromus biebersteinii, Agropyron riparum* a *Elymus dauricus*. Only four last mentioned species were also able to metabolize 2-CBA from soil with efficiency from 32 to 42% in 60 days [83].

From these examples of CBAs metabolisation in plants is evident, that this process is not only influenced by the structure and concentration of CBA, but it is also influenced by plant species and plant capacity to "resist" the toxicity of CBA.

4. Plant and microbial cooperation on bioremediation

There are known several mechanisms for the promotion of the contamination removal from soil by plants. One of these mechanisms is a non-specific support of growth and proliferation of microbial communities by plants [84]. Support of microbial activity in the rhizosphere leading to the protection of plants from the effects of contamination and increasing the speed of xenobiotics elimination [85]. Another mechanism is specific support of degrading microorganisms in the rhizosphere of plants [86, 87], secretion of enzymes capable of catalyzing the conversion of contaminants in the soil by plants [88].

Plants can promote the growth of microorganisms in the rhizosphere by excretion of root exudates into the soil [89]. Exudates may contain various ions, free oxygen, water, enzymes (e.g. laccase, peroxidase, dehalogenase, nitroreductase), substances of the saccharide nature to facilitate root growth and diverse mixes of primary and secondary metabolites. Organic compounds in exudates are divided into two groups. One is for substances with low molecular weights such as amino acids, organic acids, sugars, phenolic compounds and other secondary metabolites and second is for substances of higher molecular weight such as proteins and polysaccharides [90].

The problem of the use of the cooperation in bioremediation is that the increase in the number of microorganisms in the rhizosphere of plants does not always mean increased degradation of the contamination. An example of this behavior is experiment when exudates of perennial ryegrass (*Lolium Perence*) were used in the degradation of phenanthrene [89]. Results showed that despite the increased number of microorganisms caused by addition of exudates, no increased phenanthrene degradation was observed. Phenanthrene degradation in this case was even half than in the control without exudates.

Just as plants can affect the growth of microorganisms, microorganisms can also affect the growth and survival of plants in the contaminated areas. Microorganisms can help plants by production of protective biofilms or antibiotics acting as a protection against organisms potentially pathogenic for plants [91, 92]. In addition, the activity of certain microorganisms is source of nutrients for plant e.g. nitrogen compounds. Microorganisms may also increase the solubility of minerals such as phosphorus, and thus make them more accessible for plants. Another mechanism of protection of plants by microorganisms is the synthesis of low molecular weight siderophores in addition to phytosiderophores. Siderophores bind the free iron in the rhizosphere of plants. Lack of free iron negatively affects the proliferation of fungal pathogens in the root system of plants [93, 94].The presence of microorganisms can also initiate the synthesis of plant hormone such as auxins, cytokinins and gibberellins, which promote the growth of plants [91, 95]. Equally important is the mechanism for reducing stress by reduction of stress hormone ethylene concentration in plant by synthesis of microbial enzyme 1-aminocyklopropan-1-carboxylate deaminase (ACC deaminase) [96].

During the life of the plant its growth may be influenced by microorganisms with one or more of these mechanisms. Influence is most evident in plants grown under stressful conditions.

In a consortium consisting of microorganisms and plants growing in the contaminated soil, in addition to the above mentioned mechanisms, can occur production of metabolites or intermediates of degradation no matter whether by plant or microorganism. These metabolites may be by the other partner further degraded and metabolized or they may be toxic. The study of the metabolism of both biological systems and their possible influence should lead to the obtaining of such a system consisting of plant - microorganisms, which increases speed and efficiency of the xenobiotics degradation process.

4.1. Plant and microbial cooperation on chlorobenzoic acids remediation

The issue of co-operation of plants and microorganisms on degradation of CBA is solved in several publications.

Haby and Crowley [84] compared the degradation capacity of soil from the rhizosphere of perennial ryegrass (*Lolium perenne*) with non-vegetated soil. Results showed that the degradation rate of 3-CBA was at the beginning of the experiment accelerated by the rhizosphere soil, they also showed that in soil from the rhizosphere is increased amount of microorganisms [84].

Dittmann et al. [97] examined the ability of three months old pine seedlings (*Pinus sylvestris*) with mycorrhizal fungi *Suillus bovinus* to degrade 3-CBA. They found that 3-CBA is accumulated in the aboveground parts of the seedlings and removal efficiency of 3-CBA from the soil by a consortium was very low (around 25% for 4 weeks). In plant tissues were not detected elevated concentrations of Cl^- in comparisom with control plants grown in uncontaminated soil. Therefore they concluded that 3-CBA was not or only to a very small extent degraded by plants [97].

The aforementioned grass *Elimus dauricus*, together with microorganismus *Pseudomonas aeruginosa* R75 and *Pseudomonas savastanoi* CB35 was able to remove almost 50% of 2-CBA from soil during 56 days [98]. The consortium showed greater metabolic efficiency than non-vegetated soil.

In this work has been continued testing the degradation capacity of microorganisms and plants with other grasses and other CBAs or with a mixture of CBAs. The same composition of the consortium, thus *Elimus dauricus* and strains of *Pseudomonas aeruginosa* R75 and *Pseudomonas savastanoi* CB35 was able to remove 74% of 3-CBA from the soil. *Bromus biebersteinii* with strain *Alcaligenes* sp. BR60 remove 56% of 2,3-CBA from the soil and the same consortium was able to metabolize 61% of 2,3-CBA and 50% of 3-CBA if both CBAs were added together. A consortium consisting of *Elymus angitus* and strains *Pseudomonas aeruginosa* R75 and *Pseudomonas savastanoi* CB35 eliminate 46% of 2,5-CBA from soil and from the mixture of 3-CBA, 2,3-CBA and 2,5-CBA this consortium has removed around 40% of all three CBAs from soil [99]. These results were obtained by testing a large number of combinations of grasses with bacterial inoculants. Most of the combinations did not show the positive impact on reducing the tested CBA or a mixture of CBAs from soil. Inoculants did not reduce CBA phytotoxicity and caused plant death in contaminated soil.

The above examples show that the use of plants and microorganisms in bioremediation technologies has great potential. However, selection of suitable pair of plant and microorganism is very complicated. We can assume that not only depend on the type of contamination, but it requires a deeper understanding of the principles of cooperation between this two organisms.

Acknowledgements

The work was supported by the grant GAČR 525/09/1058 and EU project MINOTAURUS (FP7-KBBE-2010-4/265946).

Author details

Blanka Vrchotová*, Martina Macková, Tomáš Macek and Kateřina Demnerová

*Address all correspondence to: vrchotob@vscht.cz

Institute of Chemical Technology, Prague, Faculty of Food and Biochemical Technology, Department of Biochemistry and Microbiology, Prague, Czech Republic

References

[1] Horvath RS Cometabolism of the herbicide 2trichlorobenzoate. Journal of Agricultural and Food Chemistry. (1971). , 19(2), 291-293.

[2] Holtze, M. S, Sørensen, S. R, & Sørensen, J. Aamand J Microbial degradation of the benzonitrile herbicides dichlobenil, bromoxynil and ioxynil in soil and subsurface

environments- Insights into degradation pathways, persistent metabolites and involved degrader organisms. Environmental Pollution. (2008). , 154, 155-168.

[3] Wildung, R. E, & Chesters, G. Armstrong DE Chloramben (ambien) degradation in soil. Weed Research. (1968). , 8(3), 213-225.

[4] Kamanavalli, C. M. Ninnekar HZ Biodegradation of DDT by a *Pseudomonas* species. Current Microbiology. (2004). , 48(1), 10-13.

[5] Nováková, L, Matysová, L, & Havlíková, L. Solich P Development and validation of HPLC method for determination of indomethacin and its two degradation products in topical gel. Journal of Pharmaceutical and Biomedical Analysis. (2005). , 37(5), 899-905.

[6] Byrne, O, Williams, P. M, Walsh, R, & Gilmer, J. J. JF The aqueous stability of bupropion. Journal of Pharmaceutical and Biomedical Analysis. (2010). , 53(3), 376-381.

[7] Molnari, J. C. Myers AL Carbonyl reduction of bupropion in human liver. Xenobiotica. (2012). , 42(6), 550-561.

[8] Quintana, J. B, & Weiss, S. Reemtsma T Pathways and metabolites of microbial degradation of selected acidic pharmaceutical and their occurrence in municipal wastewater treated by a membrane bioreactor. Water Research. (2005). , 39(12), 2654-2664.

[9] Pieper DH Aerobic degradation of polychlorinated biphenylsApplied and Environmental Microbiology. (2005). , 67, 170-191.

[10] Macková, M, Vrchotová, B, Francová, K, Sylvestre, M, Tomaniová, M, Lovecká, P, & Demnerová, K. Macek T Biotransformation of PCBs by plants and bacteria- consequences of plant-microbe interactions. European Journal of Soil Biology. (2007). , 43(4), 233-241.

[11] Nováková, H, Vošahlíková, M, Pazlarová, J, Macková, M, & Burkhard, J. Demnerová K PCB metabolism by *Pseudomonas* sp. International Biodeterioration & Biodegradation. (2002). , 2.

[12] Furukawa, K. Miyazaki T Cloning of a Gene Cluster Encoding Biphenyl and Chlorobiphenyl Degradation in *Pseudomonas pseudoalcaligenes*. Journal of Bacteriology. (1986). , 166(2), 392-398.

[13] Dietrich, D, & Hickey, W. J. Lamar R Degradation of 4,4′-dichlorobiphenyl, 3,3′,4,4′-tetrachlorobiphenyl, and 2,2′,4,4′,5,5′-hexachlorobiphenyl by the white rot fungus *Phanerochaete chrysosporium*. Applied and Environmental Microbiology. (1995). , 61(11), 3904-3909.

[14] Kamei, I, & Kogura, R. Kondo R Metabolism of 4,4′-dichlorobiphenyl by white-rot fungi *Phanerochaete chrysosporium* and *Phanerochaete* sp. MZ142. Applied Microbiology and Biotechnology. (2006). , 72(3), 566-575.

[15] Adebusoye, S. A, Picardal, F. W, & Ilori, M. O. Amund OO Influence of chlorobenzoic acids on the growth and degradation potentials of PCB-degrading microorganisms. World Journal of Microbiology and Biotechnology. (2008). , 24(7), 1203-1208.

[16] Sondossi, M, & Sylvestre, M. Ahmad D Effects of chlorobenzoate transformation on the *Pseudomonas testosteroni* biphenyl and chlorobiphenyl degradation pathway. Applied and Environmental Microbiology. (1992). , 58(2), 485-495.

[17] Field, J. Sierra-Alvarez R Microbial transformation of chlorinated benzoates. Reviews in Environmental Science and Biotechnology. (2008). , 7(3), 191-210.

[18] Sharak Genthner BRPrice WA, Pritchard PH Characterization of anaerobic dechlorinating consortia derived from aquatic sediments. Applied and Environmental Microbiology. (1989). , 55(6), 1472-1476.

[19] Commandeur LCMPersons JR Degradation of halogenated aromatic compounds. Biodegradation. (1990). , 1, 207-220.

[20] Sharak Genthner BR Preliminary characterization of four 2-chlorobenzoate-degrading anaerobic bacterial consortiaBiodegradation. (1999). , 10(1), 27-33.

[21] Häggblom, M. M. Young LY Anaerobic degradation of 3-halobenzoates by a denitrifying bacterium. Archives of Microbiology. (1999). , 171(4), 230-236.

[22] Sharak Genthner BRTownsend GT, Blattmann BO Reduction of 3-chlorobenzoate, 3-bromobenzoate, and benzoate to corresponding alcohols by Desulfomicrobium escambiense, isolated from a 3-chlorobenzoate-dechlorinating coculture. Applied and Environmental Microbiology. (1997). , 63(12), 4698-4703.

[23] Van Der Woude, B. J, & Deboer, M. Vanderput NMJ, Vandergeld FM, Prins RA, Gottschal JC Anaerobic Degradation of Halogenated Benzoic-Acids by Photoheterotrophic Bacteria. FEMS Microbiology Letters. (1994).

[24] Elder DJEKelly DJ The bacterial degradation of benzoic acid and benzenoid compounds under anaerobic conditions: Unifying trends and new perspectives. FEMS Microbiology Reviews. (1994). , 13(4), 441-468.

[25] Williams, R. J. Evans WC The metabolism of benzoate by *Moraxella* species through anaerobic nitrate respiration. Evidence for a reductive pathway. The Biochemical Journal. (1975). , 148(1), 1-10.

[26] Dispensa, M, Thomas, C. T, Kim, M. K, Perrotta, J. A, & Gibson, J. Harwood CS Anaerobic growth of *Rhodopseudomonas palustris* on 4-hydroxybenzoate is dependent on AadR, a member of the cyclic AMP receptor protein family of transcriptional regulators. Journal of Bacteriology. (1992). , 174(18), 5803-5813.

[27] Ni, S, & Fredrickson, J. K. Xun L Purification and characterization of a novel 3-chlorobenzoate-reductive dehalogenase from the cytoplasmic membrane of *Desulfomonile tiedjei* DCB-1. Journal of Bacteriology. (1995). , 177(17), 5135-5139.

[28] Kasberg, T, Daubaras, D. L, Chakrabarty, A. M, & Kinzelt, D. Reineke W Evidence that operons *tcb*, *tfd*, and *clc* encode maleylacetate reductase, the fourth enzyme of the modified *ortho* pathway. Journal of Bacteriology. (1995). , 177(13), 3885-3889.

[29] Springael, D, Peys, K, Ryngaert, A, Van Roy, S, Hooyberghs, L, Ravatn, R, Heyndrickx, M, Van Der Meer, J. R, Vandecasteele, C, & Mergeay, M. Diels L Community shifts in a seeded 3-chlorobenzoate degrading membrane biofilm reactor: Indications for involvement of *in situ* horizontal transfer of the *clc*-element from inoculum to contaminant bacteria. Environmental Microbiology. (2002). , 4(2), 70-80.

[30] Chae, J. C, & Zylstra, G. J. Chlorobenzoate uptake in *Comamonas* sp. strain DJ-12 is mediated by a tripartite ATP-independent periplasmic transporter. Journal of Bacteriology. (2006). , 188(24), 8407-8412.

[31] Radice, F, Orlandi, V, Massa, V, Battini, V, Bertoni, G, & Reineke, W. Barbieri P Cloning of the *Arthrobacter* sp. FG1 dehalogenase genes and construction of hybrid pathways in *Pseudomonas putida* strains. Applied Microbiology and Biotechnology. (2007). , 75(5), 1111-1118.

[32] Zhuang, Z. H, Gartemann, K. H, & Eichenlaub, R. Dunaway-Mariano D Characterization of the 4-hydroxybenzoyl-coenzyme A thioesterase from *Arthrobacter* sp. strain SU. Applied and Environmental Microbiology. (2003). , 69(5), 2707-2711.

[33] Francisco, P. B, Ogawa, N, & Suzuki, K. Miyashita K The chlorobenzoate dioxygenase genes of *Burkholderia* sp. strain NK8 involved in the catabolism of chlorobenzoates. Microbiology. (2001). , 147(1), 121-133.

[34] Nakatsu, C. H. Wyndham RC Cloning and expression of the transposable chlorobenzoate-3,4-dioxygenase genes of *Alcaligenes* sp. strain BR60. Applied and Environmental Microbiology. (1993). , 59(11), 3625-3633.

[35] Romanov, V. Hausinger RP *Pseudomonas aeruginosa* 142 uses a three-component ortho-halobenzoate 1,2-dioxygenase for metabolism of 2,4-dichloro- and 2-chlorobenzoate. Journal of Bacteriology. (1994). , 176(11), 3368-3374.

[36] Ajithkumar, P. V. Kunhi AA Pathways for 3-chloro- and 4-chlorobenzoate degradation in *Pseudomonas aeruginosa* 3mT. Biodegradation. (2000). , 11(4), 247-261.

[37] Krooneman, J, Wieringa, E. B, Moore, E. R, Gerritse, J, & Prins, R. A. Gottschal JC Isolation of *Alcaligenes* sp. strain L6 at low oxygen concentrations and degradation of 3-chlorobenzoate via a pathway not involving (chloro)catechols. Applied and Environmental Microbiology. (1996). , 62(7), 2427-2434.

[38] Haak, B, & Fetzner, S. Lingens F Cloning, Nucleotide Sequence, and Expression of the Plasmid-Encoded Genes for the Two-Component 2-Halobenzoate 1,2-Dioxygenase from *Pseudomonas cepacia* 2CBS. Journal of Bacteriology. (1995). , 177(3), 667-675.

[39] Fetzner, S, & Muller, R. Lingens F Degradation of 2-chlorobenzoate by *Pseudomonas cepacia* 2CBS. Biological Chemistry Hoppe-Seyler. (1989). , 370(11), 1173-1182.

[40] Fetzner, S, & Müller, R. Lingens F Purification and some properties of 2-halobenzoate 1,2-dioxygenase, a two-component enzyme system from *Pseudomonas cepacia* 2CBS. Journal of Bacteriology. (1992). , 174(1), 279-290.

[41] Sylvestre, M, Mailhiot, K, & Ahmad, D. Masse R Isolation and preliminary characterization of a 2-chlorobenzoate degrading Pseudomonas. Canadian Journal of Microbiology. (1989). , 35(4), 439-443.

[42] Fetzner, S, Müller, R, & Lingens, F. A novel metabolite in the microbial degradation of 2-chlorobenzoate. Biochemical and Biophysical Research Communications. (1989). , 161(2), 700-705.

[43] Pieper, D. H, & Knackmuss, H-J. Timmis KN Accumulation of 2-chloromuconate during metabolism of 3-chlorobenzoate by *Alcaligenes eutrophus* JMP134. Applied Microbiology and Biotechnology. (1993). , 39(4), 563-567.

[44] Reineke, W. Knackmuss H-J Chemical structure and biodegradability of halogenated aromatic compounds substituent effects on 1,2-dioxygenation of benzoic acid. Biochimica et Biophysica Acta (BBA)- General Subjects. (1978). , 542(3), 412-423.

[45] Nakatsu, C. H, & Straus, N. A. Wyndham RC The nucleotide sequence of the Tn5271 3-chlorobenzoate dioxygenase genes (*cbaAB*) unites the class IA oxygenases in a single lineage. Microbiology. (1995). , 3, 4.

[46] Providenti, M. A. Wyndham RC Identification and functional characterization of CbaR, a MarR-li Applied and Environmental Microbiology. (2001). , 67(8), 3530-3541.

[47] Singh, O. V. Jain RK Phytoremediation of toxic aromatic pollutants from soil. Applied Microbiology and Biotechnology. (2003). , 63(2), 128-135.

[48] Nakatsu, C. H, & Providenti, M. Wyndham RC The *cis*-diol dehydrogenase *cbaC* gene of Tn*5271* is required for growth on 3-chlorobenzoate but not 3,4-dichlorobenzoate. Gene. (1997). , 196, 209-218.

[49] Yu, Y, & Wang, L. H. Tu SC *Pseudomonas cepacia* 3-hydroxybenzoate 6-hydroxylase: stereochemistry, isotope effects, and kinetic mechanism. Biochemistry. (1987). , 26(4), 1105-1110.

[50] Johnston, H. W, & Briggs, G. G. Alexander M Metabolism of 3-chlorobenzoic acid by a pseudomonad. Soil Biology and Biochemistry. (1972). , 4(2), 187-190.

[51] Kobayashi, K, & Katayama-hirayama, K. Tobita S Metabolic pathway of benzoic acid in an *Acinetobacter* sp. that mineralizes 4-chlorobenzoic acid. Japanese Journal of Toxicology and Environmental Health. (1998). , 44(1), 25-33.

[52] Hoskeri, R. S, Mulla, S. I, & Shouche, Y. S. Ninnekar HZ Biodegradation of 4-chlorobenzoic acid by *Pseudomonas aeruginosa* PA01 NC. Biodegradation. (2011). , 22(3), 509-516.

[53] Chang, K. H. Dunaway-Mariano D Determination of the chemical pathway for 4-chlorobenzoate:coenzyme A ligase catalysis. Biochemistry. (1996). , 35(41), 13478-13484.

[54] Gulick, A. M, & Lu, X. Dunaway-Mariano D Crystal structure of 4-chlorobenzoate:CoA ligase/synthetase in the unliganded and aryl substrate-bound states. Biochemistry. (2004). , 43(27), 8670-8679.

[55] Niedan, V. Schöler HF Natural formation of chlorobenzoic acids (CBA) and distinction between PCB-degraded CBA. Chemosphere. (1997). , 35(6), 1233-1241.

[56] Egland, P. G, & Gibson, J. Harwood CS Reductive, coenzyme A-mediated pathway for 3-chlorobenzoate degradation in the phototrophic bacterium *Rhodopseudomonas palustris*. Applied and Environmental Microbiology. (2001). , 67(3), 1396-1399.

[57] Romanov, V. Hausinger RP NADPH-dependent reductive *ortho* dehalogenation of 2,4-dichlorobenzoic acid in *Corynebacterium sepedonicum* KZ-4 and Coryneform bacterium strain NTB-1 via 2,4-dichlorobenzoyl coenzyme A. Journal of Bacteriology. (1996). , 178(9), 2656-2661.

[58] van den Tweel WJKok JB, de Bont JA Reductive dechlorination of 2,4-dichlorobenzoate to 4-chlorobenzoate and hydrolytic dehalogenation of 4-chloro-, 4-bromo-, and 4-iodobenzoate by *Alcaligenes denitrificans* NTB-1. Applied and Environmental Microbiology. (1987). , 53(4), 810-815.

[59] Hartmann, J, & Reineke, W. Knackmuss HJ Metabolism of 3-chloro-, 4-chloro-, and 3,5-dichlorobenzoate by a pseudomonad. Applied and Environmental Microbiology. (1979). , 37(3), 421-428.

[60] Baggi, G, Bernasconi, S, & Zangrossi, M. Chloro-, 2,3- and 3,5-dichlorobenzoate co-metabolism in a 2-chlorobenzoate-degrading consortium: role of 3,5-dichlorobenzoate as antagonist of 2-chlorobenzoate degradation- metabolism and co-metabolism of chlorobenzoates. Biodegradation. (2005). , 16(3), 275-282.

[61] Hickey, W. J. Focht DD Degradation of mono-, di-, and trihalogenated benzoic acids by *Pseudomonas aeruginosa* JB2. Applied and Environmental Microbiology. (1990). , 56(12), 3842-3850.

[62] Hernandez, B. S, Higson, F. K, & Kondrat, R. Focht DD Metabolism of and inhibition by chlorobenzoates in *Pseudomonas putida* Applied and Environmental Microbiology. (1991). , 111.

[63] Lang, G. H, Ogawa, N, Tanaka, Y, Fujii, T, Fulthorpe, R. R, & Fukuda, M. Miyashita K Two kinds of chlorocatechol 1,2-dioxygenase from 2,4-dichlorophenoxyacetate-degrading *Sphingomonas* sp. strain TFD44. Biochemical and Biophysical Research Communications. (2005). , 332(4), 941-948.

[64] Moiseeva, O. V, Solyanikova, I. P, Kaschabek, S. R, Groning, J, Thiel, M, Golovleva, L. A, & Schlomann, M. A new modified *ortho* cleavage pathway of 3-chlorocatechol

degradation by *Rhodococcus opacus* 1CP: genetic and biochemical evidence. Journal of Bacteriology. (2002). , 184(19), 5282-92.

[65] Mars, A. E, Kasberg, T, Kaschabek, S. R, Van Agteren, M, & Janssen, D. B. Reineke W Microbial degradation of chloroaromatics: Use of the *meta*- cleavage pathway for mineralization of chlorobenzene. Journal of Bacteriology. (1997). , 179(14), 4530-4537.

[66] Arensdorf, J. J. Focht DD Formation of chlorocatechol *meta* cleavage products by a pseudomonad during metabolism of monochlorobiphenyls. Applied and Environmental Microbiology. (1994). , 60(8), 2884-2889.

[67] Reineke, W, Jeenes, D. J, & Williams, P. A. Knackmuss HJ TOL plasmid pWW0 in constructed halobenzoate-degrading *Pseudomonas* strains: Prevention of *meta* pathway. Journal of Bacteriology. (1982). , 150(1), 195-201.

[68] Arensdorf, J. J, & Focht, D. D. A *meta* cleavage pathway for 4-chlorobenzoate, an intermediate in the metabolism of 4-chlorobiphenyl by *Pseudomonas cepacia* Applied and Environmental Microbiology. (1995). , 166.

[69] Brinkrolf, K, & Brune, I. Tauch A Transcriptional regulation of catabolic pathways for aromatic compounds in *Corynebacterium glutamicum* Genetics and Molecular Research. (2006). , 5(4), 773-789.

[70] Vrana, B, Dercová, K, & Baláž, Š. Ševčíková A Effect of chlorobenzoates on the degradation of polychlorinated biphenyls (PCB) by *Pseudomonas stutzeri*. World Journal of Microbiology and Biotechnology. (1996). , 12(4), 323-326.

[71] Bartels, I, & Knackmuss, H. J. Reineke W Suicide inactivation of catechol 2,3-dioxygenase from *Pseudomonas putida* mt-2 by 3-halocatechols. Applied and Environmental Microbiology. (1984). , 47(3), 500-505.

[72] Blasco, R, Wittich, R. M, Mallavarapu, M, & Timmis, K. N. Pieper DH From xenobiotic to antibiotic, formation of protoanemonin from 4- chlorocatechol by enzymes of the 3-oxoadipate pathway. The Journal of Biological Chemistry. (1995). , 270(49), 29229-29235.

[73] Fava, F. Di Gioia D, Bignami A, Marchetti L An attempt to control the polychlorocatechol pigment production during 3-chlorobenzoate aerobic co-metabolism in growing-cell batch culture. Chemosphere. (1994). , 29(1), 39-46.

[74] Häggblom MM Microbial breakdown of halogenated aromatic pesticides and related compoundsFEMS Microbiology Letters. (1992). , 103(1), 29-71.

[75] Fava, F, & Digioia, D. Marchetti L Characterization of a Pigment Produced by Pseudomonas-Fluorescens during 3-Chlorobenzoate Co-Metabolism. Chemosphere. (1993). , 27(5), 825-835.

[76] Soudek, P, Petrová, Š, Benešová, D, & Kotyza, J. Vaněk T Fytoremediace a možnosti zvýšení jejich účinnosti. Chemické Listy. (2008).

[77] Macek, T, & Macková, M. Káš J Exploitation of plants for the removal of organics in environmental remediation. Biotechnology Advances. (2000)., 18(1), 23-34.

[78] Morikawa, H. Erkin OC Basic processes in phytoremediation and some applications to air pollution control. Chemosphere. (2003)., 52(9), 1553-1558.

[79] Peuke, A. D. Rennenberg H Phytoremediation. EMBO Reports. (2005)., 6(6), 497-501.

[80] Suresh, B. Ravishankar GA Phytoremediation--a novel and promising approach for environmental clean-up. Critical Reviews in Biotechnology. (2004).

[81] Macek, T, Uhlik, O, Jecna, K, Novakova, M, Lovecka, P, Rezek, J, Dudkova, V, Stursa, P, Vrchotova, B, Pavlikova, D, Demnerova, K, & Mackova, M. Advances in phytoremediation and rhizoremediation In: Singh A, Kuhad RC, Ward OP, (eds.) Advances in Applied Bioremediation: Springer; (2009)., 257-277.

[82] Deavers, K, Macek, T, & Karlson, U. G. Trapp S Removal of 4-chlorobenzoic acid from spiked hydroponic solution by willow trees (*Salix viminalis*). Environmental Science and Pollution Research. (2010)., 17(7), 1355-1361.

[83] Siciliano, S. D. Germida JJ Enhanced phytoremediation of chlorobenzoates in rhizosphere soil. Soil Biology and Biochemistry. (1999)., 31(2), 299-305.

[84] Haby, P. A. Crowley DE Biodegradation of 3-chlorobenzoate as affected by rhizodeposition and selected carbon substrates. Journal of environmental quality. (1996)., 25(2), 304-310.

[85] Buyanovsky, G. A, Kremer, R. J, & Gajda, A. M. Kazemi HV Effect of corn plants and rhizosphere populations on pesticide degradation. Bulletin of Environmental Contamination and Toxicology. (1995)., 55, 689-696.

[86] Ferro, A. M, & Sims, R. C. Bugbee B Hycrest crested wheatgrass accelerates the degradation of pentachlorophenol in soil. Journal of environmental quality. (1994)., 23(2), 272-279.

[87] Walton, B. T, Hoylman, A. M, Perez, M. M, Anderson, T. A, Johnson, T. R, Guthrie, E. A, & Christman, R. F. Rhizosphere microbial communities as a plant defense against toxic substances in soils. In: Anderson T. (ed.) Bioremediation through Rhizosphere Technology: American Chemical Society; (1994)., 82-92.

[88] Alkorta, I. Garbisu C Phytoremediation of organic contaminants in soils. Bioresource Technology. (2001)., 79, 273-276.

[89] Louvel, B, & Cébron, A. Leyval C Root exudates affect phenanthrene biodegradation, bacterial community and functional gene expression in sand microcosms. International Biodeterioration & Biodegradation. (2011)., 65(7), 947-953.

[90] Bais, H. P, Weir, T. L, Perry, L. G, & Gilroy, S. Vivanco JM The role of root exudates in rhizosphere interactions with plants and other organisms. Annual Review of Plant Biology. (2006)., 57, 233-266.

[91] Vessey JK Plant growth promoting rhizobacteria as biofertilizersPlant and Soil. (2003)., 255(2), 571-586.

[92] Walker, T. S, Bais, H. P, & Grotewold, E. Vivanco JM Root Exudation and Rhizosphere Biology. Plant Physiology. (2003)., 132, 44-51.

[93] Compant, S, Duffy, B, Nowak, J, & Clément, C. Barka EA Use of Plant Growth-Promoting Bacteria for Biocontrol of Plant Diseases: Principles, Mechanisms of Action, and Future Prospects. Applied and Environmental Microbiology. (2005). , 71(9), 4951-4959.

[94] Neilands, J. B. Leong SA Siderophores in Relation to Plant Growth and Disease. Annual Review of Plant Physiology. (1986)., 37, 187-208.

[95] Khalid, A, & Arshad, M. Zahir ZA Screening plant growth-promoting rhizobacteria for improving growth and yield of wheat. Journal of Applied Microbiology. (2004)., 96(3), 473-480.

[96] Glick, B. J, Cheng, Z, & Czerny, J. Duan J Promotion of plant growth by ACC deaminase-producing soil bacteria European Journal of Plant Pathology. (2007). , 119(5), 329-339.

[97] Dittmann, J, & Heyser, W. Bucking H Biodegradation of aromatic compounds by white rot and ectomycorrhizal fungal species and the accumulation of chlorinated benzoic acid in ectomycorrhizal pine seedlings. Chemosphere. (2002)., 49(3), 297-306.

[98] Siciliano, S. D, & Germida, J. J. Headley JV Evaluation of prairie grass species as bioindicators of halogenated aromatics in soil. Environmental Toxicology and Chemistry. (1997)., 16(3), 521-527.

[99] Siciliano, S. D. Germida JJ Degradation of chlorinated benzoic acid mixtures by plant-bacteria associations. Environmental Toxicology and Chemistry. (1998)., 17(4), 728-733.

Chapter 2

Development of a Bioremediation Technology for the Removal of Thiocyanate from Aqueous Industrial Wastes Using Metabolically Active Microorganisms

Yogesh B. Patil

Additional information is available at the end of the chapter

http://dx.doi.org/10.5772/56975

1. Introduction

Contamination of water and soil environment due to the release of toxic and hazardous chemicals as a result of industrialization has taken its toll by causing environmental pollution. If not treated and managed appropriately, toxic and hazardous pollutants may cause severe detrimental (negative), reversible or irreversible, intangible and incapacitating impacts on all forms of living cells. Thiocyanate ($N{\equiv}C{-}S^-$) is one such known hazardous chemical and an important member of cyanide (CN^-) family. It is a simple inorganic and one carbon (C-1) compound. Despite its toxicity, it is introduced into the environment by natural (principally by biological cyanide detoxification processes) as well as industrial processes (Kelly and Baker, 1990; Wood, 1975). Thiocyanate (SCN^-) has some novel properties. It is linear in nature, electronegative polyatomic ion and a good example of pseudohalide; and therefore produced on a grand scale for its use in diverse industrial processes such as dyeing, acrylic fibre production, thiourea production, photo-finishing, herbicide and insecticide production, metal extraction and electroplating industries (Hughes, 1975). SCN^- is also known for its applications in soil sterilization and corrosion inhibition (Beekhuis, 1975). Consequently, these industries emanate large volumes of SCN^- bearing wastewaters. Apart from SCN^-, these effluents might contain other contaminants like heavy metals, cyanide (CN^-), metal-cyanides (M_xCN) and metal-thiocyanates (M_xSCN). Cyanide has the potential to reacts readily with sulphur to produce SCN^- and any industry with cyanide in its waste is a potential source of SCN^- contamination. Steel manufacturing, metal mining and electroplating units are some examples of such industries.

All the species of cyanide family (viz. thiocyanate, cyanide and their metal complexes) have potential to interacting with living cells and strong tendencies to connect to proteins and thereby acts as a non-competitive inhibitor (Westley, 1981). This fact necessitate the industries using and/or emanating SCN^- to adequately detoxify the effluents on priority basis before its discharge in soil and water environment; as it may pose detrimental implications on aquatic life. Moreover, in water scarce situations such untreated and partially treated wastewaters could not be recycled back into the industrial processes. The concentration of SCN^- arising from all the above mentioned sources is normally in the range of 5 - 110 mg/l (Mudder and Whitlock, 1984). Although many statutory agencies across the world have set the statutory limits for cyanide and heavy metal discharge, till date there are no such prescribed limits set or documented for SCN^- discharge. Earlier scientific studies indicate that in general, SCN^- is approximately 7 to 10 times less toxic then free cyanide species. The US-health service cites 0.01 mg/l as guideline and 0.2 mg/l as the permissible limit for cyanide species. In India, the Central Pollution Control Board (CPCB) had set a Minimum National Standard (MINAS) limit for cyanide as 0.2 mg/l. Therefore, the cyanide bearing effluents generated from industries needs suitable treatment to bring down the total cyanide levels below 0.2 mg/l. Taking into consideration the mentioned facts, standards for discharge of SCN^- could readily be deduced to 1 mg/l to be on the safer side. In order to minimize the risk of exposure to the public and aquatic ecosystems, the clean-up of SCN^- contaminated wastewaters is therefore necessary. Patil and Kulkarni (2008) have reported the environmental sensitivity and safety aspects in mining industries in regard to cyanide. Impact of cyanide species on fresh water fish Catla catla have also been reported (Prashanth and Patil, 2008).

Numerous technologies are currently employed to detoxify SCN^- bearing effluents; and the most widely being used is direct alkaline chlorination or addition of hypochlorite. However, this method produces large aggregates of chemical sludge, which does not have any further utilization and is environmentally hazardous to handle (Lanza and Bertazzoli, 2002). As per Indian regulations, such hazardous chemical sludge is transited from the industrial location to a specially designed Treatment, Storage and Disposal Facility (TSDF) thereby increasing the overall energy consumption, transportation cost and air pollution. Secondly, chlorination also fails to bring the concentration of SCN^- (and other CN^- species) within the statutory limits especially when heavy metals are present in the effluents. Thirdly, chlorination increases the total dissolved solids (TDS) content of the treated wastewaters, which makes it unfit for further use. Other physico-chemical processes like hydrogen peroxide oxidation, ozone oxidation, electrolytic decomposition, etc. are highly expensive and are rarely used for the treatment of SCN^-. Thus, there is a pressing need for the development of an alternative treatment process capable of achieving high degradation efficiency at low cost.

Bioremediation (biological treatment system) using metabolically active (live) microorganism is one such effective alternative for the detoxification of toxic chemical wastes. This process has immense potential of treating variety of pollutants (both toxic and non-toxic); has several advantages over conventional methods and therefore being explored by the researchers world-wide. Microorganisms capable of utilizing C-1 compounds like CN^- and

M_xCNs are well documented and have been studied for long time (Dash et al., 2009; Gurbuz, et al., 2004; Karavaiko et al., 2000; Patil and Paknikar, 2000a; Patil and Paknikar, 2000b; Patil and Paknikar, 2001; Patil et al., 2012). Some research papers on biodegradation of SCN^- have also been reported (Chaudhari and Kodam, 2010; Hung and Pavlostathis, 1998; Patil, 2008a; Van Zyl et al., 2011). Use of metabolically passive (dead or inactive) microorganisms for the removal and recovery of metal-cyanides and SCN^- have also been reported (Gaddi and Patil, 2011; Patil, 2012; Patil and Paknikar, 1999; Thakur and Patil, 2009). Successful efforts to set-up large scale bioremediation technology for the treatment of cyanide, metal-cyanide and SCN^- from mining effluent have been made on commercial scale (Mudder and Whitlock, 1984). However, there are very few reports on the microbial SCN^- degradation from the process development point of view (Patil, 2006; Patil, 2008a; Patil, 2008b; Patil, 2011; Sorokin et al., 2001; Stratford et al., 1994). Moreover, utilization of SCN^- by microbes as a suitable growth substrate (carbon and/or nitrogen source) is poorly understood. Lack of scientific knowledge in this regard may pose problems in the biological treatment systems. The author in the present research chapter focuses on the development of a bioremediation technology for the removal of SCN^- from aqueous industrial wastes using metabolically active microorganisms.

2. Materials and methods

2.1. Analyses, chemicals and glassware

Potassium SCN^- (KSCN) was obtained from Qualigens, Mumbai, India. SCN^- assay was carried out spectrophotometrically (Spectronic, Model-20D, India) using ferric nitrate method at 460 nm as described in Standard Methods (APHA-AWWA-WEF, 1998). Digital pH meter (Elico, Model Ll-120, India) was employed to determine pH of solutions. Bacterial population from culture media, activated sludge and soil were determined microscopically (Metzer, India, METZ-778A) using Neubauer's chamber (Fein-Optik, Blankenburg, GDR) and by total viable count (TVC). Analytical grade chemicals were used for all experiments. Reagents were prepared in glass distilled water and stored under refrigerated conditions (8-10°C).

2.2. Enrichment and isolation of SCN^- degrading bacterial consortium

Enrichment culture and growth of mixed bacterial community (bacterial consortium) was carried out using M-9 minimal salts medium (MSM) (Patil and Paknikar, 2000a). One litre of medium contained $Na_2HPO_4.2H_2O$ - 3.0 g; KH_2PO_4 - 1.5 g; NaCl - 0.25 g; distilled water - 1000 ml and 1 ml/l trace metal solution (Bauchop and Elsden, 1960; Millar, 1972). The medium pH was adjusted to 7.5 using 1 M NaOH/HCl. Glucose (10 mM) was added as the sole source of carbon and energy. SCN^- (50 mg/l) was supplemented to the enrichment medium as the sole source of nitrogen. Enrichment culture for the isolation of SCN^- degrading microorganisms were set-up in aerobic and unsterilized conditions using activated sludge (obtained from secondary treatment of sewage treatment plant) and garden soil. Both the samples were collected in clean polythene bags and carried to the laboratory. Two cylindrical glass jars (reactors) of 1200 ml capacity each were employed for the enrichment purpose. Working volume of the glass reactor was 1000 ml. 100 ml of activated sludge or 100 g of garden soil was added in 900 ml M-9 MSM containing SCN^- and glucose as the source of nitrogen and carbon, respectively in order to obtain the final concentration of SCN^- and glucose as 50 mg/l and 10 mM, respectively. Enrichment was carried out at the pH 7.5. Both the glass reactors were incubated at room temperature (30±2°C). Air was sparged continuously at the bottom of medium at the rate of 1000±50 ml/min using electrical aerator units. Seven to eight successive transfers of 10% solution enriched with bacterial flora were given periodically in the fresh M-9 MSM containing SCN^- and glucose as mentioned earlier (Patil, 2006; Patil, 2008a).

2.3. Purification and Identification of bacterial cultures

The enrichment cultures as obtained were streaked on nutrient agar medium and M-9 agar medium (containing SCN^- and glucose) plates in aseptic conditions at 35°C for 48-96 h. In all, six diverse types of bacterial colonies (three each from garden soil and activated sludge enrichment) appeared on the plates. The cultures were further purified and then transferred to nutrient agar and M-9 agar slopes. By way of periodic transfers, one set of bacterial consortium was consistently maintained in liquid medium (i.e. M-9 MSM) (Patil 2008a; Patil, 2008b). Further, the isolated bacterial cultures were subjected for microscopic examination (Gram staining and motility) and cultural characteristics on the nutrient agar plates. Bergey's Manual of Systematic Bacteriology (Holt, 1989) was used to identify the SCN^- degrading bacterial cultures up to genus level only.

2.4. SCN^- degradation efficiency of the isolated bacterial cultures

Quantitative studies on SCN^- degradation were performed to determine the efficiency of isolated cultures in their individual capacity and mixed form. Experiments were performed in 250 ml Erlenmeyer flasks containing 100 ml sterile M-9 MSM (pH 7.0) and 50 mg/L potassium thiocyanate (KSCN), which acted as a nitrogen source. 10 mM glucose was used as carbon source. Bacterial cell suspension of 0.1 ml containing 10^8 cells/ml were inoculated into the flasks and were incubated at 30°C in a rotary shaker incubator (Remi, India) at 150 rpm for 48 h. Requisite controls were used and experiments were performed in duplicates and repeated twice. SCN^- contents were determined periodically as mentioned earlier. SCN^- degradation efficiency of individual and mixed bacterial culture was expressed in terms of percent total SCN^- degraded in 48 h. Reaction rate and first order rate constant for SCN^- biodegradation were calculated experimentally using equation 1 and 2, respectively (Sellers, 1999).

$$\Delta C / \Delta t = k\, C \tag{1}$$

$$\ln C_t - \ln C_o = - k\, t\, \mu \tag{2}$$

Where, C = SCN^- concentration (mg/l); t = time (h); k = rate of reaction / first order rate constant (h^{-1}); C_o= initial SCN^- concentration (mg/l) and C_t= SCN^- concentration at time t.

2.5. Utilization of SCN^- as the sole source of cellular nitrogen by bacterial consortium

Batch mode experiments on SCN^- biodegradation were conducted in aseptic conditions with 100 ml M-9 MSM in 250 ml conical flasks. The medium was augmented with carbon and nitrogen sources with different combinations as mentioned: - (i) potassium thiocyanate - KSCN (50 mg/l) as the sole carbon and nitrogen source or (ii) KSCN (50 mg/l) and glucose (10 mM) as the sole nitrogen and carbon, respectively or (iii) KSCN (50 mg/l) and NH_4Cl (1 mM) as the sole carbon and nitrogen, respectively or (iv) KSCN (50 mg/l), NH_4Cl (1 mM) and glucose (10 mM) as the nitrogen, nitrogen and carbon sources, respectively. All experiments were conducted at pH 7.0. Flasks were incubated at 30°C on a rotary shaker incubator (Remi, CIS-24 BL) at 150 rpm for 48-72 h (Patil, 2011).

2.6. Factors influencing SCN^- biodegradation

Series of batch culture experiments were conducted to investigate the influence of various parameters on SCN^- biodegradation. Experimental conditions used were as follows: 150 ml capacity Erlenmeyer flasks with 25 ml M-9 MSM containing SCN^- (50 mg/l) and glucose (10 mM). 1 ml of previously grown culture (for 24 h) having cell density of 10^8 cells/ml was used as inoculum. The flasks were incubated in stationary conditions. Impact of pH (5.0 - 9.0), temperature (20 – 45 °C), initial cell density (10^5 – 10^9 cells/ml) and glucose (1-20 mM) were checked by running different set of experiments, wherein, one parameter was varied keeping the others constant. Periodic analyses were conducted as mentioned earlier.

2.7. Biosorption of SCN^- by bacterial consortium at high cell density

Experiments were performed in 150 ml flasks. 50 ml aliquots of SCN^- (50 mg/l) adjusted to optimum pH (7.0) was contacted with bacterial consortium (10^8 cells/ml). The culture was inactivated by boiling for the period of 10 min prior to contact. The flasks were incubated at 30°C on a rotary shaker (150 rpm) for 24 h. Bacterial cells were separated by centrifugation at 1000 rpm for 10 min and the cell free supernatant was subjected to determine the residual SCN^- concentration.

2.8. Impact of cations and anions on SCN^- biodegradation

Batch experiments were performed under optimized conditions as described earlier. Impact of diverse cations, especially heavy metals (0.1 mM each) on biodegradation of SCN^- was studied. Metals were added as sulfate salts (range 0.1-1 mM). To study the influence of sulfates, additional sulfate was added to the medium as sodium sulfate. Chlorides were added as sodium chloride in the range of 1-10 mM, while cyanide was added as sodium cyanide (0.1-1 mM).

2.9. Degradation of SCN^- from industrial effluent by bacterial consortium

SCN^- effluent was synthetically prepared in the laboratory because of the difficulty in procurement of effluent from industry. This was to test the practical applicability of the microbial process for degradation of SCN^-. Batch experiments were performed as mentioned earlier under optimized conditions (pH 7.0, temperature 30°C and bacterial cell density of 10^8 cells/ml). Thiocyanate served as nitrogen source, while sucrose (COD 500 mg/l) was used as carbon source. Parameters such as pH, SCN^-, COD and soluble metal content were measured at regular intervals for a period of 48 h.

2.10. Treatment of SCN^- waste in a Continuous Treatment System (CTS)

Thiocyanate containing simulated was treated in a continuous treatment system (CTS) as shown in Fig. 1. The CTS comprised of cylindrical glass column (height, 24 cm; diameter, 8 cm and total volume 0.2 L) containing one-litre simulated SCN^- effluent (50 mg/l SCN^-) having COD of 500-600 mg/l. The consortium culture was inoculated at the level of 10^8 cells/ml (final cell density) and the contents of the reactor were stirred by sparging air at the rate of 1000 ml/min. The pH of wastewater supplemented with nutrients was adjusted between 7.0-7.3 (using 1 M $NaOH/H_3PO_4$) and then added from the top of the reactor by manual adjustment at the flow rate of approximate 40-50 ml/h as calculated from mass balance equation. The treated effluent was removed from the bottom at the same flow rate. The CTS was operated at ambient temperature (30±2°C) in continuous mode for over a period of 30 days (720 h) by periodically checking the influent and effluent water characteristics for pH, SCN^-, COD and cell count according to the method prescribed in Standard Methods (APHA- AWWA-WEF, 1998).

3. Results and discussion

3.1. Enrichment and isolation of SCN^- degrading bacterial consortium

Both the enrichment culture (garden soil and activated sludge) elucidated that the time incurred reduced significantly with each subsequent transfer cycle for complete disappearance of SCN^-. Time taken for biodegradation in first, third and fifth cycle was 100, 80 and 70 hours, respectively. After seventh cycle the time taken for biodegradation of >98% SCN^- was stabilized around 40-45 hours. Each subsequent transfer was given in fresh M-9 MSM soon after the SCN^- concentration reached to < 1 mg/l (efficiency ≥98%) in the previous cycle. The bacterial count was consistently >10^8 cells/ml during each transfer cycle. pH and total viable count (TVC) of garden soil prior to enrichment was 8.12 and 3.5 x 10^8 cells/ml, respectively; and for activated sludge it was 7.64 and >1.2 x 10^{10} cells/ml, respectively.

The main objective of the present work was to isolate bacterial cultures capable of degrading SCN^- from the aqueous industrial wastes. In order to accomplished this objective, activated sludge and garden soil was subjected to the most powerful tool called 'enrichment culture', which is popularly being used world across by the microbiologists to selected desired type

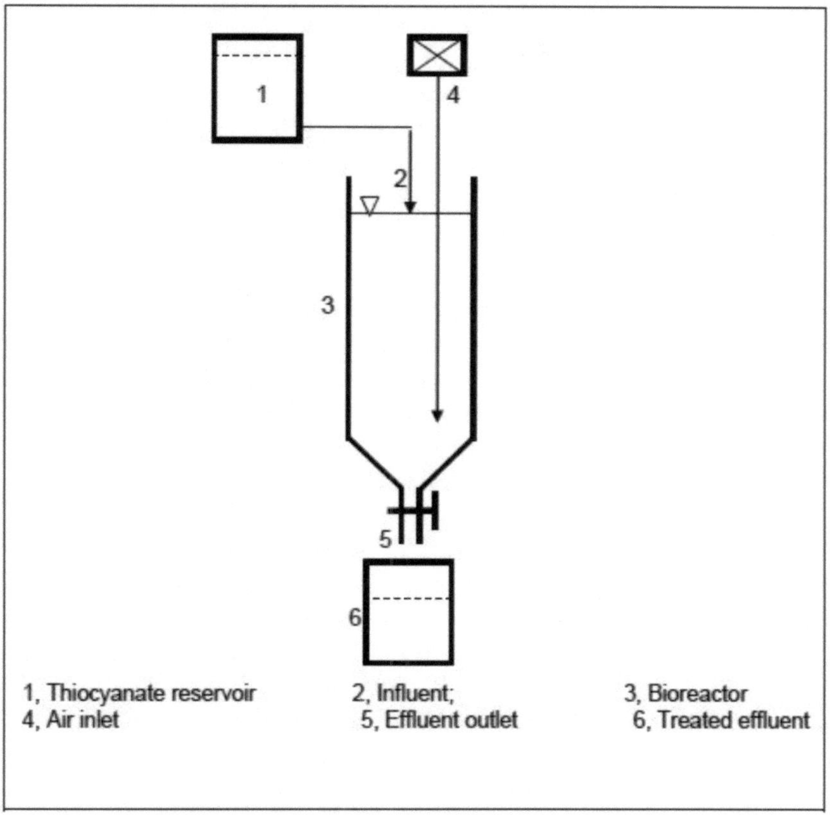

Figure 1. Schematic outline of laboratory scale Continuous Treatment System (CTS) for degradation of thiocyanate

of microorganims. There are reports of successfully utilizing this tool for the isolation of microorganism capable of degrading toxic and hazardous chemicals like SCN^- and M_xCN (Patil, 1999; Patil, 2008a; Sorokin *et al.*, 2001). Reduction in time during each subsequent transfer could be explained by the fact that bacterial flora in the enrichment medium got gradually acclimatized to the hazardous chemical environment. High bacterial population (>10^8 cells/ml) in both the procured samples indicated the presence of substantial organic matter content and nutrient availability, thus giving enhanced probability to obtain SCN^- degrading cultures.

3.2. Purification and identification of bacterial cultures

In all, six heterotrophic bacterial cultures (three each from both enrichment cultures) capable of degrading SCN^- were isolated by enrichment technique subsequently followed by streak plate and spread plate technique that were employed for purification of cultures. Microscopic examination showed that all the six bacterial culture were Gram-negative rods and motile. Detailed cultural characteristics were previously reported by Patil (2008a). Based on cultural and biochemical characteristics, all the six identified bacterial cultures belonged to the genus *Pseudomonas* as and reported by Patil (2008b).

The microbial source employed for enrichment culture for the isolation of thiocyante degrading microorganisms were garden soil and sewage sludge of STP. These sites did not have any past history of cyanide or SCN^- contamination. The prime objective was to test whether SCN^- degrading bacterial cultures could be isolated from such non-cotaminated sites and secondly to conduct a comparative assessment of the cultures isolated from two completely different niche areas. The fact that six SCN^- degrading cultures could be isolated from these samples indicates that SCN^- degradation is an intrinsic property of certain microorganisms and that no prior exposure is required to induce this property. SCN^- degrading ability of various heterotrophic and autotrophic microorganisms have been reported by few authors (Kwon *et al.*, 2002; Patil, 2006; Sorokin *et al.*, 2001; Stratford *et al.*, 1994).

3.3. SCN^- degradation efficiency of the isolated bacterial cultures

Data in Table 1 depicts the wide variation of SCN^- degradation efficiency of the bacterial isolates. However, the bacterial consortium isolated from garden soil and activated sludge showed maximum degradation of SCN^- (>99.9%) in 42 and 36 h giving the SCN^- degradation rate constant (k) of 0.0931 and 0.1086 per h, respectively. In contrast, isolate-2 degraded only 75.9% of SCN^- in 48 h (k = 0.029 per h). It was also observed that the first order rate constant of bacterial consortiums was 2-3 folds higher than their individual isolates.

Source	Bacterial Isolates	% SCN^- degradation		Rate of Reaction (mg/l/h)	First Order Rate constant (per h)
		With culture	Control (without culture)		
Bacterial cultures isolated from garden soil	*Pseudomonas* sp. # 1	78.24	0	0.815	0.0317
	Pseudomonas sp. # 2	75.91	0	0.790	0.0297
	Pseudomonas sp. # 3	82.47	0	0.859	0.0362
	Bacterial consortium (1+2+3)	>99.9 (in 42 h)	0	1.189	0.0931
Bacterial cultures isolated from activated sludge	*Pseudomonas* sp. # 4	92.07	0	0.959	0.0528
	Pseudomonas sp. # 5	87.65	0	0.913	0.0435
	Pseudomonas sp. # 6	89.41	0	0.931	0.0467
	Bacterial consortium (4+5+6)	>99.9 (in 36 h)	0	1.387	0.1086

Table 1. SCN^- degradation efficiency of pure and mixed bacterial cultures (Conditions: pH 7.0; Temperature 30°C; Inoculum size 10^5 cells/ml; SCN^- conc. 50 mg/L; Glucose 10 mM; Agitation speed: 150 rpm; Incubation time: 48 h) (Patil, 2008b)

This experiment was conducted to ascertain the efficacy of bacterial cultures in their individual capacity and in consortium form. And the results clearly revealed that that consortium of bacteria were efficient compared to individual (pure) isolates (Table 1). These results confirmed the studies carried out by Patil and Paknikar (2000a) on biodegradation of copper- and zinc-cyanide using bacterial consortium. In this study, the consortium consisted of four bacterial isolates out of which three were *Pseudomonas* sp. and one was *Citrobacter* sp. The wide variation in SCN^- degradation efficiency among the cultures tested in the present study could be a manifestation of the natural diversity. Bacterial consortium isolated from activated sludge was more efficient than the consortium isolated from garden soil. Uninoculated controls did not show any decrease in SCN^- levels confirmed that biodegradation of SCN^- was the predominant reaction taking place during SCN^- degradation by the cultures. Experimental determination of reaction rate and first order rate constants are essentially needed because such data gives valuable information regarding the time requirement for reaction completion and the size of the treatment facilities that must be provided (Patil, 2008b; Sellers, 1999).

3.4. Utilization of SCN^- as the sole source of cellular nitrogen by bacterial consortium

This experiment was carried out only on bacterial consortium isolated from activated sludge because it was more efficient than the consortium isolated from garden soil. Detailed results of this experiment could be obtained from Patil (2011). Overall results are summarized as follows. It was established that when SCN^- was supplemented in M-9 MSM as the sole carbon and nitrogen, the consortium failed to utilize SCN^-. The SCN^- concentration of 50 mg/l remained unchanged throughout the tested period of 48 h. However, it was found that the bacterial consortium was capable of utilizing SCN^- as the sole source of cellular nitrogen in the presence external carbon source like glucose (10 mM) within 40 h with an efficiency of >99.9%. This was also confirmed from the control experiments run simultaneously. In the third combination, SCN^- when supplemented as the sole carbon source in the presence of external nitrogen like ammonium chloride resulted in complete cessation of consortium growth. In the fourth combination, when SCN^- was supplied in MSM along with external carbon (glucose) and nitrogen (ammonium chloride) source, showed an interesting diauxic growth (diauxie) pattern of the bacterial consortium as shown in Fig. 2. The bacterial consortium preferentially utilized ammonium chloride first until its depletion and only then switched over to the utilization of SCN^- as nitrogen source.

All bioremediation processes essentially depends on the availability of principal nutrients in the wastes that could potentially be utilized by the microorganisms as either carbon and/or nitrogen source. Elucidating this is crucial because if SCN^- is utilized by bacterial consortium as both carbon and nitrogen source, then at practical scale external supplementation of nutrients will not be required, thereby benefiting the industries economically those using microbial technologies for the effluents containing SCN^-, cyanide and metal-cyanides. However, in the present study, the SCN^- compound posed toxic problems to the consortium for growth utilizing it as suitable growth substrate. It is well known that concentration of nitrogen required for a given amount of growth is less than the requirement for carbon it might be easier for bacterial consortium to utilize SCN^- as the source of nitrogen in the presence of a separate source of carbon and energy (Patil, 1999). Therefore, enrichment culture was de-

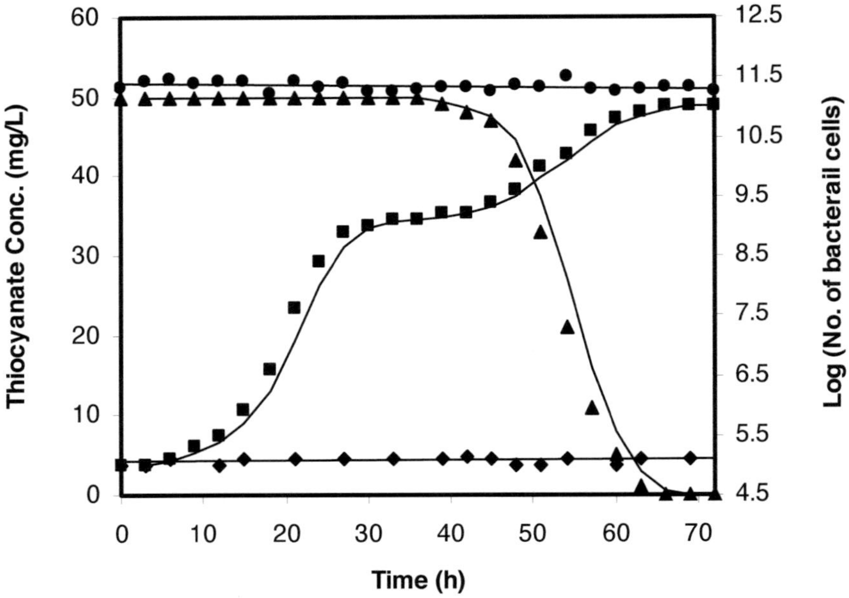

Figure 2. Diauxic growth pattern exhibited by bacterial consortium in the presence two nitrogen sources (viz. SCN^- and ammonium chloride) in the presence external carbon source (glucose). Growth of consortium (■) and SCN^- degradation (▲) in the presence of two N sources; SCN^- concentration in absence of consortium (●); Cessation of bacterial growth in absence of either nitrogen or carbon source (◆) (Patil, 2011)

signed/manipulated for the isolation of microorganisms capable of degrading SCN^- as the source of nitrogen nitrogen (Patil, 2006; Patil, 2008a). The experiments conducted explicitly proved that SCN^- is used by the consortium as nitrogen source in the presence of external carbon viz. glucose, thereby giving the C/N ratio of 10. In view of microbial process development, it is imperative to supplement some cheaper source of carbon like molasses, which is readily available in India at cheaper rate. Patil (1999) had successfully demonstrated the use of molasses as carbon source to develop a microbial technology for metal cyanide biodegradation/removal from wastes utilising it as the sole nitrogen source. There are few reports, which describe microbial SCN^- degradation utilising it as the sole nitrogen source (Bipinraj *et al.* 2003; Patil, 2011; Sorokin *et al.* 2001). The bacterial consortium ceased to grow when SCN^- was supplied as sole carbon source in the presence of external nitrogen. This could be attributed to the higher amount of available nitrogen compared to carbon (C/N ratio 0.5). Obviously the culture would find it more difficult to obtain sufficient amount of energy from low amount of carbon.

Example of diauxie pattern (biphasic growth) in *Escherichia coli* in the presence of two carbon sources (viz. glucose and lactose) is well documented (Atlas, 1997). In the present study, diauxic growth pattern was observed when two nitrogen salts (i.e. SCN^- and ammonium chloride) along with one carbon source (glucose) were supplied to the consortium. Ammonium chloride acted as preferred growth substrate by the consortium followed by SCN^- degradation. This suggests that SCN^- utilization by consortium culture is inducible. It was also revealed that biodegradation of SCN^- took place rapidly (within 25 h) in second phase of growth after the exhaustion of ammonium chloride from medium in the first phase. The biomass that built-up in the first phase of growth was readily available in the medium for SCN^- degradation in the second phase, which ultimately led to the rapid biodegradation of SCN^-. This result immediately suggests its possible application in bioreactor designing that will retain large microbial biomass. Immobilization of the biomass in bioreactor using inert material will certainly hasten the process of biodegradation of toxic SCN^-.

Further, it could be also observed from the experiments that decrease of SCN^- concentration in the MSM was concomitant with the increase in bacterial population. The fact that the final cell density obtained was considerably high (>10^8cells/ml) indicated the use of well acclimatised SCN^- tolerant culture, having high SCN^- removal efficiency and therefore has immense potential of using the microbial technology on industrial scale. The treatment of wastewater involves a number of chemical and biological reactions and conversions. The rate at which these reactions and conversions occur decides the size of the treatment facilities that must be provided (Tchobanoglous and Burton, 1997). The study also showed that SCN^- degradation by the bacterial consortium isolated from activated sludge was comparatively more efficient than the consortium isolated from garden soil. This might be due to the acclimation/tolerance of sewage microorganisms to a variety of hazardous and non-hazardous waste contaminants/components naturally existing in it, thereby making them more tolerant and efficient degraders as compared to the microbial flora prevailing in garden soil.

3.4. Factors influencing SCN^- biodegradation

Factors influencing SCN^- biodegradation was restricted to the bacterial consortium isolated from activated sludge because of its high degradation efficiency compared to the consortium isolated from garden soil as mentioned earlier. Table 2 shows that degradation of SCN^- was significantly influenced by the various factors tested. Optimum pH and temperature for maximum SCN^- biodegradation (>99.9%) was found to be 7.0 and 30°C, respectively. Under the optimized conditions of pH and temperature, the initial cell density had a substantial influence on the biodegradation efficiency of SCN^-. With initial cell density of 10^8 cells/ml, SCN^- degradation process completed within 24 h with >99.9% efficiency. As regard to carbon source, the consortium culture exhibited maximum biodegradation efficiency only above glucose concentration of 5 mM.

In general, the SCN^- containing effluents released from various industries have pH in neutral to alkaline range. Our study showed that growth of SCN^- degrading bacterial consortium occurred in wide range of pH (6.0-9.0), while the optimum being 7.0. From practical applicability point of view very little or no pH adjustment would be required for the effluents containing SCN^-. Experiments also showed the unchanged pH of the solution after SCN^- biodegradation. This may be perhaps due to the formation of ammonia as one of the by-products of SCN^- degradation, which neutralized the accumulated carboxylic acids in the medium. These results corroborate with the studies carried out by other researchers on the

Parameter	Range selected	% SCN^- biodegradation	Parameter	Range selected	% SCN^- biodegradation
pH	Control*	0	Cell Density (cells/ml)	Control	0
	5	35.3		10^5	19.5
	5.5	43.0		10^6	47.1
	6	75.3		10^7	62.8
	6.5	84.6		10^8	>99.9
	7	>99.9		10^9	>99.9
	7.5	89.2			
	8	87.6	Glucose (mM)	Control	0
	8.5	84.5		0.1	6.2
	9	84.2		1	24.7
	9.5	56.9		5	>99.9
Temperature	Control	0		10	>99.9
	10	16.3		20	>99.9
	20	51.8			
	30	>99.9			
	37	36.3			
	45	0			

* Control indicates flask without culture; All the values are the average of two readings

Table 2. Degradation of SCN^- by a bacterial consortium isolated from activated sludge as a function of pH, temperature, initial cell density and glucose

biodegradation/biodetoxification free cyanide (Babu *et al.*, 1993) and metal-cyanides (Patil and Paknikar, 2000b). The bacterial consortium used in the study was neutrophilic and mesophilic. Optimum temperature for thiocyanate degradation by bacterial consortium was 30°C. This is very important from the view point of actual applicability of the bioremediation process in a tropical country like India having ambient temperature ranging from 20-40°C. Results on the impact of inoculum size clearly showed that SCN^- degradation increased with the increase in inoculum size. Therefore, from the point of view of process development, it is essential to use a reactor capable of retaining high microbial biomass that will hasten the degradation of SCN^-. Results on the influence of glucose requirement for SCN^-degradation could possibly be explained on the basis of nutrient availability. Even though the available nitrogen in the form of SCN^- was ample the externally supplied glucose at concentrations < 5 mM limited the biodegradation process. However, at adequate glucose concentration of ≥ 5 mM complete degradation could take place. These findings confirm the utilisation of SCN^- as the sole source of nitrogen. In the previous studies carried out by Stratford *et al.* and Wood *et al.* glucose was supplemented at the concentration of 10 and 25 mM for the degradation of 3 mM (≈ 174 mg/l) and 2.5 mM (≈ 145 mg/l) of SCN^-, respectively (Stratford *et al.*, 1994; Wood *et al.*, 1998). However, these reports did not mention optimisation of this parameter, which needs to be worked out for economizing the process. In another study carried out by Patil (1999) on the biodegradation of various metal-cyanides (copper-, nickel-, zinc- and silver-cyanide), glucose was required in the range of 1-5 mM (≈ COD 100 – 500 mg/l) (Patil, 1999). Scanty information is available on the biochemical pathway involved in SCN^- biodegradation. For heterotrophic bacterium, Stratford *et al.* has proposed the conversion of SCN^- to carbon dioxide and ammonia via cyanate by an inducible enzyme; while the sulphur moiety gets hydrolysed to sulphide, which further gets oxidised to tetrathionates via formation of thiosulphate (Stratford *et al.*, 1994). It might be possible that bacterial cultures isolated in the present study also have similar SCN^- removal or tolerant mechanism.

3.5. Biosorption of SCN^- by bacterial consortium at high cell density

It can be seen from Table 3 that the bacterial consortium had low biosorption efficiency (~7-14%) at the pH values tested (6.5 to 7.5). In fact it is possible that biosorbed SCN^- also could subsequently be biodegraded by the live culture used in the biodegradation process. These observations confirmed that biodegradation of SCN^- was the predominant reaction taking place during detoxification of SCN^- by the consortium culture isolated from activated sludge.

pH	SCN^- concentration (mg/l)		% Sorption
	Initial	Final	
6.5	53.29	48.22	9.51
7.0 (optimum pH)	50.94	45.97	13.68
7.5	51.03	47.50	6.91
*Optimum pH			

Table 3. Biosorption of SCN^- by consortium culture

3.6. Impact of cations and anions on SCN^- biodegradation

Apart from SCN^- various metal cations and anions are normally present in the various industrial effluents. Therefore, the influence of some of the commonly occurring cations such as copper, cadmium, iron, lead, nickel, zinc and anions such as sulfates, chlorides and cyanide on SCN^- biodegradation was checked.

Table 4 shows the effect of various cations such as copper, nickel, zinc, cadmium, iron and lead on biodegradation of SCN^- in thiocyanate-metal system. It can be seen that biodegradation of thiocyanate was not affected in the presence of copper, nickel and zinc (degradation >90%). In presence of lead and cadmium the biodegradation efficiency was reduced by approximately 30 and 45%, respectively. Chromium and iron significantly affected the degradation of SCN^- by >80%. Anions such as sulfates and chlorides (1000 μM and 10000 μM, respectively), and cyanide (0.1-1 mM) however, did not had much impact on SCN^- degradation.

Thiocyanate + cations/anions	% Thiocyanate biodegradation
Thiocyanate (Control without culture)	0
Thiocyanate (Control with culture)	>99
Thiocyanate + Copper	95.2
Thiocyanate + Nickel	>99
Thiocyanate + Zinc	90.4
Thiocyanate + Cadmium	56.2
Thiocyanate + Iron	17.9
Thiocyanate + Lead	69.5
Thiocyanate + Chromate	11.8
Thiocyanate + Sulfate	74.0
Thiocyanate + Chlorides	90
Thiocyanate + Cyanide	>98

Table 4. Effect of cations and anions onμSCN^- biodegradation

Biodegradation of the SCN^- was adversely affected in the presence of metals such as iron and chromium. In case of free cyanide, it is known that cyanide ion has a great tendency to act as a ligand and can thus be found associated with metal-complexes (Pohlandt *et al.*, 1983). Cyanide complexes with different metals have widely varying stabilities depending on metal oxidation states (Cotton and Wilkinson, 1972), but can be broadly classified as weakly complexed cyanides and strongly complexed cyanides. The former group includes complexes with copper, cadmium, lead, nickel and zinc (Chapman, 1992) while the latter group consists of very stable hexacyanoferrates and chromium-cyanide (Lordi *et al.*, 1980). Since the chemistry of free cyanide and SCN^- almost being similar, the heavy metals are capable of forming ligands with thiocyanate to form metal-thiocyanate complexes. The high stability of iron-thiocyanate and chromium- SCN^- might be the reason for poor degradation efficiency observed in presence of these ions.

3.7. Degradation of SCN^- from industrial effluent by bacterial consortium

The SCN^- containing effluent simulated in laboratory could be effectively treated by the bacterial consortium with a degradation efficiency exceeding >99.9%. The time incurred for the complete biodegradation of SCN^- from waste waters was less than 24 h. Table 5 shows the parameters such as pH, total cyanide, COD and metal content before and after biodegradation. COD removal was more than 80%. It was also observed that the level of soluble copper, zinc, silver and nickel was reduced to less than 5 mg/l. There was no significant change in the pH after biodegradation.

Parameter	Simulated Industrial Effluent concentration (mg/l)			% Removal efficiency
	Before biodegradation	After biodegradation	Uninoculated control	
Color	Colourless	Colourless	Colourless	-
Turbidity	Clear	clear	Clear	-
pH	7.3	7.7	7.4	-
Thiocyanate	51.5	< 0.1	52.4	>99
Copper	12.5	1.92	13.1	84.6
Nickel	8.1	0.55	8.0	93.2
Zinc	18.3	2.17	16.9	88.1
Iron	3.9	0.1	4.1	97.4
Sulfates	55	57	62	-
Chlorides	42	45	50	-
Cyanide	5.2	0.13	4.7	97.5
COD	< 550	97	600	82.3

All the values given in the table are in mg/l, except pH

Table 5. Composition and biodegradation of simulated industrial effluent

The microbial process for degradation of thiocyanate was found to be highly effective in the detoxification of simulated industrial effluents. The levels of total thiocyanate, COD and metals could be brought down below the statutory limits as per Indian Standards (IS: 2490-1981). The treatment of effluent required more time as compared to the synthetic solutions. It is known that the applicability of any such process to real effluents is always complicated by the fact that effluents contain a variety of other contaminants which might interfere with or prolong the detoxification process. However, it must be emphasized that the microbial process described was highly efficient, safe and environment-friendly. In addition, the process had the potential of becoming economically attractive if scaled-up to a sufficient level, especially as a continuous operation. Therefore, it was decided to further develop the process in continuous mode and evaluate its performance.

3.8. Treatment of SCN^- waste in a Continuous Treatment System (CTS)

Studies in CTS showed that SCN^- level in the treated effluent was consistently below 0.1 mg/l for over a period of 30 days. The HRT of CTS was constant during the treatment period around 20 h. A closer monitoring of the CTS revealed that further reduction in the HRT was not possible because the bacterial cells could not be retained in the system. The COD removal efficiency after treatment was >75% for thiocyanate effluent (Table 6).

Parameter	Influent	Effluent	Bureau of Indian Standards (BIS)
pH	7.0 - 7.3	7.5 - 7.7	5.5 - 9.0
Thiocyanate	51.67 ± 2.1	0.03± 0.01	NA
COD	596 ± 103	147 ± 41	250

All the figures in the table are expressed in mg/l, except pH; *HRT of the system was ~20 h. Figures represent average values of 30 readings taken each at 24 h interval

Table 6. Treatment of metal-cyanide waste waters in CTS

The results of CTS showed that SCN^- was degraded efficiently by the bacterial consortium with the minimum hydraulic retention time (HRT) of approximately 20 h. However, there was no reduction in the HRT of CTS further. The main reason for this was the continuous loss of active biomass from the reactor, which makes it unattractive from process economics point of view. This necessitates the immobilization of the bacterial consortium in the reactor. In principle, it is possible to retain active biomass in CTS if the culture used has an inherent property of producing wall growth. Further, biomass retention is also possible by changing the reactor design, introducing inert support material or changing nutrient supplementation, etc. However, the consortium culture used in the present studies did not show wall growth. Also, in our studies during optimization of process parameters it was conclusively proved that degradation efficiency increased with the increase in cell number, which in turn hastens the degradation of SCN^-. Thus, the above results emphasize the fact that the bioremediation process developed during the course of present work is highly efficient and completely safe. After further scale-up the bacterial process developed could have the following advantages: (i) no sludge generation; (ii) no expensive chemical additives required; (iii) very little or no pH adjustment required; (iv) the process would be easy to operate and maintain. Thus, the bacterial process developed could have the potential of becoming an economical and reliable alternative to the conventional processes employed for the treatment of SCN^- bearing industrial effluents on a commercial scale.

Acknowledgements

The author gratefully acknowledges the research grant provided by University Grants Commission (UGC), WRO, Pune.

http://dx.doi.org/10.5772/56975

Author details

Yogesh B. Patil

Address all correspondence to: head_respub@siu.edu.in

Symbiosis Institute of Research and Innovation (SIRI), Symbiosis International University (SIU), Lavale, Pune, Maharashtra, India

References

[1] APHA, AWWA, WEF (1998) *"Standard Methods for the Examination of Water and Wastewater"*, 20th Ed. American Public Health Association, Washington, DC.

[2] Atlas, R.M. (1997) *Principles of Microbiology*, McGraw-Hill, New York.

[3] Babu, G.R.V., Wolfram, J.H. and Chapatwala, K.D. (1993) Degradation of Inorganic Cyanides by Immobilised *Pseudomonas putida* Cells, In: Torma, A.E., Apel, M.L. and Brierley, C.L. eds., Biohydrometallurgical Technologies, The Minerals Metals and Materials Society, Warrendale, PA, Vol. II, pp. 159-165

[4] Bauchop, T. and Elsden, S.R. (1960) The growth of microorganisms in relation to energy supply, Gen. Microbiol. 23 : 457.

[5] Beekhuis, H.A. (1975) Technology and Industrial Application In: Newmann, A.A., Chemistry and Biochemistry of Thiocyanic Acid and its Derivatives, Academic Press, London.

[6] Bipinraj, N.K., Joshi, N.R. and Paknikar, K.M. (2003) *"Biohydrometallurgy: A Sustainable Technology in Evolution"*, In: International Biohydrometallurgy Symposium, Athens, pp. 491.

[7] Chaudhari, A.U. and Kodam, K.M. (2010) Biodegradation of thiocyanate using coculture of *Klebsiella pneumoniae* and *Ralstonia* sp. *Applied Microbiology and Biotechnology* 85(4): 1167-1174.

[8] Cotton, F.A. and Wilkinson, G. (1972) *Advanced Inorganic Chemistry: A Comprehensive Text*. Wiley-Interscience, New York.

[9] Dash, R.R., Gaur, A. and Balomajumder, C. (2009) Cyanide in industrial wastewaters and its removal: A review on biotreatment. *Journal of Hazardous Materials*, 163(1):1-11.

[10] Gaddi Shivanand S and Patil Yogesh B (2011) Screening of some low-cost waste biomaterials for the sorption of silver-cyanide [Ag(CN)2-] from aqueous solutions. *International Journal of Chemical Sciences* 9: 1063-1072.

[11] Gurbuz, Fatma, Ciftci, Hasan, Akcil, Ata and Karahan, Aynur Gul. (2004) Microbial detoxification of cyanide solutions: a new biotechnological approach using algae, *Hydrometallurgy*, 72, 167-176.

[12] Holt, J.G. (1989) *Bergey's Manual of Systematic Bacteriology*. Vol. I. Williams and Wilkins, Baltimore, USA.

[13] Hughes, M.N. (1975) General Chemistry. In: Newmann, A.A., ed., Chemistry and Biochemistry of Thiocyanic Acid and its Derivatives, Academic Press, London.

[14] Hung, C.H. and Pavlostathis, S.G. (1998) Fate and transformation of thiocyante and cyanate under methanogenic conditions. *Applied Microbiology and Biotechnology* 49:112-116.

[15] Karavaiko, G.I., Kondrat'eva, T.F., Savari, E.E., Grigor'eva, N.V. and Avakyan, Z.A. (2000) Microbial degradation of cyanide and thiocyanate, *Microbiology*, 69(2):167-173.

[16] Kelly, D.P. and Baker S.C. (1990) Theorganosulphur cycle: aerobic and anaerobic processes leading to turnover of C1-sulfur compounds. FEMS Microbiology Review 87: 241-246.

[17] Kwon, Hyouk K., Woo, Seung H. and Park, Jong M. (2002) Thiocyanate degradation by *Acremonium strictum* and inhibition by secondary toxicants, *Biotechnol. Lett.*24, 1347-1351.

[18] Lanza, M.R.V. and Bertazzoli, R. (2002) Cyanide Oxidation from Wastewater in a Flow Electrochemical Reactor. *Industrial Engineering and Chemical Research*. 41, 22-26.

[19] Lordi, D.T., Lue-Hing, C., Whitebloom, S.W., Kelada, N. and Dennison, S. (1980) Cyanide problems in municipal wastewater treatment plants. *J. Wat. Pollut. Cont. Fed.* 52: 597-609.

[20] Millar, J.H. (1972) *"Experiments in Molecular Genetics"*, Cold Spring Harbour, NY: Cold Spring Harbour Laboratory.

[21] Mudder, T.I. and Whitlock, J.L. (1984) Biological treatment of cyanidation waste waters. *Minerals and Metallurgical Processing*, pp. 161-165

[22] Patil Y.B. (1999) Studies on Biological Detoxification of Metal-Cyanide Containing Industrial Effluents, Ph.D. Thesis, University of Pune, Pune, India.

[23] Patil YB and Paknikar KM (1999a) Removal and recovery of metal cyanides using a combination of biosorption and biodegradation processes. *Biotechnology Letters* 21: 913-919.

[24] Patil YB and Paknikar KM (1999b) Removal and recovery of metal cyanides from industrial effluents. *Process Metallurgy* 9: 707-716.

[25] Patil YB and Paknikar KM (2000a) Development of a process for biodetoxification of metal cyanides from wastewaters. *Process Biochemistry* 35: 1139-1151.

[26] Patil YB and Paknikar KM (2000b) Biodetoxification of silver-cyanide from electroplating industry wastewater. *Letters in Applied Microbiology* 30: 33-37.

[27] Patil YB and Paknikar KM (2001) Biological detoxification of nickel-cyanide from industrial effluents. *Process Metallurgy* 11: 391-400.

[28] Patil Yogesh B (2006) Isolation of thiocyanate degrading chemoheterotrophic bacterial consortium. *Nature Environment and Pollution Technology* 5(1): 135-138.

[29] Patil Yogesh B (2008a) Biodegradation of thiocyanate from aqueous waste by a mixed bacterial community. *Research Journal of Chemistry and Environment* 12(1): 69-75.

[30] Patil Yogesh B (2008b) Thiocyanate degradation by pure and mixed bacterial cultures. *Bioinfolet* 5(3): 308-309.

[31] Patil Yogesh B (2011) Utilization of thiocyanate (SCN-) by a metabolically active bacterial consortium as the sole source of nitrogen. *International Journal of Chemical, Environmental & Pharmaceutical Research*, 2: 44-48.

[32] Patil Yogesh B (2012) Development of an innovative low-cost industrial waste treatment technology for resource conservation - A case study with gold-cyanide emanated from SMEs. *Procedia- Social and Behavioral Sciences* 37: 379-388.

[33] Patil Yogesh B and Kulkarni Anil R (2008) Environmental sensitivity and management of toxic chemical waste in mining industry with special reference to cyanide. In: *High Performing Organizations: Needs and Challenges*, Tata McGraw Hill Publications, Part D: pp. 183-196.

[34] Patil Yogesh, Chourey Jayati and Rao Prakash (2012) Biotechnological strategy for the management of industrial waste with concurrent mitigation of global warming – A feasibility study using microalgae. In: *Inclusiveness and Innovation – Challenges for Sustainable Growth of Emerging Economies* (Edited by Rajiv Divekar and BR Londhe), Excel India Publishers, New Delhi, India.

[35] Prashanth MS and Patil Yogesh B (2006) Behavioural surveillance of Indian major carp *Catlacatla*(Hamilton) exposed to free cyanide. *Journal of Current Sciences* 9(1): 313-318.

[36] Sellers, K. (1999) *"Fundamentals of Hazardous Waste Site Remediation"*, Lewis Publishers, Washington, DC.

[37] Sorokin Dimitry Y., Tourova, Tatyana P., Lysenko, Anatoly M. and Kuenan, J. Gigs. (2001) Microbial thiocyanate utilization under highly alkaline conditions.*Applied and Environmental Microbiology* 67(2): 528-538.

[38] Stratford, J., Dias, A.E.X.O and Knowles, C.J. (1994) The utilization of thiocyanate as a nitrogen source by a heterotrophic bacterium: the degradative pathway involves formation of ammonia and tetrathionate, *Microbiology* 140: 2657-2662.

[39] Tchobanoglous, G. and Burton, F.L., (1997) Wastewater Engineering: Treatment, Disposal, Reuse, Metcalf and Eddy Inc., McGraw-Hill Publication, New Delhi.

[40] Thakur Ravindra Y and Patil Yogesh B (2009) Management of thiocyanate pollution using a novel low cost natural waste biomass.*South Asian Journal of Management Research* 1(2): 85-96.

[41] Van Zyl, Andries W., Harrison, Susan T.L., van Hille, Robert P. (2011) Biodegradation of thiocyanate by a mixed microbial population. In: Mine Water – Managing the Challenges (IMWA 2011), Aachen, Germany.

[42] Westley, J. (1981) *Cyanide and sulfane sulfur*. In: Vennesland, B., ed., Cyanide in Biology, Academic Press, London, pp. 201-212.

[43] Wood, A.P., Kelly, D.P., McDonald, I.R., Jordan, S.L., Morgan, T.D., Khan, S., Murell, J.C. and Borodina, E., (1998) A novel pink-pigmented facultative methylotroph, *Methylobacterium thiocyanatum* sp. nov., capable of growth on thiocyanate as sole nitrogen source. *Arch. Microbiol.*, 169, 148-158.

[44] Wood, J.L. (1975) Biochemistry, In: Newmann, A.A., ed., Thiocyanic Acid and its Derivatives, Academic Press, London, United Kingdom.

[45] Zargury, G.J., Oudjehani, K. and Deschenes, L. (2004) Characterisation and availability of cyanide in solid mine tailings from gold extraction plants. *Science of the Total Environment* 320: 211-224.

Chapter 3

Bioremediation of Olive Mill Wastewater by Yeasts – A Review of the Criteria for the Selection of Promising Strains

Antonio Bevilacqua, Leonardo Petruzzi,
Maria Rosaria Corbo and Milena Sinigaglia

Additional information is available at the end of the chapter

http://dx.doi.org/10.5772/56916

1. Introduction

The cultivation of olive trees and the production and use of olive oil has been a well-known and established practice in the Mediterranean region for more than 7000 years [1].

Olive is the most extensively cultivated fruit crop in the world, counting 9,2 million hectares of area harvested in 2009 and its cultivation area has tripled in the past 50 years [2].

Over the last decade, olive oil production has increased about 40% worldwide and Europe obtained an increase of 45% in production [3], due to its high dietetic and nutritional value (the high smoke point-210 °C- and an excellent lipid profile as the proportion of saturated, mono-unsaturated and poly-unsaturated fatty acids is 14:77:9) [1]. It is generally accepted that olive oil consumption brings benefits to human health, such as reduction of risk factors of coronary heart disease, prevention of several types of cancers, and modifications of immune and inflammatory responses [4].

Mediterranean Countries produce more than 98% of the world's olive oil, which is estimated at over 2.5 million metric tons *per* year. Three quarters of the annual production in the world comes from European Union, in particular Spain (36% of the worldwide production), Italy (24% of the world's total), and Greece (17% of the global production) [3].

These data reflect the importance of olive oil sector in the Mediterranean area and consequently the magnitude of the problems related with the disposal of large amounts of olive mill wastewaters (OMW). Many studies report that OMW is a major pollutant to surface and ground water resources in the Mediterranean basin [5]. Moreover, olive oil production is no longer restricted to the Mediterranean basin, and new producers such as Australia, USA and South America will also have to face the environmental problems posed by OMW [6].

OMW (*acque reflue* in Italy; *alpechin* in Spain; *katsigaros* in Greece; *zebar* in Arab countries) is a dark red to black-coloured, mildly acidic liquid of high conductivity, obtained from mechanical olive processing during olive oil production [7]. Only in the Mediterranean area, OMW generation varies between 10×10^6 and 30×10^6 m^3 [3]. In general, the quality and quantity of OMW, and consequently the environmental impact, depends on many factors, such as the type of olives, the type of soil, the cultivation system and the production process [8]. The traditional cold press method typically generates about 50% of OMW relative to the initial weight of the olives, while the continuous centrifugation process generates 80–110% of OMW due to the continuous washing of the olive paste with warm water prior to oil separation from the paste [9].

The problems connected with OMW depend on their high chemical oxygen demand (COD) (up to 100 g/L), biological oxygen demand (BOD) (13-46 g/L), low pH (4–5), and other recalcitrant organic compounds, such as water-soluble phenols (hydroxytyrosol, tyrosol, catechol, methylcatechol, caffeic acid, vanillic acid, *p*-coumaric acid, etc.) and polyphenols originating from the olives [1]; the conductivity of OMW is around 18.0 mmhos/cm, while the average value of TSS (total suspended solids) and VSS (volatile suspended solids) are respectively 40-60 g/L and 30-50 g/L, with a TOC (total organic matter) of 10-30 g/L and TN (total nitrogen) of 0.6-1.4 g/L [1]. OMW contains also other mineral elements (P_2O_5, K_2O, Na, Mg, Fe, Cu etc.), but the amount of these compounds is greatly variable.

OMW is one of the most complex agro-industrial effluent [10]. Most of the problems associated with OMW pollution can be attributed to the phenolic fraction [11]. Monomeric phenols of OMW have been associated with the phytotoxic and antimicrobial properties of these wastewaters, while the dark brownish color of OMW, which is particularly recalcitrant to decolorization, has been attributed to the polymerization of tannins and low molecular weight phenolic compounds [12].

OMW are often poured into the soil (up to 50 m^3 *per* hectar in Italy) or disposed of in sewage, causing soil and water pollution. In fact untreated OMW are able to change the microbial composition of the soil through their antibacterial activity and produce phytopathogenic effects due to their high toxicity [13] (i.e. 1 m^3 of OMW is equivalent to 100-200 m^3 of domestic sewage) [1]. Due to the high organic load of OMW, it may contribute significantly to eutrophication of recipients in which fluid exchange rates are low (closed gulfs, estuaries, lakes, etc.). An additional adverse impact of OMW on the environment is the aesthetic degradation caused by its strong odour and dark coloration [14]. Furthermore, environmental regulations and enforcements have become more and more strict [15], thus there is the need of new guidelines to manage these wastes; in fact the most olive oil is produced in Countries that are deficient in water and energy [3].

For these reasons, in recent years, several disposal methods have been proposed such as thermal processes (combustion and pyrolysis), physico-chemical treatments (e.g. precipitation/flocculation, ultrafiltration and reverse osmosis, adsorption, chemical oxidation processes and ion exchange), extraction of valuable compounds (e.g. antioxidants, residual oil, sugars), agronomic applications (e.g. land spreading), animal-breeding methods (e.g. direct utilisation as animal feed or following protein enrichment) and biological treatments [8]. Among the different options, biological treatments are considered the most environmentally compatible and the least expensive methods [9].

Two different approaches have been developed for OMW biological treatment: aerobic and anaerobic processes [16]. Some drawbacks of OMW bioremediation under anaerobic conditions are the difficulties to remove high-molecular weight phenols [16], the need for a long period for the adaptation of microorganisms, the high costs for the storage [17]; on the other hand, the aerobic protocols do not show these limits.

Early studies focused on the use of specific bacterial species, including *Bacillus pumilus* [18], *Arthrobacter* sp. [19], *Azotobacter vinelandii* [20], *Azotobacter chroococcum* [21], *Pseudomonas putida* and *Ralstonia* sp. [22] and various bacterial consortia [23-25]. In general, aerobic bacteria appeared to be very effective against some low-molecular-mass phenolic compounds but are relatively ineffective against the more complex polyphenolics responsible for the dark colouration of OMW [3].

Several strains of filamentous fungi have revealed interesting capacities for the removal of problematic OMW compounds [26]. A variety of white-rot fungi have been used including *Phanerochaete chrysosporium* [27], *Trametes versicolor* [28], *Pleurotus* spp. [29], *Funalia trogii* [30-31], *Lentinus edodes* [3]. Although Garcia et al. [32] studied the ability of *Aspergillus niger* and *Aspergillus terreus* to remove phenol compounds from OMW, the use of *Aspergillus* spp. is not so common as the application of white rot fungi.

According to a recent review, fungi - including white rot fungi - are more effective than bacteria for the degradation of the phenols of OMW [6]. The high efficiency of fungi relies upon the structure of the aromatic compounds present in OMW; they are analogous to those of many lignin monomers, and only a few microorganisms, mainly white rot fungi, are able to efficiently degrade lignin by producing ligninolytic enzymes such as lignin peroxidases, manganese peroxidases and laccases [6]. However, there is usually a need to employ a heat pre-treatment to facilitate establishment of introduced fungi [26, 33]. Starter cultures for bioremediation usually requires aeration, and the duration of treatment is ca. 8-24 days, depending on some process variables such as degree of dilution, aeration and supplementation [6]. In addition, only some white-rot fungi were reported as able to perform decolorization and COD reduction in OMW when the active COD is >50 g/L [34]. Finally, the application of fungi for OMW treatment on a large scale was limited by the difficulty of achieving continuous culture because of the formation of filamentous pellets and mycelia [16].

To overcome this limitation, the use of yeasts could be a promising way. In fact, among the mentioned microbiota, yeasts are the more adapted and resistant to high concentrations of phenols and low pH values of mill wastes, allowing them to dominate this environment [35].

Some genera have already been tested successfully to detoxify and/or decolourise OMW, including *Candida, Geotrichum, Pichia, Saccharomyces, Trichosporon* and *Yarrowia* (table 1). Little information is now available on the indigenous yeasts present in the OMW and their possible use for performing biodegradation of the waste.

Yeasts	Method	Results		Reference
		Phenol Reduction	COD Reduction	
Candida boidinii	Fed-batch microcosm	42.2%	-	[36]
	Culture in OMW	40%	45%	[37]
C. cylindracea	Culture in OMW	27%	45.8-70.2%	[38]
	Bioreactor batch culture with OMW	12.8-31.3%	27.4-55.9%	[39]
	Culture in OMW	36.2%	48.4%	[40]
C. diddensiae	Culture in OMW	32.14-43.56%	55.40-64.84%	[41]
	Culture in OMW	10-72%	-	[42]
C. ernobii	Culture in OMW	34.09-35.23%	51.85-62.65	[41]
C. holstii	Culture in OMW	39%	57.93%	[41]
C. oleophila	Bioreactor batch culture with OMW	20.3% (tannins)	-	[43]
	Culture in OMW	83%	55%	[12]
C. rugosa	Culture in OMW	12.2-20.4%	20.4-62.2%	[38]
	Culture in OMW	15.3%	31.1-62.2%	[38]
C. tropicalis	Culture in OMW	51.7%	62.8%	[44]
	OMW from industrial mills	25%	18%	[45]
	Culture in OMW	12-36.5%	39.4-69.7%	[46]
	Culture in bioreactors with a mixture of OMW (75%) and pig slurry (25% v/v)	51%	62%	[47]
Geotrichum sp.	Culture in OMW	46.6%	55%	[44]
G. candidum	Culture in bioreactors with OMW	-	12.4-62%	[11]
	Fed-batch microcosm	42.9%	-	[36]
	OMW from industrial mills	25-31%	20-23%	[45]
	Culture in OMW	47%	77%	[48]
	Culture in bioreactors with OMW	-	25-65%	[16]
	Culture in OMW	20-41%	25-56%	[49]
	Culture in OMW	46%	51%	[37]

Yeasts	Method	Results		Reference
		Phenol Reduction	COD Reduction	
Pichia guilliermondii	Culture in OMW	25.09-33.52%	34.47-53.21%	[41]
Pichia sp.	Culture in OMW	40%	41.04%	[41]
P. fermentans	OMW from industrial mills	26%	18%	[45]
P. holstii	OMW from industrial mills	17%	15%	[45]
Saccharomyces sp.	Fed-batch microcosm	38.8%	-	[36]
Trichosporon cutaneum	Culture in OMW	> 80%	>80%	[50]
	Culture in OMW	64%	88%	[48]
Yarrowia lipolytica	Culture in OMW	≤78.2%	1.47-41.22%	[51]
	Culture in OMW	-	67-82%	[52]
	Culture in bioreactors with OMW	-	80%	[53]
	Culture in OMW	19.2-31.3%	21.6-52.6%	[38]
	Culture in OMW	25.3%	23.5-51.3%	[38]
	Culture in OMW	20%	23.1-50.9%	[38]
	Culture in OMW	43-72%	54-79%	[54]
	Culture in OMW	39-68%	75-80%	[54]

Table 1. Phenol removal and COD decrease in OMW by yeasts. A review of the literature. -, data not available.

2. Yeast selection: A step-by step protocol

The selection of yeasts intended as functional starter for the bioremediation of OMW is a quite complex process, involving different steps; figure 1 proposes a possible scheme.

Namely, after strain isolation from OMW, yeasts should be characterized (step 1) and identified (2); then, some promising isolates could be studied in relation to their functional properties (phenol removal and COD/BOD decrease). Finally, a multivariate approach could be used to choose the best strains for the final validation under laboratory and factory-scale conditions.

In the following sections, there are some details on the most important assays for the selection of promising yeasts.

2.1. Isolation

This is a critical step as it important to recover yeasts and many times they are not able to grow under laboratory conditions.

Generally, OMW are stored under controlled conditions (for example at 25 °C) and let to ferment; for example, authors of reference [55] analyzed OMW for 90 days. Periodically, the samples are serially diluted and plated on opportune media, like acidified Malt Extract Agar [55], Potato Dextrose Agar and Yeast Malt Agar [56], YEPD agar (Yeast Extract Potato Dextrose) supplemented with 50 μg/mL of ampicillin [45]. Then, yeasts are selected on the basis of colony morphology.

An interesting approach was proposed by other authors [57]; they optimized the protocol for the isolation of bacteria able to remove phenols from wastewater and slurry, but their method, with some modifications, could be used successfully for yeasts. OMW should be added to a mineral salt medium (MSM) containing (g/L): Na_2HPO_4, 1.6; KH_2PO_4, 0.4; NH_4NO_3, 0.5; $MgSO_4*7H_2O$, 0.2; $CaCl_2$, 0.025; $FeCl_3$, 0.0025 with and without 1% glucose (w/v) as an additional carbon source.

Different concentrations of phenol (100, 200, 300, 500, 700, and 900 mg/L) should be added to the medium; after adjusting the pH to 7.0, the samples can be stored at 25°C for at least 5 days and then plated onto MSM agar plates, with and without glucose.

2.2. Technological characterization

The technological characterization of yeasts deals with both the taxonomic assays and the technological traits (growth requirements and enzymatic traits).

The most important trait is the effect of phenols on yeast growth/survival; this characteristic includes both the ability to use phenols as carbon sources and the growth/survival in OMW. Phenol assimilation can be assessed on Yeast Nitrogen Base (a laboratory medium without carbon source), added with either caffeic, vanillic or *p*-coumaric acid [55]. Another way to assess the suitability of yeasts for bioremediation is the evaluation of growth in OMW or in solid/liquid media containing OMW [58].

A modification of this last assay was proposed by Aissam et al. [37], who cultured yeasts into lab media containing increasing amounts of OMW (from 50 to 100%) to induce yeast adaptation to such a stressful environment.

Although the assimilation of phenols and the growth in OMW are the most important traits for the selection of promising yeasts for bioremediation, some other interesting characteristics are the thermal profile (*i.e.* the minimal and maximum temperatures of growth, the optimal temperature), the effect of pH, and nitrogen assimilation. The effects of temperature and pH can be evaluated through a spectrophotometric measurement, followed by the calculation of Growth Index, as proposed for yeasts intended as starters for table olives [59].

On the other hand, nitrogen assimilation should be assessed in a poor medium, containing a single nitrogen source (for example KNO_3 or ethylamine) [55]; this assay, as well as spore production, has also a taxonomic potential: for example *Saccharomyces cerevisiae* is not able to use nitrate as the only nitrogen source, whereas other yeasts do it.

http://dx.doi.org/10.5772/56916

2.3. Enzymatic traits

Yeasts intended for bioremediation should be assessed for different enzymatic traits; some of them (pectinolytic, lipolytic and protease activities) rely on the ability to persist in a stressful environment, whereas other traits are strongly related with the ability to remove phenols.

For example, Taccari and Ciani [60] reported that ligninolytic enzymes lignin peroxidase (LiP), manganese peroxidase (MnP) and laccase, characterized by a low substrate specificity, are involved in the degradation of polyphenols in OMW. Reference [61] reports the most common protocols to assess enzymatic traits.

2.4. Identification

For long time yeasts have been identified through the fermentation/assimilation profiles of sugars; a good profile should include the assays for the following sugars: D-glucose, D-galactose, maltose, α-methyl-D-glucoside, sucrose, trehalose, melibiose, lactose, cellobiose, melezitose, raffinose, inulin, starch and D-xylose. Nowadays, these assays are usually run through some commercial miniaturized kits [61].

It is well known that the phenotypic identification shows some limits and drawbacks, therefore yeast identification should be performed through genotypic method. One of the most used approach is PCR amplification of the region spanning ITS1 and ITS2 and the 5.8S rRNA gene (5.8S–ITS region) and subsequent restriction analysis, following the protocols by the references [62, 63]; the results of amplification and restriction are used as input data for an analysis through a specific database (for example Yeast-id database, CECT, University of Valencia, Spain).

2.5. Functional characterization

For yeasts intended for bioremediation, phenol removal, the decrease of BOD/COD and OMW decolorization could be referred to as the functional traits, as they are strictly related to the decrease of the pollutant impact of OMW.

Focusing on phenol removal, yeasts should be inoculated onto aliquots of OMW under laboratory conditions (static temperature and agitation) for some days [55]. Thereafter, the amount of residual phenols can be assessed through HPLC equipments or simply using the method by Folin-Ciocalteau [64]. Other authors [12,65] evaluated indirectly phenol removal through toxicity attenuation, thus they studied the phytoxicity of OMW towards seeds and the microbial toxicity towards *Bacillus cereus*.

Other traits are the reduction of COD and BOD [45, 66], as well as waste decolorization; for this last assay, OMW should be diluted with distilled water and then analyzed through absorbance measurement at 390 nm.

2.6. Selection of promising strains and validation

Choosing the most promising strains is the final step for a starter selection; as reported elsewhere [59], the management of a such large amount of data (many strains, each of them studied for different parameters) is quite difficult and complex. A possible solution could be the use of multivariate statistical approaches, like the Principal Component Analysis, Cluster Analysis or Multiple Correspondence Analysis or all of them in a sequence.

The main result of the multivariate approach is the choice of the best strains (3-10) for an *in vivo* validation; however, yeasts require a preliminary optimization and/or validation in small volumes and under controlled conditions.

Some variables that should be assessed are:

1. **Use of coadjutants**. It has been reported that yeast metabolism could be enhanced by the addition of some ingredients; for example, Sinigaglia et al. [55] proposed the use of $(NH4)_2SO_4$ (1.5-6.0 g/L), while authors of the reference [46] used hexadecane and yeast extract.

2. **Temperature and shaking**. Some authors [46, 58] proposed a bioremediation with agitation (100-150 rpm) and at relatively high temperatures to increase the yield of the process.

3. **State of cells (free or immobilized in a bioreactor)**. OMW can be detoxified by free cells, as proposed by many authors or using the novel method proposed in the reference [46], who loaded a strain of *Geotrichum candidum* in Na-alginate beads and increased by 2-2.2 fold the yield of removal.

4. **Use of a single strain or a multiple strain starter**. The use of a multiple strain starter could be a promising way to enhance the yield and avoid a stop in the detoxification; thus, validation should focus on the composition of the starter (amounts of the different strains) and the way of inoculation (single inoculum or step-by-step inoculum).

5. **OMW dilution**. It was proposed a 10-fold dilution to increase fungal bioremediation by *Aspergillus wentii, A niger* and *Pleurotus ostreatus* [67] and these data were confirmed by a preliminary investigation performed on yeasts on our laboratory with a 3-fold dilution.

6. **Kind of process (aerobic or anaerobic)**.The use of an aerobic step could increase the yield [68]. The authors of the reference a preliminary and aerobic step with *G. candidum* to reduce COD and phenolic and fatty acid contents and increase substrate up-taking during the anaerobic treatment.

3. Conclusions

The use of yeasts for the bioremediation of OMW is a promising and open way; the starting question of this paper was: why yeasts?

We can try to point out some-key elements/benefit for the use of yeasts in OMW:

1. yeasts represent the dominant microflora of OMW and many strains are well adapted and able to grow in this stressful environment;

2. yeasts can be used for the aerobic and anaerobic treatment of wastes;
3. the yield of moulds in phenol removal is high, many times higher than for yeasts; however, micelia could absorb phenols and release them again in the case of a prolonged storage;
4. some yeasts could be used to produce biomass and useful metabolites (for example lipases) using OMW as medium;
5. yeasts can be used in continuous or in batch cultures, while moulds do not;
6. some yeasts are able to remove both low and high molecular weight phenols, whereas bacteria do not.

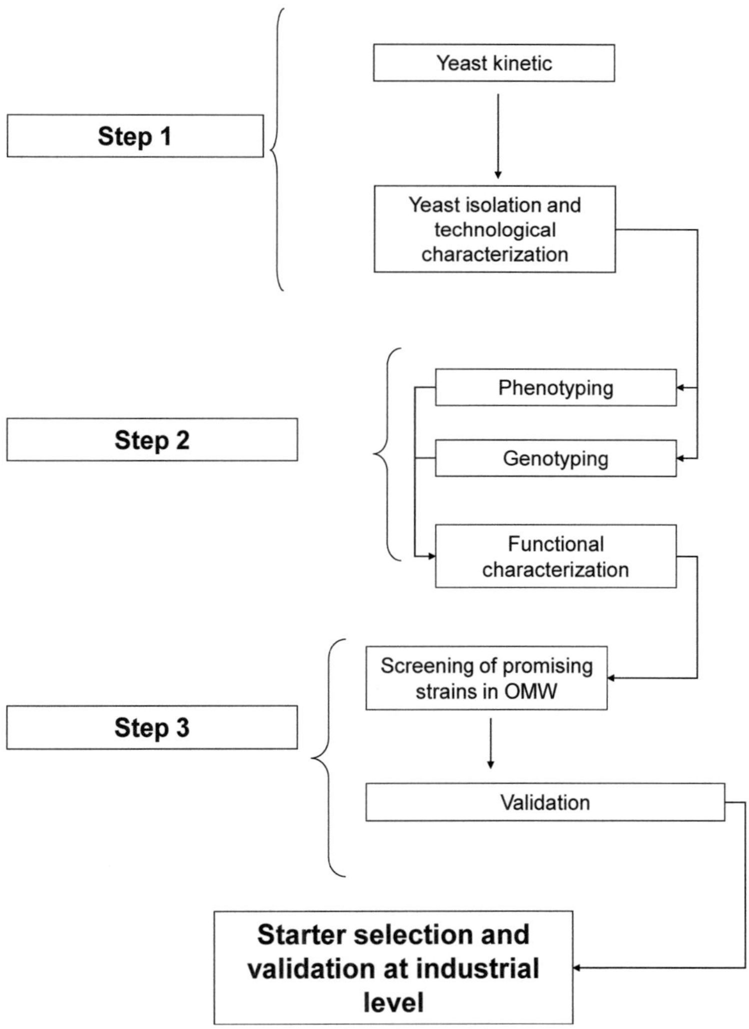

Figure 1. Selection of yeasts for phenol removal in OMW

Taxonomy
Spore production
Growth requirements
Nitrogen assimilation
Phenol assimilation
Growth in OMW
Effect of temperature and pH
Enzymatic traits
Catalase activity
Hydrolysis of pectins and xylans
Cellulolytic activity
Lipolytic activity
Protease activity
Polyphenoloxydase activity
Peroxidase activity

Table 2. Technological and taxonomic characterization

Acknowledgements

This research was supported by project ECO_P4 PON02_00186_2866121 (Promotion of ECO-friendly processes for the enhancement of quality of Apulian food productions), granted by the Italian Ministry of Education, University and Research.

Author details

Antonio Bevilacqua*, Leonardo Petruzzi, Maria Rosaria Corbo and Milena Sinigaglia

*Address all correspondence to: antonio.bevilacqua@unifg.it

Department of the Science of Agriculture, Food and Environment, University of Foggia, Foggia, Italy

References

[1] Tsagaraki E, Harris N, Lazarides, and Konstantinos B. Petrotos. Olive Mill Wastewater Treatment. In: Oreopoulou V., Russ W. (eds.) Utilization of By-Products and Treatment of Waste in the Food Industry. New York: Springer; 2007. p. 133-158.

[2] Migliorini P. Development of Organic Olive Cultivation and its Importance for the Sustainability in the Mediterranean. In: Migliorini P., Minotou C., Lusic D., Hashem Y., Martinis A. (eds.) Book of Abstract. International Conference on ORGANIC AGRICULTURE and AGRO-ECO TOURSIM in the Mediterranean, International Conference AgriBioMediterraneo, 16-18 September 2011, Zakynthos, Greece.

[3] McNamara CJ, Anastasiou CC, O'Flaherty V, Mitchell R. Bioremediation of Olive Mill Wastewater. International Biodeterioration & Biodegradation 2008; 61 (1) 127–134.

[4] Justino CI, Pereira R, Freitas AC, Rocha-Santos TA, Panteleitchouk TS, Duarte AC. Olive Oil Mill Wastewaters before and after Treatment: a Critical Review from the Ecotoxicological Point of View. Ecotoxicology 2012; 21 (2): 615-29.

[5] Hashwa F, Mhanna E. Aerobic and Anaerobic Biotreatment of Olive Oil Mill Wastewater in Lebanon. In: Al Baz I, Otterpohl R, Wendland C. (Eds.) Efficient management of wastewater. Berlin: Springer-Verlag; 2008. p. 187-203.

[6] Morillo JA, Antizar-Ladislao B, Monteoliva-Sanchez M, Ramos-Cormenzana A, Russell NJ. Bioremediation and Biovalorization of Olive-Mill Wastes. Applied Microbiology and Biotechnology 2009; 82 (1): 25–39.

[7] Kapellakis IE, Tsagarakis KP, Crowther JC. Olive Oil History, Production and By-Product Management. Review Environmental Science and Biotechnology 2008; 7 (1): 1-26.

[8] Niaounakis M, Halvadakis CP. Olive-Mill Waste Management: Literature Review and Patent Survey. Typothito: Greece; 2004.

[9] Mantzavinos D, Kalogerakis N. Treatment of Olive Mill Effluents: Part I. Organic Matter Degradation by Chemical and Biological Processes—an Overview. Environment International 2005; 31 (2): 289–295.

[10] Azbar N, Bayram A, Ayes F, Ayesn M, Fusun S, Ozer A. A Review of Waste Management Options in Olive Oil Production. Critical Reviews on Environmental Science and Technology 2004; 34 (3): 209–247.

[11] Asses N, Ayed L, Bouallagui H, Ben Rejeb I, Gargouri M, Hamdi M. Use of *Geotrichum candidum* for olive mill wastewater treatment in submerged and static culture. Bioresource Technology 2009; 100 (7): 2182–2188.

[12] Amaral C, Lucas MS, Sampaio A, Peres JA, Dias AA, Peixoto F, Anjos M, Pais C. Biodegradation of Olive Mill Wastewaters by a Wild Isolate of *Candida oleophila*. International Biodeterioration & Biodegradation 2012; 68 (3): 45-50.

[13] Laconi S, Molle G, Cabiddu A, Pompei R. Bioremediation of Olive Oil Mill Wastewater and Production of Microbial Biomass. Biodegradation 2007; 18 (5): 559–566.

[14] Arvaniti EC, Zagklis DP, Papadakis VG, Paraskeva CA. High-added Value Materials Production from OMW: a Technical and Economical Optimization. International Journal of Chemical Engineering 2012; doi:10.1155/2012/607219.

[15] Panizza M, Cerisola G. Olive Mill Wastewater Treatment by Anodic Oxidation with Parallel Plate Electrodes. Water Research 2006; 40 (6): 1179-1184.

[16] Assas N, Ayed L, Marouani L, Hamdi M. Decolorization of Fresh and Stored-Black Olive Mill Wastewaters by *Geotrichum candidum*. Process Biochemistry 2002; 38 (3): 361-365.

[17] Marques IP. Anaerobic Digestion Treatment of Olive Mill Wastewater for Effluent Re-use in Irrigation. Desalination 2001; 137(1-3): 233–239.

[18] Ramos-Cormenzana A, Juarez-Jimenez B, Garcia-Pareja MP. Antimicrobial Activity of Olive Mill Wastewaters (Alpechin) and Biotransformed Olive Oil Mill Wastewater. International Bioterioration and Biodegradation 1996; 38 (3-4): 283-290.

[19] Knupp G, Rucker G, Ramos-Cormenzana A, Hoyos SEG, Neugebauer M, Ossenkop T. Problems of Identifying Phenolic Compounds During the Microbial Degradation of Olive Mill Wastewater. International Biodeterioration and Biodegradation 1996; 38 (3-4): 277-282.

[20] Papadelli M, Roussis A, Papadopoulou K, Venieraki I, Chatzipavlidis P, Katinakis P, Balis K. Biochemical and Molecular Characterization of an *Azotobacter vinelandii* Strain with Respect to its Ability to Grow and Fix Nitrogen in Olive Mill Wastewater. International Biodeterioration and Biodegradation 1996; 38 (3-4): 179-181.

[21] Borja R, Martin A, Alonso V, Garcia I, Banks CJ. Infuence of Different Aerobic Pre-treatments on the Kinetics of Anaerobic Digestion of Olive Mill Wastewater. Water Research 1995; 19 (2): 489–495.

[22] Di Gioia D, Bertin L, Fava F, Merchetti L. Biodegradation of Hydroxylated and Methoxylated Benzoic, Phenylacetic and Phenylpropenoic Acids Present in Olive Mill Wastewaters by Two Bacterial Strains. Research in Microbiology 2001; 152 (1): 83-93.

[23] Ranalli A. Microbiological Treatment of Oil Mill Wastewaters. Grasas Aceites 1992; 43 (1): 12-19

[24] Zouari N, Ellouz R. Microbial Consortia for the Aerobic Degradation of Aromatic Compounds in Olive Oil Mill Effluent. Journal of Industrial Microbiology 1996; 16 (3): 155-162.

[25] Benitez J, Belrtan-Heredia J, Torregrosa J, Acero JL, Cercas V. Aerobic Degradation of Olive Mill Wastewaters. Applied Microbiology Biotechnology 1997; 47 (2): 185-188.

[26] Dias A, Bezerra RM, Periera AN. Activity and Elution Profile of Laccase During Biological Decolorization and Dephenolization of Olive Mill Wastewater. Bioresource Technology 2004; 92 (1): 7–13.

[27] Dhouib A, Aloui F, Hamad N, Sayadi S. Pilot-plant Treatment of Olive Mill Wastewaters by *Phanerochaete chrysosporium* Coupled to Anaerobic Digestion and Ultrafiltration. Process Biochemistry 2006; 41 (1): 159–167.

[28] Bourbonnais R, Paice M, Reid I, Lanthier P, Yaguchi M. Lignin Oxidation by Laccase Isozymes from *Trametes versicolor* and Role of the Mediator 2, 2′-azinobis(3-ethylbenzthiazoline-6-sulfonate) in Kraft Depolymerization. Applied and Environmental Microbiology 1995; 61 (5): 1876–1880.

[29] Tsioulpas A, Dimou D, Iconomou D, Aggelis G. Phenolic Removal in Olive Oil Mill Wastewater by Strains of *Pleurotus* spp. in Respect to their Phenol Oxidase (Laccase) Activity. Bioresource Technology 2002; 84 (3): 251-257.

[30] Yesilada O, Fiskin K, Yesilada E. The Use of White Rot Fungus *Funalia trogii* (Malatya) for the Decolourization and Phenol Removal from Olive Mill Wastewater. Environmental Technology 1995; 16 (1): 95–100.

[31] Yesilada O, Sik S, Sam M. Biodegradation of Olive Oil Mill Wastewater by *Coriolus versicolor* and *Funalia trogii*: Effects of Agitation, Initial COD Concentration, Inoculum Size and Immobilization. World Journal of Microbiology and Biotechnology 1998; 14 (1): 37-42.

[32] Garcia IG, Peña PRJ, Venceslada JLB, Martin AM, Santos MAM, Gómez ER. Removal of Phenol Compounds from Olive Mill Wastewater Using *Phanerochaete chrysosporium, Aspergillus niger, Aspergillus terreus* and *Geotrichum candidum*. Process Biochemistry 2000; 35 (8): 751–758.

[33] Ahmadi M, Vahabzadeh F, Bonakdarpour M, Mofarrah E. Phenolic Removal in Olive Oil Mill Wastewater Using loofah-immobilized *Phanerochaete chrysosporium*. World Journal of Microbiology and Biotechnology 2006; 22 (2): 119–127.

[34] Jaouani A, Sayadi S, Vanthournhout M, Penninckx MJ. Potent Fungi for Decolourisation of Olive Oil Mill Wastewaters. Enzyme and Microbial Technology 2003; 33 (6): 802–809.

[35] Ben Sassi A, Boularbah A, Jaouad A, Walker G, Boussaid A. A Comparison of Olive Oil Mill Wastewaters (OMW) from Three Different Processes in Morocco. Process Biochemistry 2006; 41 (1): 74–78.

[36] Giannoutsou EP, Meintanis C, Karagouni AD. Identification of Yeast Strains Isolated from a Two-phase Decanter System Olive Oil Waste and Investigation of their Ability for its Fermentation. Bioresource Technology 2004; 93 (3): 301–306.

[37] Aissam H, Penninckx MJ, Benlemlih M. Reduction of Phenolics Content and COD in Olive Oil Mill Wastewaters by Indigenous Yeasts and Fungi. World Journal of Microbiology and Biotechnology 2007; 23 (9): 1203–1208.

[38] Gonçalves C, Lopes M, Ferreira JP, Belo I. Biological Treatment of Olive Mill Wastewater by Non-Conventional Yeasts. Bioresource Technology 2009; 100(15): 3759-3763.

[39] Brozzoli V, Crognale S, Sampedro I, Federici F, D'Annibale A, Petruccioli M. Assessment of Olive-Mill Wastewater as a Growth Medium for Lipase Production by *Candida cylindracea* in Bench-top Reactor. Bioresource Technology 2009; 100 (13): 3395-3402.

[40] D'Annibale A, Sermanni GG, Federici F, Petruccioli M. Olive-Mill Wastewaters: a Promising Substrate for Microbial Lipase Production. Bioresource Technology 2006; 97 (15): 1828-1833.

[41] Ben Sassi A, Ouazzani N, Walker GM, Ibnsouda S, El Mzibri M, Boussaid A. Detoxification of Olive Mill Wastewaters by Moroccan Yeast Isolates. Biodegradation 2008; 19 (3): 337–346.

[42] Chakri M, El Haidani A, El Mzibri M, Haggoud A, Iraqui M, Houari A, Koraichi SI. Yeast Strains from the Endogenous Microflora of the Olive Flies *Bactrocera oleae* Larvae which Could Degrade the Olive Oil Mill Wastewaters Polyphenols. Annals of Microbiology 2007; 57 (2): 143-147.

[43] Peixoto F, Martins F, Amaral C, Gomes-Laranjo J, Almeida J, Palmeira CM. Evaluation of Olive Oil Mill Wastewater Toxicity on the Mitochondrial Bioenergetics after Treatment with *Candida oleophila*. Ecotoxicology and Environmental Safety 2008; 70 (2): 266-275.

[44] Fadil K, Chahlaoui A, Ouahbi A, Zaid A, Borja R. Aerobic Biodegradation and Detoxification of Wastewaters from the Olive Oil Industry. International Biodeterioration and Biodegradation 2003; 51 (1): 37-41.

[45] Bleve G, Lezzi C, Chiriatti MA, D'Ostuni I, Tristezza M, Di Venere D, Mita SG, Grieco F. Selection of Non-Conventional Yeasts and their Use in Immobilized Form for the Bioremediation of Olive Oil Mill Wastewaters. Bioresource Technology 2012; 102 (2): 982-989.

[46] Ettayebi K, Errachidi F, Jamai L, Tahri-Jouti MA, Sendide K, Ettayebi M. Biodegradation of Polyphenols with Immobilized *Candida tropicalis* under Metabolic Induction. FEMS Microbiology Letters 2003; 223 (2): 215–219.

[47] Martinez-Garcia G, Johnson AC, Bachman RT, Williams CJ, Burgoyne A, Edyvean RGJ. Two-stage biological treatment of olive mill wastewater with whey as co-substrate. International Biodeterioration and Biodegradation 2007; 59 (4): 273-282.

[48] Sollner Dragičević TL, Zanoški Hren M, Gmajnić M, Pelko S, Kungulovski D, Kungulovski I, Cvek D, Frece J, Markov K, Delaš F. Biodegradation of Olive Mill Wastewater by *Trichosporon cutaneum* and *Geotrichum candidum*. Archives of Industrial Hygiene and Toxicology 2010; 61 (4): 399-405.

[49] Ayed L, Assas N, Sayadi S, Hamdi M. Involvement of Lignin Peroxidase in the Decolourization of Black Olive Mill Wastewaters by *Geotrichum candidum*. Letters in Appied Microbiology 2005; 40 (1): 7-11.

[50] Chtourou M, Ammar E, Nasri M, Medhioub K. Isolation of a Yeast, *Trichosporon cutaneum*, Able to Use Low Molecular Weight Phenolic Compounds: Application to

Olive Mill Waste Water Treatment. Journal of Chemical Technology and Biotechnology 2004; 79 (8): 869–878.

[51] Lanciotti R, Gianotti A, Baldi D, Angrisani R, Suzzi G, Mastrocola D, Guerzoni ME. Use of *Yarrowia lipolytica* Strains for the Treatment of Olive Mill Wastewaters. Bioresource Technology 2005; 96 (3): 317-322.

[52] Wu L, Ge G, Wan J. Biodegradation of Oil Wastewater by Free and Immobilized *Yarrowia lipolytica* W29. Journal of Environmental Sciences 2009; 21 (2): 237-242.

[53] Scioli C, Vollaro L. The Use of *Yarrowia lipolytica* to Reduce Pollution in Olive Mill Wastewaters. Water Research 1997; 31(10): 2520–2524.

[54] Lopes M, Araújo C, Aguedo M, Gomes N, Gonçalves C, Teixeira JA, Belo I. The Use of Olive Mill Wastewater by Wild Type *Yarrowia lipolytica* Strains: Medium Supplementation and Surfactant Presence Effect. Journal of Chemical Technology and Biotechnology 2009; 84 (4): 533–537.

[55] Sinigaglia M, Di Benedetto N, Bevilacqua A, Corbo MR, Capece A, Romano P. Yeasts Isolated from Olive Mill Wastewaters from Southern Italy: Technological Characterization and Potential Use for Phenol Removal. Applied Microbiology and Biotechnology 2010; 87 (6): 2345-2354.

[56] Amaral C, Lucas MS, Coutinho J, Crespi AL, do Rosário Anjos M, Pais C. Microbiological and Physicochemical Characterization of Olive Mill Waastewaters from a Continuous Olive Mill in Northeastern Portugal. Bioresource Technology 2008; 99 (15): 7215-7223.

[57] Chandra R, Yadav S, Bharagava RN, Rai V. Phenol Degradation by *Paenibacillus thiaminolyticus* and *Bacillus cereus* in Axenic and Mixed Conditions. World Journal of Microbiology and Biotechnology 2011; 27 (12): 2939-2947.

[58] Taccari M, Ciani M. Use of *Pichia fermentans* and *Candida* sp. Strains for The Biological Treatment of Stored Olive Mill Wastewater. Biotechnology Letters 2011; 33 (12): 2385-2390.

[59] Bevilacqua A, Corbo MR, Sinigaglia M. Selection of yeasts as starter cultures for table olives: a step-by-step procedure. Frontiers in Food Microbiology 2012; 3: article 194.

[60] Taccari M, Ciani M. Olive Mill Wastewater: Treatments and Valorisation. In: Haghi AK (ed.) Waste Management: Research Advances to Convert Waste to Wealth. New York: Nova Publisher Inc.; 2010. p. 203-221.

[61] Suh SO, Zhang N, Nguyen H, Gross S, Blackwell M. Lab Manual for Yeast Study. 2008. http://lsb380.plbio.lsu.edu/beetlebellyfolder/Manual_for_Yeast_Work_Sept2008.pdf (accessed 20 September 2012).

[62] Esteve-Zarzoso B, Belloch C, Uruburu F, Querol A. Identification of Yeasts by RFLP Analysis of the 5.8S rRNA Gene and the Two Ribosomal Internal Transcribed Spacers. International Journal of Systematic Bacteriology 1999; 49 (1): 329-337.

[63] Granchi L, Bosco M, Vicenzini M. Rapid Detection and Quantification of Yeast Species During Spontaneous Wine Fermentation by PCR-RFLP Analysis of the rDNA ITS Region. Journal of Applied Microbiology 1999; 87 (6): 949-956.

[64] Bray HG, Thrope WV. Methods of Biochemical Analysis. London: Interscience Publishers; 1954. p. 27-50.

[65] Colarieti ML, Toscano G, Greco G. Toxicity Attenuation of Olive Mill Wastewater in Soil Slurries. Environmental Chemistry Letters 2006; 4 (2): 115-118.

[66] Hamdi M, Kadir A, Garcia JL. The Use of *Aspergillus niger* for Bioconversion of Olive Mill Wastewaters. Applied Microbiology and Biotechnology 1991; 34 (6): 828–831.

[67] Afify AS, Mahmoud MA, Emara HA, Abdelkreem KI. Phenolic Compounds and COD Removal from Olive Mill Wastewater by Chemical and Biological Procedures. Australian Journal of Basic and Applied Science 2009; 3 (2): 1087–1095.

[68] Martin A, Borja R, Chica A. (1993). Kinetic Study of an Anaerobic Fluidized Bed System Used for the Purification of Fermented Olive Mill Wastewater. Journal of Chemical Technology and Biotechnology 1993; 56 (2): 155–162.

Chapter 4

Nutrients and Organic Matter Removal in a Vertical-Flow Constructed Wetland

Silviya Lavrova and Bogdana Koumanova

Additional information is available at the end of the chapter

http://dx.doi.org/10.5772/56245

1. Introduction

Constructed wetlands are promising engineering technique that reproduce the conditions of the natural wetlands [1]. They have high water treatment capacity because of the intensive "work" of the plants and the microorganisms. Depending on the conditions various types of plants are growing: common reed (*Phragmites australis*), rush (*Typha latifolia*), iris, etc. (Fig. 1). These plants are stable toward the climatic changes and the quality parameters of the medium in which they are growing. The metabolism of the microorganisms plays an important role in the pollutants removal from wastewaters. The main chemical and physical processes are sedimentation, sorption, chemical oxidation, photo degradation, evaporation [2] as well as biotic processes such as aerobic/anaerobic degradation, plants accumulation, phytodegradation, phytoevaporation. Many publications demonstrate the removal of suspended solids, organic matter, nutrients and bacteria from wastewater in constructed wetlands. There are two types of constructed wetlands: *surface flow wetlands systems* and *subsurface flow wetland systems*. The latter are subsurface horizontal flow wetlands systems (Fig. 2) and subsurface vertical flow wetland systems (Fig. 3), [3-6]. They are characterized with the different extent of nutrients removal [7-10].

In the subsurface vertical-flow constructed wetlands (SSVFCW) the wastewater enters through the surface and flows in vertical direction slowly through the supporting material and the plant roots until reaching the bottom outlet zone. These systems are built with porous materials such as sand and gravel, that restrict the clogging. The package clogging was observed at high organic load of the system [11]. The recirculation of the wastewater is helpful to overcome this limitation.

Phragmites australis *Typha latifolia* *Scirpus lacustris* *Iris pseudacorus* *Glyceria maxima*

Figure 1. Different types of plants growing in the wetland systems

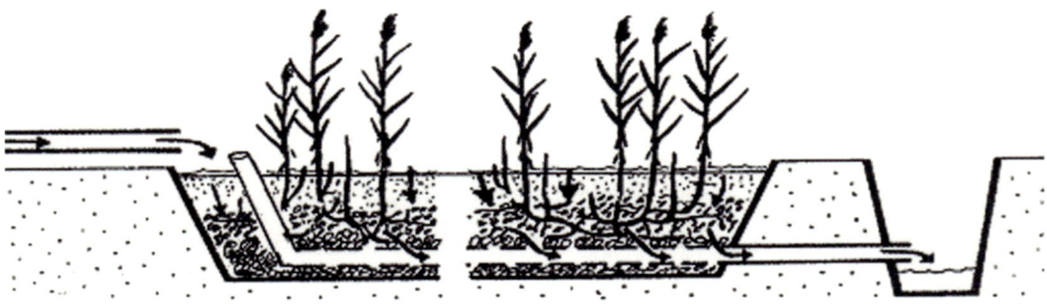

Figure 2. Subsurface horizontal-flow wetlands system

http://dx.doi.org/10.5772/56245

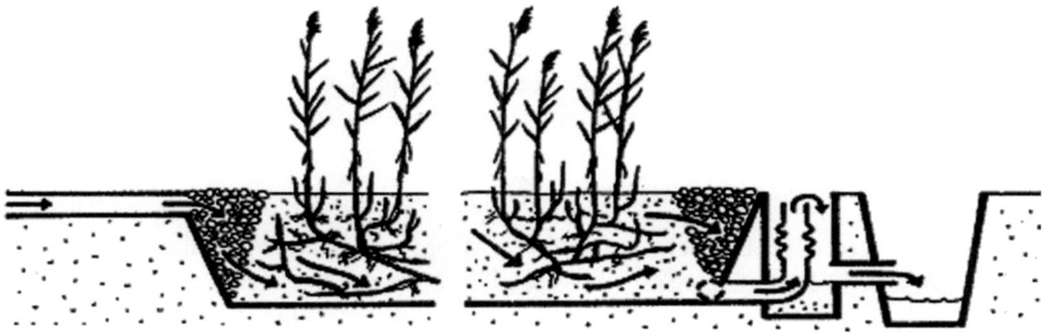

Figure 3. Subsurface vertical-flow wetlands system

Scholz and Hu [12] investigated various filter materials and macrophytes for the removal of lead and copper from wastewaters. They demonstrated the possibility to replace the expensive activated carbon and charcoal with the cheaper sand and gravel. Korkusuz [13] reports for the treatment of domestic wastewater in subsurface flow wetland with wasted granular slag and gravel. He obtained for both materials the removal of suspended solids (SS) 64 % and 62 %, COD – 49 % and 40 %, NH_4^+-N – 88% and 58 %, total nitrogen (TN) – 41 % and 44 %, total phosphorus (TP) – 63 % and 9 %, PO_4^{3-}-P – 60 % and 4 %.

SSVFCW are used for treatment of industrial wastewaters from different sources: dye-containing waters [14,15], pharmaceutically polluted waters [16], wastewaters from food industry [17], olive mill wastewaters [18], liquid waste activated sludge from a soft drink factory [19]. Gross et al. [20] reports for a novel method of recycling greywater for irrigation. The all SS and BOD were removed and about 80 % of COD after 8 h. A recirculating vertical flow constructed wetland was used also for treatment of domestic waters [21].

A combined subsurface vertical and horizontal flow constructed wetland system was designed for rular domestic wastewater treatment [22]. Several water quality parameters pH, BOD, COD, TSS, TKN, TP and faecal bacteria's number in both raw and treated wastewaters were monitored during a macrophytes life cycle. Seven mesocosm-scale constructed wetlands of different configurations were operated out-doors for 39 months to assess their ability to remove organic matter and nutrients from urban wastewaters [23]. F. Ye and Y. Li [24] have shown that nitrification/denitrification is the main mechanism for nitrogen removal from domestic wastewater in a novel constructed wetland configuration with three stages towery hybrid CW. Increased dissolved oxygen (DO) by passive aeration enhanced nitrification rates and additional organic matter supplied - for denitrification. In an installation, consisted of two settling tanks in series, a VFCW and a zeolite tank, Gicas and Tsihrintzis have studied household wastewater treatment [25]. The zeolite was found to offer additional removal of nitrogen, total phosphorus and organic matter. Significant reduction of total coliform and faecal coliform was achieved in a pilot scale VFCW in North Cairo planted with three kinds of plants [26]. The use of VFCW as a post treatment step will make possible the usage of the treated water for irrigation. The treatment effect of two pilot-scale VFCWs (one planted with *Tipha latifolia* and the second – with *Phragmites australis*) on municipal wastewaters and their suitability for irrigation reuse were studied in a 2-year experiment [27]. Zurita et al. [28] suggested that it is possible to produce commertial flowers in CW.

High rate nitrogen removal in a two-stage SSVFCW has been studied by Langergraben et al. [29]. The first stage used sand with a grain size of 2-3.2 mm and the second stage – 0.06-4 mm. Better effluent quality as compared with to conventional single-stage VFCW was obtained. The Austrian effluent standards for organic matter and ammonium nitrogen were met and the average nitrogen removal efficiency was 53 % without recirculation. A three-stage experimental CW system consisting of a vertical flow-gravel filtration bed without plants, a horizontal subsurface flow bed planted with *Iris australis* and a vertical subsurface flow bed planted with *Phragmites australis* in series were fed with primary treated domestic water [30]. The beds with plants produced effluents of better quality than that without plants. It was observed that the average removal efficiencies increased with the decrease of hydraulic loading rate. Panuvatvanich et al. studied the nitrification and denitrification potential of sand layer and the effect of percolate impounding regime on nitrogen transformation in four laboratory-scale units of vertical-flow CW fed once a week with faecal sludge [31]. Biabowiec et al. investigated the effects of reed and willow on bioremediation of landfill leachate in comparison with an unplanted control by measuring redox potential levels in the rhizosphere of microcosm systems in a greenhouse [32]. Molle et al. discussed the nitrogen removal in terms of the efficiency of the stages in a hybrid constructed wetlands plant designed for 100 person equivalent [33]. The first stage was composed of vertical filters, followed by a second stage of horizontal filters. Ouyang et al. [34] developed a model using the STELLA software for estimating nitrogen dynamics in a vertical-flow constructed wetland. It was established 18 % of TN lost due to denitrification, 6 % of TN was taken up by roots of a single plant and the rest of 22 % TN from the wastewater was removed from other mechanisms, such as volatilization, adsorption and deposition. Anaerobically pretreated domestic wastewater was treated in a hybrid CW with recirculation (first in horizontal-flow CW and then in vertical-flow CW). 98% total Kjeldahl nitrogen and 79 % total nitrogen removal was obtained [35]. Li-Hua Cui et al. demonstrated the role of *Cyperus alternifolius* for the removal of total nitrogen in a VFCW [36]. Saeed and Sun [37] conducted comparative experiments in a lab-scale hybrid system with gravel, wood mulch and zeolite as medium. Average NH_4^+-N, TN and BOD removal percentages were over 99 %, 72 % and 97 %, respectively. The removal of *Escherichia coli* was 99,9 %. Nitrogen and phosphorus removal was studied also in [38-41].

Many investigations have been done of the influence of the operational parameters on the treatment efficiency of the constructed wetlands. Giraldi and Iannelli [42] used a capacitance probe to measure water content in a vertical flow CW pilot plant. They compared field measurements with data recorded in a laboratory apparatus. The effect of various design parameters has been studied by Stefanakis and Tsihrintzis [43]. Various porous media materials (carbonate material, material from river bed, zeolite and bauxite), two vegetation types (common reeds and cattails) and three total thicknesses of the porous media were used in 10 wetlands. Organic matter removal was good in all units, since it reached on the average 71,1 % and 66,9 % for BOD and COD, respectively. Nitrogen removal was 47,1 % for TKN and 42,2 % for NH_4^+-N. J.Ye et al. studied the vertical oxygen distribution in a VFCW treating domestic wastewater [44]. The main oxygen source was the atmospheric reoxygenation and approximately 50 % of it was supplied to 0-10 cm below the water distribution system. Over 99,8 % of the oxygen consumed was used for organic degradation and nitrification. The performance response of planted and unplanted wetlands to simulated wastewater with different ratios of carbon to nitrogen (2,5:1, 5:1 and 10:1) was studied during a 9-month period in greenhouse conditions by Zhao et al. [45]. At C/N ratio 5:1 was achieved a relatively high biological nitrogen removal efficiency and a low level of greenhouse gases flux [46]. Prochaska et al. studied the influence of the season, substrate, hydraulic load and frequency of application of simulated urban sewage on the performance of pilot-scale VFCW [47]. The ANOVA statistical model was applied to analyse the relationships between the main operational factors and the effluent COD, NO_2^--N, NO_3^--N, TN and PO_4^{3-}-P. The hydrodynamics of VFCW was tested with rhodamine WT and numerical modeling was used as written in [48]. The capacity of an on-site recirculating VFCW to withstand disturbances and highly variable influent quality was studied [49]. It was found that the general recovery is reached within 24 h. Lihua Cui et al. treated domestic wastewater using three different slags, hydraulic loading rates, operational periods with and without plants for the removal of nitrogen and phosphorus [50]. Hybrid systems were compared at different C/N ratios by Zhao et al. [51]. S. Prost-Boucle and P. Molle established the dependence of nitrification on the recirculation rate and seasons (temperature effect) [52]. Effect of loading, resting period, temperature, porous media, vegetation and aeration were studied by Stefanakis and Tsihrintzis [53]. In a review Saeed and Sun [54] discussed the dependence of nitrogen and organics removal on the environmental parameters, operating conditions and supporting media.

The role of the plants was studied toward the removal of nitrogen and phosphorus [55,56]. Iamchaturapatr et al. studied nutrient removal by 21 plants (18 emergent and 3floating plants) by area-based calculation and biomass-based calculations [57].

Bacterial carbon utilization in VFCW was studied by Tietz et al.[58]. A simple mass-balance approach was applied to explain the bacterially catalysed organic matter degradation. In another paper Tietz et al. [59] made a quantitative description of the microbial biocoenosis in subsurface VFCW fed with municipal wastewater. The microbial biomass was measured at different depths of planted and unplanted systems. Sleytr et al. demonstrated the influence of the plants on the rhizosphere community [60].

Based on Life Cycle Assessment (LCA) Fuchs et al. suggested that constructed wetlands have less environmental impact in terms of resource consumption and greenhouse gas emissions [61].

Different filter materials for phosphorus removal from wastewater in treatment wetlands have been studied [62, 63]. The potential of fragmented Moleanos limestone [64], wollastonite [65], crushed brick and palygorskite [66], a mixture of river sand and dolomite (10:1 w/w), [67] was investigated.

The landfill leachate is characterized with high nitrogeneous pollutants content. Investigations have been done on its purification in constructed wetlands. Four vertical-flow wetlands under predominately aerobic conditions were used for a mass-balance study in the transformation of nitrogeneous pollutants [68]. Landfill leachate was treated in a pilot-scale sub-surface CW planted with *Cyperus haspan* and three weeks retention time. Samples were tested for 13 parameters (pH, turbidity, color, TSS, COD, BOD_5, NH_4^+-N, TP, TN, Fe, Mg, Mn, Zn) and a high removal efficiency was obtained [69]. Justin et al. present a combination of landfill leachate pre-treatment in CW and subsequent reuse for the irrigation of grass and willows [70]. Six interconnected beds with horizontal and vertical subsurface water flow and planted with *Phragmites australis* were used. According to Bulk [71] CWs as a tertiary system or as an independent system could be a low-cost alternative for the treatment of leachate from old landfill sites. Leachate from a closed landfill was treated in an integral system consisted of extraction, aeration, settling, intermittent vertical sand filtration, a surface flow wetland with recycle and discharged in a river [72]. Experiments were conducted to treat a sanitary landfill leachate with high nitrogen and bacterial contents [73]. Mass balance analysis, based on total nitrogen contents of the plant biomass and dissolved oxygen and oxidation reduction potential values, suggested that 88 % of the input total nitrogen were uptaken by the plant biomass. Lavrova and Koumanova studied the influence of recirculation in a lab-scale VFCW on the treatment efficiency of landfill leachate [74]. Comparison of horizontal and vertical CW systems for landfill leachate treatment with two types of material (gravel and zeolite) and planted with *Typha latifolia* was made by Yalcuk and Ugurlu [75]. Better NH_4^+-N removal performance was observed in the VF system with zeolite. Horizontal flow system was more effective in COD removal.

2. Aim

The aim of this study is to investigate treatment efficiency of the raw pig slurry and the landfill leachate in a lab-scale subsurface vertical-flow wetland (SSVFW) planted with *Phragmites australis*, in the lab-scale aerobic activated sludge bioreactor (ASR) and in an hybrid installation where the first stage includes an aerobic activated sludge bioreactor and the second stage – a subsurface vertical-flow wetland (ASR-SSVFW).

3. Material and methods

Pig slurry was taken from a farm located in south-western part of Bulgaria and the landfill leachate was taken from a landfill situated in the north-western region in Bulgaria. After collection, the wastewater was allowed to settle overnight. After that the supernatant was treated. Table 1 summarizes the main characteristics of the influent wastewaters.

Parameter	Influent	
	Pig slurry	Landfill leachate
COD, *mg* L^{-1}	1535 ± 502	2940 ± 140
BOD, *mg* L^{-1}	612 ± 419	230.5 ± 26.5
NH_4^+-N, *mg* L^{-1}	322 ± 87	206.7 ± 8.3
NO_3^--N, *mg* L^{-1}	0	1.5 ± 0.4
pH	7.2 ± 1.1	7.9 ± 0.4

Table 1. Characteristics of the influent wastewater

The water samples were taken every day. The water samples have been examined for pH, Chemical Oxygen Demand (COD), Biochemical Oxygen Demand (BOD), Ammonium-Nitrogen (NH_4^+ - N) and Nitrate-Nitrogen (NO_3^- - N) by standard methods [76].

- **Lab-scale subsurface vertical-flow wetland planted with *Phragmites australis* (SSVFW)**

The laboratory system consisted of sedimentation tank, subsurface vertical-flow wetland, peristaltic pump and storage tank of the treated water (Fig. 4). The SSVFW was made of Plexiglas with dimensions of 123 mm in diameter and 900 mm in height. The reactor was filled with 35 ÷ 55 mm round gravel with 300 mm height as bottom layer and top layer of 5 ÷ 25 mm gravel with a height of 500 mm. Young *Phragmites australis,* obtained from comparatively clean area, was planted in the top layer of the SSVFW. After collection, the wastewater was allowed to settle overnight, the supernatant was diluted with tap water and then was treated. This was done to avoid possible damage of the plant because of the significant contamination of the raw pig slurry and landfill leachate. For increasing the purification capacity effluent recirculation was used [77-82]. The SSVFW was operated continuously in recirculation regime. The recirculation was employed at ratio of 1:1, giving SSVFW 1 h of wastewater-bed matrix contact and 1 h of effluent recirculation. The flow rate of the system was 80 ml min^{-1} [80-82] and the hydraulic retention time was 0.9 h. After filling the reactor with wastewater, the laboratory peristaltic pump was turned on and the water started to flow through the system for a period of one hour. After that, controlled by programmed electronic timer, connected with peristaltic pump, the water stopped moving and remained calm in the SSVFW for one hour. After one hour the peristaltic pump started again and the water began to flow again through the SSVFW.

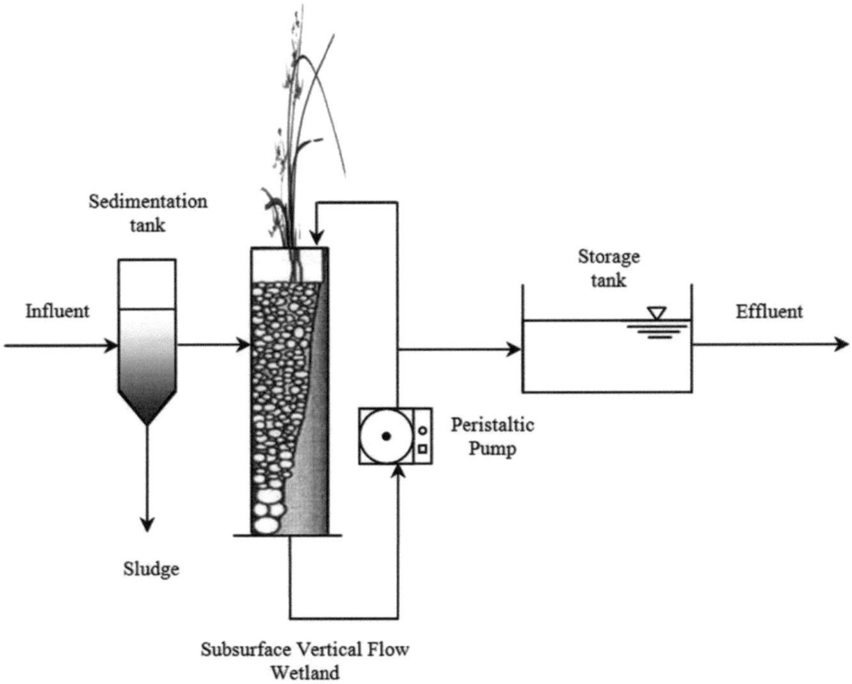

Figure 4. Flow diagram of the SSVFW

- **Lab-scale aerobic activated sludge reactor (ASR)**

ASR of 195 mm in diameter and 650 mm in height was used. The aeration system consisted of three diffusers, situated at the bottom of the bioreactor. The activated sludge (AS) was taken from a municipal wastewater treatment plant. After preliminary sedimentation in primary sedimentation tank the wastewater entered the ASR, where it was mixed with activated sludge in volume ratio 1:1. After reaching of the standard measurements of the controlled physico-chemical characteristics, wastewater flow into secondary sedimentation tank for clarifying, which leading to removal of the suspended activated sludge (Fig. 5).

- **hybrid installation consisted of an aerobic activated sludge reactor (ASR) and a subsurface vertical-flow wetland (ASR / SSVFW)**

After preliminary sedimentation in primary sedimentation tank the wastewater entered the ASR, where it was mixed with activated sludge in volume ratio 1:1. For several days the water is treated in the ASR to achieve a double reduce of pollutants concentration in terms of COD and [NH_4^+-N]. After reaching the necessary concentration, the suspension passed through the secondary sedimentation tank for clarification and then entered the SSVFW for polishing (Fig. 6). The SSVFW was operated continuously in recirculation regime. The recirculation was employed at ratio of 1:1, giving SSVFW 1 h of wastewater-bed matrix contact and 1 h of effluent recirculation. The flow rate of the system was 80 ml min^{-1} and the hydraulic retention time was 0.9 h.

http://dx.doi.org/10.5772/56245

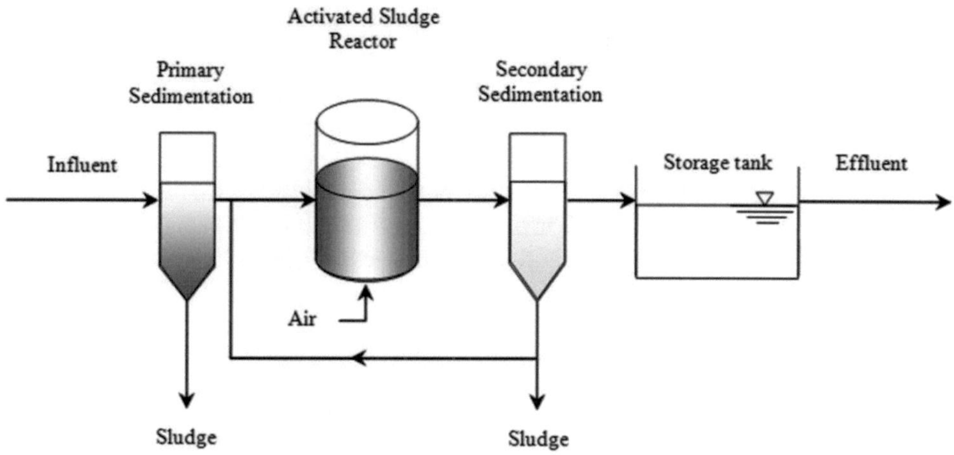

Figure 5. Flow diagram of the ASR

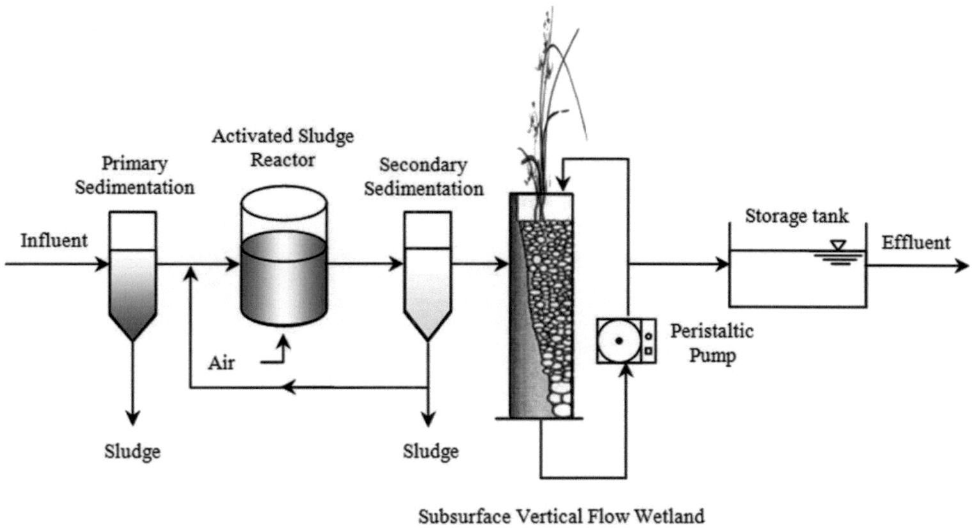

Figure 6. Flow diagramme of the hybrid installation ASR-SSVFW

4. Case study 1: Pig slurry treatment

Fig. 7 to Fig. 10 illustrate the comparison of the water characteristics during the experiments. COD, BOD and NH_4^+-N values are presented as a proportion between their concentration in certain moment and its initial concentration (COD_t/COD_0, BOD_t/BOD_0 and NH_4^+-N_t/NH_4^+-N_0).

During the first 3 days a significant decreasing of COD concentration was observed, especially in the ASR and in the hybrid installation ASR-SSVFW (fig. 7). This is a result of the additional aeration in the ASR. Until this moment the extent of COD decreasing in the ASR, SSVFW and hybrid installation ASR-SSVFW was: 58.9 %, 39 % and 49.8 %. After that the process slows down. During this period easier chemically oxidizable organic matter and biodegradable organic matter undergo changes under the influence of the oxygen and microbial activity. In the SSVFW the decreasing of the COD becomes slower because there is not additional aeration and the plants roots are the only suppliers of oxygen. During the first 3 days the COD concentration decreasing in the hybrid installation is almost identical with that in the ASR, but after wastewater inflow to the SSVFW of the hybrid installation, the tenor of the decreasing curve of the COD concentration lightly altered and reached that in the SSVFW. Due to the preliminary pig slurry dilution in the lab-scale including only subsurface vertical flow wetland, the initial concentrations of the analyzed parameters were almost twice lower than those in the other two experiments. That is why the water purification in the SSVFW is faster than that in the combined system. The reaching of the standards of measured physicochemical characteristic in the SSVFW, ASR and hybrid installation ASR-SSVFW becomes for twelve, then and sixteen days, respectively. Significant extent of the COD concentration decreasing in the three systems was achieved: in SSVFW - 93.1 %, in ASR 96.7 % and in hybrid installation ASR-SSVFW 97.1 %.

BOD decreasing in these systems becomes without significant differences and lightly in comparison with that of the COD decreasing (fig. 8). Under aeration BOD values in ASR and ASR-SSVFW, are decreasing a little bit fully in comparison with that in the SSVFW. During the first 3 days the decreasing extents of the BOD values in the three installations are: SSVFW - 43.3 %, ASR – 61.8 % and hybrid installation ASR-SSVFW – 54.2 %. The reaching of the standards in the SSVFW, ASR and ASR-SSVFW becomes for twelve, ten and sixteen days, respectively.

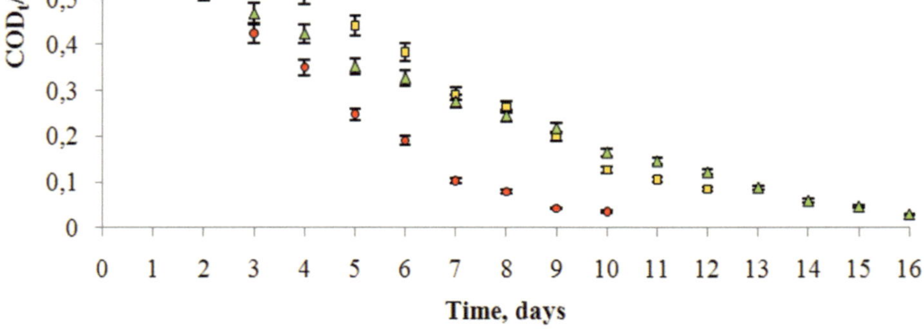

Figure 7. COD decreasing of the water samples in SSVFW, ASR and hybrid installation ASR-SSVFW

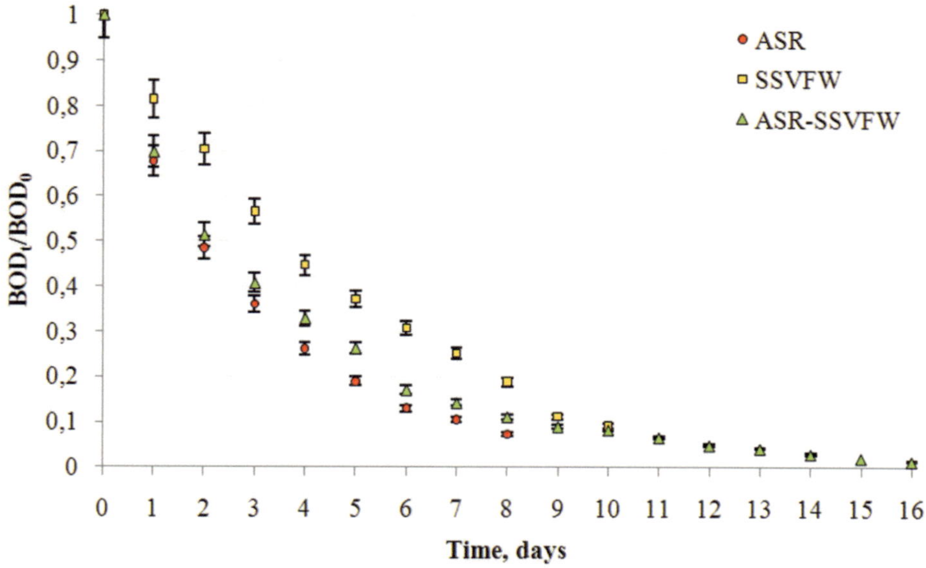

Figure 8. BOD decreasing of the water samples in SSVFW, ASR and hybrid installation ASR-SSVFW

The nitrogen is one of the main pollutants in wastewater that can cause eutrophication, affects dissolved oxygen levels of receiving water, and may cause toxicity (depending on the nitrogen form) to the aquatic organisms [83]. In these systems the transformation and removal of nitrogen are accomplished by both classical and newly discovered routes. The classical pathways include biological i.e. ammonification, nitrification, denitrification, plant uptake, biomass assimilation, dissimilatory nitrate reduction), and physicochemical routes (e.g. ammonia volatilization and adsorption). The newly discovered nitrogen removal routes are solely dependent on microbiological metabolism such as partial nitrification-denitrification.

The decreasing of ammonium - nitrogen in the systems is shown on Fig. 9. The similar effect was observed. During the first three days of the purification process in these three systems was observed elimination of significant part of the ammonium – nitrogen: SSVFW – 53.7 %, ASR – 100 % and in the hybrid installation ASR-SSVFW – 93.7 %. This process is slower in the SSVFW because of insufficient oxygen concentration in the column matrix. Fully elimination of the ammonium - nitrogen in ASR was achieved for three days and for twelve days in the SSVFW. In the hybrid installation this was achieved for seven days. The decreasing of ammonium – nitrogen can be common result of volatilization, nitrification, plant uptake in wetland system and immobilization onto microbial cells. It is believed that nitrogen constitutes a major part of biomass, e.g. 12.4 % of $C_5H_7O_2N$ mass being nitrogen [84].

In parallel with the decreasing of ammonium-nitrogen concentration, an increasing of nitrate-nitrogen is observed, as a result of the nitrification (fig. 10). In the ASR that becomes faster in comparison with the process in the SSVFW and hybrid installation ASR-SSVFW. During the 5th day was observed equalization of the nitrate-nitrogen concentration in the ASR and SSVFW. Elimination of the nitrate-nitrogen was not achieved during the wastewater purification in these systems. The reason for impossibility of nitrate-nitrogen removal is the lack of anoxic conditions in the systems. It is also well established that carbon availability plays an important role in both synthesis and activity of denitrifying enzymes as well as general support of the denitrifying population [85]. The lack of organic carbon source is supposed to keep insignificant denitrification. There are two required conditions for denitrification: anoxic environment and sufficient organic carbon source. In the ASR this process does not occur, because there is aeration. In constructed wetland system the bottom layer provides anoxic conditions for achieving a denitrification. Hence, initial high ammonium-nitrogen concentrations and deficient of organic carbon source were the reasons to depress the denitrification in the wetland system treated the pig wastewater.

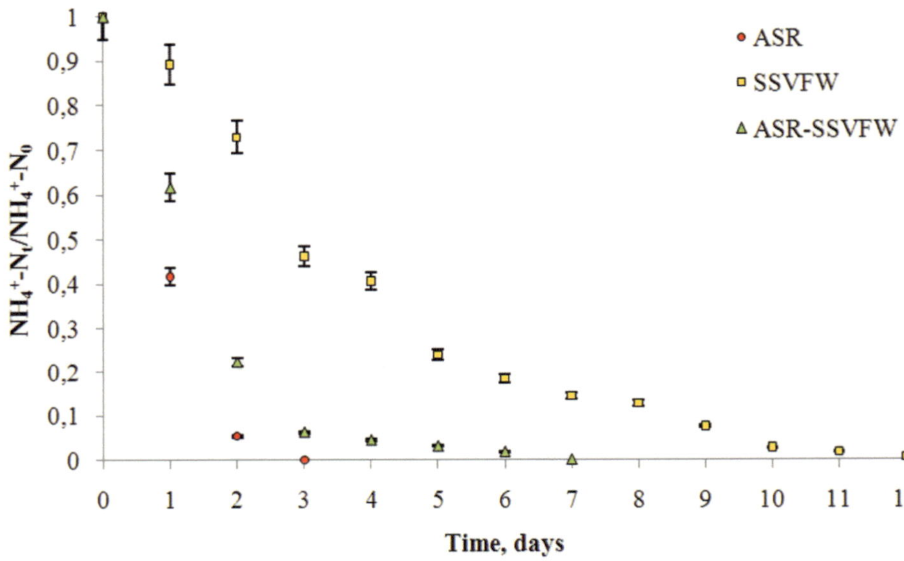

Figure 9. Ammonium-nitrogen in the water samples from SSVFW, ASR and hybrid installation ASR-SSVFW

During the first two days the odour of the treated wastewater in the systems was eliminated. The pH values of the wastewater are in the neutral zone (6.3 – 8.3).

5. Case study 2: Landfill leachate treatment

Fig. 11 to Fig. 14 illustrate the comparison of the water characteristics during the experiments. COD, BOD and NH_4^+-N values are presented like proportion between their concentration in certain moment and its initial concentration (COD_t/COD_0, BOD_t/BOD_0 and NH_4^+-N_t/NH_4^+-N_0).

During the experiments the COD decreasing in these three systems is similar (fig. 11). The reduction of the COD was slightly faster with the preliminary aerobic treatment of the landfill leachate. During the first three days there was a significant decrease of COD. The removal

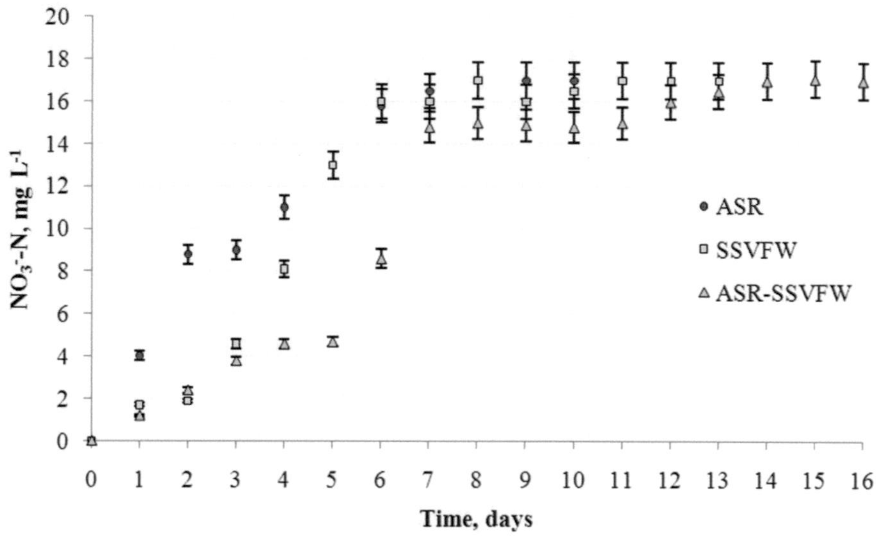

Figure 10. Nitrate-nitrogen in the water samples from SSVFW, ASR and integrated installation ASR-SSVFW

efficiency in the landfill leachate on the third day in the SSVFW is 24 %, in the ASR it is 51.5 % and in the hybrid system ASR-SSVFW it is 51 %. The COD reduction is smooth and without significant fluctuations. The limit concentration of COD in the SSVFW was reached for 15 days, in the ASR for 9 days and in the hybrid installation ASR-SSVFW for 12 days. The removal efficiency is 98 %, 97.8 % and 97.9 %, respectively.

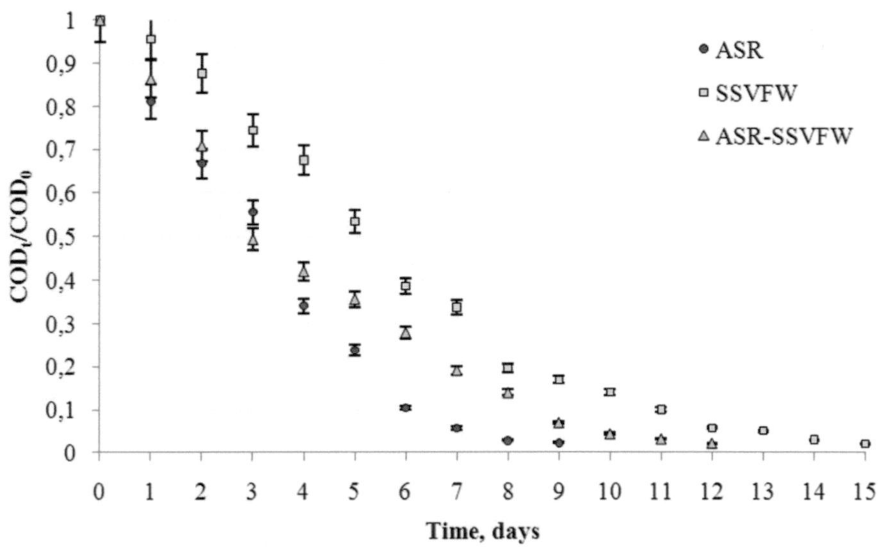

Figure 11. COD decreasing of the water samples in SSVFW, ASR and hybrid installation ASR-SSVFW

The preliminary dilution with tap water of the landfill leachate and the flowing into the SSVFW, lead to almost double decrease of the BOD. BOD decreasing in the wetland system is smoother compared to the ASR and the hybrid installation (fig. 12). The removal efficiency during the first three days is 46.8 % in the SSVFW, 72.2 % in the ASR and 71.4 % in the hybrid installation ASR-SSVFW. The limiting concentration for BOD in the SSVFW was achieved in 11 days with 92.9 % removal efficiency, in the ASR – 6 days with 95.4 % removal efficiency and in the ASR-SSVFW respectively in 9 days with 94.5 % removal efficiency.

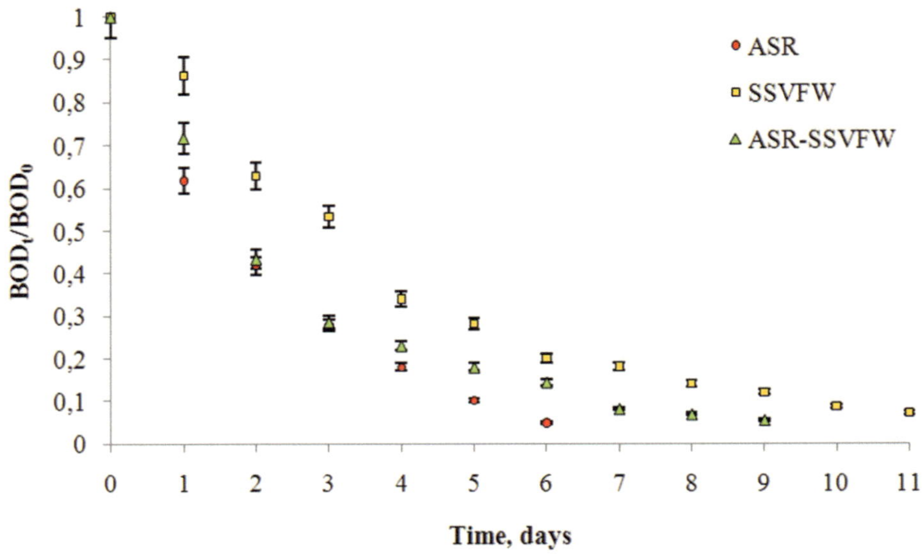

Figure 12. BOD decreasing of the water samples in SSVFW, ASR and hybrid installation ASR-SSVFW

The preliminary dilution of the wastewater and the flowing into the SSVFW, resulted in double decrease of the concentration of the [NH_4^+-N]. As shown on Fig. 13 a sharp decrease of the [NH_4^+-N] concentration in the first three days occurs in the systems where preliminary aerobic treatment of the landfill leachate was used. The removal efficiency in the SSVFW is 74.8 %, in the ASR 99.6 % and in the ASR-SSVFW 96.2 %. This is probably due to the air blowing into the aerobic reactor to support microbial activity. On the other hand this intensive aeration can cause the escape of ammonia from the system.

Complete removal of the [NH_4^+-N] in the SSVFW was achieved for 10 days while in the ASR it was for 4 days and in the ASR-SSVFW - for 5 days. This can be explained by the fact that in this reactor the oxygen is not enough for the nitrification and from the constructive point of view the separation of ammonia from the system is embarrassed.

The aerobic treatment of the landfill leachate in the reactor with suspended activated sludge results in a significant accumulation of nitrate ions in comparison with the process taking place in the wetland system where preliminary diluted landfill leachate is treated (Fig. 14). In the laboratory systems an accumulation of the nitrate ions was observed. Probably this effect is

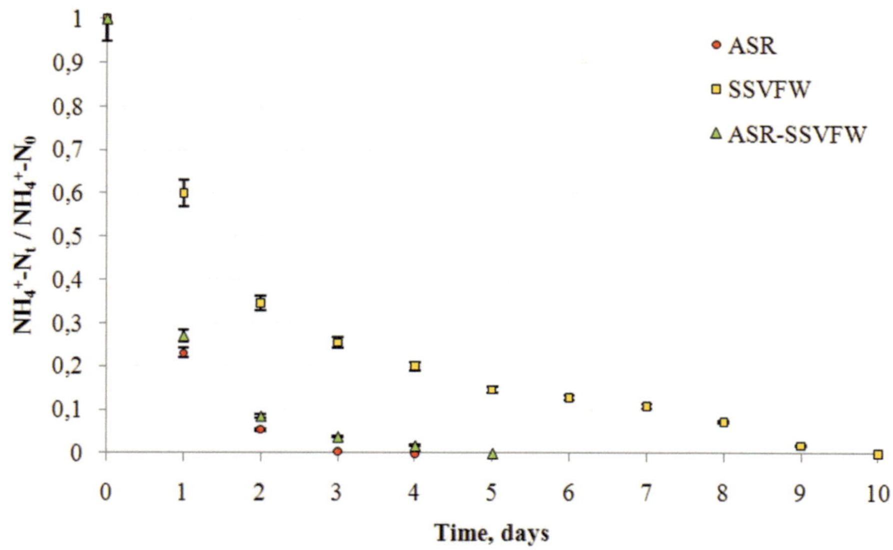

Figure 13. NH_4^+-N decreasing in the water samples from SSVFW, ASR and hybrid installation ASR-SSVFW

due to the insufficient quantity of the source of organic carbon in the wetland system [85] and on the other hand - the aerobic conditions in the aerobic bioreactor.

During the experiments pH decreased slightly from 8,3 to 7,5. The neutralization of the landfill leachate was achieved during the preliminary dilution with tap water, while in the hybrid installation it was achieved during the third day of operation.

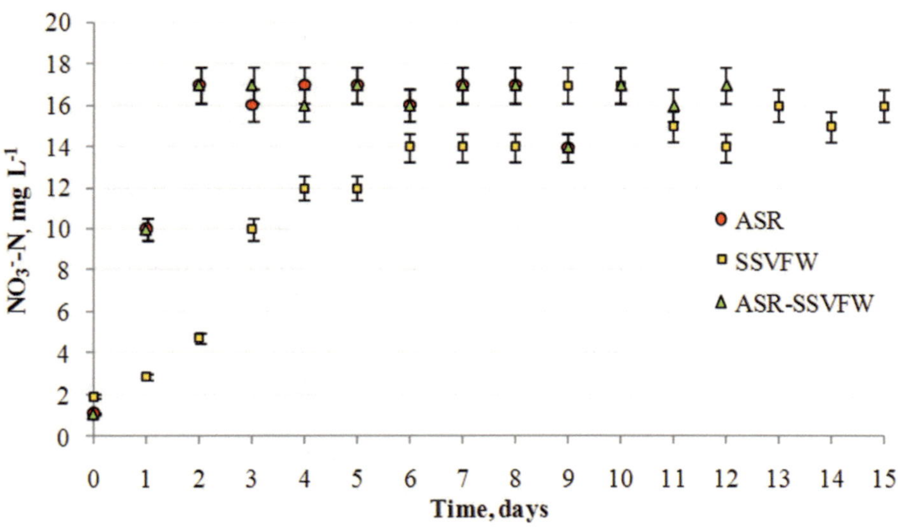

Figure 14. [NO_3^--N] decreasing of the water samples in SSVFW, ASR and hybrid installation ASR-SSVFW

In Table 2 are compared the data characterized the two studied wastewaters that are treated in ASR, SSVFW and hybrid installation ASR-SSVFW. The data are compared with the standards in Bulgaria (Benchmarks). Obviously, the requirements of the national standards were met.

Parameter	Influent		Effluent						Bench marks	Average efficiency, %	Time, days					
			ASR		SSVFW		ASR-SSVFW				ASR		SSVFW		ASR-SSVFW	
	Pig slurry	Landfill leachate	Pig slurry	Landfill leachate	Pig slurry	Landfill leachate	Pig slurry	Landfill leachate			Pig slurry	Landfill leachate	Pig slurry	Landfill leachate	Pig slurry	Landfill leachate
COD, mg L^{-1}	1535 ± 502	2940 ± 140	72.4 ± 7.1	69 ± 1	69.3 ± 3.9	67.5 ± 11.5	63.6 ± 4	62.5 ± 4.5	70	95.6	11	12	12	16	10	12
BOD, mg L^{-1}	612 ± 419	230.5 ± 26.5	14.4 ± 0.4	15.6 ± 0.7	15.7 ± 1.5	14.5 ± 0.5	15.8 ± 0.4	15.7 ± 0.8	15	97.5	9	10	10	9	9	11
NH_4^+-N, mg L^{-1}	322 ± 87	206.7 ± 8.3	0	0	0	0	0	0	2	100	7	6	10	10	8	8
NO_3^--N, mg L^{-1}	0	1.5 ± 0.4	16.7 ± 0.3	16.5 ± 0.5	16.9 ± 0.1	17 ± 0.1	15.9 ± 0.6	17 ± 0.1	10	-	-	-	-	-	-	-
pH	7.2 ± 1.1	7.9 ± 0.4	7.1 ± 0.1	7.2 ± 0.1	7.2 ± 0.1	7.2 ± 0.1	7.1 ± 0.1	7 ± 0.1	6.0 – 8.5							

Table 2. Characteristics of the influent and the effluents from the three systems

The flow rate is one of the important factors, which control the performance of subsurface vertical flow wetland systems. The higher flow rate promotes faster passage of wastewater through the media, thus reducing the optimum contact time and leads to longer period needed for treatment [74]. The recirculation of the treated wastewater through the subsurface vertical-flow wetland also has a significant role in the purification efficiency [74, 87-88]. For confirmation of these statements experiments with three different wastewater flow rates (40, 60 and 82 ml/min) and three different recirculation ratios (1:1, 1:2 and 1:3) were conducted. The hydraulic retention time in the wetland system was 1.8, 1.2 and 0.9 h, respectively.

The comparison of the COD and BOD values of the treated landfill leachate during the experiments is illustrated in fig.15. The data demonstrate the influence of the recirculation at the three different flow rates on the treatment ability of the lab-scale vertical-flow wetland system. The decreasing of COD values after five days from the beginning is fast. Then the process slows down. The efficiency at the 5th day (recirculation ratio 1:1) was 67 % at flow rate 82 ml min^{-1}, 81 % - at flow rate 60 ml min^{-1} and 90 % - at flow rate 40 ml min^{-1}. The efficiency at recirculation ratio 1:2 was 78 %, 86 % and 90 %, respectively, and at recirculation ratio 1:3 it was 78 %, 90 % and 96 %, respectively. COD decreased slower when the flow rate was higher.

The elimination of BOD occurs fast in most cases during the initial five days. The efficiency of BOD removal at recirculation ratio 1:1 was 72 %, 85 % and 92 % for flow rate 82 ml min^{-1}, 60 ml min^{-1} and 40 ml min^{-1}, respectively. At recirculation ratio 1:2 it was 83 %, 92 % and 93 % for the corresponding flow rates. The efficiency was 91 % at flow rate 82 ml min^{-1} and recirculation ratio1:3. The same efficiency was obtained after 4 days at flow rate 60 ml min^{-1} and after 3 days at flow rate 40 ml min^{-1}. It was observed that the longer the water remained quiet in SSVFW, the faster COD and BOD decreased.

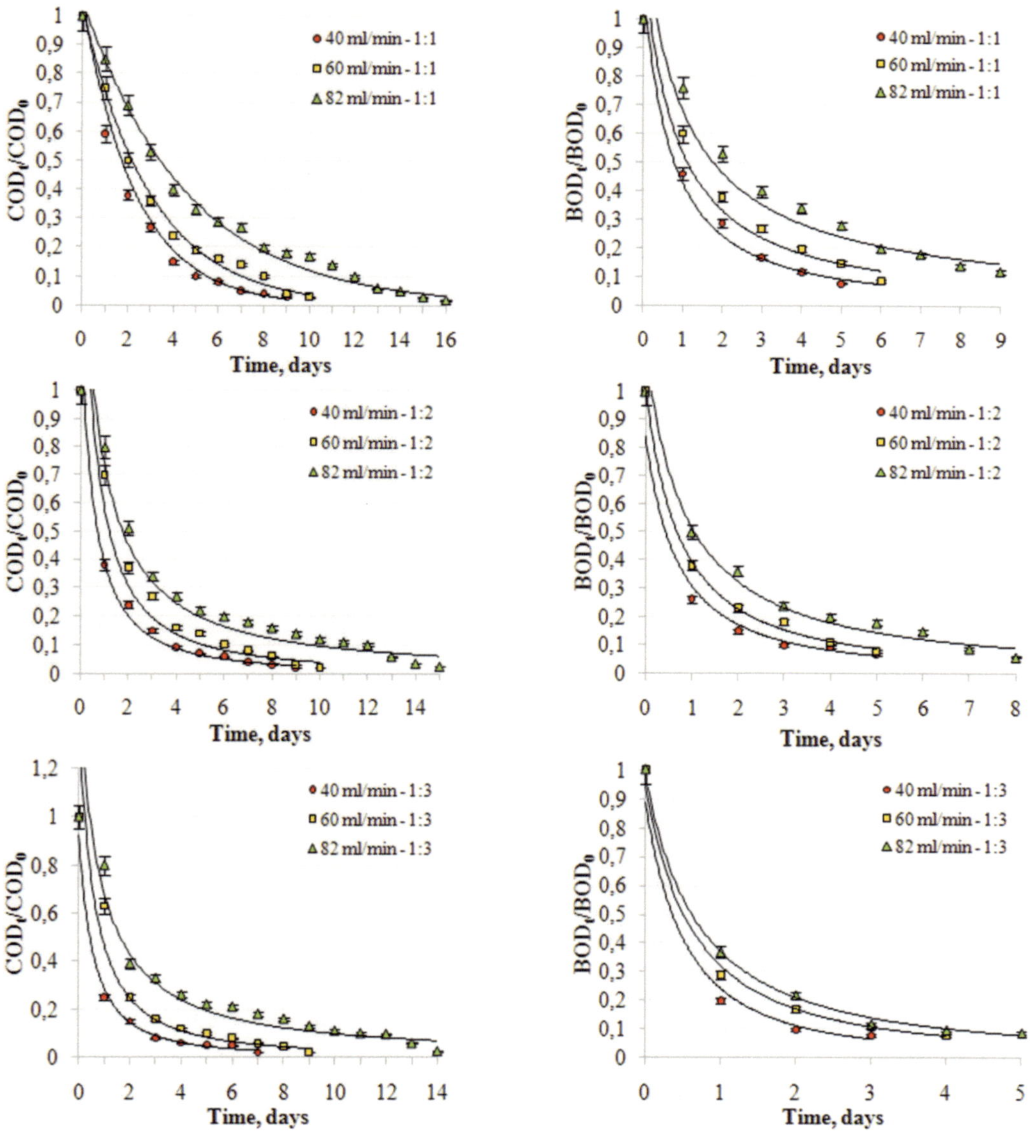

Figure 15. Comparison of the COD and BOD values of the treated landfill leachate at different flow rates

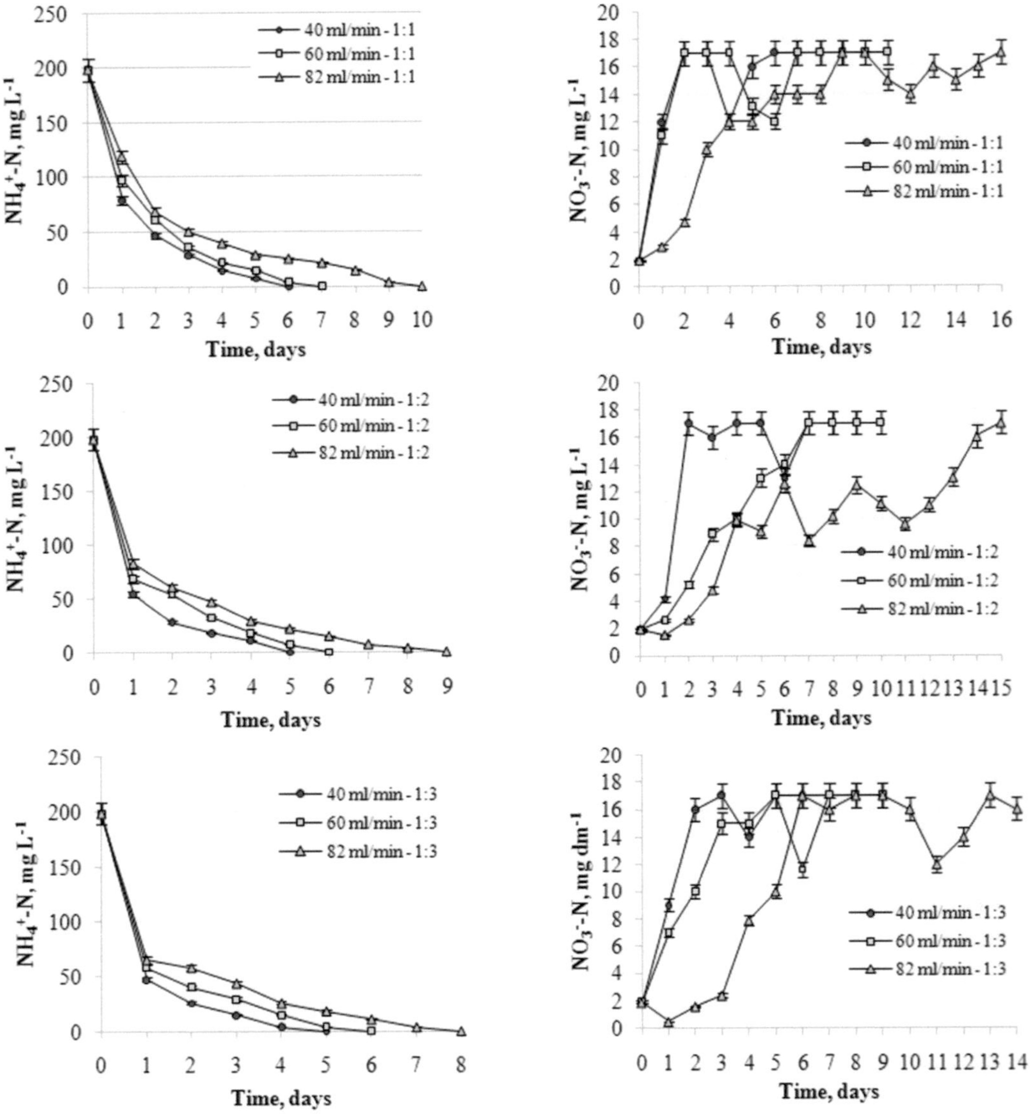

Figure 16. Comparison of the [NH_4^+-N] and [NO_3^--N] values of the treated landfill leachate at different flow rates

The importance of nitrogen removal is comparable with that for organic carbon, toxic compounds and metals removal during the leachate treatment in SSVFW. Ammonium removal by nitrification in constructed wetlands differing in design and purpose was reported [89]. It is known that autotrophic nitrification consists of two successive aerobic reactions, the conversion of ammonium to nitrite by ammonium oxidizing bacteria and the conversion of nitrite to nitrate by nitrite oxidizing bacteria. The concentration of ammonium-nitrogen in the influent used in this study was relatively high. So, it was interesting to record the changes of NH_4^+-N and NO_3^--N values during the leachate treatment at different flow rates and recirculation ratios.

The influence of these parameters on the ammonium depletion is illustrated in Fig. 16. It was exhausted completely during the experiments. The decrease of ammonium-nitrogen could be collective result of volatilization, nitrification, plant uptake in wetland system and immobilization. At the same time NO_3^--N increased depending on the flow rate and the recirculation ratios. In all cases the values increased during the first 1-2 days. After this period the curves shape depended on the experimental conditions. The concentration of NO_3^--N has increased faster when the flow rate was lower. The influence of the recirculation ratios was opposite. During the experiments at different conditions dissolved oxygen was measured and the values were from 5.2 to 8 mg L^{-1}. It is known that the concentration of 1 mg L^{-1} is sufficient for oxidation of ammonium [90].

The lack of the denitrification during the treatment can be a result of the less activity of denitrifying bacteria in the system. Vymazal reported that SSVFW removes successfully NH_4^+-N but the denitrification is very limited in these systems [91]. It was also well established that carbon availability plays an important role in both synthesis and activity of denitrifying enzymes as well as general support of the denitrifying population. The lack of organic carbon sources is thought to prevent significant levels of denitrification [85].

Phosphorus removal in wetland treatment systems occurs through adsorption, plant uptake, complexation, and precipitation [92]. The value of total phosphorus (TP) in the treated leachate was relatively low (5.5 mg L^{-1}). It was established that TP removal follows the same tendency as NH_4^+-N removal. During the first two days a significant TP elimination occured (Fig.17). The higher flow rate leads to a longer period of elimination of TP. At the same time the change of the recirculation ratio from 1:1 to 1:3 (water stays quiet in SSVFW longer) leads to shorter period of elimination (e.g. at flow rate 82 ml min^{-1} the TP removal was 41.8 %, 60 % and 67.3 % for 1:1, 1:2 and 1:3 recirculation ratios during 8, 6 and 4 days, correspondingly).

It was established that during the experiments with different flow rate and hydraulic retention time pH slightly decreased from 7.9 to 7.5 and the salinity also decreased from 2.5 to 1.9 ‰. TDS gradually decreased from 2460 to 1778 mg L^{-1}. The values of TSS varied from 1.91 to 3.96 g L^{-1}. Landfill leachate conductivity decreased from 4710 to 3408 μS cm^{-1}. TDS, TSS, salinity, phosphorus concentration as well as the conductivity have been determined only for the case of landfill leachate at different flow rate and hydraulic retention time.

It is well known that the vegetation in the wetland systems play a significant role in purification process. Aquatic plants enhance nutrient removal through biomass accumulation, fixation of inorganic and organic particles and where ammonium-N is present, by the creation of an oxidized rhizosphere [93]. In the absence of plants, the gravel substrate provided significant wastewater treatment [94, 95], although most studies report improved nutrient removal where plants are present [96]. Our experiments without vegetation in the lab-scale subsurface vertical-flow wetland system leads to lower treatment efficiency in comparison whit that where in the laboratory system has grown vegetation (Fig. 18). That confirms the important role of the plants in purification process. On the other hand the lack of plants allows the use of additional organic carbon source, which achieves a denitrification process (Fig. 19). In these experiment was added methanol in the SSVFW without vegetation. As a result was observed decreasing of the nutrition elements and their elimination from the system.

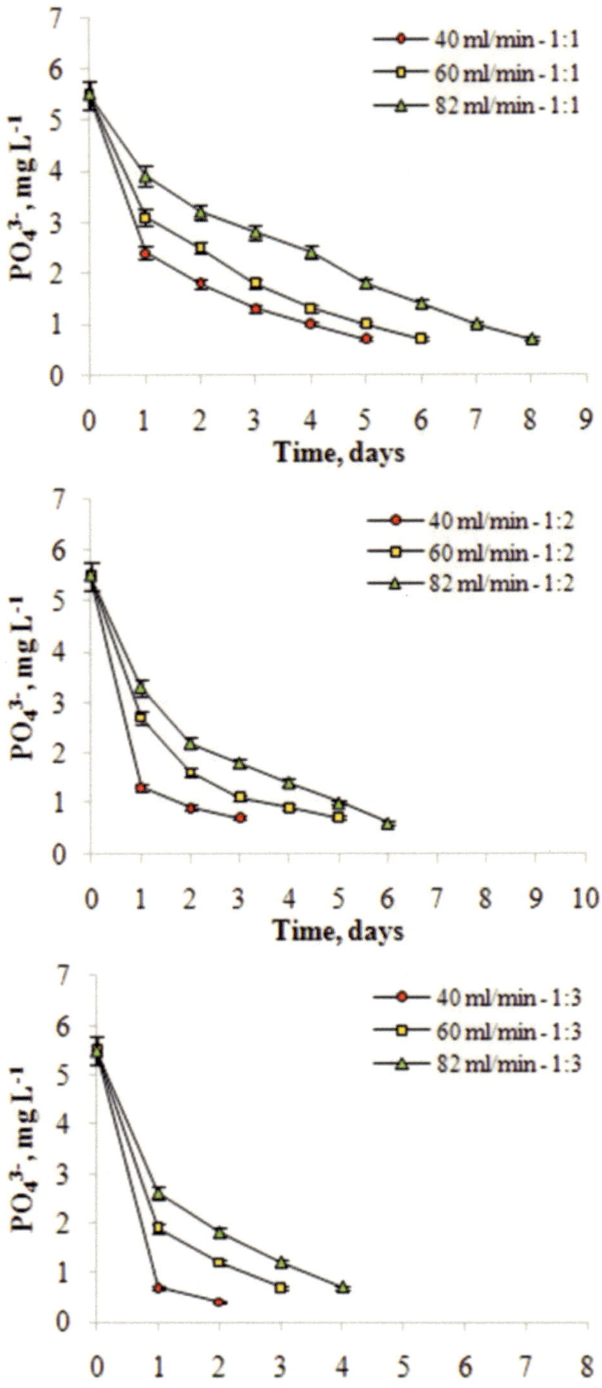

Figure 17. Comparison of the [PO_4^{3-}] values of the treated landfill leachate at different flow rates

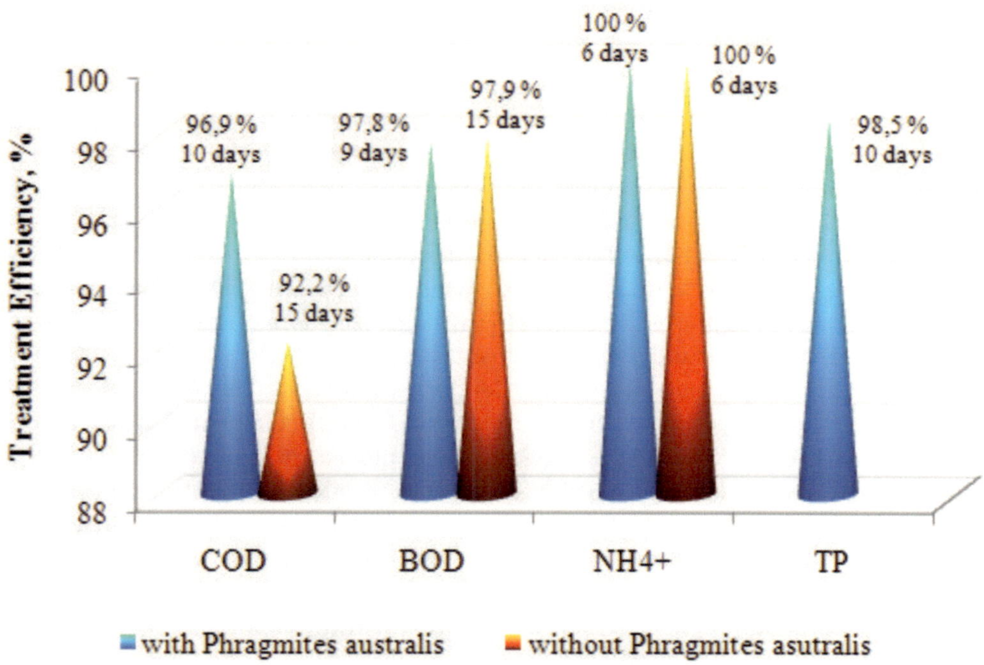

Figure 18. Treatment efficiency during the experiments with/without *Phragmites australis*

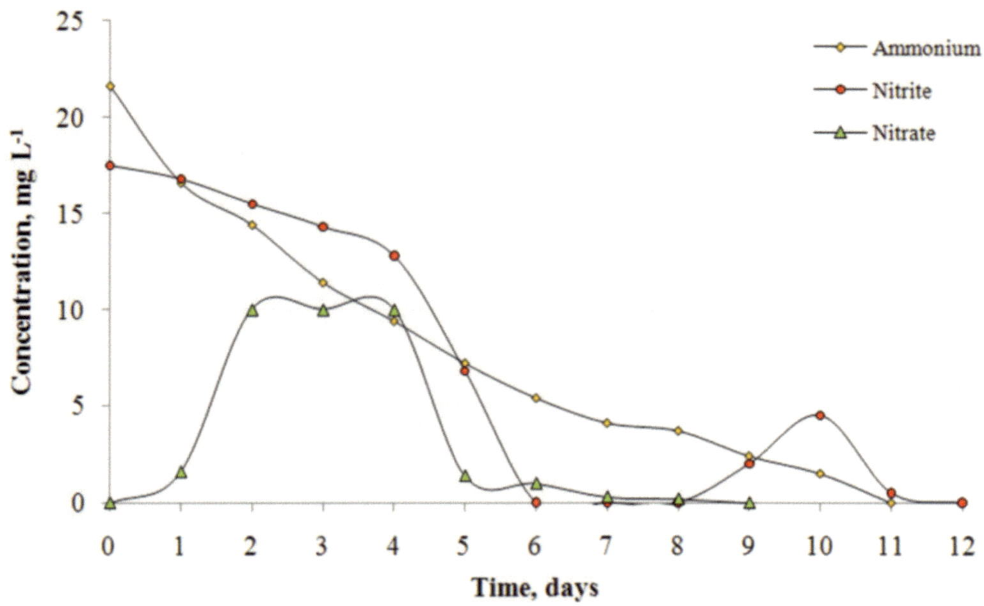

Figure 19. Nutrition concentration decreasing with applying of additional carbon source

6. Conclusions

Activated sludge reactor, susbsurface vertical-flow wetland and hybrid installation were studied for aerobic treatment and polishing of two types wastewater - pig slurry and landfill leachate. It was established that the values of the treated water characteristics significantly decreased for comparatively short time accompanied by odour elimination and neutralization of wastewater. Significant COD and BOD decreasing were attained in those cases and the aquatic standards were met. Fully elimination of the ammonium-nitrogen in the SSVFW was achieved for longer period of time in comparison whit that in the ASR. Decreasing of obtained nitrate-nitrogen was not achieved in the SSVFW with growing *Phragmites australis* because of absence of anoxic conditions and probably of insufficient organic carbon source. In the SSVFW without vegetation was achieved denitrification process. It was established that the higher flow rate leads to longer period needed for treatment. The recirculation ratios also influence the purification process. Alternating between water movement through the SSVFW and stagnant periods resulted in a varying extent of purification, and the longer the stagnant period of the water in SSVFW the shorter the period for obtaining the desired characteristics of the effluent water was. These investigations show that the use of SSVFW is also effective as ASR and combination of the processes accelerate the purification process. The SSVFW has some advantages simulating the processes occurring in the natural wetlands, easy maintenance, energy conservation and cost effectiveness.

Nomenclature

AS - Activated Sludge

ASR - Aerobic Sludge Reactor

ASR-SSVFW - Aerobic Sludge Reactor - Subsurface Vertical-Flow Wetland (hybrid installation)

BOD - Biochemical Oxygen Demand

COD – Chemical Oxygen Demand

CW – Constructed Wetland

DO – Dissolved Oxygen

LCA - Life Cycle Assessment

NH_4^+-N – Ammonium Nitrogen

NO_2^- -N – Nitrite Nitrogen

NO_3^--N – Nitrate Nitrogen

PO_4^{3-} -P - Phosphates

http://dx.doi.org/10.5772/56245

SSVFW – Subsurface Vertical Flow Wetland

TKN – Total Kjehldahl Nitrogen

TN – Total Nitrogen

TP – Total Phosphorus

TSS – Total Suspended Solids

VF – Vertical Flow

Author details

Silviya Lavrova and Bogdana Koumanova*

*Address all correspondence to: bkk@uctm.edu

University of Chemical Technology and Metallurgy, Sofia, Bulgaria

References

[1] Reed, S, & Brown, D. Constructed Wetland Design- The First Generation, Water Environment Research (1992). , 64, 776-81.

[2] Technical/Regulatory GuidelineTechnical and Regulatory Guidance Document for Constructed Treatment Wetlands. The Interstate Technology and Regulatory Council Wetlands Team December; (2003).

[3] Kadlec, R, & Knihght, R. Treatment wetlands. Lewis Publishers; (1996).

[4] Moshiri, G. Constructed Wetlands for Water Quality Improvement. Lewis Publishers, Boca Raton; (1993).

[5] Cooper, P, Smith, M, & Maynard, H. The Design and Performance of a Nitrifying Vertical Flow Reed Bed Treatment System. Water Science and Technology (1996). , 35, 215-221.

[6] Haberl, R. Constructed Wetlands: A Chance to Solve Wastewater Problems in Developing Countries. Water Science and Technology (1999). , 40, 11-17.

[7] Vymazal, J. Types of Constructed Wetlands for Wastewater Treatment: Their Potential for Nutrient Removal, In: Vymazal J. (ed.) Transformations of Nutrients in Natural and Constructed Wetlands. Backhuys Publishers; (2001). , 1-93.

[8] Vymazal, J. Constructed Wetlands for Wastewater Treatment in Europe, In: Dunne EJ, Reddy R, Carton OT (ed.) Nutrient Management in Agricultural Watersheds: a Wetland Solution. Wageningen Academic Publishers; (2005a)., 230-244.

[9] Vymazal, J. Horizontal Sub-Surface Flow and Hybrid Constructed Wetlands Systems for Wastewater Treatment. Ecological Engineering (2005b)., 25, 478-490.

[10] Van de Moortel ARousseau D., Tack F., Pauw N. A Comparative Study of Surface and Subsurface Flow Constructed Wetlands for Treatment of Combined Sewer Overflows: a Greenhouse Experiment. Ecological Engineering (2009)., 35(2), 175-183.

[11] Tuszynska, A, & Obarska-pempkowiak, H. Dependence Between Quality and Removal Effectiveness of Organic Matter in Hybrid Constructed Wetlands. Bioresource and Technology (2008)., 99, 6010-6016.

[12] Aslam, M, Murtaza, M, Baig, M, Qazi, I, & Iqbal, J. Treatment Performances of Compost-Based and Gravel-Based Vertical Flow Wetlands Operated Identically for Refinary Wastewater Treatment in Pakistan. Ecological Engineering (2007)., 30, 34-42.

[13] Korkusuz, E. A. Treatment Efficiencies of the Vertical Flow Pilot-Scale Constructed Wetlands for Domestic Wastewater Treatment. Turkish Journal of Engineering and Environmental Sciences (2004)., 28, 333-344.

[14] Ongnext, S, Uchiyama, K, Inadama, D, & Yamagiwa, K. Simultaneous removal of color, organic compounds and nutrients in azo dye-containing wastewater using upflow constructed wetland. Journal of Hazardous Materials (2009).

[15] Davies, L, Pedro, J, Novais, J, & Martins-dias, S. Aerobic Degradation of Acid Orange 7 in a Vertical-Flow Constructed Wetland. Water research (2006)., 40, 2055-2063.

[16] Zhang, D, Gersberg, R, Hua, T, Zhu, J, Tuan, N, & Tan, S. Pharmaceutical Removal in Tropical Subsurface Flow Constructed Wetlands at Varying Hydraulic Loading Rates. Chemosphere (2012)., 87, 273-277.

[17] Wang, R, Korboulewsky, N, Prudent, P, Baldy, V, & Bonin, G. can Vertical-Flow Wetland Systems Treat High Concentrated Sludge From a Food Industry? A Mesocosm Experiment testing Three Plant Species. Ecological Engineering (2009)., 35, 230-237.

[18] Herouvim, E, Akratos, C, Tekerlekopoulou, A, & Vayenas, D. Treatment of Olive mill Wastewater in Pilot-Scale Vertical Flow Constructed Wetlands. Ecological Engineering (2011)., 37, 931-939.

[19] Korboulewsky, N, Wang, R, & Baldy, V. Purification process involved in Sludge Treatment by a Vertical Flow Wetland System: Focus on the Role of the Substrate and Plants on N and P Removal. Bioresource Technology (2012)., 105, 9-14.

[20] Gross, A, Shmueli, O, Ronen, Z, & Raveh, E. Recycled Vertical Flow Constructed wetland (RVFCW)- A Novel Method of Recycling Greywater for Irrigation in Small Communities and Households. Chemosphere (2007)., 66, 916-923.

[21] Sklarz, M, Gross, A, Yakirevich, A, & Soares, M. A Recirculating vertical Flow Constructed wetland for the Treatment of Domestic Wastewater. Desalination (2009). , 246, 617-624.

[22] Kouki, S, Hiri, M, Saidi, F, Belaïd, N, & Hassen, S. A. Performances of a Constructed Wetland Treating Domestic Wastewaters During a macrophytes Life Cycle. Desalination (2009). , 246, 452-467.

[23] Hijosa-valsero, M, Sidrach-cardona, R, & Bécares, E. Comparison of Interannual Removal of Various Constructed Wetland Types. Science of the Total Environment (2012). , 430, 174-183.

[24] Ye, F, & Li, Y. Enhancement of Nitrogen Removal in Towery Hybrid Constructed Wetland to Treat Domestic Wastewater for Small Rural Communities. Ecological Engineering (2009). , 35, 1043-1050.

[25] Gikas, G, & Tsihrintzis, V. A Small-Size Vertical Flow Constructed Wetland for Onsite Treatment of Household Wastewater. Ecological Engineering (2012). , 44, 337-343.

[26] Abou-elela, S, & Hellal, M. Municipal Wastewater Treatment Using Vertical Flow Constructed Wetlands Planted with Canna, Phragmites and Cyprus. Ecological Engineering (2012). , 47, 209-213.

[27] Morari, F, & Giardini, L. Municipal Wastewater Treatment With Vertical Flow Constructed Wetlands for Irrigation Reuse. Ecological Engineering (2009). , 35, 643-653.

[28] Zurita, F, Anda, J, & Belmont, M. Treatment of Domestic Wastewater and Production of Commercial Flowers in Vertical and Horizontal Subsurface-Flow Constructed Wetlands. Ecological Engineering (2009). , 35, 861-869.

[29] Langergraber, G, Leroch, K, Pressl, A, Sleytr, K, Rohrhofer, R, & Haberl, R. High-Rate Nitrogen Removal in a Two-Stage Subsurface Vertical Flow Constructed Wetland. Desalination (2009). , 246, 55-68.

[30] Tunçsiper, B. Nitrogen removal in a Combined Vertical and Horizontal Subsurface-Flow Constructed Wetland System. Deaslination (2009). , 247, 466-475.

[31] Panuvatvanich, A, Koottatep, T, & Kone, D. Influence of Sand Layer Depth and Percolate Impounding Regime on Nitrogen Transformation in Vertical-Flow Constructed Wetlands treating Faecal Sludge. Water Research (2009). , 43, 2623-2630.

[32] Bialowiec, A, Davbies, L, Albuquerque, A, & Randerson, P. Nitrogen removal From Landfill Leachate in Constructed Wetlands With Reed and Willow: redox Potential in the Root Zone. Journal of Environmental Management (2012). , 97, 22-27.

[33] Molle, P, Prost-boucle, S, & Lienard, A. Potential for Total Nitrogen Removal by Combining Vertical Flow and Horizontal Flow Constructed Wetlands: A Full-Scale Experiment Study. Ecological Engineering (2008). , 34, 23-29.

[34] Ouyang, Y, Luo, S, & Cui, L. Estimation of Nitrogen Dynamics in a Vertical-Flow Constructed Wtland. Ecological Engineering (2011). , 37, 453-459.

[35] Ayaz, S, Aktas, Ő, Findik, N, Akça, L, & Kinaci, C. Effect of Recirculation on Nitrogen removal in a Hybrid Constructed Wetland System. Ecological Engineering (2012). , 40, 1-5.

[36] Cui Li-HuaOuyang Y., Chen Y., Zhu Xi- Zhen, Zhu Wen-Ling. Removal of Total Nitrogen by Cyperus alternifolius From Wastewaters in Simulated Vertical-Flow Constructed Wetlands. Ecological Engineering (2009). , 35, 1271-1274.

[37] Saeed, T, & Sun, G. Enhanced Denitrification and Organics Removal in Hybrid Wetland Columns: Comparative Experiments. Bioresource Technology (2011). , 102, 967-974.

[38] Huett, D, Morris, S, Smith, G, & Hunt, N. Nitrogen and Phosphorus Removal From Plant Nursery Runoff in Vegetated and Unvegetated Subsurface Flow Wetlands. Water Research (2005). , 39, 3259-3272.

[39] Zhang, D, Tan, S, Gersberg, R, Zhu, J, Sadreddini, S, & Li, Y. Nutrient Removal in Tropical Subsurface Flow Constructed Wetlands Under Batch and Continuous Flow Conditions. Journal of Environmental Management (2012). , 96, 1-6.

[40] Yousefi, Z, & Mohseni-bandpei, A. Nitrogen and Phpsphorus Removal From Wastewater by Subsurface Wetlands Planted With Iris pseudacorus. Ecological Engineering (2010). , 36, 777-782.

[41] Bustamante, M, Mier, M, Estrada, J, & Domiguez, C. Nitrogen and Potassium Variation on Contaminant Removal for a Vertical Subsurface Flow Lab Scale Constructed Wetland. Bioresource Technology (2011). , 102, 7745-7754.

[42] Giraldi, D, & Iannelli, R. Measurements of Water Content Distribution in Vertical Subsurface Flow Constructed Wetlands Using a Capacitance Probe: Benefits and Limitations. Desalination (2009). , 243, 182-194.

[43] Stefanakis, A, & Tsihrintzis, V. Performance of Pilot-Scale Vertical Flow Constructed Wetlands Treating Simulated Municipal Wastewater: Effect of Various Design Parameters. Desalination (2009). , 248, 753-770.

[44] Ye, J, Wang, L, Li, D, Han, W, & Ye, C. Vertical Oxygen Distribution and Oxygen Source Analysis for Vertical-Flow Constructed Wetlands Treating Domestic Wastewater. Ecological Engineering (2012). , 41, 8-12.

[45] Zhao, Y. Liu Bo, Zhang W., Ouyang Y., An S. Performance of Pilot-Scale Vertical-Flow Constructed Wetlands in Responding to Variation in Influent C/N Ratios of Simulated Urban Sewage. Bioresource Technology (2010). , 101, 1693-1700.

[46] Yan, C, & Zhang, H. li Bo, Wang D., Zhao Y., Zheng Z. Effects of Influent C/N Ratios on CO_2 and CH_4 Emissions From Vertical Subsurface Flow Constructed wetlands Treating Synthetic Municipal Wastewater. Journal of Hazardous Materials (2012).

[47] Prochaska, C, Zouboulis, A, & Eskridge, K. Preformance of Pilot-Scale Vertical Flow Constructed Wetlands, As Affected by Season, Substrate, Hydraulic Load and Frequency of Application of Simulated Urban Sewage. Ecological Engineering (2007). , 31, 57-66.

[48] Giraldi, D, Vitturi, M, Zaramella, M, Marion, A, & Iannelli, R. Hydrodynamics of Vertical Subsurface Flow Constructed Wetlands: Tracer Tests WITH Rhodamine WT and Numerical Modelling. Ecological Engineering (2009). , 35, 265-273.

[49] Zapater, M, Gross, A, & Soares, M. Capacity of an On-site Recirculating Vertical Flow Constructed Wetland to Withstand Disturbances and Highly Variable Influent Quality. Ecological Engineering (2011). , 37, 1572-1577.

[50] Cui, L, Ouyang, Y, Lou, Q, Yang, F, Chen, Y, Zhu, W, & Luo, S. Removal of Nutrients from Wastewater with Canna indica L. Under Different Vertical-Flow Constructed Wetland Conditions. Ecological Engineering (2010). , 36, 1083-1088.

[51] Zhao, Y, & Hui, Z. Chao Xu, Nie Er, Li H., He J., Zheng Z. Efficiency of Two-Stage Combinations of Subsurface vertical Down-Flow and Up-Flow Constructed Wetland Systems for Treating Variation in Influent C/N Ratios of Domestic wastewater. Ecological Engineering (2011). , 37, 1546-1554.

[52] Prost-boucle, S, & Molle, p. Recirculation on a Single Stage of Vertical Flow Constructed Wetland: Treatment Limits and Operation Modes. Ecological Engineering (2012). , 43, 81-84.

[53] Stefanakis, A, & Tsihrintzis, V. Effects of Loading, Resting Period, Temperature, Porous Media, vegetation and Aeration on Performance of Pilot-Scale Vertical Flow Constructed Wetlands. Chemical Engineering Journal (2012).

[54] Saeed, T, & Sun, G. A Review on Nitrogen and Organics removal Mechanisms in Subsurface Flow Constructed Wetlands: Dependency on Environmental Parameters, Operating Conditions and Supporting Media. Journal of Environmental management (2012). , 112, 429-448.

[55] Kantawanichkul, S, Kladprasert, S, & Brix, H. Treatment of High-Strength Wastewater in Tropical vertical Flow Constructed Wetlands Planted with Typha angustifolia and Cyperus involucratus. Ecological Engineering (2009). , 35, 238-247.

[56] Lio, X, Huang, S, Tang, T, Liu, X, & Scholz, M. Growth Characteristics and nutrient removal Capability of Plants in Subsurface Vertical Flow Constructed Wetlands. Ecological Engineering (2012). , 44, 189-198.

[57] Iamchaturapatr, J. Yi Su, Rhee J. Nutrient Removal by 21 Aquatic Plants for Vertical Free Surface-Flow (VFS) Constructed Wetland. Ecological Engineering (2007). , 29, 287-293.

[58] Tietz, A, Langergraber, G, Watzinger, A, Haberl, R, & Kirschner, A. Bacterial Carbon Utilization in Vertical Subsurface Flow Constructed Wetlands. Water Research (2008). , 42, 1622-1634.

[59] Tietz, A, Kirschner, A, Langergraber, G, Sleytr, K, & Haberl, R. Characterisation of Microbial Biocenosis in Vertical Subsurface Flow Constructed Wetlands. Science of the Total Environment (2007). , 380, 163-172.

[60] Sleytr, K, Tietz, A, Langergraber, G, Haberl, R, & Sessitsch, A. Diversity of Abundant Bacteria in Subsurface vertical Flow Constructed Wetlands. Ecological Engineering (2009). , 35, 1021-1025.

[61] Fuchs, V, Mihelcic, J, & Gierke, J. Life Cycle Assessment of Vertical and Horizontal Flow Constructed Wetlands for Wastewater Treatment Considering Nitrogen and Carbon Greenhouse Gas Emissions. Water research (2011). , 45, 2073-2081.

[62] Vohla, C, Kõiv, M, Bavor, H, Chazarenc, F, & Mander, Ű. Filter Materials for Phosphorus Removal from Wastewater in Treatment Wetlands- A Review. Ecological Engineering (2011). , 37, 70-89.

[63] Wang, D, Jin, X, Zhang, H, Zhang, X, Zeng, H, You, S, Zhao, W, Xu, Z, & Wu, Y. Research Progres on Phosphorus Removal in Substrates of Constructed Wetlands. Bioinformatics and Biomedical Engineering, (2009). rd International Conference on; ICBBE 2009. , 1-6.

[64] Mateus, D, Vaz, M, & Pinho, H. Fragmented limestone Wastes as a Constructed Wetland Substrate for Phosphorus Removal. Ecological Engineering (2012). , 41, 65-69.

[65] Brooks, A, Rozenwald, M, Geohring, L, Lion, L, & Steenhuis, T. Phosphorus Removal by Wollastonite: A Constructed Wetland Substrate. Ecological Engineering (2000). , 15, 121-132.

[66] White, S, Taylor, M, Albano, J, Whitwell, T, & Kleine, S. Phosphorus Retention in Lab and Field-scale Subsurface-Flow Wetlands Treating Plant Nursery Runoff. Ecological Engineering (2011). , 37, 1968-1976.

[67] Prochaska, C, & Zouboulis, A. Removal of Phosphates by Pilot Vertical-Flow Constructed Wetlands Using a mixture of Sand and Dolomite as Substrate. Ecological Engineering (2006). , 26, 293-303.

[68] Sun, G, & Austin, D. Completely Autotrophic Nitrogen-removal Over Nitrite in Lab-Scale Constructed Wetlands: Evidence from a Mass Balance Study. Chemosphere (2007). , 68, 1120-1128.

[69] Akinbile, C, Yusoff, M, & Zuki, A. Landfill Leachate Treatment Using Sub-Surface Flow Constructed Wetland by Cyperus haspan. Waste Management (2012). , 32, 1387-1393.

[70] Justin, M, & Zupancic, M. Combined Purification and reuse of Landfill Leachate by Constructed Wetland and Irrigation of Grass and Willows. Desalination (2009). , 246, 157-168.

[71] Bulc, T. Long Term Performance of a Constructed Wetland for Landfill Leachate Treatment. Ecological Engineering (2006). , 26, 365-374.

[72] Kadlec, R, & Zmarthie, L. Wetland Treatment of Leachate from a Closed Landfill. Ecological Engineering (2010). , 36, 946-957.

[73] Sawaittayothin, V, & Polprasert, C. Nitrogen Mass balance and Microbial Analysis of Constructed Wetlands Treating Municipal Landfill Leachate. Bioresource Technology (2007). , 98, 565-570.

[74] Lavrova, S, & Koumanova, B. Influence of Recirculation in a Lab-Scale Vertical Flow Constructed Wetland on the Treatment Efficiency of Landfill Leachate. Bioresource Technology (2010). , 101, 1756-1761.

[75] Yalcuk, A, & Ugurlu, A. Comparison of Horizontal and Vertical Constructed Wetland System for Landfill Leachate treatment. Bioresource technology (2009). , 100, 2521-2526.

[76] Standard Methods for the Examination of Water and Wastewaterth ed., (1992).

[77] Zhao, Y, Sun, G, & Allen, S. Purification capacity of a highly loaded laboratory scale tidal flow reed bed system with final recirculation. Science of the Total Environment (2004). , 330, 1-8.

[78] Zhao, Y, Sun, G, & Allen, S. Anti-sized reed bed system for animal wastewater treatment: a comparative study. Water Research (2004). , 38, 2907-2917.

[79] Sun, G, Zhao, Y, & Allen, S. Enhanced removal of organic matter and ammoniacal-nitrogen in a column experiment of tidal flow constructed wetland system. Journal of Biotechnology (2005). , 115, 189-197.

[80] Lavrova, S, & Koumanova, B. Polishing of aerobically treated wastewater in a constructed wetland system. Journal of the University of Chemical Technolgy and Metallurgy (Sofia) (2007). , 42(2), 195-200.

[81] Lavrova, S. Koumanova, Doprechistvane v konstruirana vlazna zona na aerobno tretirani otpaduchni vodi ot svinefermi, Proceedings of the International Conference of the Food Science, Engineering and Technologies, 19-20 October, Plovdiv 2007, iss. 1, (2007). in Bulgarian)., 34

[82] Lavrova, S, & Koumanova, B. A comparative study of the nitrogen rich pig slurry treatment in aerobic bioreactor and in constructed vertical flow wetland system. Journal of Environmental Protectection and Ecology (2009). , 4, 913-920.

[83] Saeed, T, & Sun, G. A review on nitrogen and organics removal mechanisms in subsurface flow constructed wetlands: Dependency on environmental parameters, operating conditions and supporting media. Journal of Environmental Management (2012). , 112, 429-448.

[84] Cannon, A, Gray, K, Biddlestone, A, & Thayanithy, K. Pilot-scale development of a bioreactor for the treatment of dairy dirty water. Journal of Agricultural Engineering Research (2000). , 77, 327-334.

[85] Schipper, L, Cooper, A, Harfoot, C, & Dyck, W. Regulators of denitrification in an organic riparian soil. Soil Biology and Biochemistry (1993). , 25(7), 925-933.

[86] Stefanakis, A, & Tsihrintzis, V. Effects of loading, resting period, temperature, porous media, vegetation and aeration on performance of pilot-scale vertical flow constructed wetlands. Chemical Engineering Journal (2012).

[87] Prost-boucle, S, & Molle, P. Recirculation on a single stage of vertical flow constructed wetland: Treatment limits and operation modes. Ecological Engineering (2012). , 43, 81-84.

[88] Lavrova, S, & Koumanova, B. Landfill leachate purification in a vertical flow constructed wetland with/without preliminary aerobic treatment. Journal of the University of Chemical Technology and Metallurgy (2011). , 46(3), 299-304.

[89] Faulwetter, J, Gagnon, V, Sundberg, C, Chazarenc, F, Burr, M. D, Brisson, J, Camper, A. K, & Stein, O. R. Microbial processes influencing performance of treatment wetlands: A review. Ecological Engineering (2009). , 35, 987-1004.

[90] Hammer, M. Hammer Jr. Water and Wastewater Technology, 4th ed. (2001). Prentice-Hall, Inc. Upper Saddle River, NJ, USA.

[91] Vymazal, J. Removal of nutrients in various types of constructed wetlands. Science of Total Environment (2007). , 380, 48-65.

[92] Watson, J, Reed, S, Kadlec, R, Knight, R, & Whitehouse, A. Performance expectations and loading rates for constructed wetlands. In: Hammer, D.A. (ed.), Constructed wetlands for wastewater treatment. Municipal, industrial and agricultural. Lewis publishers, Chelsea, Michigan. (1989). , 319-358.

[93] Huett, D, Morris, S, Smith, G, & Hunt, N. Nitrogen and phosphorus removal from plant nursery runoff in vegetated and unvegetated subsurface flow wetlands. Water research (2005). , 39, 3259-3272.

[94] Hamilton, H, Nix, P, & Sobolavski, A. An overview of constructed wetlands as alternatives to conventional waste treatment systems. Water Pollution Research Journal of Canada (1993). , 28, 529-548.

[95] Coleman, J, Hench, K, Garbutt, K, Sexstone, A, Bissonnelte, G, & Skousen, J. Treatment of domestic wastewater by three plant species in constructed wetland. Wter Air Soil Pollution (2001). , 128, 283-295.

[96] Hunter, R, Combs, D, & George, D. Nitrogen, phosphorus, and organic carbon removal in simulated wetland treatment systems. Archives of Environmental Health (2001). , 41, 274-281.

Chapter 5

Removal of Acrylamide by Microorganisms

Jittima Charoenpanich

Additional information is available at the end of the chapter

http://dx.doi.org/10.5772/56150

1. Introduction

Acrylamide (CH_2=$CHCONH_2$) is a well-known bifunctional monomer, appearing as a white odorless flake-like crystal. It is soluble in water, methanol, ethanol, dimethyl ether, and acetone, but insoluble in benzene and heptane. Acrylamide is incompatible with acids, bases, oxidizing agents, irons and iron salts. It decomposes non-thermally to form ammonia while thermal decomposition produces carbon monoxide, carbon dioxide, and oxides of nitrogen [1].

As a commercial conjugated reactive molecule, acrylamide has been used worldwide for the synthesis of polyacrylamide and other polymers [2, 3]. It has also been used as a binding, thickening, or flocculating agent in grout, cement, sewage, wastewater treatment, pesticide formulation, cosmetics, sugar manufacturing, and to prevent soil erosion. Polymers of this compound have been used in ore processing, food packaging, plastic products, and in scientific and medical laboratories as solid support for the separation of proteins by electrophoresis [4]. Acrylamide monomer is also widely used as an alkylating agent for the selective modification of sulfhydryl proteins and in fluorescence studies of tryptophan residues in proteins. In 2002, there was an alarming report of the occurrence of acrylamide at high levels up to 3 mg/kg in plant-derived foods and thought to form during cooking allowing the formation of Maillard browning products [5]. Many reports have suggested that acrylamide seems to be found in foods that have been processed by heat-treatment methods other than boiling [6]. One possible pathway to the formation of acrylamide is via the Maillard reaction between amino acids, particularly asparagines, and reducing sugars at high temperatures [5, 6]. Some reports suggest acrylamide could form by acrolein (2-propenal, CH=CHCHO), a three-carbon aldehyde, by either the transformation of lipids or the degradation of amino acids, proteins and carbohydrates [7-12].

Acrylamide could be absorbed through unbroken skin, mucous membranes, lungs, and the gastrointestinal tract. Human exposure to acrylamide is primarily occupational from dermal contact with the solid monomer and inhalation of dust and vapor. Although it is not toxic in polymer form, the monomer can cause peripheral neuropathy. Residual monomer in polymers is also of health concern [13]. Primary exposure occurs during the handling of monomers. Two acrylamide manufacturing factories showed breathing zone concentrations of 0.1 to 3.6 mg/m^3 [1]. During normal operations, workers at another plant were exposed to not more than 0.3 mg/m^3. Aside from occupational exposure, probable exposure to the general public is through consumption of certain foods [14]. Another source of acrylamide exposure to the general public could be through drinking water treated with polyacrylamide flocculants [13]. Acrylamide may not be completely removed in many water treatment processes with some remaining after flocculation with polyacrylamides probably due to its water solubility and is not absorbed by sediment [15].

Acrylamide is evidentially a neurogenic, terratogenic or carcinogenic toxicant in animals [16]. The neurotoxic properties of acrylamide have been studied for humans in relation to occupational exposures and, experimentally, in laboratory animals. Understanding of acrylamide-induced neuropathies is quite advanced, a consequence of more than 30 years of research on the possible mechanisms of action [17]. The mechanism underlying the neurotoxic effects of acrylamide as with other toxins are interference with the kinesin-related motor proteins in nerve cells or with fusion proteins in the formation of vesicles at the nerve terminus and eventual cell death [18]. Neurotoxicity and resulting behavioral changes in acrylamide-exposed laboratory animals can reduce reproductive fitness. Further, kinesin motor proteins are important in sperm motility, which could alter reproductive parameters. Effects on kinesin proteins could also explain some of the genotoxic effects on acrylamide. These proteins form the spindle fibers in the nucleus that function in the separation of chromosomes during cell division. This could explain the clastogenic effects of the chemical noted in a number of tests for genotoxicity and assays for germ cell damage [4].

2. Release of acrylamide in environment

Acrylamide is a synthetic monomer with a broad spectrum of industrial applications, mainly as a precursor in the production of several polymers, such as polyacrylamide [1, 19]. High molecular weight polymers can be modified to develop nonionic, anionic, or cationic properties for specific uses [1, 20]. Various grades of acrylamide are available with the industrial grade typically with a purity of 98 to 99%. Acrylamide for laboratory use ranges from routine to pure, the former for electrophoresis, the latter for molecular applications [21]. The largest demand for acrylamide polymers in industry is for flocculation of unwanted chemical substances in water arising from mining activities, pulp and paper processing, sewage treatment, and other industrial processes. Applications are based on the principles of colloidal suspensions and used to clean up liquids, particularly aqueous media, either for disposal or human consumption [20, 22-23]. Acrylamide is also used as a chemical intermediate in the production of *N*-methylol acrylamide and *N*-butoxy acrylamide and as a superabsorbent in disposable diapers, medical, and agricultural products [24]. Small amounts of acrylamide are also used in sugar beet juice clarification, adhesives, binders for seed coatings and foundry sand, printing ink emulsion stabilizers, thickening agents for agricultural sprays, latex dispersions, textile printing paste, and water retention aids [25]. An aqueous 50% solution of acrylamide is used as acrylic copolymer dispersions in surface coatings and adhesives. In surface coatings, polymers are used as dispersants and binders to provide better pigment separation and flow. Surface coatings are used on home appliances and in the automotive trade [1, 26]. In addition, polyacrylamide has been used in both paper production process and treatment of mill wastewater [27]. Emulsions of polyacrylamide, calcium carbonate and clay are applied as a white coating in the manufacture of cardboard cartons [22]. These polymers are used as thickeners in soap and cosmetic preparations, and in skin care and hair grooming products, to impart a smooth after-feel and shine [22]. For oil drilling, liquid or powder partially-hydrolyzed polyacrylamide is used as additives to water based drilling mud to provide a lubricating film and reduce friction at the drill bit, impart stability to shale and clay and increase viscosity [1, 22, 26]. Moreover, specialized gels comprised in part of acrylamide polymers are manufactured for use as lubricants in the textile dying components to which fabric or finished garments are added. The gel lubricates the cloth preventing it from clumping together and aids pigment dispersion during the dying process to ensure an even color [22, 28]. In leather processing, acrylamide is used as polymers impart a gloss or specific feel and suppleness to leather. The hide is most commonly placed in a drum with the polymer and various other constituents such as dyes, formaldehyde and pigments, then rolled for about two hours. The polymer can also be applied by brush or spray. There is no set formulation for the components of the mix and the proportion of acrylamide polymer is at the discretion of the operator seeking to obtain the properties required in the tanned product [29-30]. Another major application of acrylamide is to reduce herbicide drift during spray applications. The polymer increases the viscosity of the herbicide solution, allowing for more uniform spray applications, and also increases plant contact time [31-33].

In worldwide usage, acrylamide is released into environment as waste during its production and in the manufacture of polyacrylamides and other polymers. Residual acrylamide concentrations in 32 polyacrylamide flocculants approved for water treatment plants ranged from 0.5 to 600 ppm [13]. Acrylamide may remain in water after treatment [15] and after flocculation with polyacrylamides due to its high solubility and is not readily adsorbed by sediment [34]. Other sources of release to water are from acrylamide-based sewer grouting and recycling of wastepaper. Another important source of contamination is from acrylonitrile-acrylamide production which releases approximately 1 g acrylamide in each liter of effluent [35]. Some reports have indicated that polyacrylamide, in the presence of sunlight and glyphosate, photolytically degrades to acrylamide monomer and this is a direct introduction of acrylamide into agricultural areas [36-38]. The half-life of acrylamide monomer in rivers ranges from weeks to months [22]. However, one report indicates that polyacrylamide does not degrade to acrylamide monomer in the presence of sunlight and glyphosate. Additionally, glyphosate appears to interact with either the acrylamide monomer or polymer, decreasing the rate of monomer degradation [39]. The most important environmental contamination results from acrylamide use in soil grouting [13]. Half-life of acrylamide in aerobic soil increases with decreasing temperature [40]. Under aerobic conditions, acrylamide was readily degraded in fresh water by bacteria with a half-life of 55-70 h, after acclimatization for 33-50 h [41]. Acrylamide has been shown to remain slightly longer in estuarine or salt than fresh water [15].

Acrylamide releases to land and water from 1987 to 1993 totaled over 18.16 tons of which about 85 percent was to water, according to Toxic Chemical Release Inventory of the U.S. Environmental Protection Agency (EPA) [40]. These releases were primarily from plastic industries which use acrylamide as a monomer. In 1992, discharges of acrylamide, reported to the Toxic Chemical Release Inventory by certain US industries included 12.71 tons to the atmosphere, 4.54 tones to surface water, 1,906.8 tones to underground injection sites, and 0.44 tones to land [4]. In an EPA study of five industrial sites that produce acrylamide and polyacrylamide, acrylamide (1.5 ppm) was found in only one sample downstream from a polyacrylamide producer and no acrylamide was detected in soil or air samples [13]. Concentrations of 0.3 ppb to 5 ppm acrylamide have been detected in terrestrial and aquatic ecosystems near industrial areas that use acrylamide and/or polyacrylamides [42-43]. Cases of human poisoning have been documented from water contaminated with acrylamide from sewer grouting. The acrylamide monomer was found to remain stable for more than 2 months in tap water [22]. Atmospheric levels around six US plants were found on an average of $< 0.2\ \mu g/m^3$ (0.007 ppb) in either vapor or particulate form [15]. The vapor phase chemical should react with photochemically produced hydroxyl radicals (half-life 6.6 h) and be washed out by rain [15].

3. Microbial degradation of acrylamide

The interest in environmental problems is continuously growing and there are increasing demands to seek the sustainable and controllable process which do not burden the environment significantly. Biodegradation is one of the classic methods for removal of undesired organic compounds to concentrations that are undetectable or below limits established as acceptable by regulatory agencies.

Acrylamide is likely to partially biodegrade in water within approximately 8-12 days [13]. If released on land, acrylamide can be expected to leach readily into the ground and biodegrade within a few weeks. In five surface soils that were moistened to field capacity, 74-94% degradation occurred in 14 days in three soils and 79 to 80% in 6 days in the other two soils [44]. Acrylamide may not be completely degraded in domestic sewage and water treatment facilities if residence times are relatively short [13, 45]. Further degradation through bioremediation of acrylamide to less harmful substances would alleviate environmental concerns.

Since 1982, microbial degradation of acrylamide has been explored extensively with a diversity of isolates (Table 1), mainly *Bacillus*, *Pseudomonas* and *Rhodococcus* [3, 46-55]. Further, numerous other microorganisms including the representatives of *Arthrobacter*, *Xanthomonas*, *Rhodopseudomonas*, *Rastonia*, *Geobacillus*, and a newly family of Enterobacteriaceae [49, 56-62]. *Aspergillus oryzae*, a filamentous fungal has also been documented as an acrylamide degrader [63].

Several acrylamide degraders use a coupling reaction of nitrile hydratase (EC 4.2.1.84) and amidase (EC 3.5.1.4) for biotransformation of acrylonitrile to acrylic acid via acrylamide as an intermediate [46, 56]. For example, *R. rhodochrous* J1 changed acrylonitrile to acrylamide and subsequently to acrylic acid [47] and *R. erythropolis* utilized either 2-arylpropionamides or acrylamide to form acrylic acid and ammonia [64]. In China, *Nocardia* sp. 163, a soil derived bacterium from Taishan Mountain harboring the highest nitrile hydratase activity on acrylonitrile was also used frequently for bioconversion of acrylamide [65]. Another prominent example is *Rhodococcus* sp. AJ270 which is a powerful and robust nitrile hydratase/amidase-containing microorganism isolated by Guo et al [66]. An aliphatic amidase (amidohydrolase) has been found to be the responsive enzyme for the deamidation of acrylamide to acrylic acid and ammonia [50, 59, 62, 64-67].

In 1990, Shanker and his colleagues isolated an acrylamide-degrading bacterium, *Pseudomonas* sp., from soil using an enrichment method. This bacterium degraded high concentration of acrylamide (4 g/l) to acrylic acid and ammonia. An amidase was also found to be the relevant enzyme for the hydrolysis of acrylamide and other short chain aliphatic amides like formamide and acetamide but not on acrylamide analogues, methacrylamide and *N, N*-methylene bisacrylamide [48].

Many aerobic microorganisms utilize acrylamide as their sole source of carbon and energy including *Pseudomonas* sp. and *Xanthomonas maltophilia*. Nawaz and his team found amidase in cell free extracts of these species and suggested it was involved in acrylamide degradation [49]. This is consistent with their earlier conclusion of acrylamide degradation by *Rhodococcus* sp. [50]. Later, the denitrifying bacteria, *Pseudomonas stutzeri* was found to use acrylamide as substrate in the acrylonitrile–butadiene–styrene resin wastewater treatment system. The strain could remove acrylamide at concentrations below 440 mg/l under aerobic conditions [52]. Acclimation of microorganisms is believed to enhance acrylamide biodegradation. Complete degradation of acrylamide at 10–20 ppm in river water occurred in about 12 days with non-acclimated microorganisms, but in only 2 days with acclimation [3]. In 2009, scientists in Malaysia reported two acrylamide-degrading bacteria, *Bacillus cereus* DRY135 and *Pseudomonas* sp. DRYJ7. Acrylic acid was also detected as a metabolite in the degradation [53-54]. *Aspergillus oryzae* KBN 1010 has been the only fungi documented as an acrylamide degrader [63].

In domestic wastewater in Thailand, four novel acrylamide-degrading bacteria (*Enterobacter aerogenes, Kluyvera georgiana, Klebsiella pneumoniae,* and *Enterococcus faecalis*) were isolated. *E. aerogenes* and *K. georgiana* showed degradation potential of acrylamide up to 5000 ppm at the mesophilic temperatures and could degrade other aliphatic amides especially short to medium-chain length but not amide derivatives [60-61]. Removal of acrylamide and ammonium nitrogen from industrial wastewater by *E. aerogenes* was generally higher than that by mixed cultures of microorganisms [68].

Degradation of acrylamide under anaerobic conditions has been rarely described. Recently a new strain of *Rhodopseudomonas palustris* was found capable of using acrylamide under photoheterotrophic conditions but grew poorly under anaerobic dark or aerobic conditions. A study of acrylamide metabolism by nuclear magnetic resonance showed the rapid deamidation of acrylamide to acrylate and further to propionate [57]. More recently, the denitrifying

Microorganisms	Source	Conditions	Reference
Bacteria			
Pseudomonas chlororaphis B23	Soil	Aerobic (Enzymatic degradation)	[46]
Arthrobacter sp. J-1	Soil	Aerobic (Enzymatic degradation)	[56]
Rhodococcus rhodochrous J1	Soil	Aerobic (Free cells)	[47]
Pseudomonas sp.	Soil	Aerobic (Free cells)	[48]
Pseudomonas sp. *Xanthomonas maltophilia*	Soil	Aerobic (Immobilized cells)	[49]
Rhodococcus sp.	Soil	Aerobic (Enzymatic degradation)	[50]
Rhodococcuserythropolis MP50	Soil	Aerobic (Enzymatic degradation)	[64]
Rhodococcus sp.	Soil	Aerobic (Immobilized cells)	[51]
Pseudomonas stutzeri	Wastewater treatment system	Aerobic (Free cells)	[52]
Rhodopseudomonas palustris	Bovine slaughterhouse	Photoheterotropic (Free cells)	[57]
Pseudomonas aeruginosa	Soil	Aerobic (Free and immobilized cells)	[3]
Ralstonia eutropha TDM-3	Wastewater treatment system	Anaerobic (Free cells)	[58]
Bacillus cereus DRY135	Soil	Aerobic (Free cells)	[53]
Pseudomonas sp. DRYJ7	Antarctic soil	Aerobic (Free cells)	[54]
Natural microbial populations	Rocky Ford Highline Canal, Colorado USA	Aerobic and anaerobic (Free cells)	[69]
Ralstonia eutropha AUM-01	Soil	Aerobic (Free cells)	[59]
Enterobacter aerogenes	Domestic wastewater	Aerobic (Free and immobilized cells)	[60]
Kluyvera georgiana *Klebsiella pneumoniae* *Enterococcus faecalis*	Domestic wastewater	Aerobic (Free cells)	[61]
Geobacillus thermoglucosidasius AUT-01	Soil	Aerobic (Free cells)	[62]
Pseudomonas aeruginosa DS-4	Soil	Aerobic (Free cells)	[55]
Fungi			
Aspergillus oryzae KBN 1010	Filamentous fungi used in food and beverage industries	Aerobic (Free cells)	[64]

Table 1. Acrylamide-degrading microorganisms.

bacterium, *Ralstonia eutropha* TDM-3 isolated from the wastewater treatment system associated with the manufacture of polyacrylonitrile fiber consumed acrylamide to concentration of 1446 mg/l, above which it was toxic [58]. This report is similar with the potential of soil bacteria, *Ralstonia eutropha* AUM-01 and *Geobacillus thermoglucosidasius* AUT-01 [59, 62]. One report, and perhaps most interesting, removal of acrylamide has been found potentially with the natural microbial populations in Rocky Ford Highline Canal, Colorado USA [69]. Degradation of acrylamide occurs under aerobic or anaerobic conditions, with nitrate serving as the most favorable anaerobic electron acceptor. Phylogenetic analysis of these cosmopolitan microorganisms suggest the potential for biodegradation in similar lotic systems such as *Pseudomonas*, *Rhodococcus*, and *Bacillus*. New proteobacterial genera (*Pectobacterium*, *Citrobacter*, *Delftia*, *Comomonas*, and *Methylobacterium*) were also found [69]. Microbial degradation of a lipid in conjunction with acrylamide was also report with *Pseudomonas aeruginosa* DS-4. Salad oil was believed to be an essential factor for acrylamide biodegradation by this bacterium. The degradation rate of acrylamide was affected by the incubation time of the acclimated strain DS-4. Longer incubation time with acrylamide resulted in more efficient degradation [55].

4. Metabolism of acrylamide

Until now, we can not deny possible routes for acrylamide other than deamination via amidase [50, 59, 62, 64, 67]. The subsequent fate of acrylate is not well understood but probably involves pathways and enzymes that have been characterized to various degrees for other acrylate utilizing bacteria (Figure 1). Acrylate metabolism is believed to proceed via hydroxylation to β-hydroxypropionate, then oxidized to CO_2 [48] or reduced to propionate [57]. Another plausible pathway for mineralization of acrylamide is via formation of acrylyl CoA which eliminates lactate as a final product [48].

A powerful tool that also enables unraveling acrylamide metabolic pathways is the sequential induction of catabolic enzymes and intermediatary metabolites. Further, insight into degradative pathways is also provided from assaying the probable key proteins that are synthesized at sufficient levels when acrylamide is present. Using proteome analysis, fifteen proteins differentially expressed from *Enterobacter aerogenes* grown on acrylamide were identified. Six protein homologues with amidohydrolase, urease accessory protein, quaternary ammonium compound resistance proteins, dipeptide transport protein, Omp36 osmoporin and large conductance mechanosensitive channel proteins (MscL) are seemingly involved in acrylamide stress response and its degradation. Five proteins identified as GroEL-like chaperonin, ArsR-transcriptional regulator, Ts- and Tu-elongation factor and trigger factor and four proteins (phosphoglycerate kinase, ATP synthase β-subunit, malate dehydrogenase and succinyl-CoA synthetase α-subunit) are expected to be relevant to adaption of *E. aerogenes* in the presence of acrylamide [70]. Based on the results, Charoenpanich and Tani have proposed acrylamide may be assimilated using Omp36 osmoporin and dipeptide transport proteins. Acrylamide is toxic, indeed lethal, to most microorganisms, however some bacteria have adapted their metabolism to use this substance as an energy source. Important to this adaptation is the evolution of genes that encode amidohydrolase (amidase) and other synthesis proteins that deaminate acrylamide to acrylic acid and ammonium [48, 50-51, 60-61]. With this, acrylic acid can be changed to propionate and subsequently succinyl CoA [57, 71-72] to generate energy. Potentially harmful ammonium is detoxified and with MscL protein and released from the cell [70].

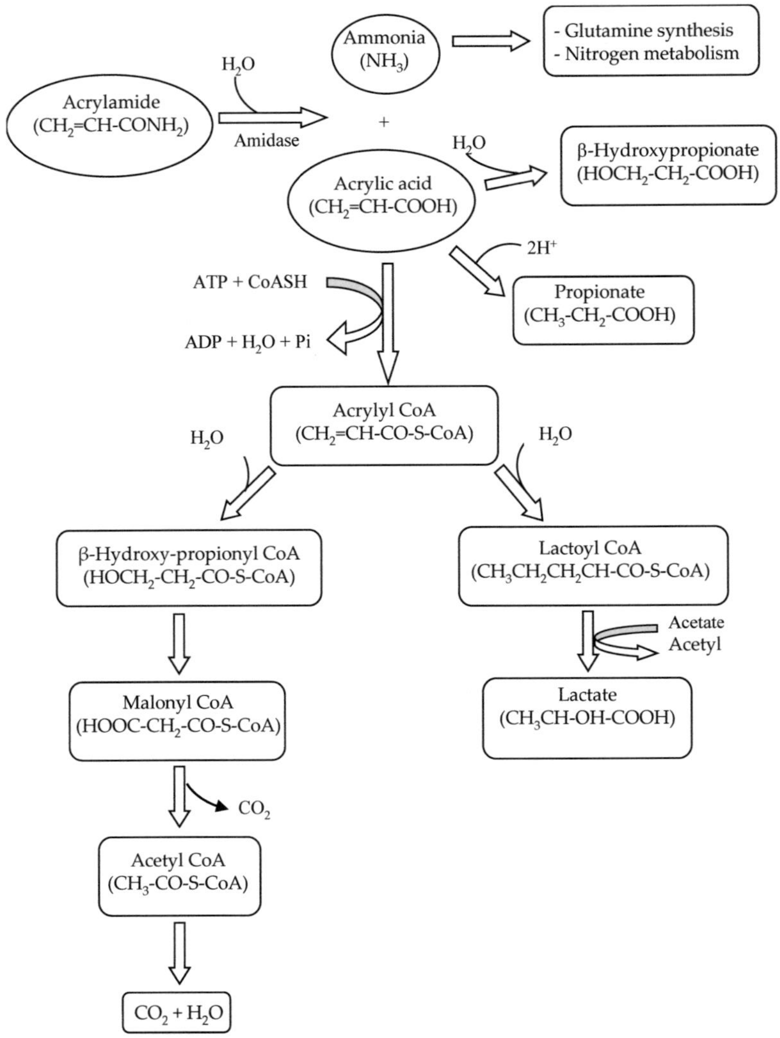

Figure 1. Possible biological fates of acrylate produced from acrylamide deamidation.

5. Bioremediation of acrylamide and future prospects

Bioremediation is viewed as a sustainable process for wastewater treatment, which under appropriate conditions, can promote an efficient reduction of organic matter with minimal energy requirements and, therefore, low costs. Major limitations are the bioavailability of the organic matter and the finding of efficient biodegraders. Physico-chemical environmental conditions also greatly influence the rate and extent of degradation. In general, degradation efficiency is dependent on three overall factors (i) microorganisms that can degrade the specific chemical structure (ii) environmental conditions that allow the microorganisms to grow and express their degradation enzymes and (iii) good physical contact between the organic substrate and the organism.

Rapid degradation of acrylamide coupled with growth requires not only amidase or microorganism producing amidase, but also a whole pathway, i.e. a set of enzymes that are differentially synthesized in the presence of acrylamide. Although a complete catabolic pathway for acrylamide does not exist, recombination and mutation processes and exchange of genetic information between microorganisms may lead to the development of organisms with improved catabolic activities. Alternatively, microorganisms can cooperate by combining their catabolic potential in mixed cultures and in this way may completely mineralize acrylamide. Wang and Lee elucidated the effectiveness of *Ralstonia eutropha* TDM-3 and mixed cultures of wastewater from the manufacture of polyacrylonitrile fiber in treating acrylamide in synthetic wastewater. They found that mixed culture and *R. eutropha* TDM-3 can jointly consume acrylamide up to concentrations of 1446 mg/l and completely remove acrylamide with a sufficient supply of nitrate as electron acceptors [58]. A similar result has been found in *E. aerogenes*. If grown with mixed cultures from a municipal wastewater treatment plant, they can completely and rapidly convert acrylamide to acrylic acid [68]. Acrylamide up to 100 mg/L can efficiently be removed from amended canal water and sediment slurries under aerobic conditions. Using natural nitrate-reducing microorganisms in a canal environment, potential fate of acrylamide (70.3-85%) was found after 60 days [69].

Microorganisms typically require sufficient water, inorganic nutrients, carbon sources, and trace elements for maintenance and growth. Besides growth substrates, other specific organic compounds such as vitamins or other growth factors are essential for some microorganisms. Monosaccharides like glucose and fructose have been reported as support elements for the growth and degradation potential of acrylamide-degrading bacteria [53-54]. However, in some cases supplementation of acrylamide containing growth medium with glucose or succinate as additional carbon source demonstrated a severe repression in degrading ability [48, 71-75]. Addition of glutamate or ammonium sulfate as an additional nitrogen source to the growth medium demonstrated an increase in degradation potential compared to the cells grown only on acrylamide [48]. One interesting study found that *Pseudomonas aeruginosa* DS-4 isolated from lipid wastewater required salad oil for growth and acrylamide degradation [55].

Toxic compounds (e.g. heavy metals) should not be present at high concentrations, since they can inactivate essential enzymes. As explained in [51] iron (<10 mM) enhanced the rates of acrylamide degradation of *Rhodococcus* sp. but copper, cobalt and nickel inhibited the degradation. Mercury and chromium inhibited acrylamide degradation by *Pseudomonas aeruginosa* while nickel at lower concentrations (200 and 400 ppm) improved the degrading ability [3].

Optimum conditions for acrylamide biodegradation are achieved if pH and temperature are in the range of pH 6-8 and mesophilic temperature (15-30ºC), respectively [3, 45-48, 53-55].

Most microorganisms consume considerably less energy for the maintenance of basic functions under neutral conditions. This means that more energy is available for growth. It has been known that metabolic activity of tropical soils typically is high and fosters several processes such as carbohydrate fermentation and carbon dioxide production leading to the lowering of pH. Thus, for successful bioremediation of pollutants including acrylamide pH control may be essential. Addition of an inexpensive chemical such as calcium carbonate to neutralize soil pH during bioremediation can optimize remediation [76].

Studies on acrylamide biodegradation are mainly concerned with the isolation and identification of suitable microbial strains. Most studies use either free or immobilized cells for acrylamide removal. Of these, immobilized cells are advantageous because the immobilized cells are less likely than free cells to be adversely affected by predators, toxin, or parasites [77-78]. Additionally, they can be reused, saving resources and time. However, the implementation of immobilized cells may be sensitive to pH, temperature and acrylamide concentration. Moreover, large accumulations of the metabolic intermediate, acrylic acid, may affect some microbial activity [3, 51, 60]. Hence, the attempt to biotransform acrylamide with amidase or nitrile-converting enzymes via hydrolysis.

Microbial degradation of nitriles proceeds through two enzymatic pathways. Nitrilase (EC 3.5.5.1) catalyzes the direct cleavage of nitriles to the corresponding acids and ammonia, and nitrile hydratase (NHase) catalyzes the hydration of nitriles to amides. Both nitrile-converting enzymes have increasingly attracted attention as catalysts for processing many organic chemicals [79-81]. Nitrile hydratase is commonly used as the catalyst in the production of acrylamide and is known as one of the most important industrial enzymes [82-83]. Generally, the gene operon of nitrile hydratase consists of the genes for the alpha and beta subunits of NHase, the NHase activator and amidase. The amides produced by NHase are degraded to their corresponding free carboxylic acids and ammonia by the action of amidases [84]. Thus, nitrile-converting enzymes are of broad use as alternatives for acrylamide biotransformation.

Acrylic acid, the intermediate product in acrylamide catabolism, is a commodity chemical with an estimated annual production capacity of 4.2 million metric tons [85]. Acrylic acid and its esters can be used in paints, coatings, polymeric flocculants, paper and so on. It is conventionally produced from petrochemicals. Currently, most commercial acrylic acid is produced by partial oxidation of propene which produces undesirable by-products and large amount of inorganic wastes [86]. Currently, there is an innovative manufacturing method using nitrile-amide converting enzymes. For acrylamide degraders, it is initially degraded to ammonia and acrylic acid (acrylate), a process catalyzed by amidase. Then acrylate is reduced to generate energy for growth. Until now, the acrylate-utilizing enzyme has not been well characterized but believed to be acrylate reductase [48, 57]. The identification of the gene encoding this enzyme remains a challenge. Moreover, from an economic aspect, the acrylate reductase-deficient strains created by a gene-disruption method, lead to acrylic acid accumulation in wastewater and are recommended for acrylamide bioremediation in the future.

Sequence similarities have been identified using computer methods for database searches and multiple alignment, between several nitrilases, cyanide hydratase, β-alanine synthase and the first type of aliphatic amidases which hydrolyze only short-chain aliphatic amides [87]. All these enzymes involving the reduction of organic nitrogen compounds and ammonia production exhibited several conserved motifs. One of which contains an invariant cysteine that is part of the catalytic site in nitrilases. Another highly conserved motif includes an invariant glutamic acid that might also be involved in catalysis. Sequence conservation over the entire length of these enzymes, as well as the similarity in the reactions constitutes a definite family which points to a common catalytic mechanism [88]. Chemical mutagenesis and X-ray crystallography have been analyzed for three-dimensional structures of amidases. Only a few crystal structures of nitrilase-related amidases have been reported with *Pseudomonas aeruginosa* amidase the first [89-90]. The three dimensional-structures showed a conserved α-β-β-α sandwich fold resembling the conserved structural fold of the nitrilase superfamily structures. Analysis of the three dimension-structures identified E59, K134, and C166 as a catalytic triad [89]. Similar catalytic triad residues were also reported in the three dimensional structural models of amidase from *Rhodococcus erythropolis*, *Helicobacter pylori*, and *Bacillus stearothermophilus* [89] and also in the amidase of novel acrylamide-degrading *Enterobacter aerogenes* [91]. The crystal structure of *Xanthomonas campestris* XC1258 amidase showed a monomeric structure of globular α/β protein comprising mainly six α helices and two six-stranded β-sheet (Figure 2). This is the typical nitrilase-superfamily α-β-β-α fold. The hexamer preserving the eight-layered α-β-β-αα-β-β-α structure in holoenzyme across an interface has also been reported [92]. The analysis of small asymmetric catalytic site of the *Geobacilus pallidus* RAPc8 amidase suggested that access of a water molecule to the catalytic triad (C, E, K) side chains would be impeded by the formation of the acyl intermediate. The conserved E142 in the catalytic site acts as a general base to catalyze the hydrolysis of this intermediate [93]. This confirmed the conservation of the E, K, C catalytic triad across the nitrilase superfamily members and also supported the classification of the amidases in the nitrilase superfamily.

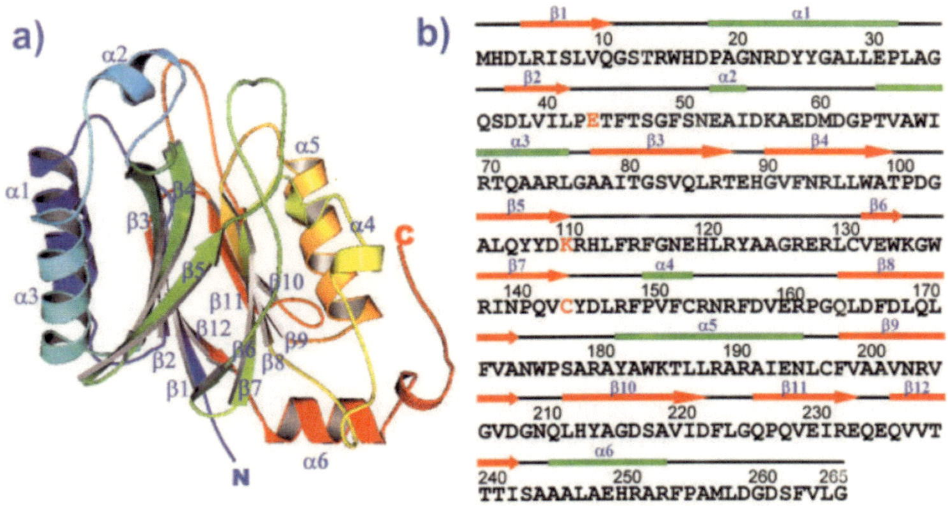

Figure 2. (a) The monomeric tertiary structure of amidase from *Xanthomonas campestris* XC1258, color-coded from blue (N-terminal) to red (C-terminal), and (b) the primary sequence of XC1258 amidase. Reprinted from Ref. [92].

Acrylamide amidases have similar sequences with nitrilases and seem to have descended from a common ancestry along with members of the sulfhydryl enzyme family. In these amidases an invariant cysteine residue was reported to act as the nucleophile in the catalytic mechanism and is confirmed by the three dimensional structural model of the amidase of *Pseudomonas aeruginosa*. This was built by comparative modeling using the crystal structure of the worm nitrilase fusion protein, NitFhit as the template. The putative catalytic triad C-E-K is conserved in all members of the nitrilase superfamily [89]. The signature amidases possesses two real active site residues D191 and S195 among the various conserved residues within the signature sequence common to all enantioselective amidases. D191N and S195A substitutions in *Rhodococcus* amidase has been shown to completely suppress amidase activity [94-95]. These sequences are also present within the active site sequences of aspartic proteinases. Thus, amide bond cleaving enantioselective amidases that are coupled with nitrile hydratases are evolutionary related to aspartic proteinases. Further structural characterization of the amidase produced by acrylamide-degrading bacteria should reveal what other differences are present. It may be possible to use this information to aid protein engineering of the enzymes in order to improve their efficiency and specificity.

Development of thermostable amidase is also important. Based on the three-dimensional structure of amidase, additional disulfide bridges can be engineered by site-directed mutagenesis for enzyme stabilization. Novel amidases that show broad substrate specificity may be developed to biodegrade the toxic environmental pollutants, acrylamide and amides. Random approaches such as directed evolution, reverse engineering and site-directed mutagenesis could be applied to achieve such ends.

Our understanding of the biochemistry and molecular biology of amidase is advancing rapidly and already providing information that is of use today. Moreover, recent developments in amidase studies have broadened the scope of potential applications of the enzyme in acrylamide bioremediation as well as that of acrylic acid production. I predict that these developments combined with progress in genetic engineering and enzyme crystallography will have a major effect on the practical applications of acrylamide bioremediation.

6. Concluding remarks

A huge demand for acrylamide as an ubiquitous monomer for industry led to its environmental presence, however the International Agency for Research on Cancer has classified this compound as a probable human carcinogen. Bioremediation seems to be the only efficient and environmentally friendly process to decompose this monomer. The first step in developing acrylamide bioremediation is to choose high potent microorganisms. Choice of microorganisms is challenging owing to the large scale degradation of acrylamide and elucidation of the intermediate in catabolic pathways is the first important step. Nevertheless, the main problem is the rapid conversion of intermediate acrylic acid to other metabolites. Research on the relationship between degradation mechanisms and membrane structure of acrylamide-utilizing bacteria awaits further characterization. It is noteworthy that successful remediation of acrylamide depends on the ability of microbes to adapt to new environmental conditions and the availability of active and stable chemical degrading bacteria. Indigenous predators, parasites and toxicants are known to severely restrict biodegradation and should be a concern.

Nomenclature

Amino acids

E: Glutamic acid

K: Lysine

C: Cysteine

D: Aspartic acid

N: Asparagine

S: Serine

A: Alanine

Acknowledgements

The author is grateful to Dr. N. Kurukitkoson for his encouragement to write this review and would like to thank F.W.H. Beamish for proofreading the manuscript.

Author details

Jittima Charoenpanich*

Address all correspondence to: jittima@buu.ac.th

Department of Biochemistry, Faculty of Science, Burapha University, Bangsaen, Chonburi, Thailand

References

[1] IARCSome Chemicals Used in Plastics and Elastomers. In: IARC Monographs on the evaluation of carcinogenic risk of chemicals to human. France: International Agency for Research on Cancer; (1986). , 403.

[2] Prasad, D. Y. Polyacrylamide as a coagulant aid in water treatment. Chemical Age of India (1982). , 34-387.

[3] Prabu, C. S, & Thatheyus, A. J. Biodegradation of acrylamide employing free and immobilized cells of *Pseudomonas aeruginosa*. International Biodeterioration and Biodegradation (2007). , 60-69.

[4] Exon, J. H. A review of the toxicology of acrylamide. Journal of Toxicology and Environmental Health (2006). , 9-397.

[5] Mottram, D. S, Wedzicha, B. L, & Dodson, A. T. Acrylamide is formed in the Maillard reaction. Nature (2002). , 419-448.

[6] Stadler, R. H, Blank, I, Varga, N, Robert, F, Hau, J, Guy, P. A, Robert, M. C, & Riediker, S. Acrylamide from Maillard reaction products. Nature (2002). , 419-449.

[7] Yasuhara, A, Tanaka, Y, Hengel, M, & Shibamoto, T. Gas chromatographic investigation of acrylamide formation in browning model systems. Journal of Agricultural and Food Chemistry (2003). , 51-3999.

[8] Zhang, Y, Dong, Y, Ren, Y, & Zhang, Y. Rapid determination of acrylamide contaminant in conventional fried foods by gas chromatography with electron capture detector. Journal of Chromatography A (2006). , 1116-209.

[9] Zhang, Y, Ren, Y, Zhao, H, & Zhang, Y. Determination of acrylamide in Chinese traditional carbohydrate-rich foods using gas chromatography with micro-electron capture detector and isotope dilution liquid chromatography combined with electrospray ionization tandem mass spectrometry. Journal of Analytical Chimica Acta (2007). , 584-322.

[10] Zamora, R, & Hidalgo, F. J. Contribution of lipid oxidation products to acrylamide formation in model systems. Journal of Agricultural and Food Chemistry (2008). , 56-6075.

[11] Capuano, E, Oliveiro, T, Açar, O. Ç, Gökmen, V, & Fogliano, V. Lipid oxidation promotes acrylamide formation in fat-rich model systems. Food Research International (2010). , 43-1021.

[12] Komthong, P, Suriyaphan, O, & Charoenpanich, J. Determination of acrylamide in Thai-conventional snacks from Nong Mon market, Chonburi using GC-MS technique. Food Additives and Contaminants Part B (2012). , 5(1), 20-28.

[13] Howard, P. H. Large production and priority pollutants. Handbook of Environmental Fate and Exposure Data for Organic Chemicals. Chelsea: Lewis; (1989).

[14] Tareke, E, Rydberg, P, Karlsson, P, Eriksson, S, & Törnqvist, M. Analysis of acrylamide, a carcinogen formed in heated foodstuffs. Journal of Agricultural Food Chemistry (2002). , 50(17), 4998-5006.

[15] Croll, B. T, & Arkell, G. H. Hodge RPJ. Residues of acrylamide in water. Water Resource (1974)., 8(11), 989-993.

[16] Shairashi, Y. Chromosome aberrations induced by monomeric acrylamide in bone marrow and germ cells of mice. Mutation Research (1978)., 57(3), 313-324.

[17] Tilson, H. A. The neurotoxicity of acrylamide. An overview. Neurobehavior Toxicological Teratology (1981)., 3-455.

[18] Sickles, D. W, Brady, S. T, Testino, A, Friedman, M. A, & Wrenn, R. W. Direct effect of the neurotoxicant acrylamide on kinesin-based microtubule motility. Journal of Neuroscience Research (1996)., 46(1), 7-17.

[19] IARCSome Industrial Chemicals. In: IARC Monographs on the Evaluation of Carcinogenic Risk of Chemicals to Humans. France: International Agency for Research on Cancer; (1994).

[20] Kirk, O. Kirk-Othmer Encyclopedia of Chemical Technology (3rd ed.). New York: John Wiley and Sons; (1978).

[21] Bachmann, M, Myers, J, & Bezuidenhout, B. Acrylamide monomer and peripheral neuropathy in chemical workers. American Journal of Industrial Medicine (1992)., 21(2), 217-222.

[22] Brown, L, & Rhead, M. M. Bancroft KCC, Allen N. Case studies of acrylamide pollution resulting from industrial use of acrylamides. Water Pollution Control (1980)., 79(4), 507-510.

[23] Sax, N. I, & Lewis, R. J. Hawley's Condensed Chemical Dictionary (11th ed.). New York: Van Nostrand Reinhold; (1987).

[24] Agency for Toxic Substance and Disease Registry (ATSDR)Toxicological profile for acrylamide: draft for public comment. Atlanta: U.S. Public Health Service; (2009).

[25] Mannsville, M. Chemical Products Synopsis, Acrylamide. Mannsville Chemical Products Corporation; (1993).

[26] Sittig, M. Handbook of Toxic and Hazardous Chemicals and Carcinogens (2nd ed.). Park Ridge: Noyes; (1985).

[27] Wong, S. S, Teng, T. T, Ahmada, A. L, Zuhairi, A, & Najafpour, G. Treatment of pulp and paper mill wastewater by polyacrylamide (PAM) in polymer induced flocculation. Journal of Hazardous Materials (2006)., 135-378.

[28] Kirk, O. Kirk-Othmer Encyclopedia of Chemical Technology (3rd ed.) New York: John Wiley and Sons; (1979).

[29] Mark, H. F, Gaylord, G. N, & Bikales, N. M. Acrylamide. Encyclopedia of Polymer Science and Technology. (1962)., 177-195.

[30] Jallageas, J. C, Arnaud, A, & Galzy, P. Bioconversion of nitriles and their applications. Advances in Biochemical Engineering and Biotechnology (1980). , 14-1.

[31] Bouse, L. F, Carlton, J. B, & Jank, P. C. Use of polymers for control of spray droplet size. American Society of Agricultural Engineering (1986). AA-86-005: 1-18.

[32] Downer, R. A, Wolf, T. M, Chapple, A. C, Hall, F. R, & Hazen, J. L. Characterizing the impact of drift management adjuvants on the dose transfer process. Proceeding Fourth International Symposium on Adjuvants Agrochemistry (1995). , 193-138.

[33] Underwood, A. K, Clark, A, Mack, R. E, Thomas, J, Roberts, J. R, & Volgas, G. C. Dry concentrate spray adjuvants. Proceeding Fourth International Symposium on Adjuvants Agrochemistry (1995). , 193-391.

[34] Brown, L, & Rhead, M. M. Bancroft KCC, Allen N. Model studies of the degradation of acrylamide monomer. Water Research (1980). , 14(7), 775-778.

[35] Rogacheva, S. M, & Ignatov, O. V. The respiratory activity of *Rhodococcus rhodochrous* M8 cells producing nitrile-hydrolyzing enzymes. Applied Biochemistry and Microbiology (2001). , 37-282.

[36] Smith, E. A, Prunes, S. L, & Oehme, F. W. Environmental degradation of polyacrylamides I. Effects of artificial environmental conditions: temperature, light and pH. Ecotoxicology and Environmental Safety (1996). , 35-121.

[37] Smith, E. A, Prunes, S. L, & Oehme, F. W. Environmental degradation of polyacrylamides II. Effects of environmental exposure. Ecotoxicology and Environmental Safety (1997). , 37-76.

[38] Smith, E. A, & Oehme, F. W. Rapid direct analysis of acrylamide residue in polyacrylamide thickening agents by HPLC. Journal of Chromatographic Science (1993). , 31-192.

[39] Ver Vers LMDetermination of acrylamide monomer in polyacrylamide degradation studies by high-performance liquid chromatography. Journal of Chromatographic Science (1999). , 37-486.

[40] U.S. Environmental Protection Agency (U.S. EPA). Integrated Risk Information System (IRIS) Online. Coversheet for Acrylamide. In: Office of Health and Environmental Assessment, U.S. EPA, Cincinnati; 1994.

[41] Conway, E. J, Petersen, R. J, Collingsworth, R. F, Graca, J. G, & Carter, J. W. Assessment of the need for and character of limitations on acrylamide and its compounds, Washington DC, US Environmental Protection Agency. (Report prepared for the Offic of Pesticides and Toxic Substances, Contract (1979). (68-10)

[42] Weideborg, M, Källqvist, T, Ødegård, K. E, Sverdrup, L. E, & Vik, E. A. Environmental risk assessment of acrylamide and methyloacrylamide from a grouting agent used

in the tunnel construction of Romeriksporten, Norway. Water Research (2001). , 35-2645.

[43] Chang, L, Bruch, M. D, Griskowitz, N. J, & Dentel, S. K. NMR spectroscopy for determination of cationic polymer concentrations. Water Research (2002). , 36-2555.

[44] Lande, S. S, Bosch, S. J, & Howard, P. H. Degradation and leaching of acrylamide in soil. Journal of Environmental Quality (1979). , 8-133.

[45] Brown, L, Rhead, M. M, & Hill, D. Bancroft KCC. Rapid screening technique utilizing high-performance liquid chromatography for assessing acrylamide contamination in effluents. Analyst (1982). , 107(1276), 749-754.

[46] Asano, Y, Tachibana, M, Tani, Y, & Yamada, H. Purification and characterization of amidase which participates in nitrile degradation. Agricultural and Biological Chemistry (1982a). , 46(5), 1175-1181.

[47] Nagasawa, T, & Yamadam, H. Microbial transformation of nitriles. Trends Biotechnology (1989). , 7-153.

[48] Shanker, R, Ramakrishna, C, & Seth, P. K. Microbial degradation of acrylamide monomer. Archives Microbiology (1990). , 154-192.

[49] Nawaz, M. S, Franklin, W, & Cerniglia, C. E. Degradation of acrylamide by immobilized cells of a *Pseudomonas* sp. and *Xanthomonas maltophilia*. Canadian Journal of Microbiology (1993). , 39-207.

[50] Nawaz, M. S, Khan, A. A, Seng, J. E, Leakey, J. E, Siitonen, P. H, & Cerniglia, C. E. Purification and characterization of an amidase from an acrylamide-degrading *Rhodococcus* sp. Applied and Environmental Microbiology (1994). , 60-3343.

[51] Nawaz, M. S, Billedeau, S. M, & Cerniglia, C. E. Influence of selected physical parameters on the biodegradation of acrylamide by immobilized cells of *Rhodococcus* sp. Biodegradation (1998). , 9-381.

[52] Wang, C. C, & Lee, C. M. Denitrification with acrylamide by pure culture of bacteria isolated from acrylonitrile-butadiene-styrene resin manufactured wastewater treatment system. Chemosphere (2001). , 44-1047.

[53] Shukor, M. Y, Gusmanizar, N, Azmi, N. A, Hamid, M, Ramli, J, Shamaan, N. A, & Syed, M. A. Isolation and characterization of an acrylamide-degrading *Bacillus cereus*. Journal of Environmental Biology (2009). , 30-57.

[54] Shukor, M. Y, Gusmanizar, N, Ramli, J, & Shamaan, N. A. MacCormack WP, Syed MA. Isolation and characterization of an acrylamide-degrading Antarctic bacterium. Journal of Environmental Biology (2009). , 30-107.

[55] Shen, S. M, Wan, T. J, & Hwang, H. Y. Enhancement of degradation of acrylamide coupled with salad oil by *Pseudomonas aeruginosa* DS-4 using incubation periods. Biocatalysis and Agricultural Biotechnology (2012). , 1-110.

[56] Asano, Y, Yasuda, T, Tani, Y, & Yamada, H. A new enzymatic method of acrylamide production. Agricultural and Biological Chemistry (1982b). , 46-1183.

[57] Wampler, D. A, & Ensign, S. A. Photoheterotrophic metabolism of acrylamide by a newly isolated strain of *Rhodopseudomonas palustris*. Applied and Environmental Microbiology (2005). , 71-5850.

[58] Wang, C. C, & Lee, C. M. Isolation of the acrylamide denitrifying bacteria from a wastewater treatment system manufactured with polyacrylonitrile fiber. Current Microbiology (2007). , 55-339.

[59] Cha, M, & Chambliss, G. H. Characterization of acrylamidase isolated from a newly isolated acrylamide-utilizing bacterium, *Rastonia eutropha* AUM-01. Current Microbiology (2011). , 62(2), 671-678.

[60] Buranasilp, K, & Charoenpanich, J. Biodegradation of acrylamide by *Enterobacter aerogenes* isolated from wastewater in Thailand. Journal of Environmental Sciences (2011). , 23(3), 396-403.

[61] Thanyacharoen, U, Tani, A, & Charoenpanich, J. Isolation and characterization of *Kluyvera georgiana* strain with the potential for acrylamide biodegradation. Journal of Environmental Science and Health, Part A (2012). , 47-1491.

[62] Cha, M, & Chambliss, G. H. Cloning and sequence analysis of the heat-stable acrylamidase from a newly isolated thermophilic bacterium, *Geobacillus thermoglucosidasius* AUT-01. Biodegradation (2012). DOIs10532-01209557-6.

[63] Wakaizumi, M, Yamamoto, H, Fujimoto, N, & Ozeki, K. Acrylamide degradation by filamentous fungi used in food and beverage industries. Journal of Bioscience and Bioengineering (2009). , 108(5), 391-393.

[64] Hirrlinger, B, Stolz, A, & Knackmuss, H. J. Purification and properties of an amidase from *Rhodococcus erythropolis* MP50 which enantioselectively hydrolyzes arylpropionamides. Journal of Bacteriology (1996). , 2.

[65] Liu, M, Jiao, P, & Cao, Z. A. Advances of nitrile hydratase used in bioconversion for acrylamide. Journal of Chemical Industry and Engineering (2001). , 52-847.

[66] Guo, X. L, Deng, G, Xu, J, & Wang, M. X. Immobilization of *Rhodococcus* sp. AJ270 in alginate capsules and its application in enantioselective biotransformation of trans-2-methyl-3-phenyl-oxiranecarbonitrile and amide. Enzyme and Microbial Technology (2005). , 39(1), 1-5.

[67] Nawaz, M. S, Khan, A. A, Bhattacharayya, D, Siitonen, P. H, & Cerniglia, C. E. Physical, biochemical, and immunological characterization of a thermostable amidase from *Klebsiella pneumonia* NCTR 1. Journal of Bacteriology (1996). , 178-2397.

[68] Charoenpanich, J, Buranasilp, K, Thanyacharoen, U, Mangkalad, O, Suktanarag, S, Sriwiriyarat, T, & Tani, A. Expression of the recombinant aliphatic amidase from Enterobacter aerogenes in the soluble form and characterization of its activity. In: Sum-

mary Book Asian Core Program (2008-2012) on Capacity Building and Development of Microbial Potential and Fermentation Technology Towards New Era (Yamada, M, Leelavatcharamas, V, Xuan, V-T, Bunphanmy, S. eds.) International Exchange Core Program in Yamaguchi University, Japan, ISBN: 978-4-9906826-0-6. November 18-21, Shimonoseki, Yamaguchi, Japan.

[69] Labahn, S. K, Fisher, J. C, Robleto, E. A, Young, M. H, & Moser, D. P. Microbially mediated aerobic and anaerobic degradation of acrylamide in a western United States irrigation canal. Journal of Environmental Quality (2010)., 39(5), 1563-1569.

[70] Charoenpanich, J, & Tani, A. Proteomic analysis of acrylamide-degradating *Enterobacter aerogenes*. In: The Science Society of Thailand: Proceedings of the 3rd Biochemistry and Molecular Biology (BMB) international conference From basic to translational researches for a better life, April (2011). The Empress Convention Centre, Chiangmai, Thailand., 6-8.

[71] Stams AJMVan Dijk JB, Dijkema C, Plugge CM. Growth of syntrophic propionate-oxidizing bacteria with fumarate in the absence of methanogenic bacteria. Applied and Environmental Microbiology (1993)., 59(4), 1114-1119.

[72] Ansede, J. H, Pellechia, P. J, & Yoch, D. C. Metabolism of acrylate to beta-hydroxypropionate and its role in dimethylsulfoniopropionate lyase induction by a salt marsh sedioment bacterium, *Alcaligenes faecalis* M3A. Applied and Environmental Microbiology (1999)., 65-5075.

[73] Clarke, P. H. The utilization of amides by microorganisms. In: Payne JW (ed) Microorganisms and nitrogen sources. John Wiley and Sons, New York, (1980)., 537-562.

[74] Potts, J. R, & Clarke, P. H. The effect of nitrogen limitation on catabolite repression of amidase, histidase and urocanase in *Pseudomonas aeruginosa*. Journal of General Microbiology (1976)., 93-377.

[75] Smyth, P. F, & Clarke, P. H. Catabolite repression of *Pseudomonas aeruginosa* amidase: The effect of carbon source on amidase synthesis. Journal of General Microbiology (1975)., 90-81.

[76] Johnston, J. J, Borden, R. C, & Barlaz, M. A. Anaerobic biodegradation of alkylbenzenes and trichloroethylene in aquifer sediment down gradient of a sanitary landfill. Journal of Contaminants Hydrolysis (1996)., 23-263.

[77] Mallory, L. M, Yuk, C. S, & Alexander, M. Alternative prey: a mechanism for elimination of bacterial species by protozoa. Applied and Environmental Microbiology (1983)., 46-1073.

[78] Murakami, Y, & Alexander, M. Destruction and formation of toxins by one bacterial species affect biodegradation by a second species. Biotechnology and Bioenergetics (1989)., 33-832.

[79] Gao, M, Wang, D. X, Zheng, Q. Y, Huang, Z. T, & Wang, M. X. Remarkable electronic and steric effects in the nitrile biotransformations for the preparation of enantiopure

functionalized carboxylic acids and amides: implication for an unsaturated carbon-carbon bond binding domain of the amidase. Journal of Organic Chemistry (2007). , 72-6060.

[80] Song, L, Wang, M, Shi, J, Xue, Z, Wang, M. X, & Qian, S. High resolution X-ray molecular structure of the nitrile hydratase from *Rhodococcus erythropolis* AJ270 reveals posttranslational oxidation of two cysteines into sulfinic acids and a novel biocatalytic nitrile hydration mechanism. Biochemical and Biophysical Research Communication (2007). , 362-319.

[81] Wang, J. Y, Wang, D. X, Pan, J, Huang, Z. T, & Wang, M. X. Nitrile and amide biotransformations for the synthesis of enantiomerically pure arylaziridine-2-carboxamide derivatives and their stereospecific ring-opening reactions. Journal of Organic Chemistry (2007). , 3.

[82] Endo, I, Nojiri, M, Tsujimura, M, Nakasako, M, Nagashima, S, Yohda, M, & Odaka, M. Fe-type nitrile hydratase. Journal of Inorganic Biochemistry (2001). , 83-247.

[83] Yamada, H, & Kobayashi, M. Nitrile hydratase and its application to industrial production of acrylamide. Bioscience Biotechnology and Biochemistry (1996). , 60-1391.

[84] Mayaux, J. F, Cerbelaud, E, Soubrier, F, Yeh, P, Blanche, F, & Petre, D. Purification, cloning, and primary structure of a new enantionmer-selective amidase from a *Rhodococcus* strain: structural evidence for a conserved genetic coupling with nitrile hydratase. Journal of Bacteriology (1991). , 173-6694.

[85] Wang, C. C, Lee, C. M, & Wu, A. S. Acrylic acid removal from synthetic wastewater and industrial wastewater using *Ralstonia solanacearum* and *Acidovorax avenae* isolated from a wastewater treatment system manufactured with polyacrylonitrile fiber. Water Science and Technology (2009). , 60(11), 3011-3016.

[86] Straathof, A. J, Sie, S, Franco, T. T, & Wielen, L. A. Feasibility of acrylic acid production by fermentation. Applied Microbiology and Biotechnology (2005). , 67-727.

[87] Schomburg, D, & Salzmann, M. Enzyme Handbook 4. Berlin, Springer-Verlag; (1991).

[88] Bork, P, & Koonin, E. V. A new family of carbon-nitrogen hydrolases. Protein Science (1994). , 3-1344.

[89] Novo, C, Farnaud, S, Tata, R, Clemente, A, & Brown, P. R. Support for a three dimensional structure predicting a Cys-Glu-Lys catalytic triad for *Pseudomonas aeruginosa* amidase comes from site-directed mutagenesis and mutations altering substrate specificity. Biochemical Journal (2002). , 365-731.

[90] Andrade, J, Karmali, A, Carrondo, M. A, & Frazao, C. Structure of amidase from *Pseudomonas aeruginosa* showing a trapped acyl transfer reaction intermediate state. Journal of Biology and Chemical (2007). , 282-19598.

[91] Suktanarag, S, & Charoenpanich, J. Sequence analysis and structural modeling of an amidase from a novel acrylamide-degrading bacterium *Enterobacter aerogenes*. The 13th FAOBMB Congress, November 25-29, (2012). Bangkok, Thailand.

[92] Chin, K. H, Tsai, Y. D, Chan, N. L, & Huang, K. F. Wang AHJ, Chou SH. The crystal structure of XC1258 from *Xanthomonas campestris*: A putative prokaryotic Nit protein with an arsenic adduct in the active site. Proteins (2007). , 69-665.

[93] Kimani, S. W, Agarkar, V. B, & Cowan, D. A. Sayed MFR, Sewell BT. Structure of an aliphatic amidase from *Geobacillus pallidus* RAPc8. Acta Crystallography (2007). , 63-1048.

[94] Kobayashi, M, Fujiwara, Y, Goda, M, Komeda, H, & Shimizu, S. Identification of active sites in amidase: Evolutionary relationship between amide bond-and peptide bond-cleaving enzymes. Proceedings of the National Academy of Science USA (1997). , 94(11), 986-991.

[95] Novo, C, Tata, R, Clemente, A, & Brown, P. R. *Pseudomonas aeruginosa* aliphatic amidase is related to the nitrilase/cyanide hydratase enzyme family and Cys166 is predicted to be the active site nucleophile of the catalytic mechanism. FEBS Letters (1995). , 367-275.

Chapter 6

Biodegradation and Leaching of Surfactants During Surfactant-Amended Bioremediation of Oil-Polluted Soil

Aare Selberg, Kätlin Juuram, Jana Budashova and Toomas Tenno

Additional information is available at the end of the chapter

http://dx.doi.org/10.5772/56908

1. Introduction

1.1. Theoretical overview

1.1.1. Bioremediation as green-tech solution for soil decontamination

Pollution of soil (and water) is quite hot problem in the different places in the world due to its generation through variety of sources (waste dumps, petrol stations, underground storage, and accidents by pipelines) [1] and soil acts as a permanent residence for pollutants and the dynamic movements of hydrological cycle transports them to groundwater aquifer [2]. The pollutants can be removed using different physical methods, among them mechanical recovery of oil by the sorbents is one of the most promising countermeasures [3]. Washing with surfactant solutions has been shown to be effective for the removal of hydrophobic organic contaminants (PAH, hydrocarbons, PCB, chlorinated solvents) from soil [4, 5] but the degradation of pollutants is not carried out through it. Generally, removal of pollutants from soil using physical and chemical processes is quite expensive [6]. The different chemicals used for the chemical treatment can induce the secondary pollution of soil and/or groundwater.

The using of indigenous microorganisms is the greener solution for soil (and water) decontamination. The enhancement of natural biological degradation processes can be a preferred cost-effective method of removing contaminants from the contaminated environments and the role of microorganisms has been shown to be essential in the remediation of organic pollution [7 - 9].

Bioremediation technologies exploit the natural ability of microorganisms to degrade organic chemical contamination in soil and ground water [7, 10] and the goal of active bioremediation of soil is to enhance the microbial systems to efficiently remediate contaminated area and to decrease the impact of natural heterogeneity of environmental conditions [11]. The most widely used bioremediation procedure is the biostimulation of indigenous microorganisms by the addition of nutrients because the input of large quantities of carbon sources (i.e. organic pollutants) tends to result in a rapid depletion of the available pools of major inorganic nutrients such as nitrogen and phosphorus [12]. The most reliable way to achieve the successful bioremediation is to ensure that appropriate microorganisms are present in adequate numbers and that the physicochemical environmental conditions are optimized through appropriate site engineering design, to enhance microbial growth and activities [13].

The activity of the indigenous microorganisms can be inhibited through the non optimal environmental conditions (content of oxygen, moisture and/or nutrients, pH) or insufficient availability (hydrophobic contaminants) or degradability (xenobiotics) of pollutants. Therefore it is necessary to optimize the conditions to activate the work of microorganisms [12, 14, 15]. Water activity is considered to be proportional to free water that is able to transport soluble nutrients to the biomass, in contrast with moisture which includes all water content (available and non-available) [14]. It is important to note that the contamination by hydrocarbons increased biochemical and microbial activities and improved soil structure [16].

It would be good to use commercial products of specified microorganisms to enhance the biodegradation of xenobiotic pollutants, but it needs monitoring for the influence of added microorganisms on the diversity of local (indigenous) organisms [17, 18]. During bioremediation, the rate of degradation of hydrocarbons in the field is largely controlled by the rate of supply of nutrients and oxygen [19], which makes it difficult to extrapolate directly the results from the laboratory to bioremediation in the field [20].

The mineralization of pollutants is the best solution for the decontamination of polluted soil, but the level of the residual amounts, which is degraded very slowly, is very important for the applicability of the bioremediation process [10]. However, with a global political shift towards sustainable and green bioremediation technologies, the use of plant-associated bacteria to degrade toxic synthetic organic compounds in environmental soil may provide an efficient, economic, and sustainable green remediation technology for future environment [21].

1.1.2. Environmentally friendly enhancement of bioremediation

The addition of surfactants can increase the bioavailability of hydrophobic pollutants [22-25] and the toxicity of organic compounds is strongly linked to their bioavailability in the soil [26], but it is necessary to monitor the leaching of surfactants. Temperature is a very critical factor in determining microbial activity, although the optimum temperature for biodegradation of petroleum products has generally been found to be in the range of 20-30°C, local environmental conditions may select for a population with a lower optimum temperature [27-29]. The decontaminations of contaminated soils achieved at 10°C showed that, even at low temperatures, abiotic processes and cold-adapted indigenous soil microorganisms contribute to a great extent to diesel oil contaminations in alpine soils [28, 30]. Cold-adapted indigenous hydrocarbon degraders also play an essential role in the in situ decontamination of deep soil horizons in temperate climates [12]. In the sub-arctic soil with the higher level of contaminants (8,100 mg/kg) low soil temperature was probably the primary factor limiting field biodegradation and the maximum microbial activity occurred at about 21°C [31]. The CO_2 production of tundra soils below 0°C is caused, at least partly, by a specialized ecological group of microorganisms adapted to extreme Arctic conditions and life at sub-zero temperatures [32]. The optimization of temperature could be good way to enhancement biodegradation of pollutants. For in situ remediation it is generally too expensive to increase the soil temperature, but for ex situ remediation, increasing temperature provides realistic option [33, 34, 35].

Biostimulation of the contaminated soil with air and mineral nutrients showed that both respiratory activity and the number of hydrocarbon-degrading bacteria sharply increased in the first two days of the treatment, in parallel with deep changes in the structure of the bacterial community as it was shown by population fingerprinting [36]. Biostimulation of microorganisms can be carried out through addition of different fertilizers, but chemical properties of used fertilizer have influence on the results: water-soluble fertilizers can be leached into the deeper layer if they are not consumed quite fast, but water insoluble fertilizers can be insufficiently available to consumers. For example, the experiment showed that combination of fast release and slow release phosphate sources may be the best for immediate treatment and long-term maintenance [37].

The addition of nutrients could be quite easy to activate bioremediation, but the real biodegradation of pollutants is carried out by microorganisms. The lack of microbial growth factors [28] can be the reason for the low ultimate degradation rate of contaminants. The results of analysis showed that the biodegradation of petroleum products is more related to the inherent biodegradability of the constituting compounds (a low bioavailability of the contaminants [38]) than to the specific enzymatic capabilities of the microbiota [39]. At the same time the sites with high percentage of total petroleum hydrocarbon removal were characterized by the presence of high levels of estimated biomass and consisted of a large diversity of microorganisms [40].

The analysis of different soil environments confirmed that vegetation has been enhanced the rate and extent of biodegradation in the rhizosphere as root exudates provide carbon and energy, which increases the diversity and number of microbial species [41, 42]. The carbon and energy sources supplied by plants help sustain the microbial population that is degrading the contaminants, for example root exudates associated with vegetation can provide sufficient carbon and energy to support $\approx 10^8$-10^9 vegetative microbes per gram of soil in the rhizosphere [41]. Through the exudation of a wide variety of compounds, roots may regulate the soil microbial community in their immediate vicinity, cope with herbivores, encourage beneficial symbioses, change the chemical and physical properties of the soil and inhibit the growth of competing plant species [43, 44, 45]. The efficient attachments of microorganisms to host plants as well as the improvement of the degradation activity of microorganisms are critical factors for the efficient bioremediation by plant-microbe systems [46].

The addition of simple organic carbon sources (the simpler fatty acids [47] or methyl derivatives of vegetable oils called as biodiesel [48, 49]) has been identified as a useful technique for further stimulating the biodegradation of petroleum hydrocarbons by an indigenous microbial biomass. Even in the optimal conditions the biodegradation of pollutants takes place more readily when the target contaminants are dissolved in an aqueous solution and the improvement of the bioavailability of hydrocarbons is performed through the use of surfactants and both non ionic and anionic surfactants increase the solubility of hydrocarbons by forming micelles [22, 24, 25, 38].

More often the enhancing of solubility of organic hydrophobic contaminants in soil is carried out by using synthetic surfactants [50], but problems can be associated with reduced availability of compounds sequestered into micelles, their toxicity and ultimate resistance to biodegradation leading increased secondary pollution [23, 51-53]. The toxicity of the surfactant and its potential degradation products is one of the most important criteria for the selection of surfactant in soil clean-up [24, 25, 33]. There are many advantages of biosurfactants compared to their chemically synthesized counterparts and it is discussed by different authors [50, 52, 54 - 56].

The use of surfactants is widely studied to increase the bioavailability of PAH which solubility is very low [7, 57-59] and a significant PAH amount (>94%) was removed from the soil solid phase to surfactants solution of Tween 80 and approximately 0.8 mg of Tween 80 was needed per milligram of total PAH removal from the soil solid phase with five successive washing [58]. Soil organic matter (such as humic acids) is a very important adsorbent for cationic surfactants and the sorbed surfactant obviously enhanced the soil sorption of PAHs [60].

The experiments showed that the biosurfactants had similar PAH bioavailability-enhancing properties to synthetic surfactants but with the advantage of being biodegradable and non-toxic to PAH-degrading bacteria and additional advantages of some biosurfactants were the lack of a true micellar formation at high surfactant concentrations [52].

The main factor that works against the wide production of truly natural surfactants (biosurfactants) is the cost of their work-up (separation and purification) [54, 56]. Nowadays the approach to the use of renewable resources increases the interest in the ability of humic acids to sequester organic pollutants, because humic acids can be used as natural surfactants for technological purposes [23]. The effectiveness of surfactants in improving contaminant biodegradation is a combination of the solubilizing power of the surfactant and the bioavailability of micellar contaminant [33, 52].

It is mentioned that a residual fraction of contaminants remained undegraded in the soil even when optimal biodegradation conditions are provided [61]. The influence of the nature of the soil and of the presence of specialized microorganisms on both the degradation rate and the residual concentration was tentatively singled out [62].

1.1.3. Dual application of surfactants

Surfactants are amphiphilic molecules that tend to partition preferentially at the interface between phases of different polarity and water bonding. Surfactants increase the aqueous solubility of non-aqueous-phase liquids (NAPLs) by reducing their surface/interfacial tension at air-water and water-oil interfaces [24, 63]. Many microbes are capable to synthesize different types of biosurfactants and insoluble substrates (including hydrocarbons) induce the biosurfactant production [64].

Soil washing is a commonly utilized soil remediation technology which is performed on excavated soil and it can be adapted to fit different masses and different contaminant content by changing the composition of the soil washing fluid [65, 66, 67]. The ecotoxicological evaluation (by Microtox®) demonstrated that all soil washing treatments (chelating agent and non ionic surfactant) increased the toxicity of soil leachates, possibly due to increased availability of contaminants and toxicity of soil washing solutions to the test organism [66].

One of the main limitations for a wider application for surfactants in soil remediation is the lack of knowledge about environmental fate and toxicity of surfactant itself [22, 25, 68]. The application of surfactants to enhance soil remediation requires precise knowledge of soil microbial ecology as well as of the fate and transport of contaminants and surfactants in environmental systems [52] including the use of biodegradable surfactants [69]. Biosurfactants have been increasingly used in soil washing and oil removal from contaminated areas [70, 71] and in enhancing the biodegradation of oil by increasing the bioavailability of hydrophobic pollutants [55, 72].

In the absence of toxicity, the net effect of addition of surfactant to a contaminated soil depends on the benefits that result from enhanced solubility of target compounds versus the reduction in direct adhesion of bacteria to the NAPL [38] and several surfactants were toxic to the test bacteria [73].

Beside of concentration the influence of surfactants depends on the composition and properties of surfactants as well. The commercial linear alkylbenzene sulfonate (LAS) which is widely used anionic surfactant is primary the Na salt, but occurrence of Ca and Mg salts in the environment might cause precipitation of Ca and Mg salts, which have lower bioavailability and thereby influence the toxicity and biodegradation [74].

The growing use of non ionic surfactants is related to their good detergent performance, which is not affected by water hardness, and to their low toxicity and they can be mixed with ionic and amphoteric surfactants [22, 25, 75]. The experiments with endosulfan proved that *Pseudomonas aeruginosa* combined with non ionic surfactant Tween 80 was able to achieve 94% degradation of endosulfan in contaminated soil [76]. Tween 80 influenced the sorption of pyrene on the bacterial cell and thus enhanced the degradation of pyrene [77].

Surfactant biodegradability is a factor that can have negative and positive effects in the use of surfactants for bioremediation [22, 25]. The negative effects can be connected with depletion of minerals and/or oxygen, toxicity of surfactants or their intermediates and preferential degradation of the surfactant [24]. The most obvious positive effect of surfactant degradation is the removal of the surfactant from the polluted site. Another positive effect is fact that a degradable surfactant might be used as a primary substrate when the pollutant is degraded co-metabolically [33].

The study of biodegradation of petro-anionic sulfonate X-100 showed that the presence of organic contaminants apparently enhanced surfactant biodegradation, suggesting a probable synergistic effect of organic contaminants [78]. Environmental risk assessment revealed that after 23 days in the case of LAS and 56 days in the case of nonylphenolic compounds (NPE) potential toxic effects could not be expected [79]. Cyclodextrins are natural, non-toxic compounds that are harmless to microorganisms and free enzymes, and their employment to form inclusion complexes with hydrophobic molecules gives opportunity to use cyclodextrins for the enhancement of biodegradation of aliphatic and polycyclic aromatic hydrocarbons [80].

The biosurfactant BS-UC produced by *Candida antarctica* from *n*-undecane had the capacity of changing the hydrophobicity and the zeta potential of the cell surface, but the biosurfactant enhanced the biodegradation of the organic compound whose structure was similar to that of the biosurfactant produced from [81].

In respirometric experiments, oxidation of all parts of the chemical will lead to oxygen consumption, which is the parameter used to quantify ultimate biodegradation [82].

Although the sorption of surfactants is very complex, especially for anionic surfactants, it is partly correlated with the organic carbon content in soil [83]. The adsorption of surfactants by soil components may lead to a significant reduction in their effectiveness to remove the contaminants from the soil and because it is important to consider surfactant tendency to be adsorbed by the soil or sediment to be decontaminated [84, 85].

1.1.4. Polluted soil composting — The easiest technological approach

Composting is an aerobic process that relies on the actions of microorganisms to degrade organic materials, resulting in the thermogenesis and production of organic and inorganic compounds [6]. Whereas the primary benefit from composting of household waste is the reduction of volume, hygienization and stabilization for recycling or ultimate disposal, the objective of composting hazardous materials is solely to convert these substances into an innocuous end-product [86]. Microorganisms consume the supplied substrates as well as the contaminants, possibly by co-metabolisms and many hazardous wastes have been also converted into innocuous end-products in this manner [87].

There are variety of composting systems and most of them utilize bulking agents (such as bark chips, straw and chopped sugar beet), which increases the porosity and, therefore, aerobicity of medium under treatment and decrease the moisture levels [88]. Composting bioremediation strategy relies on mixing the primary ingredients of composting with the contaminated soil, wherein as the compost matures, the pollutants are degraded by the active microflora within mixture [6].

Soil composting pile is a promising approach of oil-contaminated soil and it was possible to degrade up to 60% of total petroleum hydrocarbon of heavy-contaminated soil during the first eight months [89].

Contaminated soil is often poor in organic matter and has a general low microbial activity, but usually the indigenous bacterial community is adapted to the presence of the contaminant. By adding an organic matrix to contaminated soil the general microbial activity is enhanced and also the activity of specific degraders, which may be found in the contaminated soil or

http://dx.doi.org/10.5772/56908

introduced along with the organic material [10]. The mixing of polluted soil with different organic material will be the better solution to enhance the biodegradation of xenobiotic pollutants and that technique is called cocomposting as it is the process of simultaneous stabilizing organic matter and degrading toxic compounds (pollutants) [14, 90, 91].

In contrast to composting, compost can be added to polluted soil after its maturation for remediation purposes as composts are capable of sustaining diverse populations of microorganisms (bacteria including bacilli, pseudomonads, mesophilic and thermophilic actinomycetes and lignin-degrading fungi), all with the potential to degrade a variety of aromatic pollutants, but the use of composts has not been widely applied as a method for bioremediation [6, 87]. The amount and nature of soil organic matter has been proposed by many workers as being one of the most significant factors dominating organic compound interactions within soil [92, 93].

Biodegradation pathways of organic pollutants may vary in accordance with the chemical structure of the pollutant and the particular degrading microbial species present and that pathways are described by Neilson and Allard [94].

Composting and the use of composted materials have both been successfully applied to the bioremediation of PCP-contaminated soil and the principal vector of PCP loss under both bioremediation regimes was mineralization and the changes in molecular weight distribution during the composting indicated that most of the removal of organic chlorine compounds occurred in the low molecular weight fractions [95, 96]. Whereas 90% losses of TNT from contaminated soil during composting was reported although no mineralization took place [97].

Nutrient provision is an important factor for the degradation of pollutants, but low C/N ratios should favor carbon utilization and thus enhance the total mineral oil and grease degradation as it was proved by substantially increased degradation of soil-derived mineral oil and grease as the C/N ratio was reduced from 49 to 17 [87]. Maintenance of thermophilic conditions in traditional composting is generally considered to be necessary, but more total mineral oil and grease was degraded when temperature profile was maintained at 23°C rather than imposing a 5-day plateau at 50°C [87]. The kinetic parameters of composting showed that the composting temperature (in concert with the microorganisms) accounted for 82.5% of the observed reduction in the composting treatment, while only 17.5% of this reduction could be attributed to additional composting-related factors (nutrients, organic matter, abundance of microorganisms) [90].

Rhizodegradation of pollutants in soil becomes promising for the areas where plants can grow because plant rhizospheres are the most biologically active microsites in soil [98, 99] and it can be used to remediate heavily oil-polluted soils where composting with optimal organic amendments amounts (2:1) and C/N ratio (15:1) effectively reduced total petroleum hydrocarbon concentration and allowed dominant plant species to grow in the remediated soil [100].

The experiments of diesel oil degradation during contaminated soil composting showed that volatilization loss of total petroleum hydrocarbons was only about 2% of initial hydrocarbons [91]. No significant effect of the bioaugmendation on the composting process of petroleum-contaminated soil was observed as well as none of the introduced organisms were re-isolated by traditional plating techniques and because it is more important to create suitable conditions for the indigenous bacteria than to introduce new species [10]. The germination tests showed that no immediate phytotoxic properties could be attributed to the compost after 12 weeks of composting fresh biowaste together with the diesel-contaminated soil (10%) [90].

1.1.5. Objectives of experimental work

The main objective of the experimental work was to study the behavior of surfactants in oil-polluted soil and soil compost. The sub-objectives for achieving the main goal are the following:

a. To study leaching and biodegradation of anionic surfactants and their influence on the behavior of petroleum hydrocarbons in oil-polluted sandy soil;

b. To study leaching and biodegradation of non ionic surfactant Tween 80 and its influence on the properties of leachate and on the behavior of petroleum hydrocarbons in soil compost.

2. Experimental

2.1. Experimental setup and analyses

The leaching experiments were carried out in thermostated Plexiglas columns (length 50 cm, inner diameter 6 cm) at temperature 20.0±0.2°C (Figure 1.) Each column contained 1.4 L (2.2 kg) of soil, which was added into the column and tightened by manual shaking. In the experiments the columns with soil were treated once with 80 cm^3 of the diluted solution of bioremediation agent SR-100 (E-Tech, USA), which contained 9.18% of anionic surfactants as MBAS [101]. Every week, 30 cm^3 of aerated distilled water was added to the column to moisten the soil and supply the soil with oxygen in order to model natural conditions (rain). After the experiments the content of the columns was divided into four equal fractions by volume (Figure 2A).

The experiments with soil compost were carried out in the same columns (Fig. 1), which contained 1.4 L (1.9 kg) of soil compost. In the experiments the columns with soil compost were treated twice per week with aerated distilled water or 0.02% solution of Tween 80 and the amount of added liquid was calculated by mean annual precipitation in Estonia. After the experiments the content of the columns was divided into three equal fractions by volume (Figure 2B). All experiments continued two months.

First series of the experiments were carried out with contaminated coarse-grained (diameter 2-8 mm) sandy soil from Ämari Airport (northwestern Estonia). The soil was contaminated with jet and diesel fuel as well as lubricating oil. To achieve a higher porosity in the test the polluted soil was mixed with unpolluted sand and the final concentration of the hydrocarbons was about 1000 mg HEM kg^{-1} DS. Another series of experiments was carried out with fine (diameter 0.2-4 mm) natural sandy soil from Kloogaranna beach (northwestern of Estonia). It

http://dx.doi.org/10.5772/56908

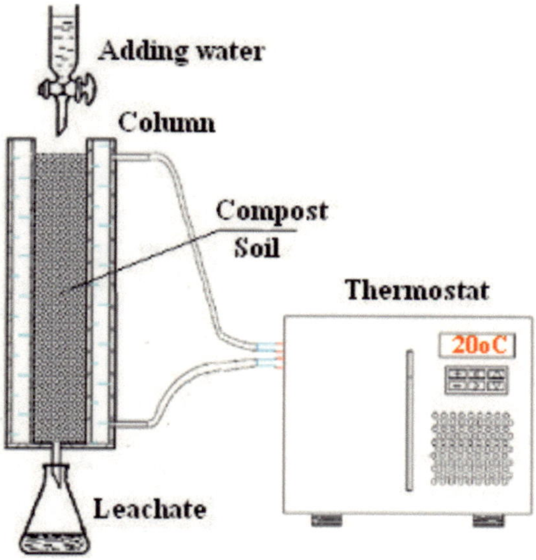

Figure 1. Experimental setup for leaching experiments

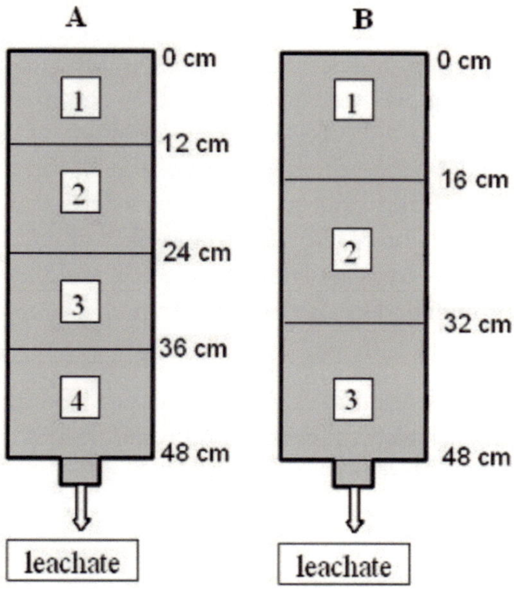

Figure 2. Fractions of soil (A) and soil compost (B)

was artificially contaminated with used diesel oil and the concentration of hydrocarbons was about 1400 mg HEM kg^{-1} DS. The column with the unpolluted sand was used to compare the leaching of hydrocarbons and surfactants.

In the latter experiments the coarse-grained soil was mixed with the hydrophobic porous adsorbent (10% v/v) prepared by modifying of the urea-formaldehyde resins. The leachate was collected and the concentrations of the leached surfactants and hydrocarbons were determined.

Soil compost was sampled from the composting windrows (composting facility of Ragn-Sells Ldt, Estonia) and for composting the polluted soil, which consist different petroleum hydrocarbons was mixed with gravel (to enhance the porosity), horse manure (as source of microorganisms), wood and bark chips and peat (as bulking agents). The concentration of hydrocarbons was about 77 g HEM kg^{-1} DS.

2.2. Chemical analysis

The pH of the leachate and water extracts was measured by a pH-meter (SensION1, Hach, USA and Model 3320, Jenway, UK). Soil pH was determined by extracting the soil samples with 5 volumes of distilled water and measured with a glass electrode. The conductivity (EC) of water extracts was measured by a conductivity meter (Model 4320, Jenway, UK) and the values have been corrected to a constant temperature of 25 °C [102].

Colorimetric methods are quite widely used for the determination of the concentrations of surfactants [103 - 106]. Anionic surfactants form ion pairs with Methylene Blue (MB) and extracted with chloroform. The concentration of anionic surfactants in the soil and leachate was determined by the spectrophotometric method using MB [105]. The chloroform phase with the dissolved colored complex was separated and its absorbance was measured at 654 nm by the spectrophotometer KFK-3 (USSR). The concentration of the anionic surfactants in the leachate was calculated by the calibration curve as methylene blue active substances (MBAS). The concentration of non ionic surfactants was determined as cobalt thiocyanate active substances (CTAS) by method 512C [102]. The surfactant-cobalt complex will partition into methylene chloride from excess aqueous cobalt thiocyanate by a single extraction and measurement of CTAS in the methylene chloride at 620 nm by spectrophotometer (Model 6300, Jenway, UK).

The concentration of total petroleum hydrocarbons (abbreviated as TPH) in the solids and leachate was determined gravimetrically as HEM by the USEPA method 1664 [107]. The microbial activity of soil compost samples was characterized through the oxygen uptake over a 7-days period determined by manometric OxiTop system (WTW, Germany) at temperature 20±0.2 °C [108-110].

2.3. Results of column studies

2.3.1. Sandy soil

Concentrations of anionic surfactants were measured regularly in the upper layer (0-5 cm) of the soil in the columns [101, 111]. The determined concentrations of anionic surfactants as MBAS in the experiments with polluted sandy soil are presented in Table 1.

http://dx.doi.org/10.5772/56908

Time, days	Unpolluted soil		Oil-polluted soil		Oil-polluted soil with adsorbent	
	Coarse soil	Fine soil	Coarse soil	Fine soil	Coarse soil	Fine soil
0	172	292	172	260	552	612
8	121	176	91.4	142	461	549
15	95.7	108	85.4	97.5	213	401
29	51.7	46.6	48.5	2.1	109	254
49	8.9	2.3	8.4	1.1	64.8	119
60	8.1	1.5	8.1	0.93	43.7	87

Table 1. Concentrations of anionic surfactants (mg MBAS g^{-1}) in the upper layer (0-5 cm) of the columns with soil during the leaching experiments.

The columns with unpolluted or polluted coarse-grained soil had a negligible difference in the concentration of surfactants in the upper soil layer (0-5 cm) during the experiment. The higher concentration of surfactants in the upper soil layer in case of a mixture of soil and adsorbent could be explained by the specific higher surface of adsorbent which adsorbed the anionic surfactants by the soaking of the solution of surfactants in the soil.

For fine soil, the initial concentration of surfactants and its reduction in the upper layer (Table 1) was similar in all soil samples. The surfactants were washed out of the upper layer in the columns of unpolluted and polluted soil during the first 30 days. The concentration of anionic surfactants decreased linearly during the last 40 days of the experiment with the mixture of polluted soil and adsorbent and the surfactants were not completely washed out of the upper layer during the 60 days in the column. The experiments continued for 60 days after which all four fractions of soil samples and leachate were analyzed and the calculated masses of anionic surfactants are presented in Table 2.

Fraction of the column	Unpolluted soil		Oil-polluted soil		Oil-polluted soil with adsorbent	
	Coarse soil	Fine soil	Coarse soil	Fine soil	Coarse soil	Fine soil
1 (0-12 cm)	22.7	10.4	8.5	19.6	61.1	74.4
2 (12-24 cm)	52.6	17.1	75.9	75.5	70.7	81.3
3 (24-36 cm)	64.3	93.8	4.4	88.4	79.9	84.8
4 (36-48 cm)	271	617	4.1	351	71.3	97.5
leachate	125	40.4	20.2	50.4	0	0
total	535	779	113	584	283	338
Reduction (%)	34.4	4.5	86.2	28.4	65.3	58.6

Table 2. The cumulative masses of anionic surfactants (mg MBAS) in soil fractions and leachate of the columns after the experiments.

In the case of unpolluted soil the anionic surfactants were washed mainly deeper than 36 cm and the higher cumulative masses of surfactants indicated the higher level of degradation in the more porous coarse soil. Due to the low porosity of fine soil the amount of oxygen that diffused into the soil was not enough to achieve the aerobic conditions needed for the surfactants degradation [25].

Fraction of the column	Oil-polluted soil		Oil-polluted soil with adsorbent	
	Coarse soil	Fine soil	Coarse soil	Fine soil
Initial soil	1039	1463	882	1245
		Soil + H_2O		
1 (0-12 cm)	216	318	154	246
2 (12-24 cm)	220	326	160	251
3 (24-36 cm)	229	340	175	259
4 (36-48 cm)	218	333	169	248
leachate	42	20	18	10
total	925	1337	676	1014
Reduction (%)	11.0	8.6	23.4	18.6
		Soil + SR-100		
1 (0-12 cm)	107	186	60	152
2 (12-24 cm)	69	193	53	164
3 (24-36 cm)	82	258	63	144
4 (36-48 cm)	128	183	61	139
leachate	30	59	10	18
total	416	879	247	617
Reduction (%)	60.0	39.9	72.0	50.4

Table 3. The cumulative masses of TPH (mg HEM) in soil fractions and leachate of the columns after the experiments.

Addition of porous urea-formaldehyde adsorbent increased the mass of generated leachate (up to 25%) due to the higher porosity of soil but anionic surfactants were no leached out. The residual content of anionic surfactants was about 35-40% from the added surfactants independently on the soil properties and fractions. The analysis of soil fractions indicated no significant accumulation of anionic surfactants in the upper layer of mixture of soil and adsorbent.

The analysis of residual TPH (Table 3) showed that porosity was quite important factor for the biodegradation of hydrocarbons as coarse soil contained less hydrocarbon in comparison with fine soil. Added anionic surfactants accelerated the degradation of petroleum hydrocarbons, but porosity of soil was important factor, too. Analysis of leachate showed very low content of hydrocarbons (less than 1% from dry residual) indicating quite active biodegradation of petroleum hydrocarbons in the soil columns.

2.3.2. Polluted soil compost

The profile of leachate generation is presented in Figure 3 and the generation of leachate had no significant difference by adding of non ionic surfactants Tween 80. The leaching experiments in the columns with the oil-polluted soil compost continued for 60 days. The measurements of leachate masses showed that 64% of added water and solution of Tween 80 leached out during the experiments and therefore the non ionic surfactant did not increase the amount of leachate during 2 months [112]. The results of leachate analysis are presented in Table 4.

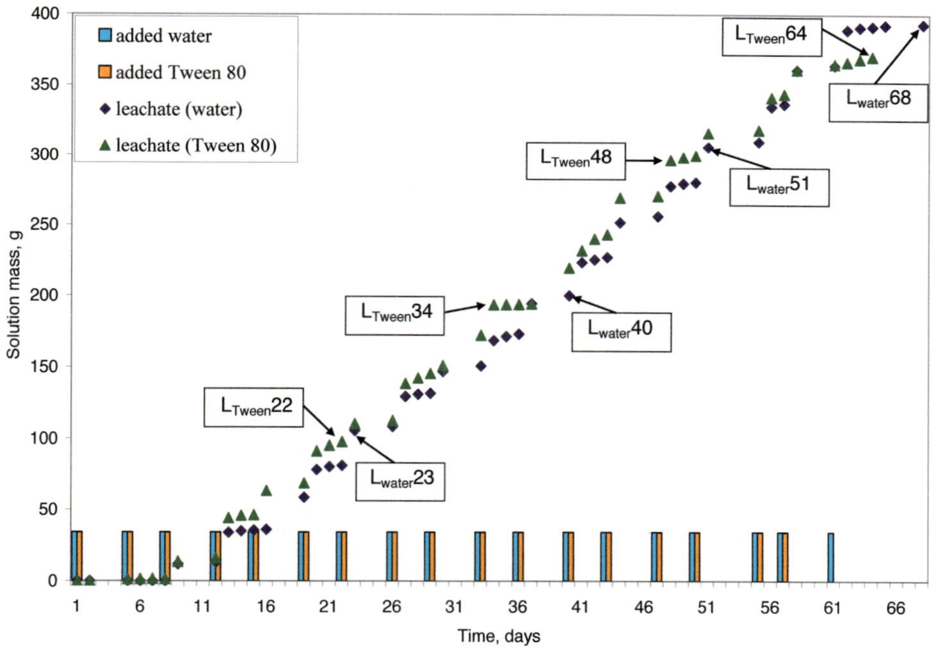

Figure 3. Profile of generation of leachate from the column of soil compost. The sampling of leachate is also presented.

In general, the leachate had quite low (below 15 mg HEM L^{-1}) content of TPH, but it was increased significantly (up to 70 mg HEM L^{-1}) for the last period through the added Tween 80 (Table 4). It means that the added non ionic surfactant can generate the leaching of petroleum hydrocarbon during longer period of treatment.

Similar trend was observed for the non ionic surfactant as well and leachate had more than 3 times higher concentration of non ionic surfactant as CTAS. The solution can be the turning of

H_2O				Tween 80				
Time, days	Added water, g	Leachate, g	TPH, mg HEM L^{-1}	Time, days	Added water, g	Leachate, g	TPH, mg HEM L^{-1}	Surfactant, mg CTAS L^{-1}
1-23	238	105.4	15	1-22	238	97.7	10	2.9
24-40	136	94.8	11	23-34	102	96.0	10	20.9
41-51	136	105.6	10	35-48	136	102.7	10	25.1
52-68	102	87.4	13	49-64	102	74.1	70	85.1
1-68	**612**	**393.2**	**12**	**1-64**	**578**	**370.5**	**22**	**30.2**

Table 4. Leachate from the column of composted soil

soil compost which was impossible in the columns. The compost windrows were turned at least once per month in the composting facility and therefore the leaching of petroleum hydrocarbons and non ionic surfactant would not occur. After the experiments the fractions of soil compost were analyzed and the results are presented in Table 5.

Compost fraction, cm	Solid soil compost			Water extract of soil compost	
	Rate of oxygen demand, mg O_2 kg $^{-1}$ h^{-1}	TPH, g HEM kg^{-1}	Surfactant, mg CTAS kg^{-1}	pH	EC, mS cm^{-1}
Initial	462±3	77.3±2.1	0	6.92	0.27
		Compost + H_2O			
0–16	358±11	61.6±5.2	0	7.13	0.46
16–32	307±8	55.4±4.2	0	7.12	0.33
32–48	375±14	64.2±4.2	0	7.47	0.48
		Compost + Tween 80			
0–16	213±48	61.6±5.7	8.7±0.8	8.30	0.33
16–32	239±16	61.1±5.2	26.1±2.5	8.01	0.32
32–48	299±25	55.9±5.3	52.3±4.2	7.87	0.30

Table 5. Characteristics of fraction of soil compost after leaching experiments in the columns

The analysis of compost fractions showed that both water and solution of Tween 80 reduced the content of petroleum products in the compost mixture about 20%. The reduction of hydrocarbon content was generated through biodegradation as less than 0.01% of hydrocarbons were leached out from the soil columns. It is important to note that addition of surfactant did not increase the leaching of petroleum products [112]. In the column of soil compost which was treated with water the lowest TPH content as well as rate oxygen demand and conductivity were determined for the centre fraction (16-32 cm). In the case of treatment with Tween 80 the rate of oxygen demand was the highest for the lower fraction, but content of TPH was lightly higher for the upper layer. The analyses of soil compost fractions showed that non ionic surfactant was leached into the lower fractions of soil (Table 5) and only up to 10% of non ionic surfactant was leached out from the soil columns during 60 days.

2.3.3. *Fluxes of pollutants and surfactants*

Calculation of mass balance of anionic surfactants showed the highest degradation of anionic surfactants (about 86% of added surfactants) in the column of coarse-grained oil-polluted soil, while only 28.4% of anionic surfactants were degraded in the column of fine polluted soil (Table 2). Quite low degradation of anionic surfactants was determined in the columns of unpolluted soil (34.4% for coarse soil and 4.5% for fine soil) while added adsorbent significantly accelerated the degradation of anionic surfactants (65.3% for coarse soil and 58.6% for fine one).

The highest degradation of TPH (72%) was determined for the mixture of polluted coarse soil with adsorbent (Table 3) treated with SR-100 containing anionic surfactants. The treatment of polluted coarse soil with SR-100 reduced the content of TPH to 60%. In the case of fine soil the reduction of content of TPH was about 20% lower. Without added anionic surfactants the reduction of content of TPH was below 25%.

The cumulative amount of added Tween 80 was 116 mg CTAS and residual amount of non ionic surfactant was 66 mg CTAS and therefore 43% of non ionic surfactant was degraded during 60 days of experiments, while the leachate contained 9.7% of surfactant and therefore main part of Tween 80 biodegraded in soil compost during experiments. The other experiment also showed not so good biodegradation of Tween 80 due to its high molecular weight [113]. From biodegradability tests, both in liquid and in solid phase, Tween 80 resulted to be, in tested conditions, extremely biodegradable and mineralisable also by not specialized soil bacteria [68].

Calculation of mass balance of TPH showed that 22% of TPH was biodegraded in the column of soil compost treated with water and 23% of TPH was degraded through addition of solution of Tween 80. It means that non ionic surfactant had no accelerating effect on the biodegradation of TPH in soil compost.

3. Conclusion

Surfactants are used to increase the bioavailability of hydrophobic pollutants (hydrocarbons), but the added anionic surfactants were washed out of the upper layer of soil columns during the first 30 days regardless of the type of soil. After the experiments (60 days), the analysis of different soil fractions showed that up to 86% of anionic surfactants had degraded in the column of coarse-grained polluted soil while up to 28% of the added surfactants had degraded in the column of fine sandy soil. Higher amount of anionic surfactants was determined in the leachate of columns of coarse-grained soil and the higher porosity of soil could increase the leaching of surfactants due to the lower surface area of coarse-grained soil.

Addition of porous urea-formaldehyde adsorbent increased the mass of generated leachate (up to 25%) due to the higher porosity of soil but anionic surfactants were no leached out and the analysis of soil fractions indicated no significant accumulation of anionic surfactants in the upper layer of mixture of soil and adsorbent.

During the experiments similar amounts of leachate were generated from the soil compost through the addition of non ionic surfactant Tween 80 or water and therefore the non-ionic surfactant did not increase the amount of leachate during 2 months. The leachate had quite low content both of TPH and non ionic surfactant, but it was increased significantly for the last period through the added Tween 80, therefore the added non ionic surfactant can generate the leaching of petroleum hydrocarbon during longer period of treatment.

The analysis of compost fractions showed that both water and solution of Tween 80 reduced the content of petroleum products in the compost mixture about 20% and the reduction of hydrocarbon content was generated through biodegradation as less than 0.01% of hydrocarbons were leached out from the soil columns.

Calculation of mass balance of anionic surfactants showed 86% of degradation of anionic surfactants in the column of coarse-grained oil-polluted soil, while 72% degradation of TPH was determined for the mixture of polluted coarse soil with adsorbent treated solution of anionic surfactants. Calculation of mass balance indicated that 43% of non ionic surfactant Tween 80 was degraded during 60 days of experiments in soil compost, but non ionic surfactant had no accelerating effect on the biodegradation of TPH in soil compost.

Acknowledgements

A support for this study was provided by the target-financed project of the Estonian Ministry of Education and Research No. SF0180135s08. Thanks are due to Mr Olav Ojala (Ragn-Sells Ldt) for sampling of soil compost.

Author details

Aare Selberg, Kätlin Juuram, Jana Budashova and Toomas Tenno

*Address all correspondence to: aare.selberg@mail.ee

Institute of Chemistry, University of Tartu, Tartu, Estonia

References

[1] van Straalen, N.M. Assessment of soil contamination – a functional perspective. Biodegradation 2002; 13 41-52.

[2] Nadim, F., Hoag, G.E., Liu, S., C, Carley, R.J., Zack, P. Detection and remediation of soil and aquifer systems contaminated with petroleum products: an overview. Journal of Petroleum Science and Engineering 2000; 26 169-178.

[3] Wei, Q.F., Mather, R.R., Fotheringham, A.F. Oil removal from used sorbents using a biosurfactant. Bioresource Technology 2005; 96 331-334.

[4] Underwood, J., Debelak, K., Wilson, D, Means, J. Soil cleanup by in-situ surfactant flushing. Separation Science and Technology 1993; 28 1527-1537.

[5] Zhang, C., Valsaraj, K.T., Constant, D., Roy, D. Surfactant screening for soil washing: comparison of foamability and biodegradability of a plant-based surfactants with commercial surfactants. Journal of Environmental Science and Health A 1998; 33(7) 1249-1273.

[6] Semple, K.T., Reid, B.J., Fermor, T.R. Impact of composting strategies on the treatment of soils contaminated with organic pollutants. Environmental Pollution 2001; 112 269-283.

[7] Alvarez, P.J.J., Illman, W.A. Bioremediation and natural attenuation: process fundamentals ans mathematical models. Wiley-Interscience; 2006.

[8] Bento, F.M., Camargo, F.A.O., Okeke, B.C., Frankenberger, W.T. Comparative bioremediation of soils contaminated with diesel oil by natural attenuation, biostimulation and bioaugmentation. Bioresource Technology 2005; 96 1049-1055.

[9] Christofi, N., Ivshina, I.B. Microbial surfactants and their use in field studies of soil remediation. A review. Journal of Applied Microbiology 2002; 93 915-929.

[10] Jørgensen, K.S., Puustinen, J., Suortti, A.-M. Bioremediation of petroleum hydrocarbon-contaminated soil by composting in biopiles. Environmental Pollution 2000;107 245-254.

[11] Korda, A., Santas, P., Tenente, A., Santas, R. Petroleum hydrocarbon bioremediation: sampling and analytical techniques, in situ treatments and commercial microorganisms currently used. Applied Microbiology and Biotechnology 1997; 48 677-686.

[12] Margesin, R. Potential of cold-adapted microorganisms for bioremediation of oil-polluted Alpine soils. International Biodeterioration and Biodegradation 2000; 46 3-10.

[13] Balba, M.T., Al-Daher, R., Al-Awadhi, N., Chino, H., Tsuji, H. Bioremediation of oil-contaminated desert soil: the Kuwaiti experience. Environment International 1998; 24(1/2) 163-173.

[14] Aguilera-Vazguez, L., Soto-Cruz, N.O., Saucedo-Castaneda, G., Gutierrex-Rojas, M. A model system for cocomposting hydrocarbon contaminated soil by using water activity and porosity as response variables. Chemical Engineering Journal 2001; 81 197-202.

[15] Russell, D.L. Remediation manual for contaminated sites. CRC; 2012.

[16] Caravaca, F., Roldan, A. Assessing changes in physical and biological properties in a soil contaminated by oil sludges under semiarid Mediterranean conditions. Geoderma 2003; 117 53-61.

[17] Mohammed, D., Ramsubhag, A., Beskles, D.M. An assessment of the biodegradation of petroleum hydrocarbons in contaminated soil using non-indigenous, commercial microbes. Water Air Soil Pollution 2007; 182 349-356.

[18] Cho, B.-H., Chino, H., Tsuji, H., Kunito, T., Makishima, H., Uchida, H., Matsumoto, S., Oyaizu, H. Analysis of oil components and hydrocarbon-utilizing microorganisms during laboratory-scale bioremediation of oil-contaminated soil of Kuwait. Chemosphere 1997; 35(7) 1613-1621.

[19] Huesemann, M.H., Hausmann, T.S., Fortman, T.J. Does bioavailability limit biodegradation? A comparison of hydrocarbon biodegradation and desorption rates in aged soils. Biodegradation 2004; 15 261-274.

[20] Balba, M.T., Al-Awadhi, N., Al-Daher, R. Bioremediation of oil-contaminated soil: microbiological methods for feasibility assessment and field evaluation. Journal of Microbiological Methods 1998; 32 155-164.

[21] Divya, B., Kumar, M.D. Plant-microbe interaction with enhanced bioremediation. Research Journal of Biotechnology 2011; 6(4) 72-79.

[22] Paria, S. Surfactant-enhanced remediation of organic contaminated soil and water. Advances in Colloid and Interface Science 2008; 138 24-58.

[23] Salati, S., Papa, G., Adani, F. Perspective on the use of humic acids from biomass as natural surfactants for industrial application. Bioresource Technology 2011; 29 913-922.

[24] Tadros, T.F. Applied surfactants: principles and applications. Wiley-VCH; 2005.

[25] Ying, G.-G. Fate, behavior and effects of surfactants and their degradation products in the environment. Environment International 2006; 32 417-431.

[26] de Jonge, H., Freijer, J.I., Verstraten, J.M., Westerveld, J., van der Wielen, F.W.M Relation between bioavailability and fuel oil hydrocarbon compostion in contaminated soils. Environmental Science and Technology 1997; 31 771-775.

[27] Westlake, D.W.S., Jobson, A.M., Cook, F.D. In situ degradation of oil in a soil of the boreal region of the Northwest Territories. Canadian Journal of Microbiology 1978; 24 254-260.

[28] Margesin, R., Schinner, F. Bioremediation of diesel-oil-contaminated alpine soils at low temperatures. Applied Microbiology and Biotechnology 1997; 47 462-468.

[29] Belousova, N.I., Baryshnikova, L.M., Shkidchenko, A.N. Selection of microorganisms capable of degrading petroleum and its products at low temperatures. Applied Biochemistry and Microbiology 2002; 38(5) 437-440.

[30] Jobson, A., Cook, F.D., Westlake, D.W.S. Microbial utilization of crude oil. Applied Microbiology1972; 23 1082-1089.

[31] Walwort, J., Braddock, J., Wooland, C. Nutrient and temperature interactions in bioremediation of cryic soils. Cold Region Science and Technology 2001; 32 85-91.

[32] Panikov, N.S., Flanagan, P.W., Oechel, W.C., Mastepanov, M.A., Christensen, T.R. Microbial activity in soils frozen to below –39°C. Soil Biology and Biochemistry 2006; 38 785-794.

[33] Volkering, F., Breure, A.M., Rulkens, W.H. Microbial aspects of surfactant use for biological soil remediation. Biodegradation 1998; 8 401-417.

[34] Ferguson, S.H., Franzmann, P.D., Snape, I., Revill, A.T., Trefry, M.G., Zappia, L.R. Effects of temperature on mineralization of petroleum in contaminated Antarctic terrestrial sediments. Chemosphere 2003; 52 975-987.

[35] Mohn, W.W., Stewart, G.R. Limiting factors for hydrocarbon biodegradation at low temperature in Arctic soil. Soil Biology and Biochemistry 2000; 32 1161-1172.

[36] Zucchi, M., Angiolini, L., Borin, S., Brusetti, L., Dietrich, N., Gigliotti, C., Barbieri, P., Sorlini, C., Daffonchio, D. Response of bacterial community during bioremediation of an oil-polluted soil. Journal of Applied Microbiology 2003; 94 248-257.

[37] Knox, A.S., Kaplan, D.I., Paller, M.H. Phosphate sources and their suitability for remediation of contaminated soils. Science of Total Environment 2006; 357 271-279.

[38] Stelmack, P.L., Gray, M.R., Pickard, M.A. (1999) Bacterial adhesion to soil contaminants in the presence of surfactants. Applied and Environmental Microbiology, 65(1), 163-168.

[39] Chaineau, C.-H., Morel, J.-L., Oudot, J. Microbial degradation in soil microcosms of fuel oil hydrocarbons from drilling cuttings. Environmental Science and Technology 1995; 29 1615-1621.

[40] Claassens, S., van Rensburg, L., Riedel, K.J., Bezuidenhout, J.J., van Rensburg, P.J.J. Evaluation of the efficiency of various commercial products for the bioremediation of hydrocarbon contaminated soil. The Environmentalist 2006; 26 51-62.

[41] Erickson, L.E., Davis, L.C., Narayanan, M. Bioenergetics and bioremediation of contaminated soil. Thermochimica Acta 1995; 250 353-358.

[42] Mortensen, G.K., Egsgaard, H., Ambus, P., Jensen, E.S., Grøn, C. Influence of plant growth on degradation of linear alkylbenzene sulfonate in sludge-amended soil. Journal of Environmental Quality 2001; 30 1266-1270.

[43] Shukla, K.P., Sharma, S., Singh, N.K., Singh, V., Tiwari, K., Singh, S. Nature and role of root exudates: efficiency in bioremediation. African Journal of Biotechnology 2011; 10(48) 9717-9724.

[44] Cohen, M.F., Yamasaki, H., Mazzola, M. Bioremediation of soils by plant-microbe systems. International Journal of Green Energy 2004; 1(3) 301-312.

[45] Kuiper, I., Lagendijk, E.L., Bloemberg, G.V., Lugtenberg, B.J.J. Rhizoremediation: a beneficial plant-microbe interaction. Molecular Plant-Microbe Interactions 2004; 17(1) 6-15.

[46] Motoyama, T., Kadokura, K., Tatsusawa, S., Arie, T., Yamaguchi, I. Application of plant-microbe systems to bioremediation. RIKEN Review 2001; 42 35-38.

[47] Obbard, J.P., Ng, K.L., Xu, R. Bioremediation of petroleum contaminated beach sediments: use of crude palm oil and fatty acids to enhance indigenous biodegradation. Water, Air, and Soil Pollution 2004; 157 149-161.

[48] Nelson, E.C., Walter, M.V., Bossert, I.D., Martin, D.G. Enhancing Biodegradation of Petroleum Hydrocarbons with Guanidinium Fatty Acids. Environmental Science and Technology 1996; 30, 2406-2411.

[49] Pereira, M.G., Mudge, S.M. Cleaning oiled shores: laboratory experiments testing the potential use of vegetable oil biodiesels. Chemosphere 2004; 54(3) 297–304.

[50] Kosaric, N. Biosurfactants and their application for soil bioremediation. Food Technology and Biotechnology 2001; 39(4) 295-304.

[51] Cort, T.L., Song, M.S., Bielefeldt, A.R. Non ionic surfactant effect on pentachlorophenol degradation. Water Research 2002; 36 1253-1261.

[52] Makkar, R.S., Rockne, K.J. Comparison of synthetic surfactants and biosurfactants in enhancing biodegradation of polycyclic aromatic hydrocarbons. Environmental Toxicology and Chemistry 2003; 22(10) 2280-2292.

[53] Mulligan, C.N., Yong, R.N., Gibbs, B.F. Surfactant remediation of contaminated soil: a review. Engineering Geology 2001; 60 371-380.

[54] Holmberg, K. Natural surfactants. Current Opinion in Colloid and Interface Science 2001; 6 148-159.

[55] Ron, E.Z., Rosenberg, E. Biosurfactants and oil bioremediation. Current Opinion in Biotechnology 2002; 13(3) 249-252.

[56] Rosenberg, E., Ron, E.Z. High- and low-molecular-mass microbial surfactants. Applied Microbiology and Biotechnology 1999; 52 154-162.

[57] Ahn, C.K., Kim, Y.M., Woo, S.H., Park, J.M. Soil washing using various non ionic surfactants and their recovery by selective adsorption with activated carbon. Journal of Hazardous Material 2008; 154 13-60.

[58] Zheng, Z., Obbard, J.P. Polycyclic aromatic hydrocarbon removal from soil by surfactant solubilization and *Phanerochaete chrysosporium* oxidation. Journal of Environmental Quality 2002; 31 1842-1847.

[59] Volkering, F., Breure, A.M., van Andel, J.G., Rulkens, W.H. Influence of non ionic surfactants on bioavailability and biodegradation of polycyclic aromatic hydrocarbons. Applied and Environmental Microbiology 1995; 61 1699-1705.

[60] Lu, L., Zhu, L. Effect of soil components on the surfactant-enhanced soil sorption of PAH. Journal of Soils and Sediments 2012; 12 161-168.

[61] Huesemann, M.H., Moore, K.O.J. Compositional changes during landfarming of weathered Michigan crude oil contaminated soil. Journal of Soil Contamination 1993; 2 245-264.

[62] Nocentini, M., Pinelli, D., Fava, F. Bioremediation of a soil contaminated by hydrocarbon mixtures: the residual concentration problem. Chemosphere 2000; 41 1115-1123.

[63] Urum, K., Pekdemir, T. Evaluation of biosurfactants for crude oil contaminated soil washing. Chemosphere 2004; 57 1139-1150.

[64] Veenanadig, N.K., Gowthaman, M.K., Karanth, N.G.K. Scale up studies for the production of biosurfactant in packed column bioreactor. Bioprocess Engineering 2000; 22 95-99.

[65] Harwell, J.H., Sabatini, D.A., Knox, R.C. Surfactants for ground water remediation. Colloids and Surface A 1999; 151 255-268.

[66] Elgh-Dalgren, K., Arwidsson, Z., Camdzija, A., Sjöberg, R., Ribe, V., Waara, S., Allard, B., von Kronhelm, T., van Hees, P.A.W. Laboratory and pilot scale soil washing of PAH and arsenic from a wood preservation site: Changes in concentration and toxicity. Journal of Hazardous Materials 2009; 172 1033-1040.

[67] Yuan, T., Marshall, W.D. Optimizing of washing procedure to mobilize polycyclic aromatic hydrocarbons (PAHs) from a field-contaminated soil. Industrial and Engineering Chemistry Research 2007; 46 4626-4632.

[68] Franzetti, A., Gennaro, P.D., Bevilacqua, A., Papacchini, M., Bestetti, G. Environmental features of two commercial surfactants widely used in soil remediation. Chemosphere 2006; 62 1474-1480.

[69] Suchanek, M., Kostal, J., Demnerova, K., Kralova, B. Use of sodium dodecyl sulphate for stimulation of biodegradation of n-alkanes without residual contamination by the surfactant. International Biodeterioration and Biodegradation 2000; 45 27-33.

[70] Mulligan, C.N., Yong, R.N., Gibbs, B.F. Removal of heavy metals from contaminated soil and sediments using the biosurfactant surfactin. Journal of Soil Contamination 1999; 8(2) 231-254.

[71] Mulligan, C.N., Yong, R.N., Gibbs, B.F. Remediation technologies for metal-contaminated soils and groundwater: an evaluation. Engineering Geology 2001; 60(1-4) 193-207.

[72] Zhang, G.-L., Wu, Y.-T., Qian, X.-P., Meng, Q. Biodegradation of crude oil by *Pseudomonas aeruginosa* in the presence of rhamnolipids. Journal of Zhejiang University SCIENCE 2005; 6B(8) 725-730.

[73] Roch, F., Alexander, M. Biodegradation of hydrophobic compounds in the presence of surfactants. Environmental Toxicology and Chemistry 1995; 14(7) 1151-1158.

[74] Berna, J.L., Ferrer, A., Moreno A., Prats, D., Bevia, F.R. The fate of LAS in the environment. Tenside Surfactants Detergents 1989; 26 101-107.

[75] Mezzanote, V., Castiglioni, F., Todeschini, R., Pavan, M. Study on anaerobic and aerobic degradation of different non-ionic surfactants. Bioresource Technology 2003; 87 87-91.

[76] Jayashree, R., Vasudvan, N. Effect of Tween 80 added to the soil on the degradation of endosulfan by *Pseudomonas aeruginosa*. International Journal of Environmental Sience and Technology 2007; 4 203-210.

[77] Zhang, D., Zhu, L. Effects of Tween 80 on the removal, sorption and biodegradation of pyrene by *Klebsiella oxytoca* PYR-1. Environmental Pollution 2012; 164 169-174.

[78] Abd-Allah, A.M.D., Srorr, T. Biodegradation of anionic surfactants in the presence of organic contaminants. Water Research 1998; 32(3) 944-947.

[79] Gonzales, M.M., Martin, J., Camacho-Munoz, D., Santos, J.L., Aparicio, I., Alonso, E. Degradation and environmental risk of surfactants after the application of compost sludge to the soil. Waste Management 2012; 32 1324-1331.

[80] Bardi, L., Mattei, A., Steffan, S., Marzona, M. Hydrocarbon degradation by a soil microbial population with β-cyclodextrin as surfactant to enhance bioavailability. Enzyme and Microbial Technology 2000; 27 709-713.

[81] Hua, Z., Chen, J., Lun, S., Wang, X. Influence of biosurfactants produced by *Candida antarctica* on surface properties of microorganism and biodegradation of *n*-alkanes. Water Research 2003; 37 4143-4150.

[82] Gejlsbjerg, B., Madsen, T., Andersen, T.T. Comparison of biodegradation of surfactants in soils and sludge-soil mixtures by use of ^{14}C-labelled compounds and automated respirometry. Chemosphere 2003; 50 321-331.

[83] Di-Toro, D.M., Dodge, L.J., Hand, V.C. A model for anionic surfactant sorption. Environmental Science and Technology 1990; 24 1013-1020.

[84] Rodriguez-Cruz, M.S., Sanchez-Martrin, M.J., Sanchez-Camazano, M. A comparative study of adsorption of an anionic and a non-ionic surfactant by soils based on physicochemical and mineralogical properties of soils. Chemosphere 2005; 61 56-64.

[85] Jacobsen, A.M., Mortensen, G.K., Hansen, H.C.B. Degradation and mobility of linear alkylbenzene sulfonate and nonylphenol in sludge-amended soil. Journal of Environmental Quality 2004; 33 232-240.

[86] Kirchmann, H., Ewnetu, W. Biodegradation of petroleum-based oil wastes through composting. Biodegradation 1998; 9 151-156.

[87] Beaudin, N., Caron, R.F., Legros, R., Ramsay, J., Ramsay, B. Identification of the key factors affecting composting of a weathered hydrocarbon-contaminated soil. Biodegradation 1999; 10 127-133.

[88] Diaz, L.F., de Bertoldi, M., Bidlingmaier, W., Stentiford, E. Compost science and technology. Elsevier; 2007.

[89] Al-Daher, R., Al-Awadhi, N., El-Nawawy, A. Bioremediation of damaged desert environment using the windrow soil pile system in Kuwait. Environment International 1998; 24(1/2) 175-180.

[90] van Gestel, K., Mergaert, J., Swings, J., Coosemans, J., Ryckeboer, J. Bioremediation of diesel oil-contaminated soil by composting with biowaste. Environmental Pollution 2003; 125 361-368.

[91] Namkoong, W., Hwang, E.-Y., Park, J.-S., Choi, J.-Y. Bioremediation of diesel-contaminated soil with composting. Environmental Pollution 2002; 119 23-31.

[92] Brusseau, M.L., Jessup, R.E. Rao, S.C. Nonequilibrium sorption of organic chemicals: elucidation of rate-limiting processes. Environmental Science and Technology 1991; 25 134-142.

[93] Piatt, J.J., Brusseau, M.L. Rate limiting sorption of hydrophobic organic compounds by soils with well characterized organic matter. Environmental Science and Technology 1998; 32 1604-1608.

[94] Neilson, A.H., Allard, A.-S. Environmental degradation and transformation of organic chemicals. CRC Press; 2008.

[95] Laine, M.M., Ahtiainen, J., Wagman, N., Öberg, L.G., Jørgensen, K.S. Fate and toxicity of chlorophenols, polychlorinated dibenzo-p-dioxins, and dibenzofurans during composting of contaminated sawmill soil. Environmental Science and Technology 1997; 31 3244-3250.

[96] Laine, M.M., Jørgensen, K.S. Effective and safe composting of chlorophenol-contaminated soil in pilot scale. Environmental Science and Technology 1997; 31 371-378.

[97] Bruns-Nagel, D., Dryzyzga, O., Steinbach, K., Schmidt, T.C., von Löw, E., Gorontzy, T., Blotevogel, K.-H., Gemsa, D. Anaerobic/aerobic composting of 2, 4, 6-trinitrotoluene-contaminated soil in a reactor system. Environmental Science and Technology 1998; 32 1676-1679.

[98] Kirk, J.L., Klironomos, J.N., Lee, H., Trevors, J.T. The effects of perennial ryegrass and alfalfa on microbial abundance and diversity in petroleum contaminated soil. Environmental Pollution 2005; 133 455-465.

[99] Willey, N. Phytoremediation: Methods and Reviews. (Methods in Biotechnology, 23) Humana Press; 2007.

[100] Wang, Z., Xu, Y., Zhao, J., Li, F., Gao, D., Xing, B. Remediation of petroleum contaminated soils through composting and rhizosphere degradation. Journal of Hazardous Materials 2011; 190 677-685.

[101] Selberg, A., Budashova, J., Tenno, T. Column study of the leaching and degradation of anionic surfactants in oil-polluted soil. Proceeding of Estonian Academy of Science. Chemistry 2007; 56 87-97.

[102] Standard Methods for the Examination of Water and Wastewater. APHA, 1985, 16th Edition

[103] Salanitro, J.P., Diaz, L.A. Anaerobic biodegradability testing of surfactants. Chemosphere 1995; 30 813-830.

[104] Zhang, C., Valsaraj, K.T., Constant, W.D., Roy, D. Aerobic biodegradation kinetics of four anionic and non ionic surfactants at sub- and supra-critical micelle concentrations (CMCs). Water Research 1999; 33 115-224.

[105] Koga, M., Yamamichi, Y., Nomoto, Y. et al. Rapid determination of anionic surfactants by improved spectrophotometric method using Methylene Blue. Analytical Science 1999; 15 563-568.

[106] Hayachi, K. A rapid determination of sodium dodecyl sulfate with methylene blue. Analytical Biochemistry 1975; 67 503-506.

[107] n-Hexane Extractable Material (HEM) and Silica Gel Treated n-Hexane Extractable Material (SGT-HEM) by Extraction and Gravimetry (Oil and Grease and Total Petroleum Hydrocarbons). USEPA Method 1664. EPA-821-B-94-004b.

[108] Strotmann, U., Reuschenbach, P., Schwarz, H., Pagga, U. Development and evaluation of an online CO_2 evolution test and a multicomponent biodegradation test system. Applied and Environmental Microbiology 2004; 70 4621-4628.

[109] Reuschenbach, P., Pagga, U., Strotmann, U. A critical comparison of respirometric biodegradation tests based on OECD 301 and related test methods. Water Research 2003; 37 1517-1582.

[110] Binner, E., Böhm, K., Lehner, P. Large scale study on measurement of respiration activity (AT_4) by Sapromat and OxiTop. Waste Management 2012; 32 1752-1759.

[111] Selberg, A. Evaluation of environmental quality in Northern Estonia by the analysis of leachate. PhD Thesis. University of Tartu; 2010.

[112] Juuram, K. Assessment of the windrow composting of oil polluted soil and its enhancement. Master's Thesis. University of Tartu; 2012. (In Estonian, Summary in English)

http://dx.doi.org/10.5772/56908

[113] Kim, I.S., Park, J.-S., Kim, K.-W. Enhanced biodegradation of polycyclic aromatic hydrocarbons using non ionic surfactants in soil slurry. Applied Geochemistry 2001; 16 1419-1428.

Chapter 7

Biodegradation of the Organophosphate Pesticide Profenofos by Marine Fungi

Natália Alvarenga da Silva, Willian Garcia Birolli,
Mirna Helena Regali Seleghim and
André Luiz Meleiro Porto

Additional information is available at the end of the chapter

http://dx.doi.org/10.5772/56372

1. Introduction

Pesticides play an important role in modern agriculture. Synthetic pesticides are recognized as a cost-effective method of controlling pests, improving productivity and food quality. However, while pesticides may have a beneficial effect on agricultural productivity, their indiscriminate use causes many serious problems to the environment and human health, since these compounds are toxic to non-target species (Diez, 2010; Coutinho *et al.*, 2005).

The fate of pesticides in the environment is influenced by many processes (biological, chemical and physical) that determine their persistence and mobility (Gravilescu, 2005). Millions of tons of pesticides are applied annually, but it is believed that only a small fraction of these products effectively reaches the target organisms, and the remainder are deposited on the soil, contaminating non-target organisms and moving into the atmosphere and water (Eerd *et al.*, 2003). Since many pesticide types are recalcitrant, they remain for a long time in soils and sediments, where they can enter the food chain directly or percolate into the groundwater (Rissato *et al.*, 2004; Gravilescu, 2005).

Detoxification of pesticides *in situ* has been achieved by treatment of the contaminated soil with certain microorganisms or plants, a technology known as bioremediation or more specifically, phytoremediation in the case of plants (Sutherland *et al.*, 2002). These microorganisms are the main biological agents capable of removing and degrading waste materials, to enable their recycling in the environment (Chowdhury *et al.*, 2008). Since the conventional treatment options for the pesticide residues clean-up in the environment include removal of the contaminated material to be incinerated or disposed in landfills, *in situ* biological remediation is seen as a safer, less disruptive and more cost-effective alternative treatment (Sutherland *et al.*, 2004).

Effective techniques for soil bioremediation are bioaugmentation, biostimulation, phytoremediation and enzymatic bioremediation. However, the three first techniques are limited by their dependence upon the growth rate of the remediating plants and microbes, which will vary with nutrients, aeration, pH and other factors relating to the contaminated soil (Scott *et al.*, 2008; Sutherland *et al.*, 2004). A successful bioremediation technique requires efficient organisms that can degrade pollutant to a minimum level. In the case of pesticides, an adequate rate of biodegradation is required to attain the acceptable level of a pesticide residue or its metabolites at the contaminated site in a limited period of time (Singh, 2008).

Organophosphate pesticides (OPs) are used worldwide in agriculture, municipal hygiene, disease vector control and against household pests; they were also a group of compounds used historically as chemical warfare agents (Yang *et al.*, 2008; Zheng *et al.*, 2007; Edwards and Tchounwou, 2005). OPs are phosphorus-containing pesticides whose insecticidal qualities were first observed in Germany during World War II (Edwards and Tchounwou, 2005). The principal types are phosphotriesters, thiophosphotriesters, and phosphorothiolesters. Phosphotriesters contain a phosphate center with three *O*-linked groups, thiophosphotriesters have the phosphoryl oxygens replaced by sulfur and in phosphorothiolesters, one or more of the ester oxygen are replaced by sulfur (Figure 1) (Bigley and Raushel, 2013).

This chemical class of pesticides has been used to replace the organochlorine pesticides, banned in the United States since the 1970s (Jauregui *et al.*, 2003). However, the OPs are also highly toxic pesticides, since they are potent irreversible acetylcholinesterase (AChE) inhibitors that have a profound effect on the nervous system of exposed organisms, including human beings (Edwards and Tchounwou, 2005).

The hydrolysis mechanism normally catalyzed by AChE depends on the attack of a serine residue at the active site on the carbonyl group in ACh, but in the presence of organophosphates, this residue is readily phosphorylated, as follows: a histidine residue at the active site captures a proton from the serine residue, increasing its nucleophilic character, so that it readily attacks the electrophilic phosphorus atom, releasing the leaving group (X) (Figure 2). Unlike the acetylated enzyme, the phosphorylated enzyme reacts slowly with water, allowing the dealkylation of the alkoxy substituent (R_2) attached to the phosphorus atom. The organophosphate compounds thus inactivate acetylcholinesterase by phosphorylation of the serine at the enzyme active site. The result is the formation of a strong hydrogen bond between a protonated histidine residue of the catalytic site and the negatively charged oxygen atom of the inhibitor. Therefore, the protonated histidine cannot function as a general base catalyst for the hydrolysis of the phosphorylated enzyme, which is a necessary step for the reactivation of AChE (Figure 2) (Mileson *et al.*, 1998; Santos *et al.*, 2007)

According the Brazillian Food, Drug and Sanitary Surveillance Agency (ANVISA), analysis of pesticide waste in food showed that OPs are those with the greatest number of occurrences in unsatisfactory samples. Among then, chlorpyrifos, methamidophos and acephate are the main active ingredients responsible for food contamination. Profenofos appeared to be the 12th

http://dx.doi.org/10.5772/56372

APPENDIX

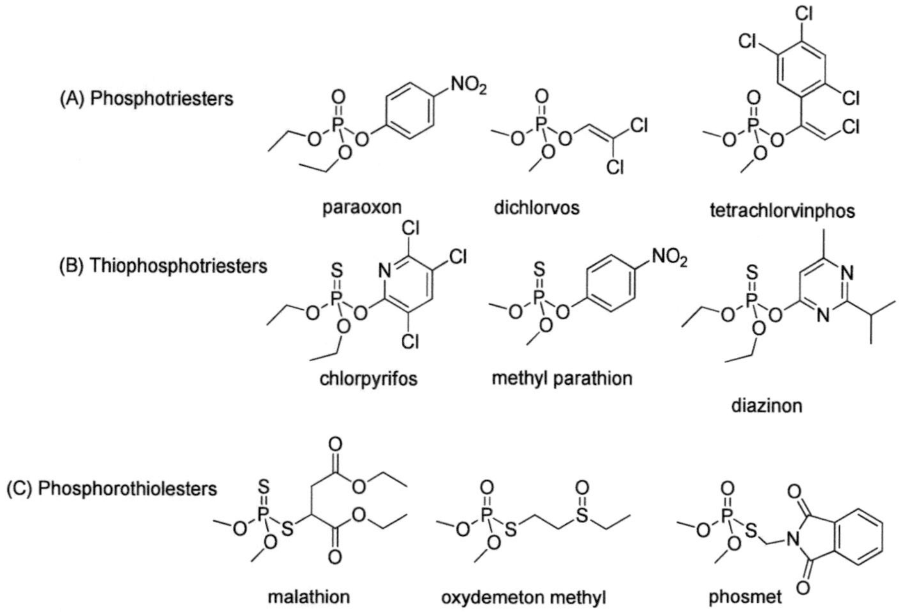

Figure 1. Chemical structures of main class of OPs. (A) Phosphotriesters; (B) Thiophosphotriesters; (C) Phosphorothiolesters

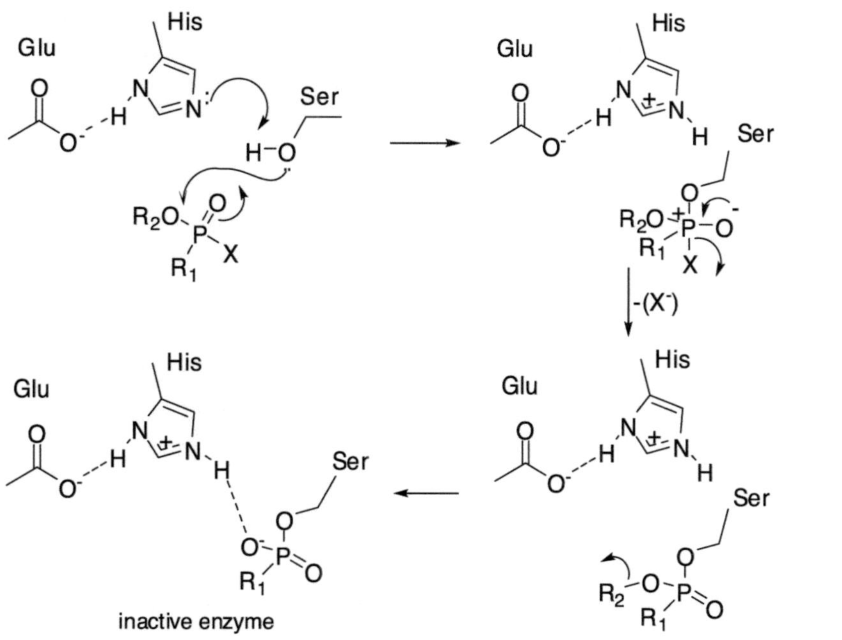

Figure 2. Mechanism of inhibition of acetylcholinesterase by organophosphate pesticides (Santos *et al.*, 2007)

commonest active ingredient in irregular samples of food, being found in samples of orange, strawberry and pepper (ANVISA, 2012).

Profenofos, *O*-(4-bromo-2-chlorophenyl) *O*-ethyl *S*-propyl phosphorothioate, is a broad spectrum, non-systemic foliar insecticide and acaricide. It is effective against a wide range of chewing and sucking insects and mites on various crops (Reddy and Rao, 2008). In the United States, profenofos is a "restricted use" pesticide sprayed only on cotton crops (McDaniel and Moser, 2004; EPA, 2012). However, in Brazil, this pesticide can also be used for foliar application on cotton, peanuts, potatoes, coffee, onions, peas, beans, green beans, watermelon, corn, cucumber, cabbage, soybean, tomato and wheat (ANVISA, 2011).

Classified as a moderately hazardous (Toxicity class II) pesticide by the World Health Organization (WHO) (Abass *et al.*, 2007; Malghani *et al.*, 2009), profenofos has a moderate order of acute toxicity following oral and dermal administration (McDaniel and Moser, 2004; Abass *et al.*, 2007). According to US Environmental Protection Agency (EPA), profenofos was first registered in the United States in 1982 and about 775,000 pounds (lbs.) of active ingredient are applied to cotton each year (EPA, 2012).

Chemical decontamination of organophosphates relies on bleach treatment, alkaline hydrolysis or incineration, but these conditions are harsh and the byproducts can be toxic (Ghanem and Raushel, 2005). Specific bioremediation of OPs requires highly specialized enzymes, so genetic engineering has been used to improve the properties of enzymes from various sources to enhance catalytic rates, stability and substrate range (Sutherland *et al.*, 2004).

A number of enzymes capable of detoxifying OPs have been discovered and the majority of them belong to the class of phosphotriesterases (PTE). Various PTEs have been identified: organophosphate hydrolase (OPH), methyl parathion hydrolase (MPH), organophosphorus acid anhydrolase (OPAA), diisopropylfluorophosphatase (DFP), and paraoxonase 1 (PON1) (Bigley and Raushel, 2013). All of these enzymes are found to promote the hydrolysis of organophosphate compounds. The most frequently cited enzyme in the literature, OPH, isolated from the bacteria *Pseudomonas diminuta* or *Flavobacterium* ATCC 27551, catalyzes the hydrolysis of a wide range of OP pesticides (Rogers, 1999; Chen and Mulchandani, 1988). Another enzyme reported involving the hydrolysis of OPs are carboxylesterases (CbEs), although the hydrolysis of OPs by PTEs is more efficient in the detoxification than the CbEs (Sogorb and Vilanova, 2002).

Zheng *et al.* showed that when OPH was coexpressed with CbE, the mixed enzymes degraded a variety of P-O bond containing OPs (chlorpyrifos, methyl parathion, dichlorvos and phoxim), whereas OPH had a very low catalytic activity for P-S bond containing OPs (malathion) (Figure 3). Thus, the hydrolase activities usually vary among structurally different OPs, ranging from the nearly diffusion-controlled limit for paraoxon to several orders of magnitude lower for phosphothiolesters, such as malathion (Zheng *et al.*, 2007).

Some degradation pathways are described in the literature for profenofos. The metabolic pathway of profenofos in cotton plants involves the cleavage of the phosphorothioate ester bond to yield 4-bromo-2-chlorophenol, followed by conjugation with glucose (Capps *et al.*, 1996). In the literature there are some cases of (bio)degradation of organophosphate pesticides

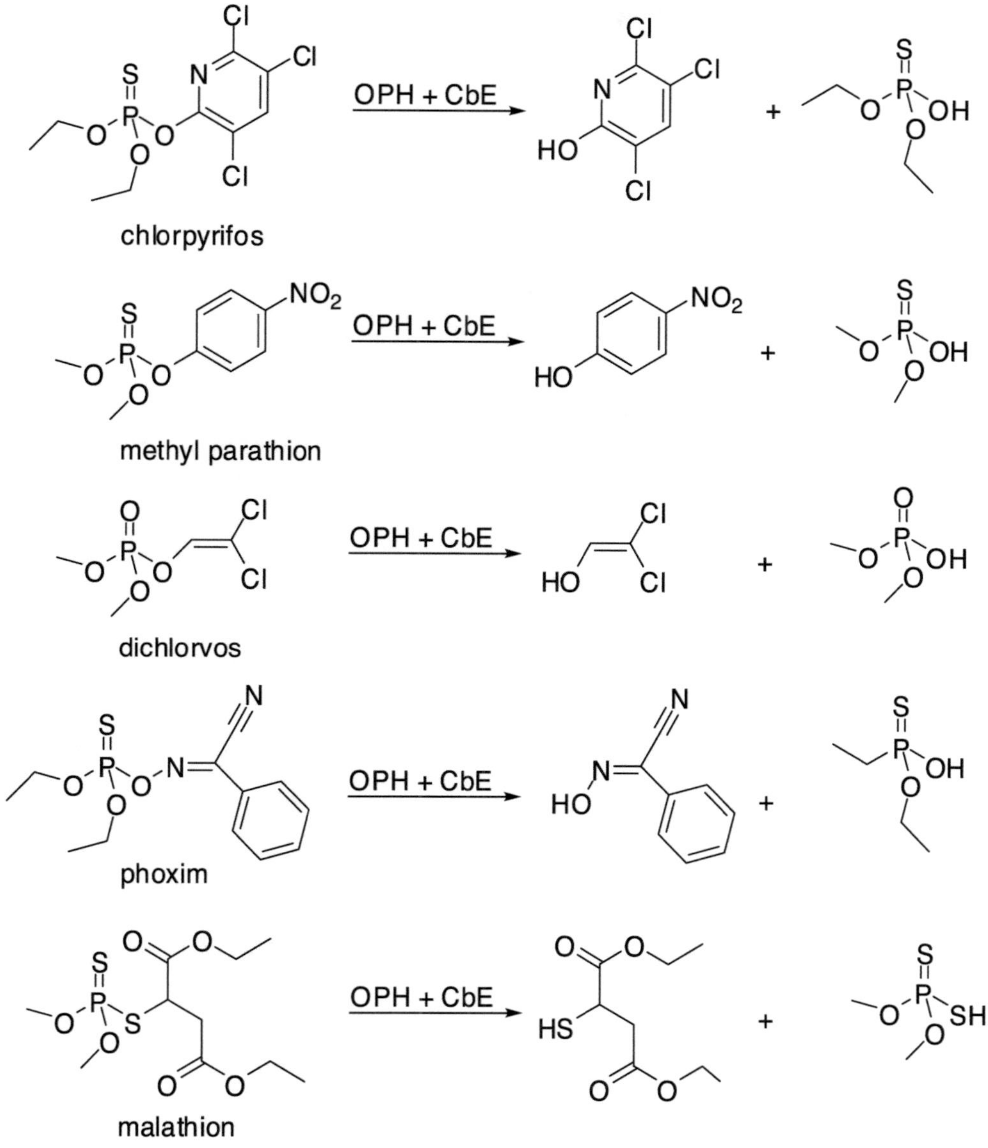

Figure 3. Hydrolysis of different OPs pesticides by OPH + CbE enzymes

that occur by subsequent biotransformations, yielding novel polar metabolites, such as glycosylated and sulfated derivatives. According to the Food and Agriculture Organization of the United Nations (FAO), in aerobic soil conditions, profenofos degraded rapidly, with mineralization and formation of unextracted residues. In sterilized soil, cleavage of the phenol-phosphorus ester bond in profenofos proceeded via chemical hydrolysis, with accumulation of 4-bromo-2-chlorophenol and formation of unextracted residues. The metabolic biotransformations of profenofos in plants and animals are similar and occur via hydrolysis to 4-

bromo-2-chlorophenol which is then conjugated by several enzymatic reactions (Figure 4) (FAO, 2012).

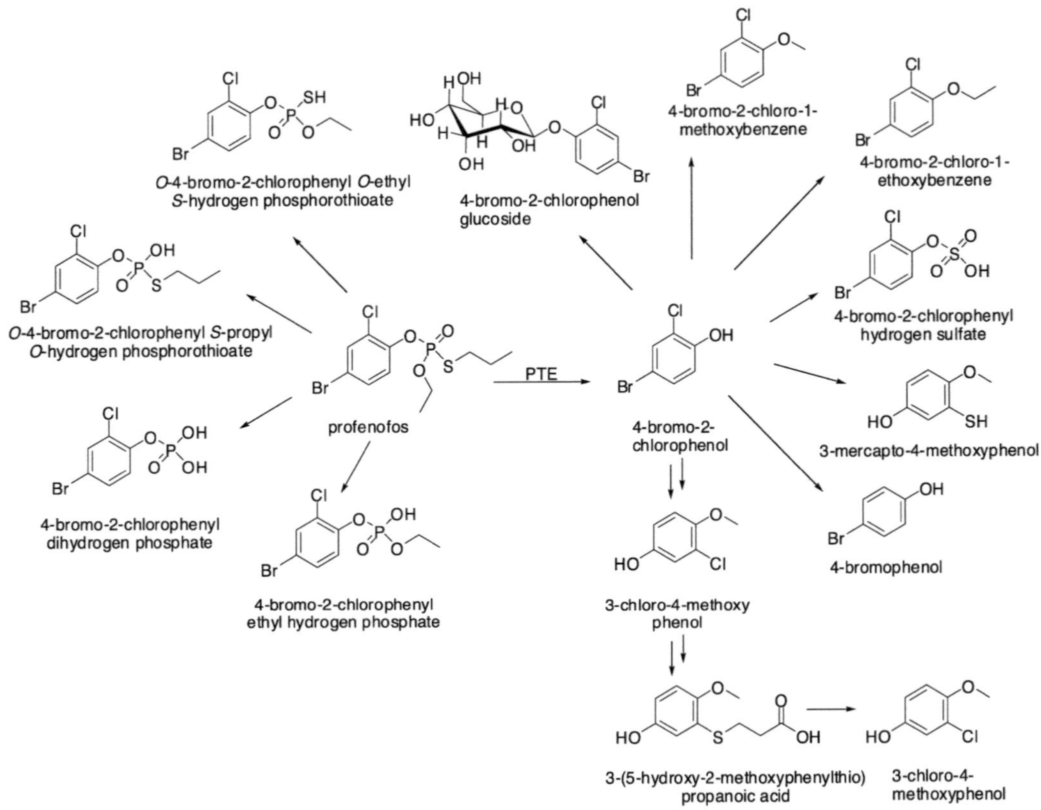

Figure 4. Proposed metabolic pathway of profenofos in soil, plant and enzymatic reactions

Fungi degrade a wide variety of compounds, a process known as mycodegradation. This process involves degradation to smaller molecules which may be toxic or non-toxic, as well as the removal of the pesticide molecule through a simple absorption or adsorption mechanism (Ramadevi *et al.*, 2012).

The ability of bacterial species to degrade organophosphates is well established and researches have even proposed possible degradation mechanisms for the OPs (Van Eerd *et al.*, 2003). However, the mechanisms of fungal degradation of these compounds are less established than those used by bacteria, since there are few studies on fungal degradation of OPs.

There are few studies on the biodegradation of profenofos by microorganisms. Malghani *et al.* reported the successful biodegradation of this compound by bacteria, and at the time of writing, the author stated that no studies on bacterial degradation of profenofos had been reported earlier

(Malghani *et al.*, 2009). Filamentous fungi of the genus *Aspergillus* have been used in the biodegradation of OPs. For instance, *Aspergillus niger* showed high biodegradation of malathion pesticide (Ramadevi *et al.*, 2012), *Aspergillus flavus* and *Aspergillus sydowii* were capable of degrading pirimiphos-methyl, pyrazophos and malathion, even at high concentrations (1,000 ppm), utilizing these compounds as sole phosphorus and carbon sources, releasing the phosphorus moiety from these pesticides by means of their phosphatases (Hasan, 1999).

Marine enzymes have a great potential for use in biocatalytic reactions, as in biodegradation of pesticides, due to the peculiar characteristics of the marine environment. As the sea covers more than three quarters of the Earth's surface and provide abundant resources for biotechnological research and development (Rush *et al.*, 2007), marine organisms offer a dramatically different environment for the biosynthesis of molecules than terrestrial organisms, and are a vast untapped source of enzymes (Venter *et al.*, 2004; Venter *et al.*, 2010). In recent years, a variety of new enzymes with specific activities have been isolated from bacteria, fungi and other marine organisms; moreover, some can produce a considerable number of molecules with potential to be transformed into commercial drugs (Ghosh *et al.*, 2005; Haefner, 2003). In fact, the marine environment is a very rich source of extremely potent compounds exhibiting significant activities in anti-tumor, anti-inflammatory, analgesic, immunomodulatory, allergic and anti-viral assays (Newman and Cragg, 2004; San-Martín *et al.*, 2008).

Marine organisms in general (fungi, bacteria, algae, sponges, fish, prawns and other crustaceans) can be rich sources of novel enzymes, but most of the current bioprospecting activity focuses on microbial ones. A marine enzyme is a protein molecule with unique properties as it is derived from an organism whose natural habitat is saline or brackish water (Trincone, 2010; Sarkar *et al.*, 2010). These enzymes can be biocatalysts with properties such as high salt tolerance, hyperthermostability, barophilicity and cold adaptability. Microorganisms isolated from ocean sediment and seawater are the most widely studied sources of marine enzymes, especially proteases, carbohydrases and peroxidases (Ghosh *et al.*, 2005).

Enzymatic reactions catalyzed by marine fungi can be used when the fungi are cultured in media based on artificial seawater. The filamentous marine fungi *Aspergillus sydowii* CBMAI 933, *Penicillium raistrickii* CBMAI 931, *Penicillium miczynskii* CBMAI 930 and *Trichoderma* sp. CBMAI 932, grown in artificial seawater were able to catalyze the hydrolysis of benzyl glycidyl ether (Martins *et al.*, 2011). Similar results were observed in a study of ligninolytic enzyme production by the marine fungi *Aspergillus sclerotiorum* CBMAI 849, *Cladosporium cladosporioides* CBMAI 857 and *Mucor racemosus* CBMAI 847 (Bonugli-Santos *et al.*, 2010). Other studies have shown that marine bacteria and fungi cultured in the laboratory have specific requirements for salts, especially sodium, potassium, magnesium and chloride ions (Martins *et al.*, 2011; MacLeod, 1965; Kogure, 1998; Rocha *et al.*, 2009).

In this chapter, the first results obtained in the biodegradation of profenofos by whole cells of marine fungi are presented. Marine fungi were selected by us, with high potential to biodegrade profenofos and its main metabolite. The results presented in this chapter explore the potential of marine fungi in biotransformation and biodegradation of a xenobiotic (pesticide profenofos). The fungal biodegradation of OPs is still underexplored by researches, especially with regard to the biodegradation of profenofos, making this work extremely relevant. The main objective of this study was the screening of Brazilian marine fungi with the enzymes required for detoxification of organophosphate pesticides (phosphotriesterases-PTEs and /or carboxylesterases-CBEs). The biodegradation of profenofos in the presence of these selected fungi was evaluated, assessing the degradation of the pesticide, as well as the formation of the metabolite, 4-bromo-2-chlorophenol. This results are environmentally important, because the pesticides applied to crops can be leached into rivers, lakes and seas under these different conditions, where they may suffer different biodegradation processes.

2. Materials and methods

2.1. General

Ethyl acetate (PA), used to extract the reaction mixtures and the salts used to prepare artificial sea water were purchased from a commercial source (Synth, Vetec, Brazil). Ethyl acetate (HPLC grade) for the analytical curve was purchased from a commercial source (Tedia, Rio de Janeiro, Brazil). The malt extract and agar used in solid and liquid culture media were purchased from commercial sources (Acumedia and Himedia, Brazil).

2.2. Pesticides

The analytical standards of chlorpyrifos and profenofos were purchased from Sigma-Aldrich, Brazil. Commercial pesticide containing profenofos was purchased from Syngenta® under the name Polytrin 400/40 CE. The commercial profenofos used in the marine fungi biodegradation test was donated by Professor Marcos R. de V. Lanza (IQSC-USP). The 4-bromo-2-chlorophenol was purchased from Sigma-Aldrich, Brazil.

2.3. Marine fungi

The Brazilian marine-derived fungal strains *Aspergillus sydowii*-CBMAI 934, *Aspergillus sydowii*-CBMAI 935 and *Penicillium raistrickii* CBMAI 931 were isolated from the sponge *Chelonaplysilla erecta; Aspergillus sydowii* CBMAI 1241, *Penicillium decaturense* CBMAI 1234 and *Penicillium raistrickii* CBMAI 1235 were isolated from the sponge *Dragmacidon reticulata; Trichoderma* sp. CBMAI 932 was isolated from the sponge *Geodia corticostylifera*. The sponges were collected in the South Atlantic Ocean at São Sebastião in São Paulo state, Brazil, by Professor Roberto G. S. Berlinck (Chemistry Institute of Sao Carlos, University of São Paulo, IQSC-USP). The marine fungi were isolated and purified in the microbiology laboratory of the Department of Ecology and Evolutionary Biology supervised by Professor Mirna H. R. Seleghim (UFSCar-Brazil). The fungi were identified by both conventional and molecular methods at the Chemical, Biological and Agricultural Multidisciplinary Research Center at UNICAMP, São Paulo, Brazil (http://www.cpqba.unicamp.br/). The isolated and identified marine fungi were deposited in the Brazilian Collection of Environmental and Industrial Microorganisms (CBMAI).

2.4. Composition of marine fungi growth media

Composition of Artificial Sea Water (ASW) (1L): $CaCl_2.2H_2O$ (1.36g), $MgCl_2.6H_2O$ (9.68g), KCl (0.61g), NaCl (30.0g), Na_2HPO_4 (0.014 mg), Na_2SO_4 (3.47g), $NaHCO_3$ (0.17g), KBr (0.1g), $SrCl_2.6H_2O$ (0.040g), H_3BO_3 (0.030g).

Solid medium for stock cultures: agar (20 g. L^{-1}) and malt extract (20 g.L^{-1}) in ASW (1L) and adjusted to pH 8 by addition of 3M KOH.

Solid medium for fungal screening: agar (20 g. L^{-1}) and malt extract (20 g.L^{-1}) dissolved in ASW (1L) and adjusted to pH 5 by addition of 3M KOH or 1M HCl.

Liquid medium: malt extract (20 g.L^{-1}) in ASW (1L), adjusted to pH 7 by addition of 3M KOH or 1M HCl.

Liquid mineral medium supplemented with KNO_3(12.5 ppm): KNO_3 (12.5 mg.L^{-1}) dissolved in ASW (1L), adjusted to pH 7 by addition of 3M KOH or 1M HCl.

The culture media were sterilized in autoclave for 20 minutes (at 121 °C, 1.5 kPa). All manipulations involving marine fungi were carried out under sterile conditions in a Veco laminar flow cabinet. The stock cultures of the marine microorganisms were stored on solid culture medium (25 mL), in Petri dishes, maintained at 4°C in the refrigerator.

2.5. Cultivation of marine fungi on solid medium in the presence of profenofos

Marine fungi were screened by culturing on Petri dishes containing 25 mL of solid culture medium (2.0 g of malt extract, 2.0 g of agar and 100 mL of ASW) with the addition of profenofos and without (control culture). After the medium sterilization in the autoclave, the agar was cooled to 40-45°C and the profenofos was added at three different concentrations: 5.0, 10.0 and 15.0 μL per plate, solubilized in 100.0, 200.0 and 300.0 μL of dimethyl sulfoxide (DMSO), respectively. At room temperature, fungal mycelia from recent cultures were transferred to the surfaces of the agar plates with an inoculating loop. The fungi were incubated for 10 days at 35°C. Tolerance of profenofos was estimated by the size of the colony formed on the surface of the plates, relative to the control culture.

2.6. Analytical curve

Stock solutions of 500.0 ppm of profenofos, 4-bromo-2-chlorophenol (main metabolite) and chlorpyrifos (used as internal standard) were prepared.

Profenofos 500.0 ppm: 3.4 μL (1.3 mmol) of profenofos analytical standard and ethyl acetate (10.0 mL).

4-bromo-2-chlorophenol 500.0 ppm: 5.0 mg (2.4 mol) of 4-bromo-2-chlorophenol and ethyl acetate (10.0 mL).

Chlorpyrifos 500.0 ppm: 5.0 mg (1.4 mol) of chlorpyrifos and ethyl acetate (10.0 mL).

All standard solutions were prepared in a volumetric flask and made up to the containing 10.0 mL mark with ethyl acetate (HPLC grade). From these stock solutions were prepared the working solutions, at concentrations of 5.0, 10.0, 20.0, 30.0 and 50.0 ppm of profenofos and 4-bromo-2-chlorophenol, in ethyl acetate (HPLC grade). The concentration of internal standard, chlorpyrifos, was maintained at 30.0 ppm in all assays. Next, 1.0 mL aliquots from the stock solutions were transferred into 1.5 mL vials. Triplicates samples, for each concentration of analyte, were analyzed by GC-MS-SIM.

2.7. Determination of profenofos concentration in commercial sample

The sample was prepared with 20.0 μL of commercial pesticide profenofos and 60.0 μL (2.4 mmol, 30.0 ppm internal standard) of stock solution of chlorpyrifos 500.0 ppm in ethyl acetate (HPLC grade), in a 100.0 mL volumetric flask. Duplicate samples of 1.0 mL were prepared and analyzed by GC-MS-SIM.

2.8. Biodegradation of profenofos by *A. sydowii* CBMAI 935 and *P. raistrickii* CBMAI 931 in liquid medium

The fungi *A. sydowii* CBMAI 935 and *P. raistrickii* CBMAI 931 were inoculated in solid culture medium (2.0 g of malt extract, 2.0 g of agar) in the presence of profenofos (50.0 ppm) and incubated for 7 days at 35°C. Two small slices of solid medium (1.2 cm x 1.2 cm) bearing the marine fungal mycelia were transferred to 250 mL Erlenmeyer flasks containing liquid medium (2.0 g of malt extract and 100.0 mL of ASW, pH 7). The liquid medium containing the inocula were incubated in an orbital shaker (Technal TE-421 or Superohm G-25) for 4 days (130 rpm, 32°C). After that, 50.0 ppm (18.6 μL, Section 2.6) of commercial profenofos was added. The reaction was incubated in the orbital shaker for 30 days (130 rpm, 32°C).

Extractions were performed at 10, 20 and 30 days. The reaction culture was filtered on a Buchner funnel to separate the mycelia from liquid medium. The mycelial mass obtained was rinsed and suspended in water and ethyl acetate (1:1). The mixture was stirred magnetically for 30 minutes and filtered again using Buchner funnel.

For the 10-days reaction, the extractions were then analyzed separately on the first mycelial extract and filtered medium. The liquid medium and mycelial extract were acidified to pH 6 and extracted separately, three times with ethyl acetate (3 x 25 mL) (Table 3).

For reactions at 20 and 30 days, the liquid medium and mycelial extract (after the Buchner filtration and extraction of mycelia with water and ethyl acetate, 1:1) were put together in an Erlenmeyer flask, acidified to pH 6 and extracted three times with ethyl acetate (3 x 25 mL). The filtered mycelial cells were dried in an oven (35 °C, 24 h) and then weighed (Tables 4-5).

After extractions, the organic phase was dried over anhydrous Na_2SO_4, followed by solvent filtration and evaporation, resulting in a final volume of 100.0 mL. The residual (no degraded) profenofos and the 4-bromo-2-chlorophenol released were analyzed by a gas cromatography coupled to a mass spectrometer in single-ion monitoring mode (GC-MS-SIM). Under these conditions, the concentration of pesticide and metabolite were determined by comparing the peak area of the samples with an analytical curve. The biodegradation results are summarized in Tables 6-10. Further degradation and growth experiments were performed, to test some parameters (Sections 2.8.1-2.8.4).

2.8.1. Biodegradation of 4-bromo-2-chlorophenol by A. sydowii CBMAI 935 and P. raistrickii CBMAI 931 in liquid medium

These reactions were prepared in 250-mL Erlenmeyer flasks containing 100.0 mL of liquid medium at pH 7 (Section 2.8) in which the 4-bromo-2-chlorophenol (50.0 ppm, 2.4 mmol) was added. The reaction was incubated in an orbital shaker for 30 days (130 rpm, 32°C). The inoculations, extractions and analyses proceeded as described in Section 2.8. The results are summarized in Table 6.

2.8.2. Biodegradation of profenofos by A. sydowii CBMAI 935 and P. raistrickii CBMAI 931 in liquid minimal medium supplemented with KNO_3

Two small slices of solid medium (1.2 cm x 1.2 cm) bearing the marine fungal mycelia were transferred to 250 mL Erlenmeyer flasks containing 100 mL of liquid mineral medium (1.25 g of KNO_3 and 100.0 mL of ASW, pH 7), previously sterilized in autoclave for 20 minutes at 121°C. Next, 100.0 ppm (37.2 μL) of commercial profenofos was added to the medium. The reaction was incubated in an orbital shaker for 30 days (130 rpm, 32°C). The extractions and analyses proceeded as described in Section 2.8. The results are summarized in Table 7.

2.8.3. Degradation of profenofos in the absence of marine fungi

To a 250 mL Erlenmeyer flask containing liquid medium (2.0 g of malt extract and 100.0 mL of ASW, pH 7), previously sterilized in the autoclave for 20 minutes at 121°C, 50.0 ppm (18.6 μL, Section 2.5) of commercial profenofos was added. The reaction was incubated in an orbital shaker for 30 days (130 rpm, 32°C). The extractions and analyses proceeded as described in Section 2.8. The results are summarized in Table 8.

2.8.4. Growth of marine fungi A. sydowii CBMAI 935 and P. raistrickii CBMAI 931 in the absence of profenofos

In a 250 mL Erlenmeyer flasks containing liquid medium (2.0 g of malt extract and 100.0 mL of ASW, pH 7), previously sterilized in the autoclave for 20 minutes at 121°C, two small slices of solid medium (1.2 cm x 1.2 cm) bearing the marine fungi was added, without any profenofos. The culture medium was incubated in an orbital shaker for 30 days (130 rpm, 32°C). The extractions and analyses proceeded as described in Section 2.8. The results are summarized in Table 9.

2.9. Biodegradation of the pesticide profenofos at various concentrations by *P. raistrickii* CBMAI 931

In four 250 mL Erlenmeyer flasks containing liquid medium (2.0 g of malt extract and 100.0 mL of ASW, pH 7), previously sterilized in the autoclave for 20 minutes at 121°C, commercial profenofos was added separately at 15.0 ppm (18.6 μL), 30 ppm (11.2 μL), 50.0 ppm (18.6 μL) and 60.0 ppm (22.3 μL). In the flasks were inoculated two small slices of solid medium (1.2 cm x 1.2 cm) bearing mycelium of *P. raistrickii* CBMAI 931. These reactions were incubated in an orbital shaker for 30 days (130 rpm, 32°C). The extractions and analyses proceeded as in Section 2.8. The results are summarized in Table 10.

2.10. GC-MS analyses

The GC-MS system was a Shimadzu GC2010plus gas chromatograph coupled to a mass-selective detector (ShimadzuMS2010plus) in electron ionization (EI, 70 eV) mode. The GC-MS oven was fitted with a DB5 fused silica column (J&W Scientific 30m x 0.25mm x 0.25 μm). The chromatographic conditions were: initial oven temperature 100 °C (for 5 min), increased to 250 °C (for 10 min) at 5 °C/min; run time 45.0 min; injector temperature 200 °C; detector temperature 200 °C; injector split ratio 1:1; helium carrier gas at a pressure of 60 kPa. The analytes were first analyzed in SCAN mode in order to select the ion and the retention time for each compound. The selected-ion mode (SIM) analyses were performed to measure the biodegradation of profenofos. Table 1 shows the retention time and selected ion for each compound, used in the SIM-mode analyses.

Compounds	Retention time (min)	Selected ion (*m/z*)
4-Bromo-2-chlorophenol	10.605	207.85
Chlorpyrifos	27.730	313.90
Profenofos	31.190	138.95

Table 1. Method data of the SIM-mode analyses.

3. Results and discussion

3.1. Screening marine fungi on solid medium

The strains studied were the filamentous marine fungi *Aspergillus sydowii* CBMAI 934, *Aspergillus sydowii* CBMAI 935, *Aspergillus sydowii* CBMAI 1241, *Penicillium decaturense* CBMAI 1234, *Penicillium raistrickii* CBMAI 931, *Penicillium raistrickii* CBMAI 1235 and *Trichoderma* sp. CBMAI 932. These are multicellular microorganisms, which grow as mycelia, composed by branching microscopic filament named hyphae. Fungi were grown on solid medium at pH 5, which is a good pH for the cultivation of most fungi, while the optimum may vary from 3.8 to 5.6. These pH values favor fungi growth and inhibit growth of most bacteria, which optimal culture condition is at higher pH (Pelczar *et al.*, 1997).

Initially, the biotransformation of profenofos by marine fungi was conducted on solid culture media. The microorganisms were grown on Petri dishes containing 2% malt extract and artificial seawater (ASW). All the strains investigated were analyzed in the presence and absence of the profenofos pesticide, in duplicate tests. Fungi with biocatalytic potential to degrade profenofos were screened by comparing the growth of fungal colonies on Petri dishes at several concentrations of the pesticide and in its absence (control). Volumes of profenofos added to the solid cultures were 5.0, 10.0 and 15.0 μL per Petri dish, corresponding to concentrations of 80.0, 160.0 and 240.0 ppm, respectively (Table 2).

After 10 days of growth at 35 °C, the colony diameters were measured and the average diameter (cm) of the colonies formed on each Petri dish was recorded. Since most of the colonies showed non-circular radial growth (Figure 5), they were measured between the furthest points. Figure 5 summarizes the qualitative results of the marine fungi growth on solid culture media in the absence and presence of profenofos, for the strains which growth-better.

When several colonies grow in a Petri dish, one colony can compete and/or inhibit the growth of another. In this experiment on solid medium, it was important to assess fungal growth on the plate surface to detect the presence or absence of microbial growth. However, the measurement of colonies had no quantitative purpose, and the test was done only to estimate the fungal growth.

Marine fungi	Colony diameter (cm)			
	Control culture	80.0 ppm[a]	160.0 ppm[b]	240.0 ppm[c]
Aspergillus sydowii CBMAI 934	4.0 x 3.0	1.0 x 1.5	1.0 x 1.0	1.0 x 1.0
Aspergillus sydowii CBMAI 935	Whole plate*	3.0 x 2.5	3.0 x 2.5	3.0 x 2.5
Aspergillus sydowii CBMAI 1241	3.5 x 2.5	3.0 x 2.5	3.0 x 2.5	2.0 x 2.0
Penicillium decaturense CBMAI 1234	2.5 x 2.5	1.0 x 1.5*	3.0 x 2.0	1.0 x 1.0*
Penicillium raistrickii CBMAI 931	3.5 x 3.0	3.0 x 3.0	3.0 x 2.5	2.5 x 2.5
Penicillium raistrickii CBMAI 1235	4.0 x 3.0	2.0 x 2.0	1.0 x 1.0	1.5 x 1.0
Trichoderma sp. CBMAI 932	Whole plate*	3.5 x 2.5	3.0 x 2.0	2.0 x 2.0

*Estimated measure, because the number of spores did not allow observation of the set of colonies
[a]5.0 μL profenofos and 100.0 μL DMSO
[b]10.0 μL profenofos and 200.0 μL DMSO
[c]15.0 μL profenofos and 300.0 μL DMSO

Table 2. Growth of marine fungi on solid agar medium of malt extract 2% with absence and addition of profenofos (35 °C, 10 days, pH 5).

Fungal development and growth requires a variety of inorganic and organic nutrients in the medium. Carbon is one of the most important elements for microbial growth, as carbon compounds provide energy for cell growth and serve as the basic units to build the cell materials. Nitrogen is also essential to the organisms, as well as other elements (hydrogen, oxygen and phosphorus) (Pelczar *et al.*, 1997). Thus, fungal growth in the presence of pesticides may indicate fungal tolerance to the pesticide toxicity; pesticide metabolism as a mechanism of defense of the microorganism to eliminate the xenobiotic compound; or even pesticide use as a source of nutrient for fungal growth, since the organophosphate pesticide profenofos has carbon, oxygen, sulfur and phosphorus in its structure.

In the screening of fungal strains on solid medium, in the presence of profenofos, excepting by the marine fungi *A. sydowii* CBMAI 934 and *P. raistrickii* CBMAI 1235, all other microorganisms (*P. raistrickii* CBMAI 931, *A. sydowii* CBMAI 935, *A. sydowii* CBMAI 1241 and *Tricho-*

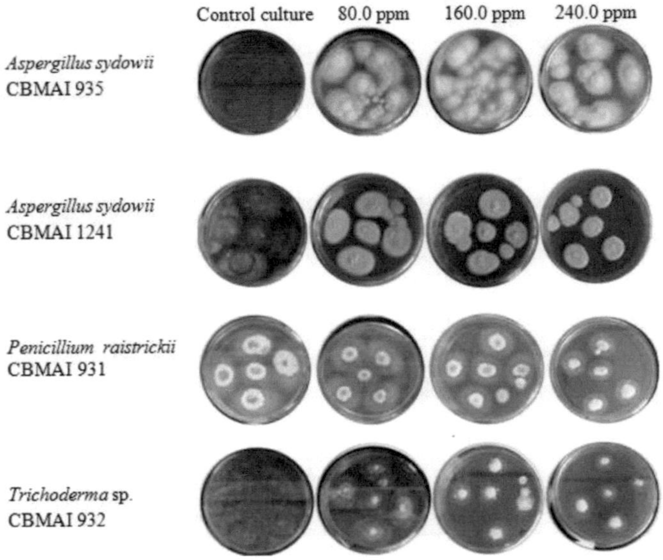

Figure 5. Marine fungi growing on solid culture medium containing various concentrations of profenofos pesticide (10 days at 35°C)

derma sp. CBMAI 932) showed excellent growth in the presence of the pesticide at all concentrations after 10 days, as shown in Table 2. Compared to the control culture, there was a slight inhibition of the cultures by the largest amount of pesticide (15.0 μL profenofos) (Figure 5). Since fungal growth was satisfactory in the highest concentration of the pesticide, it was possible to suggest that these strains showed good potential for biocatalytic degradation of profenofos.

There was a difference between the growth on the plate with 10.0 μL of pesticide and the other amounts, for the fungus *P. decaturense* CBMAI 1234. In this Petri dish, there was no sporulation and colonies were well defined, with no significant inhibition of growth in comparison with the control plate. Strains of *A. sydowii* CBMAI 934 and *P. raistrickii* CBMAI 1235 showed only a slight growth, compared to the other fungi. However, these last three fungi were able to grow even at higher concentrations of the pesticide (Table 3).

Finally, after this screening on solid culture medium, the strains of *P. raistrickii* CBMAI 931 and *A. sydowii* CBMAI 935 were selected to investigate and quantify the biodegradation of profenofos in liquid culture medium.

3.2. Analytical curves to determine the concentration of profenofos and 4-bromo-2-chlorophenol by GC-MS-SIM analysis

The OPs are particularly amenable to biodegradation because they are susceptible to hydrolysis by enzymes (Chen and Mulchandani, 1998). The best known enzymes that promote hydrolysis of OPs are phosphotriesterases (Ghanem and Raushel, 2005). The expected metabolite from hydrolysis of profenofos is 4-bromo-2-chlorophenol. Therefore, if this metabolite is a product of the mycelial reaction, enzymes were possibly active in the mycelial mass of the marine fungi *P. raistrickii* CBMAI 931 and *A. sydowii* CBMAI 935. Analytical curves by the internal standard method of GC-MS-SIM analysis were constructed in order to determine the concentration of the active ingredient profenofos in the commercial pesticide, the expected degradation product (4-bromo-2-chlorophenol) and the residual profenofos during the reaction. By analyzing, in the SCAN mode, a sample containing 4-bromo-2-chlorophenol (metabolite of profenofos), chlorpyrifos (internal standard), the ion selected and retention time for each compound was determined (Figure 6, Table 1). For 4-bromo-2-chlorophenol (*m/z* 207.85) and chlorpyrifos (*m/z* 313.90), the molecular ions of each analyte was selected, while the base peak ion (*m/z* 138.95) was selected for profenofos. All samples were analyzed in SIM mode for quantification measurements and SCAN mode in the mass range of 50-550 u.m.a. for confirmation of the molecules identities.

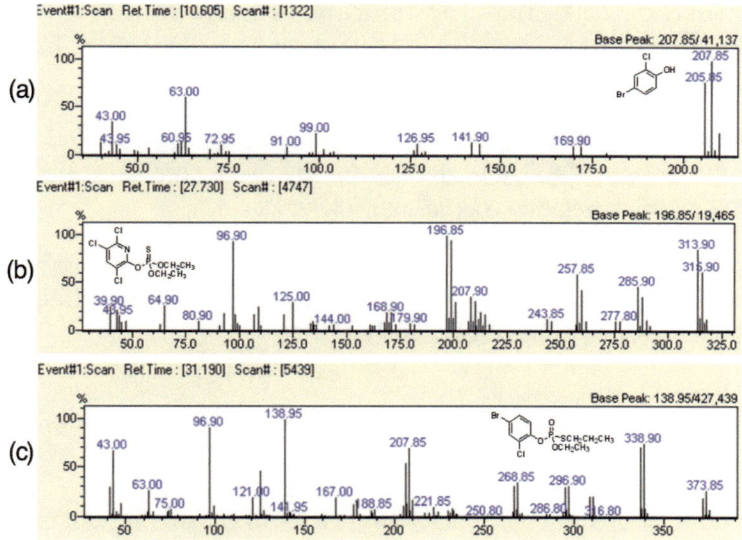

Figure 6. Mass spectra for analyses of the fragmentation patterns to select ions in the SCAN mode. (a) 4-Bromo-2-chlorophenol, (b) Chlorpyrifos (c) Profenofos

The internal standard technique is a useful method for minimizing errors due to variations in the used equipments. A substance used as an internal standard should be similar to the analyte, with a similar retention time, not react with another substance or matrix component, not be a part of the test sample and have a retention time different from those of the other substances in the sample (Ribani *et al.*, 2004). The pesticide chlorpyrifos (analytical grade) was used as the internal standard for the determination of profenofos and its metabolite. A graph was produced, the area ratio (area of the substance / area of the internal standard) versus the

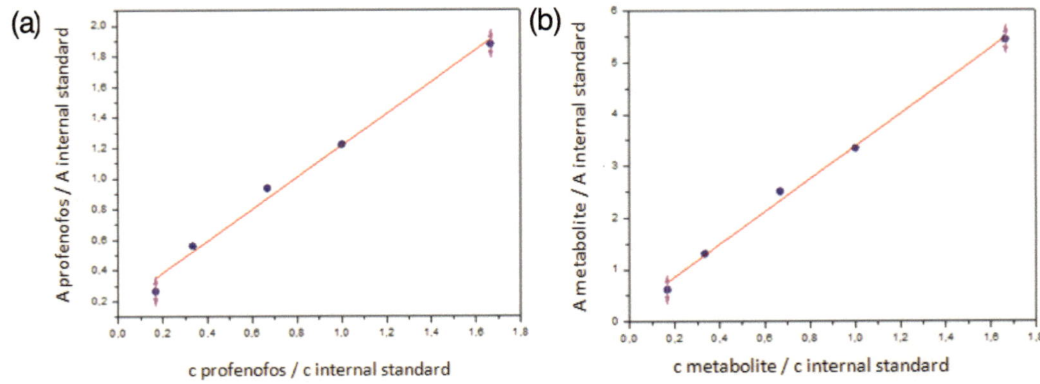

Figure 7. Analytical curves for (a) profenofos, (b) 4-bromo-2-chlorophenol

concentration ratio (variable concentration of substance / constant concentration of the internal standard) (Ribani *et al.*, 2004). This analytical curve was constructed for profenofos and 4-bromo-2-chlorophenol (metabolite of profenofos) at concentrations of 5.0, 10.0, 15.0, 20.0, 30.0 and 50.0 ppm (Figure 7).

The analytical curve for profenofos fitted by the linear equation $y = 1.03965\,x + 0.17522$, with correlation coefficient $r = 0.9955$, and the one for the metabolite was fitted by the line $y = 3.15088x + 0.2272$, with correlation coefficient $r = 0.99811$.

The Brazil´s regulatory agency ANVISA recommends a correlation coefficient of 0.99; thus, the correlation coefficients obtained for the two analytical curves are within the parameters established in the literature (Ribani *et al.*, 2004).

3.3. Determination of the active ingredients concentration in the profenofos commercial sample

According to information from Syngenta® (Syngenta, 2012), the composition of the pesticide Polytrin, used in this study, was:

Inert ingredients: 560.00 g. L^{-1} (56.0% w/v)

Profenofos: 400.00 g. L^{-1} (40% w/v)

Cypermethrin: 40.00 g. L^{-1} (4% w/v)

The amount of active ingredient present in the working sample was measured, in order to develop reactions of biodegradation with the commercial sample of profenofos. To determine the volume of pesticide profenofos required to give a concentration of 50.0 ppm in the reaction, analyses were performed in duplicate with an arbitrary amount of pesticide (20.0 μL). The analytical data yielded 54.0 ppm for the concentration of active ingredient in 100 mL of medium.

The results were in good agreement and showed that, to obtain a final concentration of 50.0 ppm in 100.0 mL of liquid culture medium, 18.6 μL of commercial profenofos must be added. According to these data, the total concentration of active ingredient (profenofos) in the working sample was approximately 320.0 $g.L^{-1}$.

3.4. Biodegradation of profenofos by marine fungi *P. raistrickii* CBMAI 931 and *A. sydowii* CBMAI 935

The inocula used for the profenofos biodegradation reactions were activated in Petri dishes containing 2% of malt extract solid medium and 50.0 ppm of the pesticide, in order to induce the production of phosphotriesterases or other enzyme classes (e.g., CbE = carboxylesterase) capable of degrading the OP. Enzyme induction occurs at the gene transcription level. Gene transcription is the first step in the flow of genetic information and, for this reason, gene expression is relatively easily affected at this point (Madigan *et al.*, 2010; Tortora *et al.*, 1995). An inducible enzyme is synthesized only when its substrate is present in the sample; hence, the inoculum was grown in the presence of profenofos, so that the enzymes of interest were already being expressed when the fungi were transferred to the liquid medium, to catalyze the biodegradation reactions.

The pH of the liquid medium was adjusted to 7, bearing to reports in the literature indicating that phosphotriesterases exhibit enhanced catalytic activity at neutral to basic pH. According to Eivazi and Tabatabal, hydrolysis of the pesticide paraoxon with animal enzymes showed good catalytic activity at pH 7.3. Assays of activity by the release of *p*-nitrophenol showed optimal activity at pH 7-11 (Eivazi and Tabatabai, 1977). The hydrolysis of organophosphates in the environment (in the absence of enzymes) is also affected by pH: the more alkaline the medium, the faster is the hydrolysis. According to Zamy *et al.* the half-life of profenofos in phosphate buffer at pH 8 is fifteen days, the pesticide being hydrolyzed in this buffer. Thus, the period in which the pesticide should be completely biodegraded by hydrolysis is considerable; however, the presence of enzymes would accelerate the process of degradation (Zamy *et al.*, 2004).

The step of mycelium extraction was important because the fungi, as well as bacteria, can absorb compounds with the aid of enzymes secreted into medium, which break or carry the complex organic molecules into the cells (Pelczar *et al.*, 1997). Thus, the extraction with magnetic stirring was used to extract both the pesticide that may be inside the mycelium (since this extraction causes the cell disruption) and adhered to surface of the cell membrane.

3.4.1. Biodegradation of profenofos by marine fungi P. raistrickii CBMAI 931 and A. sydowii CBMAI 935 after 10 days of reaction

At 10 days of reaction, the extracts of the mycelium and liquid medium were subjected to separate analysis by GC-MS-SIM. In Figures 8 and 9 chromatograms of each extraction are shown, with the analyses of the superimposed duplicates. The data concerning biodegradation of profenofos by fungal strains of *A. sydowii* CBMAI 935 and *P. raistrickii* CBMAI 931 are summarized in Table 3.

The duplicate reactions in the experiments with the fungus *A. sydowii* CBMAI 935 agreed well with each other: 53.0% and 45.0% of profenofos degraded. However, the duplicate reactions with *P. raistrickii* CBMAI 931, showed a significant discordance between for each reaction (81.4% and 48.0% degraded). As the growth behavior of the fungus varies for each experiment, other factors may have interfered and caused the difference between the results. It should also be noted that although the reactions are performed in duplicate, they occur independently and are, therefore, unique reactions.

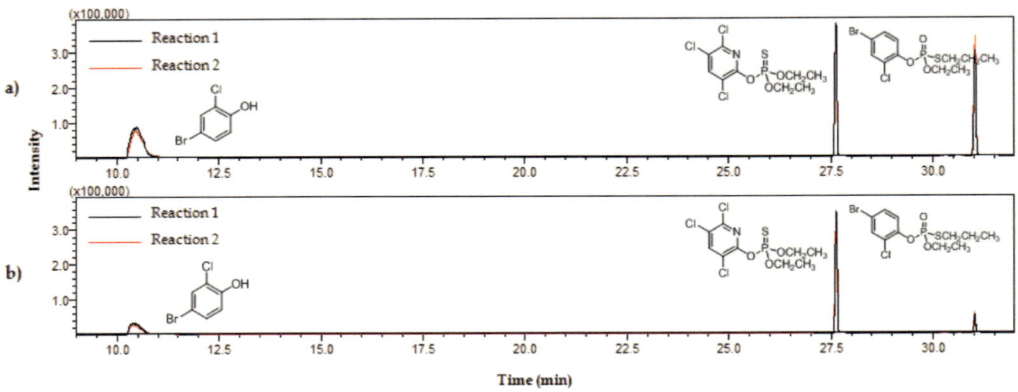

Figure 8. GC-MS-SIM analyses: Chromatogram of biodegradation of 50.0 ppm profenofos by *A. sydowii* CBMAI 935, at 10 days. a) Extract of the mycelium. b) Extract of the liquid medium

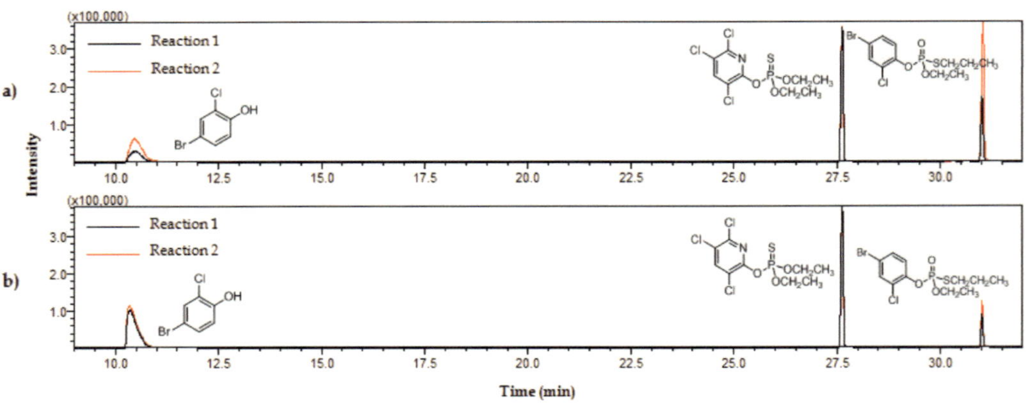

Figure 9. GC-MS-SIM analyses: Chromatogram of biodegradation of 50.0 ppm profenofos by *P.raistrickii* CBMAI 931, at 10 days. a) Extract of the mycelium. b) Extract of the liquid medium

Duplicate reaction	Fungal dry mass (g)	c[a] metabolite[b]	c[a] profenofos	% of profenofos degraded*
A. sydowii CBMAI 935 (50.0 ppm of profenofos)				
Reaction 1 (Extraction of liquid medium)	-	2.1	2.7[c]	53.0
Reaction 1 (Extraction of mycelium)	0.96	8.5	20.8	
Reaction 2 (Extraction of liquid medium)	-	1.4	2.4[c]	45.0
Reaction 2 (Extraction of mycelium)	0.64	7.7	25.1	
P. raistrickii CBMAI 931 (50.0 ppm of profenofos)				
Reaction 1 (Extraction of liquid medium)	-	8.4	1.1	81.4
Reaction 1 (Extraction of mycelium)	0.29	1.7	8.2	
Reaction 2 (Extraction of liquid medium)	-	9.3	3.0	48.0
Reaction 2 (Extraction of mycelium)	0.36	1.6	23.0	

[a]c (ppm) = concentration (data duplicates) determined by GC-MS-SIM

[b]4-bromo-2-chlorophenol

[c]estimated concentration

*total of profenofos degraded (mycelium + liquid medium)

Table 3. Quantitative biodegradation of profenofos by *A. sydowii* CBMAI 935 and *P. raistrickii* CBMAI 931, at 10 days, in liquid medium and mycelium (32 °C, 130 rpm, pH 7).

The concentrations of profenofos in the extracts of the liquid culture medium from the *A. sydowii* CBMAI 935 reaction were estimated by straight line extrapolation, since, owing to the low concentrations in the extracts, the peak areas for profenofos were not sufficient to be applied to the analytical curve; thus these values are only estimates.

In the GC-MS-SIM analyses, superimposing the mycelium and liquid culture medium extract profiles for each fungus, it was noted a higher concentration of the pesticide in the mycelium extract than in the liquid medium one. In GC-MS-SIM analyses of *P. raistrickii* CBMAI 931 it was also clear that the highest concentration of the metabolite (4-bromo-2-chlorophenol) was in the liquid medium. These data may suggest that the fungal mycelium are absorbing profenofos molecules and, after metabolization, excreting a part of the metabolite into the liquid medium. Rather than being absorbed it is also possible that these molecules are adsorbed to fungal cells membranes. A previous study conducted by us showed that the pesticide DDD was accumulated in the mycelium of the marine fungus *Trichoderma* sp. CBMAI 932 (Ortega *et al., 2011*). The higher concentration of profenofos in the mycelium could be explained by an intracelullar enzyme degrading profenofos. According to Chen and Mulchandani, the organophosphate hydrolase enzyme is found within cells and, for biodegradation to occur, the pesticide should be transported into the interior of the cell. However, this kind of enzyme may limit biodegradation, because for microorganisms containing a high intracellular activity of degradative enzymes, total detoxification may be limited by the transport mechanism (Chen and Mulchandani, 1998).

3.4.2. Biodegradation of profenofos by marine fungi P. raistrickii CBMAI 931 and A. sydowii CBMAI 935 after 20 days of reaction

The reactions that were performed for 20 days were analyzed by making a single extract from the liquid medium along with the mycelia extract that was subjected to GC-MS-SIM analyses. Table 4 shows data regarding the biodegradation of profenofos for 20 days by strains *A. sydowii* CBMAI 935 and *P. raistrickii* CBMAI 931 in liquid medium.

The results for both marine fungi showed a good percentage of biodegradation of the pesticide profenofos at 20 days of reaction. However, *P. raistrickii* CBMAI 931 was more efficient than *A. sydowii* CBMAI 935, reaching an approximately average of 82% profenofos degradation, whereas *A. sydowii* CBMAI 935 showed degradation of approximately 71%.

***A. sydowii CBMAI 935* (50.0 ppm of profenofos)**				
Duplicate reaction	**Fungal dry mass (g)**	**c[a] metabolite[b]**	**c[a] profenofos**	**% of profenofos degraded**
Reaction 1 (Extraction of liquid medium and mycelium)	0.36	5.5	13.0	74.0
Reaction 2 (Extraction of liquid medium and mycelium)	0.33	4.9	16.3	67.4
***P. raistrickii CBMAI 931* (50.0 ppm of profenofos)**				
Reaction 1 (Extraction of liquid medium and mycelium)	0.19	13.8	11.0	78.0
Reaction 2 (Extraction of liquid medium and mycelium)	0.21	11.0	7.4	85.2

[a]c (ppm) = concentration (data duplicates) determined by GC-MS-SIM

[b]4-bromo-2-chlorophenol

Table 4. Quantitative biodegradation of profenofos by *A. sydowii* CBMAI 935 and *P. raistrickii* CBMAI 931, at 20 days in liquid medium (32 °C, 130 rpm, pH 7).

3.4.3. Biodegradation of profenofos by marine fungi *P. raistrickii* CBMAI 931 and *A. sydowii* CBMAI 935 after 30 days of reaction

Finally, profenofos biodegradation reactions using marine fungi *A. sydowii* CBMAI 935 and *P. raistrickii* CBMAI 931 were performed for 30 days (Table 5, Figure 10). In this study, the following experiments were also carried out:

a. 30-day reaction containing only 4-bromo-2-chlorophenol (main metabolite) as the substrate, with the objective of assessing the biocatalytic potential of these marine fungi for complete degradation of the pesticide into non-toxic metabolites (Table 6, Figure 11);

b. 30-day reaction in mineral medium supplemented with potassium nitrate in order to assess whether the fungi are able to grow on medium with the pesticide as the sole carbon source, and nitrate as the nitrogen source (Table 7);

c. 30-day reaction for profenofos, in the absence of fungal mycelium, control experiment in order to determine the spontaneous rate of hydrolysis of the pesticide in the liquid medium (Table 8);

d. 30-day reaction with the fungi, in the absence of pesticide, control experiment in order to determine the growth of the marine fungi by measuring the mycelial mass produced without the pesticide influence (Table 9).

***A. sydowii* CBMAI 935 (50.0 ppm of profenofos)**				
Duplicate reaction	**Fungal dry mass (g)**	**c[a] metabolite[b]**	**c[a] profenofos**	**% of profenofos degraded**
Reaction 1 (Extraction of liquid medium and mycelium)	0.38	12.0	12.5	75.0
Reaction 2 (Extraction of liquid medium and mycelium)	0.39	11.0	16.0	68.0
***P. raistrickii* CBMAI 931 (50.0 ppm of profenofos)**				
Reaction 1 (Extraction of liquid medium and mycelium)	0.22	17.8	2.3*	95.4
Reaction 2 (Extraction of liquid medium and mycelium)	0.20	21.4	0.6	98.8

[a]c (ppm) = concentration (data duplicates) determined by GC-MS-SIM

[b]4-bromo-2-chlorophenol

*estimated concentration

Table 5. Quantitative biodegradation of profenofos by *A. sydowii* CBMAI 935 and *P. raistrickii* CBMAI 931, at 30 days in liquid medium (32 °C, 130 rpm, pH 7).

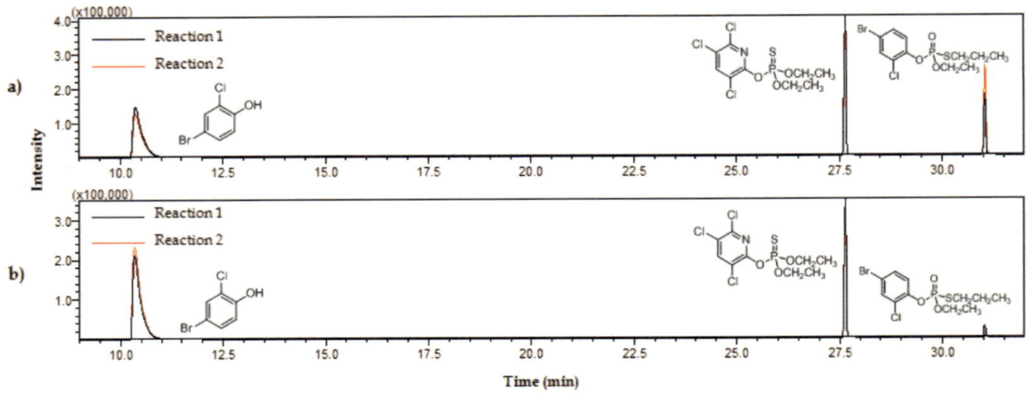

Figure 10. GC-MS-SIM analyses: Chromatogram of biodegradation of profenofos (50.0 ppm) at 30 days. a) *A. sydowii* CBMAI 935. b) *P. raistrickii* CBMAI 931.

Duplicate reaction	Fungal dry mass (g)	c[a] metabolite[b]	% of metabolite degraded
A. sydowii CBMAI 935 (50.0 ppm of 4-bromo-2-chlorophenol)			
Reaction 1 (Extraction of liquid medium and mycelium)	0.39	0.8	98.4
Reaction 2 (Extraction of liquid medium and mycelium)	0.37	0.9	98.2
P. raistrickii CBMAI 931 (50.0 ppm of 4-bromo-2-chlorophenol)			
Reaction 1 (Extraction of liquid medium and mycelium)	0.30	1.8	96.4
Reaction 2 (Extraction of liquid medium and mycelium)	0.35	1.2	97.6

[a]c (ppm) = concentration (data duplicates) determined by GC-MS-SIM

[b]4-bromo-2-chlorophenol

Table 6. Quantitative biodegradation of 4-bromo-2-chlorophenol (main metabolite) by *A. sydowii* CBMAI 935 and *P. raistrickii* CBMAI 931, at 30 days in liquid medium (32 °C, 130 rpm, pH 7).

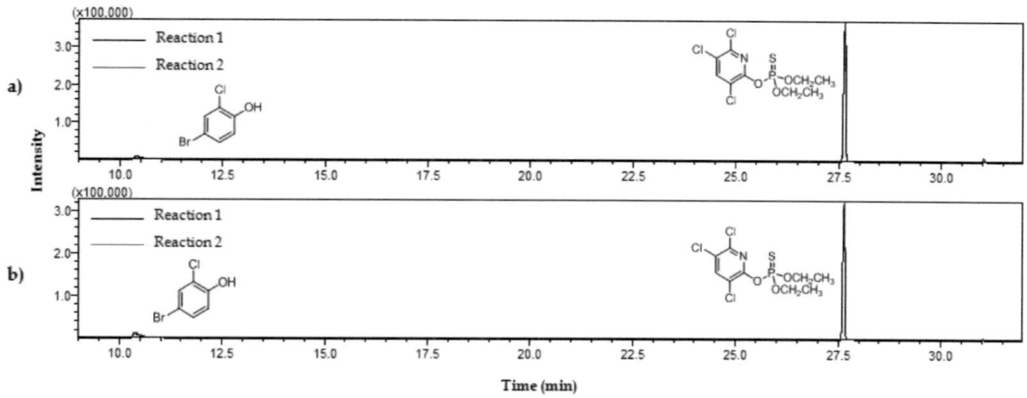

Figure 11. GC-MS-SIM analyses: Chromatogram of biodegradation of 4-bromo-2-chlorophenol (50.0 ppm) at 30 days. a) *A. sydowii* CBMAI 935. b) *P. raistrickii* CBMAI 931

Profenofos concentrations for reaction in liquid medium with 2% malt extract and reaction in mineral medium, in the presence of *P. raistrickii* CBMAI 931, were estimated by straight line extrapolation of the analytical curve for profenofos. The same was done for all concentrations of the metabolite, in the reaction with the fungi *P. raistrickii* CBMAI 931 and *A. sydowii* CBMAI 935.

A. sydowii CBMAI 935 (100.0 ppm of profenofos)				
Duplicate reaction	**Fungal dry mass (g)**	**c[a] metabolite[b]**	**c[a] profenofos**	**% of profenofos degraded**
Reaction 1 (Extraction of liquid medium and mycelium)	*	39.5	3.7	92.6
Reaction 2 (Extraction of liquid medium and mycelium)	*	38.0	5.7	88.6
P. raistrickii CBMAI 931 (100.0 ppm of profenofos)				
Reaction 1 (Extraction of liquid medium and mycelium)	*	46.3	2.2[c]	95.6
Reaction 2 (Extraction of liquid medium and mycelium)	*	44.8	1.8[c]	96.4

[a]c (ppm) = concentration (data duplicates) determined by GC-MS-SIM

[b]4-bromo-2-chlorophenol

[c]estimated concentration

*mycelial mass not obtained

Table 7. Quantitative biodegradation of profenofos by *A. sydowii* CBMAI 935 and *P. raistrickii* CBMAI 931, at 30 days, in liquid mineral medium supplemented with KNO_3 (32 °C, 130 rpm, pH 7).

Duplicate reaction	c[a] metabolite[b]	c[a] profenofos	% of profenofos degraded
Reaction 1 (Extraction of liquid medium and mycelium)	3.3	30.2	39.6
Reaction 2 (Extraction of liquid medium and mycelium)	4.2	31.0	38.0

[a]c (ppm) = concentration (data duplicates) determined by GC-MS-SIM

[b]4-bromo-2-chlorophenol

Table 8. Quantitative degradation of profenofos by spontaneous hydrolysis, at 30 days in liquid medium (32 °C, 130 rpm, pH 7).

Duplicate reaction	Fungal dry mass (g)
A. sydowii CBMAI 935	
Reaction 1	1.50
Reaction 2	1.49
P. raistrickii CBMAI 931	
Reaction 1	1.40
Reaction 2	1.23

Table 9. Quantitative mycelia mass produced by *A. sydowii* CBMAI 935 and *P. raistrickii* CBMAI 931, in absence of profenofos, at 30 days in liquid medium (32 °C, 130 rpm, pH 7).

It was observed that the biodegradation of profenofos by *A. sydowii* CBMAI 935 exhibited stagnation between 20 and 30 days of reaction, as can be seen by comparing Tables 4 and 5 data. However, between 20 and 30 days, there was a significant increase in the metabolite concentration. According to literature data, the organophosphates hydrolysis proceeds by breaking the bond between the phosphorus atom and the leaving group (Sogorb and Vilanova, 2002; Bigley and Raushel, 2012). Thus, the concentration of profenofos degraded should be about the same as the concentration of metabolite formed, where there is total degradation of the pesticide to the metabolite. However, according the United Nations Food and Agriculture Organization (FAO), after the hydrolysis and formation of 4-bromo-2-chlorophenol, the latter can be conjugated with another molecule, can react with a molecule from the fungal metabolism or even be metabolized (Figure 4) (FAO, 2012).

So, a possible explanation for the incomplete conversion of reactants to products may be the 4-bromo-2-chlorophenol further degradation or the conversion to other metabolites by *A. sydowii* CBMAI 935 and *P. raistrickii* CBMAI 931. However, at 30 days of reaction, the biodegradation of the pesticide was stagnant (compared to 20 days), so that stagnation in the degradation or conversion of the metabolite may also have occurred, causing a slight accumulation at 30 days of reaction. This stagnation may be caused either by fungal death or reaching its stationary growth phase, after one of the nutrients in the liquid medium became scarce.

The final concentrations of profenofos in 30 days biodegradation reactions with *P. raistrickii* CBMAI 931 were estimated because of the very low concentration obtained, as previously mentioned. The biodegradation of the pesticide in the presence of this fungus was almost complete, suggesting that this has a greater biocatalytic potential than *A. sydowii* CBMAI 935 for profenofos degradation.

The results for the biodegradation of the metabolite were satisfactory, since there was an almost complete degradation (or conversion) of 4-bromo-2-chlorophenol by both fungi (Table 6). However, it was not possible to identify the other metabolites formed in this degradation.

Complementing the biodegradation studies, an experiment was conducted in liquid mineral medium supplemented with potassium nitrate that demonstrated, through the high percentage of degradation of profenofos, that the fungi could be using the pesticide as a source of carbon, since this was the sole carbon source present in the reaction medium. The concentration of pesticide in the mineral medium was 100.0 ppm, higher than in the earlier tests, since the pesticide was the sole source of carbon, it was needed a high concentration for the fungi growth. There was a greater final concentration of the metabolite, which could be partially degraded / converted in other molecules, since only 50% of the pesticide was converted to this product.

Through the control reaction of profenofos, in the absence of the fungus (and hence without enzymes), the spontaneous hydrolysis of the pesticide in the medium was assessed. This experiment revealed degradation of about 40%, indicating that this pesticide is not persistent in the environment, since a half-life of about a month is relatively low compared to other pesticides, such as organochlorines. However, approximately 60% of the pesticide was not degraded, showing that the enzymatic process is highly effective for promoting the biodegradation of profenofos. It should be noted, also, that the spontaneous hydrolysis does not promote the degradation of the metabolite, as does the enzymatic system (Table 8).

Aly and Badway discussed the hydrolysis of profenofos at 20°C with buffered solutions at pH 5, 7 and 9. A loss of 50%occurred in 106 days at pH5, 43 days at pH7 and 0.7 days at pH 9. Rate constants and half-lives (t1/2) revealed that this insecticide was relatively stable in acid medium and its stability decreased in higher pHs. The studies showed that the mode of decomposition of profenofos in acidic and neutral media is dealkylation, but in an alkaline medium it undergoes hydrolysis, resulting in substituted phenol and dialkyphosphoric acid compounds (Figure 12) (Ali and Badawy, 1982; Ahmed, 2012).

In the control reaction of *A. sydowii* CBMAI 935 and *P. raistrickii* CBMAI 931, in the absence of the pesticide, there was a much greater mycelia growth indicating that the pesticide can partially inhibit the growth of fungi, as noted previously in the screenings carried out on solid medium (Table 9).

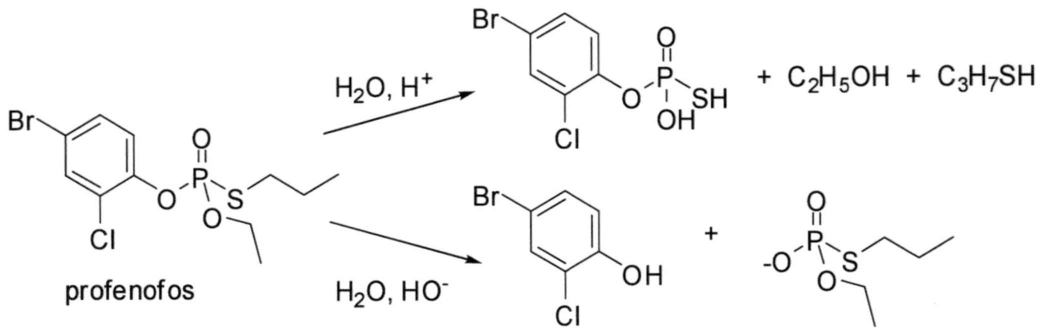

Figure 12. Degradation of profenofos in aqueous medium

3.4.4. Biodegradation of the pesticide profenofos by P. raistrickii CBMAI 931 at several concentrations

Aiming to evaluate the biodegradation of the pesticide profenofos at various concentrations, this test was carried out with variations of the initial concentration under standard biodegradation reaction (liquid culture medium with malt + profenofos + fungal inoculum for 30 days). The reactions were performed in duplicate, at concentrations of 15.0, 30.0, 50.0 and 65.0 ppm, and also a pesticide control (liquid culture medium + profenofos) was carried out and average results are shown in Table 10.

$c_{initial}$ [a] profenofos	c_{final} [a] profenofos*	c[a] metabolite[b]	%of profenofos degraded
15.0	0.5	4,90	98
30.0	0.6	20,40	98
50.0	2.5	24,1	95
65.0	3.2	27,9	95
30.0 (control)	14.9	10,8	50

[a]c (ppm) = concentration (data in duplicates) determined by GC-MS-SIM

[b]4-bromo-2-chlorophenol

Table 10. Quantitative biodegradation of profenofos by *P. raistrickii* CBMAI 931 in different concentrations, at 30 days in liquid medium (32 °C, 130 rpm, pH 7).

Almost complete biodegradation was observed at lower initial concentrations of the pesticide (15.0 and 30.0 ppm). At higher concentrations (50.0 and 65.0 ppm), the results were also satisfactory, with 95% degradation of profenofos. This test proved that in liquid medium, as well as on solid media, the fungus *P. raistrickii* CBMAI 931 was resistant to higher concentrations of pesticide and showed an excellent potential for biodegradation of profenofos.

In the profenofos control, in the absence of the fungus, at a concentration of 30.0 ppm was observed a degradation of only 50% in a period of 30 days, confirming that the presence of fungi accelerates degradation reaction, possibly through the action of phosphotriesterases enzymes. The fungus also promoted the degradation or conversion of the part of the metabolite.

4. Conclusion

The growth of fungal strains on profenofos was promising, even at the highest tested concentration. The fungi *P. raistrickii* CBMAI 931, *A. sydowii* CBMAI 935, *A. sydowii* CBMAI 1241 and *Trichoderma* sp. CBMAI 932 may have a good biocatalytic potential in the presence of profenofos, according to the results of screening. Marine fungi should be further explored as sources of enzymes capable of degrading OPs, since studies in fungal bioremediation of pesticides has shown great potential, albeit much less explored than bacterial bioremediation.

The fungi *P. raistrickii* CBMAI 931 and *A. sydowii* CBMAI 935 were efficient in profenofos biodegrading in liquid medium, as well as promoting the transformation or degradation of the metabolite, 4-bromo-2-chlorophenol, and showing that the enzymatic system of these fungi effectively expressed the necessary enzymes for a complete degradation process. Further research is under way to assess the biodegradation of methyl parathion and chlorpyrifos.

Liquid medium reactions using *P. raistrickii* CBMAI 931 with increasing concentration of the pesticide leaded to the almost complete biodegradation (99.0 to 95.0%) of the pesticide profenofos at all concentrations (15.0, 30.0, 50.0 and 65.0 ppm), showing that this fungus is resistant to high concentrations of this pesticide. The fungus *P. raistrickii*-CBMAI 931 may be a good source of phosphotriesterases, which could be isolated and purified for applications in biotechnology, biodegradation reactions in soil and water contaminated with pesticides.

Acknowledgements

NAS thanks FAPESP and WGB thanks CNPq for the scholarships. The authors wish to thank Prof. R.G.S. Berlinck (Instituto de Química de São Carlos - USP) for providing the marine fungal strains. ALMP is grateful to the Conselho Nacional de Desenvolvimento Científico e Tecnológico (CNPq) and Fundação de Amparo a Pesquisa do Estado de São Paulo (FAPESP) for financial support. The authors also wish to thank Professor Marcos Roberto de Vasconcellos Lanza (IQSC/USP, São Paulo, Brazil) for donating the commercial pesticides (profenofos).

Author details

Natália Alvarenga da Silva[1], Willian Garcia Birolli[1], Mirna Helena Regali Seleghim[2] and André Luiz Meleiro Porto[1*]

*Address all correspondence to: almporto@iqsc.usp.br

1 Instituto de Química de São Carlos, Universidade de São Paulo, São Carlos, SP, Brazil

2 Departamento de Ecologia e Biologia Evolutiva, Universidade Federal de São Carlos, São Carlos, SP, Brazil

References

[1] Abass K, Reponen R, Jalonen J, Pelkonen O. In vitro metabolism and interaction of profenofos by human, mouse and rat liver preparations. Pesticide Biochemistry and Physiology 2007; 87: 238–247.

[2] ANVISA, 2011 - Agência Nacional de Vigilância Sanitária. http://www4.anvisa.gov.br/base/visadoc/CP/CP[4708-1-0].PDF. (Accessed 2011/09).

[3] ANVISA, 2012 - Agência Nacional de Vigilância Sanitária. http://portal.anvisa.gov.br/wps/wcm/connect/ b380fe004965d38ab6abf74ed75891ae/Relat%C3%B3rio +PARA+2010+-+Vers%C3%A3o +Final.pdf?MOD=AJPERES (Accessed 2012/08).

[4] Ahmed RMAH. Chemical and biochemical studies on the residuals of the organophosphorus insecticide (prothiofos) in potato plant. Thesis. Department of Chemistry Faculty of Science Zagazig University. http://lib.bioinfo.pl/files/theses/thesis_11.pdf. (Accessed 2012/08)

[5] Ali OA, Badawy MI. Hydrolysis of organophosphate insecticides in aqueous media. Environmental International 1982; 7:373-377.

[6] Bigley, A. N, Raushel, F. M. Catalytic mechanisms for phosphotriesterases. Biochimica et Biophysica Acta (BBA)-Proteins & Proteomics (2013). , 1, 443-453"

[7] Bonugli-Santos RC, Durrant LR, Da Silva M, Sette LD. Production of laccase, manganese peroxidase and lignin peroxidase by brazilian marine-derived fungi. Enzyme and Microbial Technology 2010; 46: 32-37.

[8] Capps TM, Barringer VM, Eberle WJ, Brown DR, Sanson DR. Identification of a unique glucosylsulfate conjugate metabolite of profenofos in cotton. Journal of Agricultural and Food Chemistry 1996; 44:2408-2411.

[9] Chen W, Mulchandani A . The use of live biocatalysts for pesticide detoxification. Tibtech 1988; 16:71-76.

[10] Chowdhury A, Pradhan S, Saha M, Sanyal N. (2008) Impact of pesticides on soil microbiological parameters and possible bioremediation strategies. Indian Journal of Microbiology 2008; 48:114-127.

[11] Coutinho CFB, Tanimoto ST, Galli A, Garbellini GS, Takayama M, Amaral RB, Mazo LH, Avaca LA, Machado SAS. Pesticidas: mecanismo de ação, degradação e toxidez. Pesticidas: Revista de Ecotoxicologia e Meio Ambiente 2005; 15:65-72.

[12] Diez MC. Biological aspects involved in the degradation of organic pollutants. Journal of Plant Nutrition and Soil Science 2010;10: 244-267.

[13] Edwards FL, Tchounwou PB. Environmental toxicology and health effects associated with methyl parathion exposure – A scientific review. International Journal of Environmental Research and Public Health 2005; 2:430-441.

[14] Eerd LAV, Hoagland RE, Zablotowicz RM, Hall JC. Pesticide metabolism in plants and microorganisms. Weed Science 2003; 51: 472–495.

[15] Madigan, M. T, Dunlap, P.V, Clark, D.P. Microbiologia de Brock. Porto Alegre: Artmed (2010).

[16] EPA, 2012 - Environmental Protection Agency Reregistratione eligibility decision for profenofos. http://www.epa.gov/ oppsrrd1/REDs/profenofos_red.pdf (Accessed 2012 /08).

[17] FAO, 2012 - Food and Agriculture Organization of the United Nations (2012) http://www.fao.org/fileadmin/templates/ agphome/documents/Pests_Pesticides/JMPR/Evaluation08/Profenofos.pdf. Accessed 2012/08.

[18] Gavrilescu M. Fate of pesticides in the environment and its bioremediation. Engineering in Life Sciences 2005; 5:497-525.

[19] Ghanem E, Raushel FM. Detoxification of organophosphate nerve agents by bacterial phosphotriesterase. Toxicology and Applied Pharmacology 2005; 207:S459 – S470.

[20] Ghosh D, Saha M, Sana B, Murkherjee J. Marine enzymes. Advances in Biochemical Engineering/Biotechnology 2005; 96:189-218.

[21] Haefner B. Drugs from the deep: Marine natural products as drug candidates. Drug Discovery Today 2003; 8:536-544.

[22] Hasan HA. Fungal utilization of organophosphate pesticides and their degradation by *Aspergillus flavus* and *A. sydowii* in soil. Folia Microbiology (Praha) 1999; 44:77-84.

[23] Jauregui J, Valderrama B, Albores A, Vazquez-duhalt R. Microsomal transformation of organophosphorus pesticides by white rot fungi. Biodegradation 2003;14:397–406.

[24] Kogure K. Bioenergetics of marine bacteria. Current Opinion in Biotechnology 1998; 9:278-282.

[25] MacLeod RA. The question of the existence of specific marine bacteria. Bacteriological Review 1965; 29: 9-23.

[26] Eivazi, F, Tabatabai, M. Phosphatases in soils. Soil Biology and Biochemistry (1977)., 9,167-172.

[27] Malghani S, Chatterjee N, Yu HX, Luo Z. Isolation and identification of profenofos degrading bacteria. Brazilian Journal of Microbiology 2009;40: 893-900.

[28] Martins MP, Mouad AM, Boschini L, Seleghim MHR, Sette LD, Porto ALM. Marine fungi *Aspergillus sydowii* and *Trichoderma* sp. catalyze the hydrolysis of benzyl glycidyl ether. Marine Biotechnology 2011; 13:314-320.

[29] McDaniel KL, Moser VC. Differential profiles of cholinesterase inhibition and neurobehavioral effects in rats exposed to fenamiphos or profenofos. Neurotoxicology and Teratology 2004; 26:407–415.

[30] Mileson BE, Chambers JE, Chen WL, Dettbarn W, Enrich M, Eldefrawi AT, Gaylor DW, Hamernik K, Hodgson E, Karczmar AG, Padilla S, Pope CN, Richardson RJ, Saunders DR, Sheets LP, Sultatos LG, Wallace KB. Common mechanism of toxicity: A case study of organophosphorus pesticides. Toxicological Sciences 1998; 41:8-20.

[31] Newman DJ, Cragg GM. Marine natural products and related compounds in clinical and advanced preclinical trials. Journal of Natural Products 2004; 67: 1216-1238.

[32] Ortega SN, Nitschke M, Mouad AM, Landgraf MD, Rezende MOO, Seleghim MHR, Sette LD, Porto ALM. Isolation of Brazilian marine fungi capable of growing on DDD pesticide. Biodegradation 2011; 22:43-50.

[33] Pelczar MJ, Chan ECS, Krieg NR. Microbiologia. 2ªEd. São Paulo: Pearson Makron Books; 1997.

[34] Ramadevi C, Nath MM, Prasad MG. Mycodegradation of malathion by a soil fungal isolate, *Aspergillus niger* International Journal of Basic and Applied Chemical Sciences 2012; 2:2277-2073.

[35] Reddy NC, Rao JV. Biological response of earthworm, *Eisenia foetida* (Savigny) to an organophosphorous pesticide, profenofos. Ecotoxicology and Environmental Safety 2008; 71:574–582.

[36] Ribani M, Bottoli CBG, Collis CH, Jardim ICSF, Melo LFC. Validação em métodos cromatográficos e eletroforéticos. Quimíca Nova 2004; 27:771-780.

[37] Rissato SR, Libânio M, Giafferis GP, Gerenutti M. Determinação de pesticidas organoclorados em água de manancial, água potável e solo na região de Bauru (SP). Quimíca Nova 2004; 27:739-743.

[38] Rocha LC, Ferreira HV, Pimenta EF, Berlinck RGS, Seleghim MHR, Javaroti DCD, Sette LD, Bonugli RC, Porto ALMBioreduction of α-chloroacetophenone by whole cells of marine fungi. Biotechnology Letters 2009; 3:1559-1563.

[39] Rogers K R. Organophosphorus hydrolase-based assay for organophosphate pesticides. Biotechnology Progress 1999;15:517-521.

[40] Rusch DB, Halpern AL, Sutton G, Heidelberg KB, Williamson S, Yooseph S, Wu D, Eisen JA, Hoffman JM, Remington K, Beeson K, Tran B, Smith H, Baden-Tillson H, Stewart C, Thorpe J, Freeman J, Andrews-Pfannkoch C, Venter JE, Li K, Kravitz S, Heidelberg JF, Utterback T, Rogers YH, Falcón LI, Souza V, Bonilla-Rosso G, Eguiarte LE, Karl DM, Sathyendranath S, Platt T, Bermingham E, Gallardo V, Tamayo-Castillo G, Ferrari MR, Strausberg RL, Nealson K, Friedman R, Frazier M, Venter JC. The sorcerer ii global ocean sampling expedition: Northwest Atlantic through eastern tropical pacific. PLoS Biology 2007; 5:398-431.

[41] San-Martín A, Rovirosa J, Astudillo L, Sepúlveda B, Ruiz D, San-Martín C. Biotransformation of the marine sesquiterpene pacifenol by a facultative marine fungus. Natural Product Research 2008; 22:1627-1632.

[42] Santos VMR, Donnici CL, Dacosta JBN, Caixeiro JMR. Compostos organofosforados pentavalentes: histórico, métodos sintéticos de preparação e aplicações como inseticidas e agentes antitumorais. Quimíca Nova 2007; 30:159-170.

[43] Sarkar S, Pramanik A, Mitra A, Mukherjee J. Bioprocessing data for the production of marine enzymes. Marine Drugs 2010; 8:1323-1372.

[44] Scott C, Pandey G, Hartley CJ, Jackson JC, Cheesman MJ, Taylor MC, Pandey R, Khurana JL, Teese M, Coppin CW, Weir KM, Jain RK, Lal R, Russell RJ, Oakeshott JG. The enzymatic basis for pesticide bioremediation. Indian Journal of Microbiology 2008; 48: 65–79.

[45] Singh DK. Biodegradation and bioremediation of pesticide in soil: concept, method and recent developments. Indian Journal of Microbiology 2008; 4:35–40.

[46] Sogorb MA, Vilanova E. Enzymes involved in the detoxification of organophosphorus, carbamate and pyrethroid insecticides through hydrolysis. Toxicology Letters 2002; 128:215-228.

[47] Sutherland T, Russell R, Selleck M. Using enzymes to clean up pesticide residues. Pesticide Outlook 2002; 13:149-151.

[48] Sutherland TD, Horne I, Weir KM, Coppin CW, Williams MR, Selleck M, Russell RJ, Oakeshott JG. Enzymatic bioremediation: from enzyme discovery to applications. Clinical and Experimental Pharmacology and Physiology 2004; 31: 817–821.

[49] Syngenta, 2012. http://www.syngenta.com/country/br/pt/produtosemarcas/protecao-de-cultivos/Pages/POLYTRIN.aspx (Accessed 2012/08).

[50] Tortora, G. J, Funke, B. R, Case, C. L, Johnson, T. R. Microbiology: an introduction. San Francisco: Benjamin Cummings (2004).

[51] Trincone A. Potential biocatalysts originating from sea environments. Journal of Molecular Catalysis B: Enzymatic 2010; 66: 241-256.

[52] Van Eerd LL, Hoagland RE, Zablotowicz RM, Hall JC. Pesticide metabolism in plants and microorganisms. Weed Science 2003; 51:472–495.

[53] Venter JC, Remington K, Heidelberg JF, Halpern AL, Rusch D, Eisen JA, Wu D, Paulsen I, Nelson KE, Nelson W, Fouts DE, Levy S, Knap AH, Lomas MW, Nealson K, White O, Peterson J, Hoffman J, Parsons R, Baden-Tillson H, Pfannkoch C, Rogers Y-H, Smith HO. Environmental genome shotgun sequencing of the Sargasso sea. Science 2004; 304:66-74.

[54] Venter JC, Remington K, Heidelberg JF, Halpern AL, Rusch D, Eisen JA, Wu D, Paulsen I, Nelson, Trincone A. Potential biocatalysts originating from sea environments. Journal of Molecular Catalysis B: Enzymatic 2010; 66:241-256.

[55] Yang C, Cai N, Dong M, Jiang H, Li J, Qiao C, Mulchandani A, Chen W Surface display of MPH on *Pseudomonas putida* JS444 using ice nucleation protein and its application in detoxification of organophosphates. Biotechnology and Bioengineering 2008; 99:30-37.

[56] Zamy, C, Mazellier, P, Legube, B. Analytical and kinetic study of the aqueous hydrolysis of four organophosphorus and two carbamate pesticides. International Journal of Environmental Analytical Chemistry (2004)., 84, 1059-1068.

[57] Zheng YZ, Lan WS, Qiao CL, Mulchandani A, Chen W Decontamination of vegetables sprayed with organophosphate pesticides by organophosphorus hydrolase and carboxylesterase (B1). Applied Biochemistry and Biotechnology 2007; 136:233-241.

Chapter 8

Bioremediation of Radiotoxic Elements under Natural Environmental Conditions

Phalazane Johanna Mtimunye and
Evans M. N. Chirwa

Additional information is available at the end of the chapter

http://dx.doi.org/10.5772/56909

1. Introduction

1.1. Status of uranium use and pollution

The energy sector around the world is severely affected by the combined impact of population growth, industrialisation and the general movement of people into cities (urbanisation). Controlling carbon emissions from the energy sector lies at the centre of the strategy to curb the problem of global warming and its effects on changing weather patterns – which poses a threat to ecosystems and biodiversity. Several alternatives are under consideration worldwide to reverse the current trend of global warming; almost all strategies are concern in finding cleaner primary energy sources with low or no carbon emissions. The alternative energy sources include hydropower, wind power, nuclear power, solar energy, biomass energy, and geothermal energy. Among the proposed cleaner energy sources, nuclear energy has been demonstrated to be the most stable and concentrated enough to replace fossil fuels such as coal and natural gas on most national grids.

However, in spite of holding the promise of cleaner production in terms of carbon emissions, nuclear technology produces a waste which is highly radioactive and in most instances difficult to treat. The waste compounds originating from nuclear power generation include uranium and its fission products, and transuranic elements. These metallic elements account for over 95% of the total radioactivity of radioactive waste [1].

The lightweight fission products emanating from nuclear fuel processing plants and underground nuclear waste repositories contain high levels of mobile species such as caesium (Cs-137), strontium (Sr-90), and cobalt (Co-60). These elements are characterised by very high radiological decay rates and short half-lives. Due to their high decay rates and high radioactivity, fission products render the waste matrix in which they are detected highly radiotoxic and therefore hazardous [2]. Some of these elements are easily taken up by plants and other animal life forms upon reaching the environment [3]. As an example, the divalent cation strontium-90 ($^{90}Sr^{2+}$) is easily incorporated into bone tissue because its chemical properties resemble calcium (Ca^{2+}) which is a critical component of the mammalian bone structure. Calcium is responsible for the bone's structural integrity and strength and is therefore an essential component of the mammalian diet.

Currently, there are about 438 nuclear power plants in operation in 31 countries around the world providing about 14% of the world's primary energy needs [4]. The slow progression towards wider application of nuclear energy technology worldwide is mainly due to concerns over long term radiation contamination, reactor accidents such as those which occurred in Chernobyl and Fukushima, and the possibility of proliferation of atomic bomb making materials to renegade regimes and terrorists.

Nuclear power generation is not the only source of potential radioactive pollution. Other activities such as nuclear weapon testing, radioisotope manufacturing, and biomedical research have also contributed significant amounts of radioactive waste into the environment [5]. However, most radioactive waste originating from medical and radioisotope manufacturing facilities is predominantly organic and therefore can be easily degraded [6]. Radioactive waste from the power generation industry which is identified as High Level Waste (HLW), on the other hand, consists of a higher proportion of non-degradable metallic elements such as uranium and fission products, and transuranic elements. The waste tends to be "hot" (highly radiotoxic) to living organisms and therefore requires pre-treatment before disposal.

This chapter presents recent findings from research aimed at developing environmentally friendly treatment processes for radioactive waste. Attempts have recently been made to treat components of high level radioactive waste (HLW) prior disposal in specifically engineered facilities by immobilizing and extracting the radioactive elements in the waste using a combination of biological reduction and biosorption of the toxic metallic elements. In the case of U(VI), the reduced element is easily extracted either by precipitation/deposition on cell surfaces or removed by a biologically mitigated ion exchange process using live cells of bacteria.

1.2. Uranium compound toxicity

Uranium is found in the environment in many forms including as an oxide, organic or inorganic complex, and rarely as a free metallic ion. Free elemental uranium primarily exists in higher oxidation states typically bound to oxygen. In the aqueous phase, cationic uranium readily combines with oxygen to form oxy-cations of uranium (uranyl ions) which are highly mobile and highly reactive. For example U(VI) in the form of (UO_2^{2+}) is highly soluble in water. Whereas, the reduced form U(IV), existing as uraninite (UO_2) is less soluble and therefore represents a lower risk in the environment.

The toxicity of uranium compounds is closely related to its mobility. That is, the most soluble of the uranium species are associated with acute toxicity in organisms [7]. The highly soluble uranium compounds such as UF_6, $UO_2 (NO_3)_2$, UO_2Cl_2, uranyl acetate, uranyl sulphates, and uranyl carbonates exhibit high toxicity to mammalian cells whereas the less soluble uranium compounds including UO_2, U_3O_8, uranium hydrides, and carbides are less reactive and less toxic.

The permissible body level for soluble compounds is based on chemical toxicity, while the permissible body level for insoluble compounds is based on radiotoxicity. Because all uranium isotopes (U^{234}, U^{235}, U^{238}) mainly emit α-particles that have little penetrating ability, the main radiation hazard from soluble uranium compounds occurs when uranium compounds are ingested or inhaled [7]. Although, absorption of some soluble compounds through skin is possible, uptake through the skin is normally superseded by either surface damage due to exposure or accumulation to toxic levels through other routes of entry such as inhalation and ingestion.

2. Treatment and reclamation of uranium

2.1. Physical-chemical treatment

The most common treatment strategy for uranium and radioactive waste involves the extraction of the radioactive component to reduce the volume of radioactive waste followed by treatment of the bulky nonradioactive waste using conventional methods [8]. Various options have been utilised to achieve extraction by employing a combination of physical-chemical and biological methods. For areas that have already been contaminated, further migration of the pollutants is prevented by using *in situ* treatment options [9]. *Ex situ* pump-and-treat processes have also been attempted but these come with very high operational costs and are in most cases unsustainable [10]. *In situ* remediation technologies are generally the most preferred treatment technologies because they are more economical, do not require transportation of toxic materials that may lead to more spillage in transit, and also cause fewer disturbances on site [11, 12].

2.2. Adsorptive/ion exchange processes

Separation processes have been utilised to selectively remove cationic species from wastewater streams. Materials such as activated carbon, saw dust, and peat can remove pollutants from water. However, these materials are not selective and therefore may not be effective in removing metallic elements from nuclear waste. Specially designed resins can be utilised to target specific species by manipulating the composition of functional groups on surface of the resin. Several examples of uranium binding ion exchange systems are reported by several researchers [13-15]. Although proven successful on pilot scale, full implementation of ion exchange uranium separation is hindered by high cost. Additionally, the ion exchange resin surfaces are not self-regenerating, and therefore have limited capacity [13].

2.3. Membrane processes

Conventional membrane systems used in treating uranium includes, nano-pore filtration, ultrafiltration, microfiltration and reverse osmosis [16]. Nano-pore membrane filters have the potential to be used in recovery of radioisotopes from water or gas streams. Membrane technology is now regarded as established technology with predictable and reliable processing capability than most current alternatives. Membranes have become relatively cheap such that there use is no longer regarded uneconomical. In spite of being economically viable membrane processes generate large quantities of used membranes which contribute to the problem of radioactive solid waste from the nuclear industry.

2.4. Chemical extraction

Chemical extraction processes have mostly been used for remediation processes mostly on land. For example, uranium can be extracted from contaminated soil using sodium carbonate/ bicarbonate or citric acid [17-19]. Although this process effectively removes uranium from soil, it requires a careful balance during application since overloading the system with the acid agent may further migrate uranium in the environment [18, 20]. Certain chemical agents may oxidise other potentially toxic metals posing further risk to the environment. Furthermore, long-term stability of reaction products is of concern. Changing chemical conditions in future could remobilise the metal to its toxic form.

2.5. Biological treatment processes

Biological methods have been proposed to improve or substitute the conventional physico-chemical methods for the remediation of contaminated environments. Unlike organic compounds, toxic metals cannot be degraded or destroyed but can only be transformed from high oxidation state to lower oxidation state. Microbes can potentially affect the physical and the chemical state of the uranium by altering its speciation, solubility, and sorption properties. Strategies suggested for the removal of metals and radionuclides using appropriate microbes include biosorption, bioaccumulation, bioprecipitation, and bioreduction [21-23].

Biological treatment is based on the prospect of utilising processes already devised by nature in dealing with environmental hazards. During three and a half billion years of evolution, microorganisms have evolved mechanisms to survive in hostile environments and to adapt to changes in the environment [24]. One of the most conserved mechanisms in the living cell is the biochemical pathway for electron-transport through the cytoplasmic membrane to conserve energy through the oxidation of an electron donor and reduction of an electron acceptor such as oxygen. This process has been conserved over billions of years, such that, to this day, all life on earth depends on variants of this pathway [24-26]. Most biochemical processes for degradation and/or detoxification of compounds in the living cell are linked to the above process.

Environmental engineers around the world have undertaken to find ways to tap into the mysteries of nature by diligently studying the action of microorganisms as they adapt to extreme conditions. Lately, microorganisms have been isolated that are capable of reducing the toxic forms of heavy metal and transitional metal elements in transuranic waste (TRU) to less mobile precipitable forms [27]. Other researchers have found microbial cultures with the capability to resist high radiation doses.

Most of the microorganisms discovered thus far, utilise the energy derived from change in the redox potential or oxidation states of the compounds for metabolic purposes. No organism has yet been shown to utilise the energy derived from radioactive decay for metabolism. But recent research show promise that this scenario is soon to change. For example, recent studies have shown that microorganism may not only resist radiation, but may to a certain degree utilize the radiation for metabolic advantage. One example was illustrated in a recent study in which cultures of melanising fungi from the cold regions utilized ionizing radiation to derive metabolic energy [28].

2.6. Biosorption

Biosorption is used to describe the metabolism independent sorption (passive process). In this process uranium-bearing water is brought into contact with either living or dead biomass that possesses abandoned functional groups (carboxyl, hydroxyl, amine, and phosphate group) within the surface layer. The charged group within the cell surface layer is able to interact with ions or charged molecules present in the uranium-bearing water. As a result metal cations become electrostatically attracted and bound to the cell surface layer.

Polymers secreted by many metabolizing microbes can also immobilize metals. Different studies on biosorption demonstrated that uranium biosorption is reversible, species-specific, and depends upon the chemistry and pH of the solution, physiological state of cells as well as the presence of the extracellular soluble polymers (EPS) [29, 30]. In this process desorption and recovery of heavy metals and radionuclides for further reuse is easy [31]. Biosorption of radionuclides to the cell surface and polymer substance is a promising technology for remediating contaminated waters. However, the effectiveness of this process is highly affected by pH of the solution and saturation of the biosorbent when metal interactive sites are occupied. Studies were done [32, 33] to investigate the biosorption of uranium under acid conditions. It was observed in these studies that biosorption under acidic conditions is not favoured in several species as at low pH the protons (H^+) compete with UO_2^{2+} for sorption sites (surface hydroxyl groups–SOH), thus indicating poor selectivity of the biosorbent surface against competing ions.

In order to understand the interaction between the cell and metallic species in wastewater, the cell surface can be thoroughly characterised using Fourier Transform Infrared spectroscopy (FTIR) or Energy-Dispersive X-ray spectroscopy (EDX). In our research group at the University of Pretoria, we were able to demonstrate that different functional groups on the cell wall of sulphate reducing bacteria acted as ion exchange sites in different pH ranges.

Using autotitration date in the software MINTEQ, four equilibrium states were identified which were associated with carboxylic functional groups (pK_a 4-5), phosphates (pK_a 6-7), phenolics (pK_a 8-9), and hydroxyl/amines (pK_a 10-12). The most abundant reaction sites in sulphate reducing bacteria (SRB) were associated with the hydroxyl functional groups.

Adsorption of the divalent fission product (Sr^{2+}) was inhibited at higher pH, supposedly as a result of increased hydroxylation at the high OH- concentration in solution. Additionally, increased pH could increase the formation of SrII-OH precipitates which is counterproductive to the processes of adsorption to the cells.

Biosorption offers a unique advantage in that the biosorbent media (bacteria) is self-regenerating and can be safely disposed after expiry. Apart from the uranium species, the biosorbent can remove a range of other toxic heavy metals from the wastewater without creating hazardous sludge at costs much lower than conventionally used ion exchange systems. Regeneration of the biosorbent and concentration of the metal solution for eventual recovery further increase the cost effectiveness of the process.

2.7. Bioaccumulation

Bioaccumulation is an active process wherein metals are taken up into living cells and sequestrated intracellularly by complexing with specific metal-binding components or by precipitation. Intracellular accumulation of metals occurs among all classes of microorganisms by an energy-dependent transport system. Unlike metabolically essential metals such as Fe, Cu, Zn, Co, and Mn, which accumulates intracellularly via energy transport system, intracellular uranium sequestration is attributed to non-specific transport system mainly due to increased membrane permeability resulting from uranium toxicity in the living cell [34]. Therefore, intracellular accumulation of uranium is considered as metabolism-independent process. The major drawbacks associated with the use of active uptake systems is the requirement of metabolically active cells and also the challenge in metal desorption and recovery [35]. Specifically, the cells will need to be destroyed to release the metal either by lysis or by incineration. Therefore, in this case, the medium for the uptake of metals cannot be reused.

2.8. Bioprecepitation

Bioprecipitation also known as biocrystallization or biomineralization is the process by which metals and radionuclides can be precipitated with microbially generated ligands such as phosphate, sulphide or oxalate [35]. In these processes bacteria interact strongly with metal ions and concentrate them, eventually generating carbonates and hydroxide minerals at the surface of the cell. Macaskie et al. [35] investigated *Citrobacter sp.* accumulation of heavy deposits of UO_2^{2+}-phosphate, derived from enzymatically liberated phosphate ligand. Cells showed no saturation constrains and it could accumulate several times their own weight of precipitated metal. The above method showed that metal deposition occurs via an initial nucleation pathway involving phosphate groups localized within certain cell-surface lipopolysaccharides (LPS). Accumulation of metal-phosphate complex within the LPS was suggested to prevent fouling of the cell surface by the accumulated precipitate. The limitations of method during application in an industrial process could be similar to those encountered in biosorption. Firstly, the process is hindered by the formation of negatively charged uranyl carbonate complexes, U(VI)–CO_3, arising from microbial metabolism of the carbon source under anaerobic conditions and over time the U(VI) carbonate complex formed stimulate U(VI) oxidation over time [36]. Additionally, these processes precipitate metals other than uranium and forms insoluble uranyl-phosphate complex on the cell surface.

2.9. Bioreduction

Reduction of highly toxic and mobile U(VI) to sparingly soluble U(IV) using appropriate microbes has been proposed as a mechanism for preventing the migration of U(VI) with groundwater [37, 38]. The strategy is based on injecting physiological electron donors such as acetate, lactate, ethanol, or glucose to stimulate U(VI) reduction by microbial community native to contaminated aquifers [39]. Microorganisms are known to have evolved biochemical pathway for degradation or transformation of toxic compounds from their immediate environment either for survival or to derive energy by using toxic compounds as electron donors or acceptors [40, 41]. The overall transfer of electrons from the carbon sources to active uranium species can be visualised by the figure below (Figure 1).

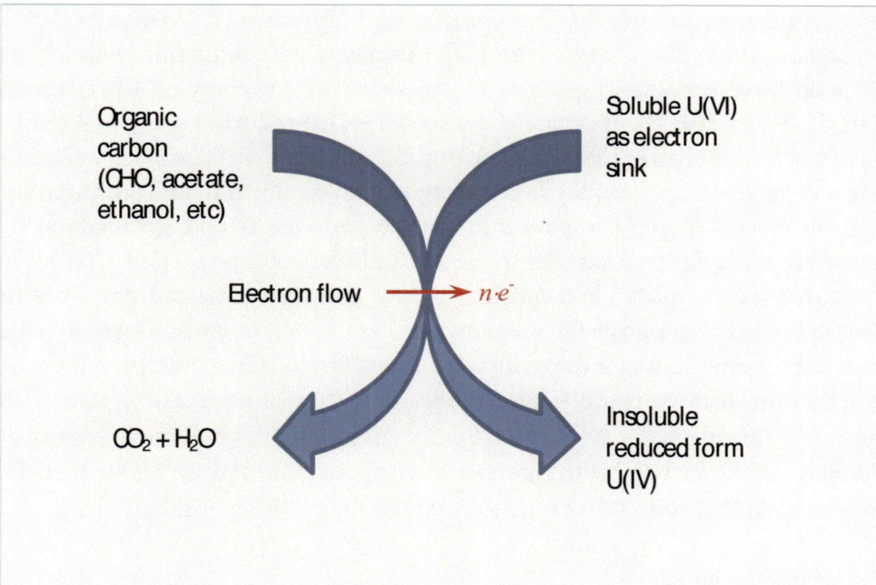

Figure 1. Electron flow during biological reduction of uranium (VI) to U(IV)

An example of a balanced stoichiometric relationship during U(VI) reduction using propanoate as an electron donor is represented by Equation 1 (below):

$$UO_2^{2+} + 1/2CH_3CH_2COO^- + 1/2H_2O \longrightarrow UO_2 + 1/2CH_3COO^- + 1/2CO_2 + 2H^+ \qquad (1)$$

Microbial reduction of U(VI) was first reported in crude extracts from *Micrococcus lactilyticus* by assaying the consumption of hydrogen which was dependent on the presence of U(VI) [42].

To date, U(VI) reduction capability has been identified in more than 25 species of phylogenetically diverse prokaryotes. Examples of these are the mesophilic sulphate-reducing bacteria (*Desulfovibro sp.*) [43], Fe(III)-reducing bacteria (*Geobacter* and *Shawanella sp.*) [44], fermentative bacteria from *Clostridium sp.*, [45], *Acidotolerant bacteria* [46], as well as *Myxobacteria sp.* [47]. Some of these organisms have been reported to conserve energy for growth from U(VI) reduction [43, 47], while others reduce U(VI) without energy gain [45, 48, 49]. Researchers such as Lloyd [50], and Wade and Di Christina [51], have demonstrated the importance of dissimilatory metal-reducing bacteria (DMRB) in reducing toxic form of uranium (U), iron (Fe), manganese (Mn), and other toxic metals. The unique physiological property of DMRD is that they are obligate anaerobes that are capable of utilising (Fe(III) and Mn(IV) oxide) as terminal electron acceptors. In the presence of the radionuclides (U(VI) and Tc(VII)), these could be mostly bioavailable since Fe(III) and Mn(IV) exist as insoluble hydroxides in the natural pH range.

2.10. Enzymatic U(VI) reduction

Members of genera *Shewanella* [43], *Desulfovibro* [52], *Clostridium* [45], *Geobacter* [50], *Thermus* [53], *Pyrobaculum* and *Desulfosporosinus* [54], display U(VI) reduction activity. The two approaches (ie. biochemical and genetic) are responsible for identifying U(VI) reductases in DMRB [46, 51, 52, 54, 55]. The mechanism by which Fe(III)-reducing bacteria (FeRB) transfer electrons to insoluble Fe(III) oxides during anaerobic growth have been extensively studied in *Shewanella* and *Geobacter* species [50]. In these organisms, an electron transfer chain containing c-type cytochromes is thought to pass through the periplasmic and terminate at the outer membrane facilitating electron transfer to the extracellular solid phase [50]. U(IV) is reported to precipitate in the periplasm and outside of both gram-negative and gram-positive cells suggesting that U(VI) complexes do not generally have access to intracellular enzyme. Thus, imply that the best candidates for the reductases would be electron carrier proteins or enzymes exposed to the outside of the cytoplasmic membrane, within the periplasm, and/or in the outer membrane. Several purified c-type cytochromes display U(VI) reductase activity in vitro. *Shewanella* and *Geobacter* has been reported to enzymatically reduce U(VI) to U(IV) via a respiratory process that conserve energy to supports their anaerobic growth [52].

2.10.1. Desulfovibrio reductase

The enzymatic system responsible for U(VI) reduction, tetraheme or periplasmic cytochrome-*c3* was characterize in the cell-free extract of the sulphate-reducing bacterium *Desulfovibrio vulgaris* in the presence of hydrogenase using H_2 gas as a physiological electron donor [43, 52]. The involvement of tetraheme cytochrome-*c3* was confirmed in the whole-cell and results on whole-cell showed that cytochrome-*c3* was oxidized during U(VI) reduction, but not during sulphate reduction [56]. Further evidence that cytochrome-*c3* was biologically important for *Desulfovibrio* U(VI) reduction came from the impairment of this process when a mutation was constructed in a related strain that eliminated the homologous cytochrome [54]. U(VI) reduction by the mutant *D. desulfuricans* strain was inhibited by at least 90% in the presence of H_2 (gas) as an electron donor and partially impaired (inhibition between 50-70%) in the presence of acetate or pyruvate as electron donors [57]. Other researchers observed a partial increase in the U(VI) reduction activity in the presence of either lactate or pyruvate, suggesting the presence of additional proteins capable of metal reduction [54].

2.10.2. *Geobacter reductase*

Genome sequence of *Geobacter sulfurreducens* revealed putative open reading frames (ORFs) for 73 multiheme c-type cytochromes [58]. The genes responsible for U(VI) reduction include, triheme periplasmic cytochrome-*c7*, diheme periplasmic cytochrome, and others [59]. Mutations have been constructed in the number of genes for these proteins and it was observed that each of the mutant constructed from a number of genes negatively affected Fe(III) reduction rates with acetate as electron donor. However, analysis over 15 cytochromes mutants of *G. sulfurreducens* showed that there was no good correlation between effects on reduction rates of Fe(III) and U(VI) [46]. Interestingly, although this organism is proficient at reducing a broad range of extracellular Fe(III) and Mn(IV) minerals, and UO_2^{2+} it was observed to be inefficient in reducing, NpO_2^{+}, the reduced species of neptunyl (NpO_2^{2+}) exiting in the spent fuel nuclear waste [60]. The authors suggested that the enzyme system responsible for uranium reduction in *G. sulfurreducens* is specific for hexavalent actinides and is capable of transferring one electron to an actinyl ion, and the instability of the resulting U(V) then generates U(IV) via disproportionation.

2.10.3. *Shawanella reductase*

Among the species that conserve metabolic energy from dissimilatory respiration utilising U(VI) an electron sink is *Shewanella oneidensis* MR-1 [61]. *S. putrefaciens* cells cultures under Fe(II) lost their orange colour indicating a major decrease in c-type cytochrome content [61]. The interpretation of these observations was that cytochromes were involved in the transfer of electrons to the terminal electron acceptor or were the terminal reductases. Mutant analysis implicated the nitrate reductase in U(VI) reduction because of the simultaneous loss of U(VI) and NO_3^- reduction observed in the absence of these reductase [51]. Further mutant studies have implicated other proteins and cytochromes to be involved in metal reduction and a model for electron transfer was proposed. The function of these electron carriers for U(VI) reduction was only recently evaluated as a part of the analysis of global transcriptional responses to U(VI) [55].

By assay of mutants, several proteins including the one involved in menaquinone biosynthesis, decaheme outer membrane cytochrome, a periplasmic decaheme cytochrome, outer membrane protein, and a tetraheme cytochrome were all shown to be needed for optimal U(VI) reduction [55]. The multiple pathways for electron delivery to U(VI) available in *Shawanella* are associated with the capability of U(VI) reduction with lactate as an electron donor in mutants lacking one or more of the above electron transfer components [55]. Comparison of uraninite (UO_2) deposition by mutants lacking outer membrane decaheme *c*-type cytochromes showed accumulation predominantly in the periplasm versus the deposition of UO_2 external to wild-type cells [61]. This result indicate that U(VI) reduction is not eliminated by any of the single mutants analysed and also supports the hypothesis that uranium reductase are likely non-specific, and that low potential electron donors are present in both the periplasm and outer membrane. It remains to be determined whether the mutants altered for U(VI) reduction are similarly affected in their ability to use U(VI) as terminal electron acceptor for growth.

2.11. Permeable reactive barriers

Waste from power generation and fuel process facilities contains high levels of uranium and transuranic elements. This type of waste, classified as high level waste (HLW) or transuranic waste (TU), is usually solidified in a concrete or bitumen before it is stored in specially engineered facilities above the ground. The chances of environmental contamination from such facilities are slight. However, most of the voluminous intermediate level and low level waste (ILW and LLW) can be packed and stored underground. The underground storage facilities pose a high risk of groundwater contamination. Where contamination has actually occurred, pump-and-treat processes are utilised to intercept the polluted groundwater for treatment above ground. The water can be treated using chemicals or using biological reactors and the clean effluent is returned to the aquifer. For toxic metals, chemical agents may be added followed by precipitation to reclaim the metals [62]. The chemical reduction process utilizes toxic reducing agents that produce toxic sludge requiring further treatment before disposal into natural waters. Biological processes have been proposed for the pump-and-treat process, but this does not eliminate the problem of disposal of the product of the precipitation stage. Several techniques for installing a biological barrier have been attempted such as construction of semi-porous walls which require a fair amount of excavation (Figure 2), injection of nutrients to encourage the growth of certain types or native species in the environment (a form of bioaugmentation), and inoculation of a region down gradient of a pollutant with specialized cultures of bacteria. Molecular *in situ* bioremediation, the process of introducing new genetic material in native species, has not been put to practice anywhere apart from small scale experimental projects on petrochemical pollution [63].

Insitu bioremediation techniques using permeable reactive barriers (PRBs) have been used to treat organic pollutants that can be completely mineralized to safe products, carbon dioxide (CO_2) and water (H_2O), by microorganisms in the environment. A wide range of toxic recalcitrant organic compounds have been treated this way. For example, case studies have been reported on the treatment of petrochemical pollutants [i.e., benzene, toluene, ethylbenzene, xylene, polychlorinated biphenyls (PCBs), and polycyclic aromatic hydrocarbons (PAHs)], and agricultural pollutants: methyl-tert-butyl-ether (MTBE) and its congeners [7, 63].

The decreasing concentration of U(VI) across the barrier is envisioned if barrier is inoculated with U(VI) reducing bacterial species. In the case of U(VI) reduction across a barrier system, we hope to utilise U(VI) as an electron sink in a dissimilatory respiration process in which the organisms introduced in the barrier (X_a) will require U(VI) to optimise their growth. If the organisms require U(VI) as a growth limiting electron sink, their survival away from the barrier zone will be limited. This will prevent increased microbial counts in the aquifer water if the aquifer downstream of the direction of flow is utilised as a drinking water supply source.

The main limitation of *in situ* bioremediation for treatment of metals such as uranium is that the element is not destroyed but rather trapped in the aquifer matrix in a reduced state. Should

http://dx.doi.org/10.5772/56909

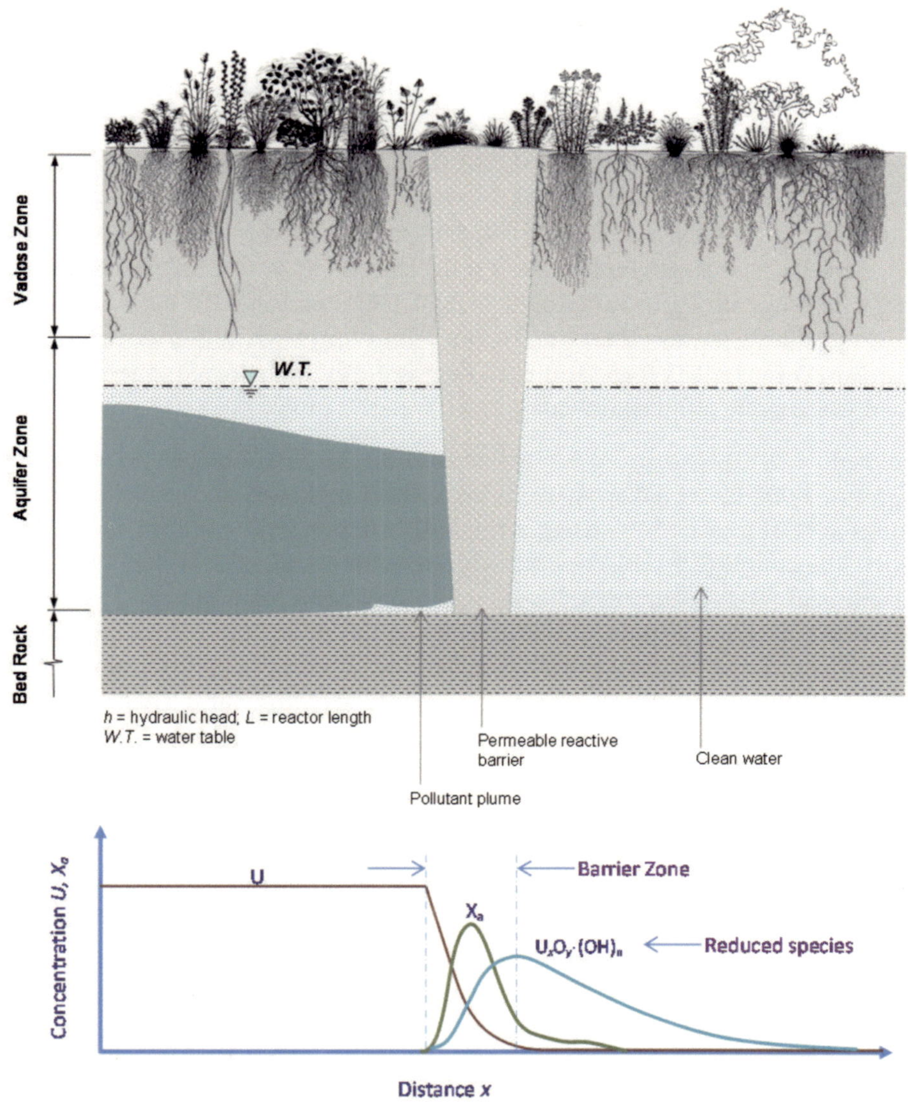

Figure 2. Theoretical representation of the microbial permeable reactive barrier system as an intervention for U(VI) pollution in an unconfined aquifer system. The graph shows the U(VI) concentration and biomass propagation under optimum operation conditions. U = uranium (VI) concentration, X_a = concentration of active biomass, and $U_xO_y(OH)_n$ = hydroxide precipitates of reduction products. The number of complexed hydroxyl ions, *n*, will depend on the charge on the uraninite group $U_xO_y^{n+}$.

the conditions in the soil change one day in future, the metal may be remobilized to its chemically toxic and mobile state and migrate down gradient to further contaminate groundwater and surface water resources. The alternative is to let the reduced form migrate without precipitating in the aquifer medium. This can then be removed by a pump-and-treat method as described above.

3. Determination of U(VI)

U(VI) reductase activity was determined by measuring the decrease in U(VI) in the solution using UV/vis spectrophotometer (WPA, Light Wave II, and Labotech, South Africa). Arsenazo III (Sigma-Aldrich, St. Louis, Missouri, USA) (1, 8-dihydroxynaphthalene-3, 6 disulphonic acid-2, 7-bis [(azo-2)-phenylarsonic acid]), a non-specific chromogenic reagent, was selected as the complexing agent for facilitating U(VI) detection. The accuracy and the precision of the method on the UV/vis spectrophotometer was determined by measuring the concentration of standard U(VI) solution in the range of (0-80 mg/L). A linearized U(VI) standard curve was generated by plotting the absorbance at 651 nm versus the known U(VI) standard concentration. Standard curve for U(VI) measurement demonstrated high degree of accuracy with R^2 = 99.7% and was used to estimate unknown U(VI) concentration.

Measurement of U(VI) was carried out by withdrawing 2 mL of homogenous solution from a 100 mL serum bottle using a disposable syringe. The sample was then centrifuged for about 10 minutes at 6000 rpm (2820 *g*) using a Minispin® Microcentrifuge (Eppendorf, Hamburg, Germany). The sample (0.5-1 mL) was then diluted with 0.4 mL of 2.5% diethylene-triamine-penta acetic acid (DTPA) and then diluted up to mark with basal mineral medium (BMM) (1:10) dilution. The homogenous solution was then mixed with 2 mL of complexing reagent (Arsenazo III) and then allowed to stand for full development of the pink-violet color prior analyses for U(VI) at a wavelength of 651 nm (10 mm light path) against a reagent blank. Total uranium in the unfiltered sample was measured using inductively-coupled plasma mass spectrometry (ICP-MS).

4. Uranium (VI) reduction capability in pure cultures

In order to achieve *ex situ* and *in situ* biological treatment of water and soil contaminated by U(VI) and transuranic elements, it is necessary to search for microorganisms capable of reducing U(VI) under natural conditions. In a current investigation at the University of Pretoria, cultures isolated from a uranium mine dump were tested for uranium (VI) reduction under anaerobic and micro-aerobic conditions. The following section presents results from detailed batch experiments conducted with non-purified and purified cultures isolated from soil samples.

Microorganisms were isolated from the soil samples collected from tailings dumps of an abandoned uranium mine. Background uranium concentration in the original sample was detected at levels as high as 29 mg/kg, much higher than values observed in natural soils (0.3-11.7 mg/kg). To select U(VI) tolerant species, microorganisms from soil were cultured overnight into a 100 mL of sterile basal mineral medium (BMM) amended with glucose as sole carbon source and a dose of U(VI) (75 mg/L uranium (VI) as uranyl nitrate). The inoculum was grown under anaerobic conditions for 24 hours at 30±2°C in 100 mL serum bottles purged with pure (nitrogen) N_2 gas (99.9% pure grade) for about 5-10 minutes to expel residual oxygen before sealing the bottle with rubber stoppers and aluminium seal. After 24 hours enriched bacterial strains were isolated by serial dilution of the cultivated culture. U(VI) reduction activity was evaluated for the isolates starting with evaluation for abiotic processes to make sure that physical-chemical processes are taken into consideration when analysing the biological process.

5. Abiotic U(VI) reduction

Heat-killed cultures and sodium azide exposed cultures were used to determine the extent of abiotic U(VI) reduction in batch experiments. For U(VI) reduction experiments cultures were grown over night in a sterile nutrient or Luria-Bettani (LB) broth under anaerobic conditions. Overnight grown cells were heat killed by autoclaving at 121°C for 20 minutes and another set of overnight grown cells were incubated with 0.1% of sodium azide (NaN_3). Cells were then harvested by centrifuging at 6000 rpm (2820 *g*) for 10 minutes. The supernatant was then decanted and the remaining pellet was washed three times with sterile 0.85% NaCl solution. The washed pellet was then re-suspended in 100 mL serum bottle containing U(VI) stock solution and sterile basal mineral medium (BMM) amended with D-glucose as sole carbon source. The serum bottles were then purged with 99.9% (N_2) for about 5-10 minutes to expel residual oxygen before sealing the bottle with rubber stoppers and aluminium seal. The experiments were all conducted at 30±2°C with continuous shaking on lateral shaker (Labotec, Gauteng, South Africa). The experimental conditions in the abiotic controls and the live cells experiments were kept the same (100 mL serum bottles containing BMM amended with D-glucose, 100 mg/L U(VI) solution, and incubated at 30±2°C under anaerobic conditions). A sample was withdrawn at regular time interval using a disposable syringe for U(VI) reduction analysis as described above.

The results showed insignificant difference U(VI) reduction between live and heat-killed cells (Figure 3). The instantaneous U(VI) reduction in heat-killed cells may be due to the presence of the cells that escaped destruction by heat. The reduction of U(VI) observed during the first 2 hours in all treatment containing biomass presented an anomaly. It was clear from these results that another mechanism rather than the direct metabolic process was involved in the U(VI) removal from solution. On the other hand abiotic control (without bacteria) showed that U(VI) reduction process is biologically mediated.

Reverting back to the biosorption studies, it is suggested that functional groups on the cell wall surfaces (-OH, -NH_2, and –COOH) may serve as ligands for U(VI)-U(IV) complexation with the cell surface. U(VI) reduction may serve as a step towards this complexation step. To evaluate these effects we conducted experiments where the pH was varied and the oxidation reduction potential (ORP) was measured with time. Results presented in Figure 4 show that that the rate of U(VI) reduction was pH dependent (Figure 4a). Electronegative conditions under anaerobic conditions created a strongly reducing environment as expected, after which the ORP increased to electropositive values (Figure 4b). As a result insignificant change over time in ORP indicated poor oxidation-reduction, while significant change in ORP over time indicated that the oxidation-reduction process approaches completeness.

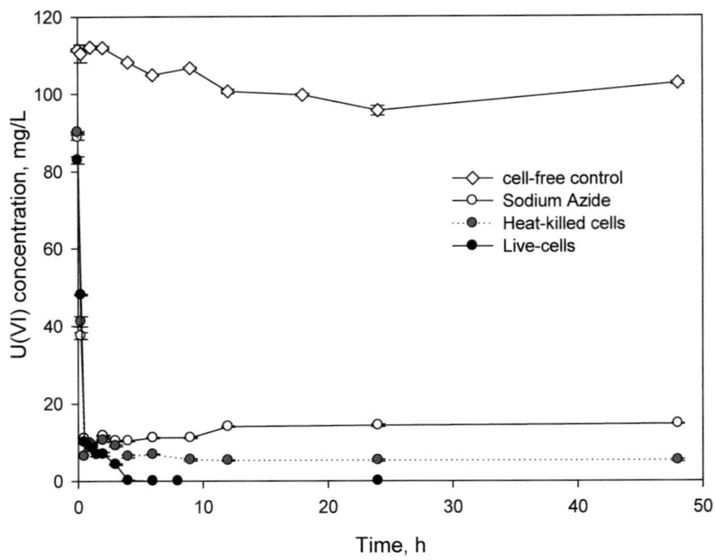

Figure 3. Evaluation of abiotic U(VI) reduction in heat-killed and azide inhibited cells

6. U(VI) Reduction by non-purified cultures (Consortium)

Preliminary experiments were conducted over a wide range of U(VI) concentration (30-400 mg/L) under similar experimental conditions (100 mL serum bottles containing BMM amended with D-glucose, U(VI) solution, and then incubated at 30±2°C under anaerobic conditions) using a reconstituted consortium culture of several identified U(VI) tolerant species. Results showed complete U(VI) reduction in batch cultures at initial U(VI) concentration up to 300 mg/L within 24 hours. In all batch studies with U(VI) concentration up to 400 mg/L (80-100%) U(VI) removal was achieved within the first 5 hours of incubation. However, inhibition of the reduction process was observed at the initial U(VI) concentration of 400 mg/L over time (Figure 5a).

U(VI) reduction trends in batches using purified cultures with single species per batch showed similar trends of U(VI) reduction. Figure 5b shows the summary of results from the best performing cultures labeled Y1, Y5, and Y6. The species characterisation results for these pure cultures are presented later in the chapter. The results in Figure 5b show that the microorganisms existing as a community possess significant stability and metabolic capabilities which can be linked to the effectiveness of synergistic interactions among members of bacterial communities [64].

http://dx.doi.org/10.5772/56909

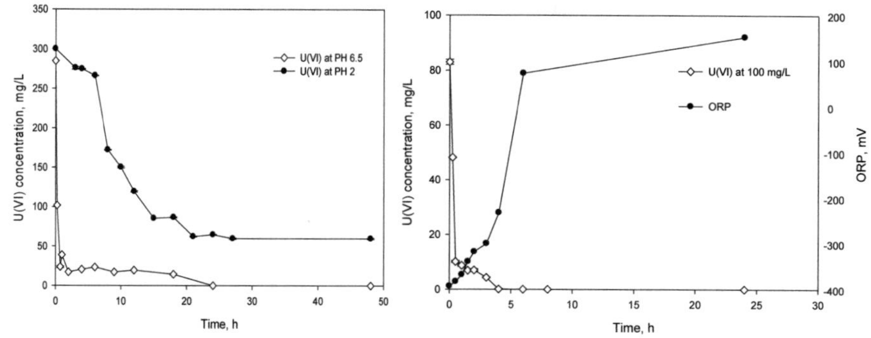

(a) (b)

Figure 4. (a) U(VI) reduction at different pH values – U(VI) reduction rate increased with increasing pH, and (b) data showing loss of U(VI) reduction capacity as ORP increases.

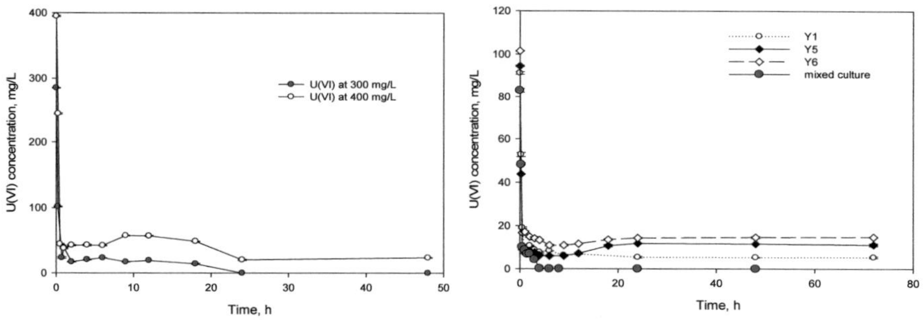

(a) (b)

Figure 5. (a) U(VI) reduction in reconstituted consortium culture from mine soil under the initial concentration of 300 and 400 mg/L, and (b) comparative performance of three pure isolates against the reconstituted consortium culture. The reconstituted consortium culture shows the best performance possibly due to symbiotic interactions within the culture system.

7. Fate of reduced uranium species in a cell

Proportional distribution of uranium precipitates in the medium and cells can be used to determine the location of U(VI) reductase activity in the culture system. This is because most the precipitates are formed from reduced uranium species. Transmission electron microscopy (TEM) was used to establish the distribution and localization of uranium deposits in the cells. The energy dispersive X-ray (EDX) spectrometer coupled to the TEM was also used for elemental characterization of the metal deposits in the medium. TEM result images show crystal structures in the medium and very little crystallisation inside the cells (Figure 6a). EDX analysis of the crystals deposited on the cell surfaces confirmed the accumulation of uranium elements in the crystal matrix. Extracellular depositions of uranium also indicate that bacteria are excellent nucleation sites for mineral formations. EDX spectra derived from the uranium deposits show that they are composed of the following elements U>Cu>P>Os>Ca>Co>Fe (according to their descending order of their weight %). The higher copper (Cu) peak results from the specimen to support grid used. Phosphorous observed in the spectrum could either be from the added phosphorous or could microbially produced. On the other hand no uranium was observed in the metal free biomass (Results not shown).

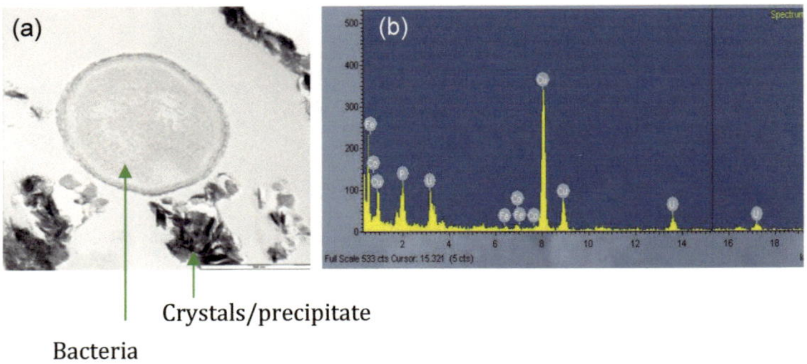

Figure 6. (a) TEM Scan of bacterial cells indicating deposition of uranium species on cell surface and (b) EDX spectrum of precipitate.

8. Microbial characterisation and activity

8.1. Correlation of U(VI) reduction to enzyme activity

Proteins make up a large fraction of the biomass of actively grown microbes. To determine microbial activity over time, protein concentration was determined using a UV/Vis Spectrophotometer (WPA, Light Wave II, and Labotech, South Africa) at the wavelength of 595 nm using Coomassie Dye as a complexing agent to facilitate protein detection. Samples required pre-treatment to reduce interferences during the spectrophotometric analyses. Cell lysis was achieved by ultrasonification of acid treated cells. Results showed that microbial activity decreased with increasing U(VI) reduction (Figure 7). These results served as a confirmation of enzymatic activity as responsible agent for U(VI) reduction.

8.2. Culture composition analysis

The phylogenetic characterization of cells from the mine dump soil was conducted after subculturing the cells in nutrient or Luria-Bettani broth. Individual colonies from a serially diluted preparation were carefully examined for colony morphology and cell morphology by Gram-staining. This process, we recognize, could eliminate a wide range of potential U(VI) reducers especially anaerobic species in the samples. But at this stage, we were targeting the species that can survive under facultative anaerobic conditions.

The purified colonies were streaked on nutrient agar followed by incubating at 30°C for 18 hours in preparation for 16S rRNA gene sequence analysis. Microbial pure cultures were grown from loop-fulls from individual colonies, transferred to fresh media containing low amounts (30-75 mg/L) of uranyl nitrate. The process was repeated at least three times for each colony type to achieve close to a pure culture of each identified species.

Genomic DNA was extracted from purified colonies according to the protocol described for the Wizard Genomic DNA purification kit (Promega Corporation, Madison, WI, USA). 16S rRNA genes were amplified by a reverse transcriptase-polymerase chain reaction (RT-PCR) using primers pA and pH1 (Primer pA corresponds to position 8-27; Primer pH to position 1541-1522 of the 16S gene under the following reaction conditions: 1 min at 94°C, 30 cycles of 30s at 94°C, 1 min at 50°C and 2 min at 72°C, and a final extension step of 10 min at 72°C). PCR fragments were then cloned into pGEM-T-easy (Promega) [Promega Wizard® Genomic DNA Purification Kit (Version 12/2010)]. The 16S rRNA gene sequences of the strains were aligned with reference sequences from *Desulfovibrio sp., Geobacter sp., Acinetobacter sp., Anthrobacter sp.*, and *Shewanella putrefaciens* using Ribosomal Database Project II programs. Sequence alignment was verified manually using the program BIOEDIT. Pairwise evolutionary distances based on an unambiguous stretch of 1274 bp were computed by using the Jukes and Cantor method [65].

U(VI) reducing colonies were identified from the genera *Bacilli, Acinobacter, Actinomycetes* and *Chrysebactreium*. Sections of phylogenetic tree diagrams with closet associations to know species are shown in Figures 8d. The associations shown Figure 8 have been reported among U(VI) reducing groups in literature. Fowle et al. [66] has shown that *Bacillus* species are effective biosorbents for uranium. Additionally, the capability of *Anthrobacter* species isolated from a uranium-contaminated site in accumulating uranium intracellularly as uranium precipitates closely associate with polyphosphate granules was also reported [21].

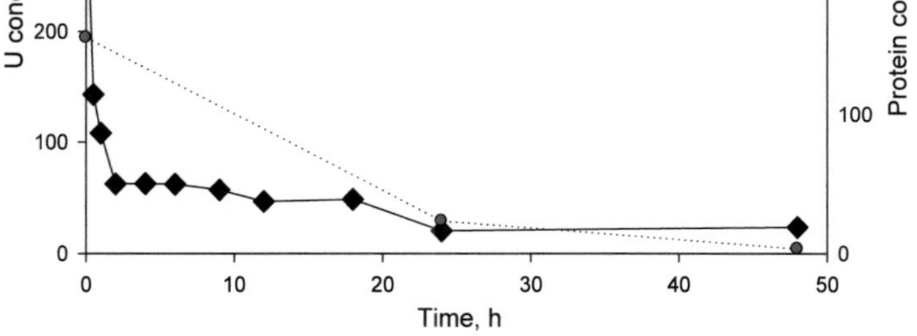

Figure 7. Evaluation of U(VI) reduction, protein concentration and total uranium under an initial concentration of 400 mg/L.

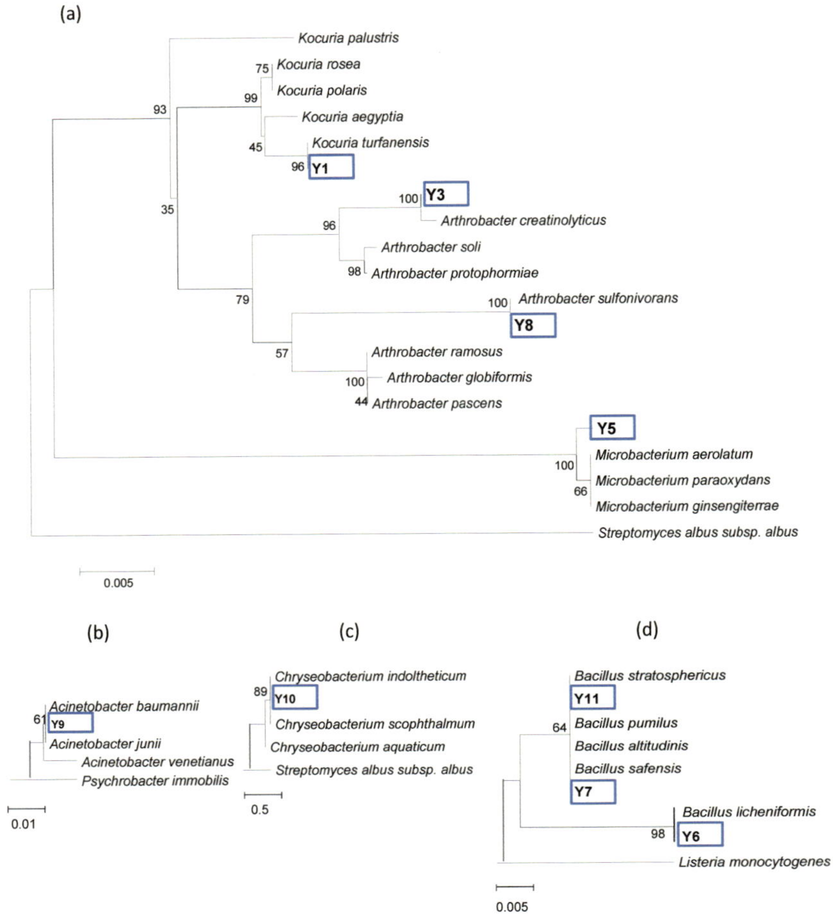

Figure 8. Phylogenetic analysis results showing the predominance of the Gram-positive (a) *Microbacterieceae* and *Anthrobacteriae*, and Gram-negative (b) *Acinetobater* under low U(VI) exposure. Colonies Y10 in figure 8c did not reduce uranium. Another uranium (VI) reducing species, Y6, was also identified among the Bacilli shown in Figure 8d.

In the phylogenetic analysis, the scale indicated at the bottom of the plots, e.g., 0.005 for Figure 8a represents the genetic distance, while the percentage numbers at the nodes indicate the level of bootstrap based on neighbour-joining analysis of 1000 replicates. The three species related to *Actinomycetes* were tolerant at 75 mg/L of U(VI) suggesting the capability of the species in reducing U(VI) in a basal mineral medium (BMM) amended with D-glucose (Figure 8a). The species related to *Acinobacter* as indicated in Figure 8b were also tolerant to U(VI) at concentration levels around 75 mg/L with D-glucose as a sole added carbon source. The group presented in Figure 8c, although tolerant to U(VI), did not reduce U(VI) under the conditions tested. Three species of 16S rRNA gene of *Bacilli* were tolerant at 75 mg/L and at least one of these, Colony Y6, was able to reduce U(VI) in a basal mineral medium (BMM) amended with D-glucose as a sole added carbon source.

9. U(VI) removal kinetics

9.1. Kinetic model adaptation

To model a biological U(VI) reducing system, the reaction scheme, rate equations and kinetic constants for the processes taking place in the batch reactor are chosen from published models on enzymatic reduction hexavalent toxic metals such as U(VI). Shen and Wang [67] demonstrated that the rate of U(VI) reduction by enzymes can be expressed as the Monod equation if viable cell concentration X is correlated to enzymes produced E:

$$-\frac{dU}{dt}=\frac{k_u\cdot U}{K_u+U}\cdot X \tag{2}$$

where: U = U(VI) concentration at time, t (mg/L); X = density of active bacterial cells at time, t (mg cells/L); k_u = specific rate of U(VI) reduction(mg U(VI)/mg cells/h); and K_u = half-velocity constant (mg/L). However, the active cell concentration, X, may be assumed to decrease in proportion to the amount of U(VI) reduced due to the toxicity of U(VI):

$$X=X_0-\frac{U_0-U}{T_u} \tag{3}$$

where: U_0 = initial U(VI) concentration (mg/L); X_0 = initial cells density of U(VI)-reducing strains (mg cells/L); and T_u = maximum U(VI) reduction capacity of cells (mg U(VI)/mg cell). Substituting Equation 3 into Equation 2 yields the following equation:

$$-\frac{dU}{dt}=\frac{k_u\cdot U}{K_u+U}\left(X_0-\frac{U_0-U}{T_c}\right) \tag{4}$$

U(VI) reduction data obtained with the pure cultures and the mixed culture were analyzed using Equation 4. Parameters in Equation 4 can be analyzed using simulation software such as AQUASIM or SigmaPlot. The model is calibrated using batch data over the incubation period. The values collected under non-inhibiting conditions are suitable for estimating the kinetic parameters k_u and K_u since these respond to cell growth dynamics. The parameter T_u is estimated under overloaded conditions since this parameter is related to the U(VI) reduction capacity of the cells.

9.2. Uranium (VI) reduction under inhibiting conditions

The inhibition model is suitable for application where the U(VI) loading per cell is very high. This is expected during startup (inoculation) of a systems with U(VI) already present. Such would be the case during the initial operation *in situ* bioremediation system. To account for toxic inhibition in such situations, a simple Monods non-competitive inhibition kinetic model incorparating inhibition term, K is suggested:

$$-r_u = \frac{k_u U}{(K_u + U)\left(K^{\left(1-\frac{U_T}{U_0}\right)}\right)}\left(X_0 - \frac{U_0 - U}{T_u}\right) \quad (5)$$

where: k_u = maximum specific rate of U(VI) reduction (1/h); K_u = half-velocity constant (mg/L); U_T = U(VI) toxicity threshold concentration (mg/L); X_o = initial biomass concentration (mg/L); K = limiting constant (mg/L); and T_u = U(VI) reduction capacity (mg U(VI) reduced/mg cells).

9.3. Continuous flow systems

Continuous flow systems better simulate actual systems especially where *in situ* bioremediation is planned. Batch systems cannot simulated the effects of diffusion, clogging of pores, advection rates. Packed columns have been used in simulating the operational conditions of a barrier system [68]. The non-steady state dynamic process of uranium removal in packed column is represented by Equation 6. The mass balance of uranium (VI) across the packed column includes U(VI) reduction rate (r_u), mass transport rate (j_u), adsorption rate (q_u). Due to the inclusion of the interstitual term u and the mass transport term j_u, the predominant process affecting performance during the transient-state in the advection and diffussion:

$$\frac{d(U \cdot V)}{dt} = A\sum_{l=0}^{l=L_i} u(U_{in} - U) - r_u \cdot \Delta V - j_u \cdot A_f - q_u \cdot \Delta V \quad (6)$$

where: U = effluent U(VI) concentration (mg/L); V = volume of the reactor (L); U_{in} = influent U(VI) concentration (mg/L); Q = influent flow rate (L/h); r_u = U(VI) reduction rate coefficient (mg/L/h); t = time (h); A_f = biomass surface area (m^2); J_u = flux of dissolved species into the biofilm (mg/m^2/h); and q_u = rate of U(VI) removal by adsorption (1/h). The interstitial velocity u (m/h) is assumed to be constant throughout the entire column. The diffusional flux (J_u) can be expresed using Fick's law:

$$\frac{\partial \mathbf{c}_i}{\partial t} = D_w \frac{\partial^2 \mathbf{c}_i}{\partial x^2} \quad (7)$$

Where: $\mathbf{c}_i$ = concentration of dissolved species (mg/L); D_w = diffusivity of dissolved species in water (m^2/h); t = time (h); and x = spatial coordinate (m).

http://dx.doi.org/10.5772/56909

The reaction term r_u is dependent on the amount of biomass accumulated in the void space of the column. However, due to space limitations, cells may only grow to a certain maximum concentration. The time at which the cells reach the maximum allowable concentration is dependent on initial cells, U(VI) toxicity, and hydraulic loading rate. These conditions cause the cells to follow a logistic curve defined by Equation 8:

$$X = X_0 + \frac{X_{\max}}{1+\left(\frac{t}{t_0}\right)^b} \tag{8}$$

where: X = viable cell density (mg/L) at any time t (h); X_{max} = maximum attainable viable cell concentration (mg/L) in the barrier column, t_0 = logistic interval (h); and b = pitch (dimensionless). The impact of the adsorptive process was determined to be minimal during continuous flow operation for an extended period of time. This is because reaction sites tend to become saturated as the system approaches equilibrium.

10. Conclusion

The chapter addresses the main feature of various U(VI) remediation techniques involving the *in situ* bioremediation using permeable reactive barrier. The technique is well known for its effectiveness for remediating organic pollutants. However, its effectiveness for removal of metallic species is hindered by possible accumulation of precipitates. In our preliminary batch studies it was observed that isolated organisms are capable of immobilizing U(VI) by means of more than one mechanism, i.e., biosorption and enzymatic reduction. These results open a new research field for understanding which of these mechanism is predominant and in what sequence does the U(VI) reduction take place under anaerobic conditions. Modelling *in situ* U(VI) bioreduction involve many uncertain parameters, including those of aqueous U(VI) speciation, surface complexation, and bioreduction kinetics. Therefore, for efficient application, sensitivity analysis is needed to simplify the models such as presented. Furthermore, re-oxidation of the biologically reduced uranium needs to be included in future models to evaluate long-term stability of bioreduction techniques.

Nomenclature

A - effective cross sectional area (m^2)

A_{f}- biomass surface area (m^2)

b - pitch factor (dimensionless)

D_w - diffusivity of dissolved species in water (m^2/h)

J_u - U(VI) flux rate (mg/m^2/h)

k_u - reaction rate coefficient (1/h)

K - inhibition coefficient (mg/L)

K_u - half-velocity constant (mg/L)

Q - inflow rate (L/h)

q_u - rate of U(VI) by adsorption (mg/L/h)

q_m - maximum specific uptake of metal corresponding to site saturation (mg/g)

r_u - U(VI) reduction rate (mg/L/h)

t - time (h)

t_0 - logistic time interval (h)

T_u - U(VI) reduction capacity (mg U(VI) reduced/mg cells)

U - effluent U(VI) concentration at time, t (mg/L)

U_{in} - influent U(VI) concentration (mg/L)

U_0 - initial value, U(VI) concentration at time zero (mg/L)

U_T - U(VI) toxicity threshold concentration (mg/L)

V - volume of the reactor (L)

ΔV - differential volume (L)

x - spatial coordinate (m)

X - biomass concentration at time t (mg/L)

X_{max} - maximum attainable viable cell concentration (mg/L)

X_0 – initial biomass concentration (mg cells/L)

Author details

Phalazane Johanna Mtimunye and Evans M. N. Chirwa

*Address all correspondence to: mtimunyepj@gmail.com

Water Utilization Division, Department of Chemical Engineering, University of Pretoria, Pretoria, South Africa

References

[1] IAEA. Nuclear Technology Review. International Atomic Energy Agency Scientific and Technical Publication Series, Vienna, Austria; 2009.

[2] Greve, K., Niesen, E., Ladefogen, O. Evaluation of health hazards by exposure to strontium in drinking water. Toxicology Letters 2007; 172(1) S210.

[3] Ajlouni, AMS. Health consequences of nuclear fission products. Journal of Applied Science and Environmental Management 2007; 11(3) 11-14.

[4] World Nuclear Association (WNA); 2008. http://www.world-nuclear.org/info/inf88.html (accessed 7 November 2012).

[5] Tikilili PV, Chirwa EMN. Characterization and Biodegradation of Polycyclic Aromatic Hydrocarbons in Radioactive Wastewater. Journal of Hazardous Materials 2011; 192(3) 1589-1596.

[6] Bouwer EJ, Zehnder AJB. Bioremediation of organic compounds-putting microbial metabolism to work. Trends Biotechnology 1993; 11(8) 360-367.

[7] Craig DK. Chemical and Radiological Toxicity of Uranium and its Compounds, Report WSRC-TR-2001-00331, Contract No. DE-AC09-96SR18500, Westinghouse Savannah River Company, Aiken/U.S. Department of Energy; 2001.

[8] Chirwa EMN. Developments in bioremediation for separation/recovery. K.L. Nash and G.J. Lumetta (Eds.), in Advanced Separation Techniques for Nuclear Fuel Reprocessing and Radioactive Waste Treatment, Woodhead Publishing, Ltd, Cambridge, UK; 2011. p436-472.

[9] Doherty R, Phillips DH, McGeough KL, Walsh KP, Kalin RM. Development of modified flyash as a permeable reactive barrier medium for a former manufactured gas plant site. Northern Ireland. Environmental Geology 2006; 50(1) 37-46.

[10] IAEA. Treatment of Liquid Effluent from Uranium Mines and Miles. Report of a Coordinated Research Project 1996-2000. International Atomic Energy Agency (IAEA), 2004; Vienna, Austria.

[11] Gavrilescu M. Overview of in situ remediation technologies for sites and groundwater. Environmental. Engineering and Management Journal 2006; 5(1) 79-114.

[12] Olexsey RA, Parker RA. Current and future *in situ* treatment techniques for the remediation of hazardous substances in soil, sediments, and groundwater I. Twardowska, H.E. Allen, M.M. Häggblom, S. Stefaniak (Eds.), Soil and Water Pollution Monitoring, Protection and Remediation, NATO Science Series, IV. Earth and Environmental Science, Springer, Netherlands 2006; 69, 211-219

[13] Zaganiaris EJ. Ion Exchange Resins in Uranium Hydrometallurgy, Books on Demand GmbH, Paris, France; 2009.

[14] Traut DE, Nichols IL, Seidel DC. Design requirements for uranium ion exchange from ammonium bicarbonate solutions in a fluidized system. Department of the Interior, Bureau of Mines, USA; 1978.

[15] Ross JR, George DR. Recovery of uranium from natural mine waters by countercurrent ion exchange.US Department of Interior, Bureau of Mines, University of Michigan; 1971. p7471-7480.

[16] Pabby AK, Rizvi SSH, Sastre AM. Membrane techniques for treatment in nuclear waste processing, a global experience. Membrane Technology 2008; 11: 9-13.

[17] Zhou P, Gu P. Extraction of oxidized and reduced forms of uranium from contaminated soils: effects of carbonate concentration and pH. Environmental Science and Technology 2005; 39(12) 4435-4440.

[18] Phillips EJP, Landa ER, Lovley DR. Remediation of uranium contaminated soils with bicarbonate extraction and microbial U(VI) reduction. Journal of Industrial Microbiology 1995; 14(3-4) 203-207.

[19] Gramss G, Voight KD, Bergmann H. Plant availability and leaching of heavy metals from ammonium-calcium, carbohydrate, and citric acid-treated uranium-mine-dump soil. Journal of Plant Nutrition and Soil Science 2004; 167, 417-427.

[20] Kantar B, Honeyman BD. Citric acid enhanced remediation of soil contaminated with uranium by soil flushing and soil washing. Journal of Environmental Engineering 2006; 132(2) 247-255.

[21] Suzuki Y, Banfield J. Resistance to, and accumulation of, uranium by bacteria from a uranium-contaminated site. Geomicrobiology Journal 2004; 21(2) 113-121.

[22] Nancharaiah YV, Joshi HM, Mohan TVK, Venugopalan VP, Narasimhan SV. Aerobic granular biomass: a novel biomaterial for efficient uranium removal. Current Science 2006; 91(4) 503-509.

[23] Nedelkova M, Merroun ML, Rossberg A, Hennig C, Selenska-Pobell S. Microbacterium isolates from the vicinity of a radioactive waste depository and their interactions with uranium. FEMS Microbiology Ecology 2007; 59(3) 694-705.

[24] Bush MB. Ecology of a Changing Planet, 3rd Edition, Prentice Hall, New Jersey, USA, 2003.

[25] Nealson KH. Post-Viking microbiology: new approaches, new data, new insights. Origins of life and evolution of the biosphere. Journal of the International Society for the Study of the Origin of Life 1999; 29(1), 73-93.

[26] Kalckar HM. Origins of the concept oxidative phosphorylation. Molecular and Cellular Biochemistry 1974; 5(1-2), 55-63.

[27] Lloyd JR. Microbial reduction of metals and radionuclides. FEMS Microbiology Reviews 2003; 27, 411-425.

[28] Dadachova E, Bryan RA, Huang X, Moadel T, Schweitzer AD, Aisen P, Nosanchuk JD, Casadevall A. Ionizing radiation changes the electronic properties of melanin and enhances the growth of melanized fungi. PLoS One, May 23, 2007.

[29] Francis AJ, Gillow JB, Dodge CJ, Harris R, Beveridge TJ, Papenguth HW. Uranium association with halophilic and non-halophilic bacteria and archaea. Radiochimica Acta 2004; 92(8) 481-488.

[30] Nakajima A, Tsuruta T. Competitive biosorption of thorium and uranium by Micrococcus luteus. Journal of Radioanalytical and Nuclear Chemistry 2004; 260(1) 13-18.

[31] Kratochvil D, Volesky B. Advances in the biosorption of heavy metals, Trends in Biotechnology 1998; 16(17) 291-300.

[32] Sar P, D'Souza SF. Biosorption of thorium (IV) by a Pseudomonas strain. Biotechnology Letters 2002; 24(3) 239-243.

[33] Jroundi F, Merroun ML, Arias JM, Rossberg A, Selenska-Pobell S, González-Muñoz MT. Spectroscopic and microscopic characterization of uranium biomineralization in*Myxococcus xanthus*. Geomicrobiology Journal 2007; 24(5) 441-449.

[34] Suzuki Y, Banfield JF. Geomicrobiology of uranium. Reviews in Mineralogy Geochemistry 1999; 38(1) 393-432

[35] Macaskie LE, Bonthrone KM, Yong P, Goddard DT. Enzymically mediated bioprecipitation of uranium by a Citrobacter sp. a concerted role for exocellular lipopolysaccharide and associated phosphatase in biomineral formation. Microbiology 2000; 146(8) 1855-1867.

[36] Reeder RJ, Nugent M, Tait CD, Morris DE, Heald SM, Beck KM, Hess WP. A Lanzirotti Coprecipitation of uranium(VI) with calcite: XAFS, micro-XAS, and luminescence characterization. Geochimica et Cosmoschimica Acta 2001; 65(20) 3491–3503.

[37] Lovley DR, Widman PK, Woodwar, JC, Phillips EJP. Reduction of uranium by cytochrome *c*3 of *Desulfovibrio vulgaris*. Applied Environmental Microbiology 1991; 59(11) 3572-3576.

[38] Gorby YA, Lovley DR. Enzymatic uranium precipitation. Environmental Science and technology 1992; 26(1) 205-207.

[39] Anderson C, Pedersen K. In situ growth of *Gallionella* biofilms and partitioning of lanthanides and actinides between biological material and ferric oxyhydroxides. Geobiology 2003; 1(2) 169-178.

[40] Rothschild LJ, Mancinelli RL. Life in extreme environments. Nature 2001; 409(6823) 1092-1101.

[41] Istok JD, Senko JM, Krumholz LR, Watson D, Bogle MA, Peacock A, Chang YJ, White DC. In situ bioreduction of Technetium and Uranium in a nitrate-contaminated Aquifer. Environmental Science and Technology 2004; 38(2) 468-475.

[42] Woolfolk CA, Whiteley HR. Reduction of inorganic compounds with molecular hydrogen by *Micrococcus lactilyticus*. I. Stoichiometry with compounds of arsenic, selenium, tellurium, transition and other elements. Journal of Bacteriology 1962; 84(4) 647-658.

[43] Lovley DR, Phillips EJP. Reduction of uranium by *Desulfovibrio desulfuricans*. Applied and Environmental Microbiology 1992; 58(3) 850-856.

[44] Coates JD, Bhupathiraju VK, Achenbach LA, McInerney MJ, Lovley DR.*Geobacter hydrogenophilus* and *Geobacter chapellei* and *Geobacter grbiciae*, three new, strictly anaerobic, dissimilatory Fe(III)-reducers. International Journal of Systematic and Evolutionary Microbiology 2001; 51(2) 581-588.

[45] Francis AJ, Dodge CJ, Lu F, Halada GP, Clayton CR. XPS and XANES studies of uranium reduction by *Clostridium sp*. Environmental Science and Technology 1994; 28(4) 636-639.

[46] Shelobolina ES, Sullivan SA, O'Neill KR, Nevin KP, Lovley DR. Isolation, characterization, and U(VI)-reducing potential of a facultatively anaerobic, acid-resistant bacterium from low-pH, nitrate- and U(VI)-contaminated subsurface sediment and description of *Salmonella subterranea* sp. nov. Applied and Environmental Microbiology 2004; 70(5) 2959-2965.

[47] Wu Q, Sanford RA, Löffler FE. Uranium reduction by *Anaeromyxobacter dehalogenans* strain 2CP-C. Applied and Environmental Microbiology 2006; 72(5) 3608-3614.

[48] Lloyd JR, Chesnes J, Glasauer S, Bunker DJ, Livens FR, Lovley DR. Reduction of actinides and fission products by Fe(III)-reducing bacteria. Geomicrobiology Journal 2002; 19(1) 103-120.

[49] Kashefi K, Lovely DR. Reduction of Fe(III), Mn(IV), and toxic metals at 100°C by *Pyrobaculum islandicum*. Applied and Environnemental Microbiology 2000; 66(3) 1050-1056.

[50] Lloyd JR, Leang C, Hodges-Myerson AL, Coppi MV, Cuifo S. Biochemical and genetic characterization of PpcA, a periplasmic *c*-type cytochrome in *Geobacter sulfurreducens*. Biochemical Journal 2003; 369(1) 153-161.

[51] Wade R, DiChristina TJ. Isolation of U(VI) reduction-deficient mutants of *Shewanella putrefaciens*. FEMS Microbiology Letters 2000; 184(2) 143-148.

[52] Lovley DR, Roden EE, Phillips EJ, Woodward JC. Enzymatic iron and uranium reduction by sulfate-reducing bacteria. Marine Geology 1993; 113(1-2) 41-53.

[53] Payne RB, Gentry DM, Rapp-Giles BJ, Casalot L, Wall JD. Uranium reduction by Desulfovibrio desulfuricans strain G20 and a cytochrome c3 mutant. Applied and Environmental Microbiology 2002; 68(6) 3129-3132.

[54] Kieft TL, Fredrickson JK, Onstott TC, Gorby YA, Kostandarithes HM. Dissimilatory reduction of Fe(III) and other electron acceptors by a *Thermus* isolate. Applied and Environmental Microbiology 1999; 65(3) 1214-1221.

[55] Bencheikh-Latmani R, Williams SM, Haucke L, Criddle CS, Wu L. Global transcriptional profiling of *Shewanella oneidensis* MR-1 during Cr(VI) and U(VI) reduction. Applied and Environmental Microbiology 2005; 71(11) 7453-7460.

[56] Elias DA, Suflita JM, McInerney MJ, Krumholz LR. Periplasmic cytochrome c3 of Desulfovibrio vulgaris is directly involved in H_2-mediated metal but not sulfate reduction. Applied and Environmental Microbiology 2004; 70(1) 413-420.

[57] Caccavo FJ, Blakemore R, Lovley D. A hydrogen-oxidizing, Fe(III)-reducing microorganism from the Great Bay Estuary, New Hampshire. Applied and Environmental Microbiology 1992; 58(10) 3211-3216.

[58] Methe BA, Nelson KE, Eisen JA, Paulsen IT, Nelson W. Genome of Geobacter sulfurreducens: metal reduction in subsurface environments. Science 2003; 302(5652) 1967-1969.

[59] Wu Q, Sanford RA, Löffler FE. Uranium reduction by Anaeromyxobacter dehalogenans strain 2CP-C. Applied and Environmental Microbiology 2006; 72(5) 3608-3614.

[60] Renshaw JC, Butchins LJ, Livens FR, May I, Charnock JM, Lloyed JR. Bioreduction of uranium: environmental implications of a pentavalent intermediate. Environmental Science and Technology 2005; 39 5657-5660.

[61] Kennedy DW, Marshall MJ, Dohnalkova AC, Saffarini DA, Culley DE. Role of *Shewanella oneidensis* *c*-type cytochromes in uranium reduction and localization. ASM 105th General Meet (Abstr.) Q-389; 2004.

[62] Mukhopadhyay B, Sundquist J, Schmitz JR. Removal of Cr(VI) from Cr-contaminated groundwater through electrochemical addition of Fe(II). Journal of Environmental Management 2007; 82(1) 66-76.

[63] Liu S-J, Jiang B, Huang G-Q, Li XG. Laboratory column study for remediation of MTBE-contaminated groundwater using a biological two-layer permeable barrier. Water Research 2006; 40(18) 3401-3408.

[64] Mukred AM, Hamid AA, Hamzah A, Yusoff WMW. Development of Three Bacterial Consoritum for the Bioremediation of Crude Petroleum-oil in Contaminated Water. Journal of Bio sciences 2008; 8(4) 73-79.

[65] Jukes TH, Cantor CR. Evolution of protein molecules, In Munro, H.N. (Ed.), Mammalian Protein Metabolism. Academic Press, New York, 1969, p21-123.

[66] Fowle DA, Fein JB, Martin AM. Experimental study of uranyl adsorption onto *Bacillus subtilis*, Environmental Science and Technology 2000; 34(17) 3737-3741.

[67] Shen H, Wang Y T. Modeling hexavalent chromium reduction in Escherichia coli ATCC 33456. Biotechnology and Bioengineering 1994; 43(4) 293-300.

[68] Mtimunye PJ. Steady-state model for hexavalent chromium reduction in simulated biological reactive barrier: microcosm analysis. Master's Thesis. University of Pretoria, Pretoria, South Africa; 2011. http://upetd.up.ac.za/thesis/available/etd-09222011-104550/ (accessed on 30/10/2012).

Chapter 9

Removal of Hexavalent Chromium from Solutions and Contaminated Sites by Different Natural Biomasses

Ismael Acosta-Rodríguez,
Juan F. Cárdenas-González,
María de Guadalupe Moctezuma-Zárate and
Víctor M. Martínez-Juárez

Additional information is available at the end of the chapter

http://dx.doi.org/10.5772/56152

1. Introduction

Chromium (Cr) toxicity is one of the major causes of environmental pollution emanating from tannery effluents. This metal is used in the tanning of hides and leather, the manufacture of stainless steel, electroplating, textile dyeing and as a biocide in the cooling waters of nuclear power plants. Consequently, these industries discharged chromium (VI) bearing effluents which are of significant environmental concerns [1]. Cr exists in nine valence states ranging from -2 to +6. From these, only the hexavalent [Cr (VI)] and trivalent chromium [Cr (III)] have primary environmental significance since they are the most stable oxidized forms in the environment.

Both are found in various bodies of water and wastewaters [2]. Cr (VI) typically exists in one of these two forms: chromate (CrO_4^{-2}) or dichromate ($Cr_2O_7^{-2}$), depending on the pH of the solution [2].These two divalent oxyanions are very water soluble and poorly adsorbed by soil and organic matter, making them mobile in groundwater. Both chromate anions represent acute and chronic risks to animals and human health, since they are extremely toxic, mutagenic, carcinogenic and teratogenic [3]. In contrast to Cr (VI) forms, the Cr (III) species are predominantly hydroxides, oxides and sulphates, less water soluble, less mobile, 100 times less toxic [4] and 1,000 times less mutagenic [5]. The principal techniques for recovering or removing Cr (VI), from wastewater are: chemical reduction and precipitation, adsorption on activated carbon, ion exchange and reverse osmosis [6]. However, these methods have certain drawbacks, namely high cost, low efficiency, generation of toxic sludge or other wastes that require disposal and imply operational complexity [7]. In this context, considerable attention has been focused in recent years upon the field of biosorption for the removal of heavy metal ions from aqueous effluents [8].

The process of heavy metal removal by biological materials is known as biosorption. Biomass viability does not affect the metal uptake. Therefore any active metabolic uptake process is currently considered to be a negligible part of biosorption. Various biosorbents have been tried, which include seaweeds, molds, yeast, bacteria, crab shells, agricultural products such modified corn stalks, [9], hazelnut shell [10], orange waste [11] and tamarind peel [12]. It has also been reported that some of these biomass can reduce chromium (VI) to chromium (III), like *Litchii cinensis* Sonn peel [13], tea fungal biomass [14], Mesquite [15], Eucalyptus bark [16], red rose's waste biomass [17] and Yohimbe bark [18]. The present study is undertaken with following objective: Investigate the use of different natural biomasses for the biosorption and reduction of Chromium (VI) in aqueous solution, and their elimination from contaminated sites.

2. Material and methods

2.1. Biosorbents used

Litchi chinensis Sonn, Citrus limonium, Mammea americana, Tamarindus indica, Citrus sinensisOsbeck, Citrus reticulate, Cucumis melo L., and *Musa cavendishii* shells were obtained from the fruits harvested between the months of June-September 2010, in the marketplace Republic of San Luis Potosi, SLP. Mexico. To obtain the natural biomass, shells rind were washed with trideionized water 72 h under constant stirring, with water changes every 12 h. Subsequently, boiled for 1 h to remove traces of the fruit and was dried at 80°C, for 12 h in the oven, ground in blender and stored in amber vials until use.

2.2. Preparation of stock solution

An aqueous stock solution (1000 mg/L) of Cr (VI) ions was prepared using $K_2Cr_2O_7$ salt. pH of the solution was adjusted using 0.1 N HCl or NaOH. Fresh dilutions were used for each study.

2.3. Biosorption studies

The biosorption capacity of shells biomasses was determined by contacting 100 mL of solution containing different concentration of Cr (VI) (100-1000 mg/L) in 250 mL Erlenmeyer glass flasks, with 1 g of biomass. The mixture was shaken in a rotary shaker at 120 rpm followed by filtration at different times (covering minutes, days and weeks). The filtrate containing the residual concentration of Cr (VI) was determined spectrophotometrically at 540 nm after complexation with 1, 5 Diphenylcarbazide, these method have a detection limit between 0.02-0.5 mg/L of Cr (VI) [19], Cr (III) with Chromazurol S [20], and Cr total by Electrothermal Atomic Absorption Spectroscopy [19]. For the determination of rate of metal biosorption by biomasses from 100 mL (at 100, 200, 300, 400, 500, and 1000 mg/L), the supernatant was analyzed for residual Cr (VI) after the contact period of 1-12 hours. The effect of pH and temperature on Cr (VI) sorption by natural biomass, was determined at pH values of 1, 2, 3, and 4, 28°C, 40°C, and 50°C, respectively. The effect of different doses of biomass ranging from 1 to 5 g/L, with 100 mg/L of Cr (VI) concentrations was determined. The values shown in the results section are the mean from three experiments carried out by triplicate.

2.4. Bioremediation assay

Four 250 mL Erlenmeyer glass flasks, with 5 g of shell biomass, were added with 20 g of contaminated earth and water with 297 mg Cr (VI)/g earth or 373 mg Cr(VI)/L water, of tannery (Celaya, Guanajuato, México), and the volume was complete to 100 mL with trideionized water. The mixture was shaken in a rotary shaker at 120 rpm followed by filtration using Whatman filter paper No. 1. The filtrate containing the residual concentration of Cr (VI) was determined with 1, 5 diphenylcarbazide [19].

2.5. Determination of hexavalent, trivalent and total Cr:

Hexavalent and trivalent chromium were quantified by a spectrophotometric method employing diphenylcarbazide and chromazurol S, respectively [19, 20], and total Chromium by Electrothermal Atomic Absorption Spectroscopy [19].

3. Results and discussion

3.1. Effect of incubation time and pH

Figure 1 shows the effect of the incubation time and pH on Cr (VI) removal by *L. chinensis* Sonn shell. The optimum time and pH for Cr (VI) removal are 10 min and pH 1.0, at constant values of biosorbent dosage (1 g/100 mL), initial metal concentration (100 mg /L) and temperature (28°C). The literature [11], report an optimum time of 60 min., for the removal of lead by orange waste, 30 min and 60 for the removal of Cr (VI) by the tamarind peel and eucalyptus bark [12, 16]. Changes in the permeability of unknown origin, could partly explain the differences found in the incubation time, providing greater or lesser exposure of the functional groups of the cell wall of biomass analyzed. Adsorption efficiency of Cr (VI) was observed maximum at pH 1.0 with Litchi shell. This was due to the dominant species (CrO_4^{2-} and $Cr_2O_7^{2-}$) of Cr ions in solution which were expected to interact more strongly with the ligands positively charges [21]. These results are like for tamarind peel [10], but the most of authors report an optimum pH of 2.0 like tamarind seeds [10], eucalyptus bark [16], bagasse and sugarcane pulp, coconut fibers and wool, [22], for the tamarind fruit shell treated with oxalic acid [23], at pH of 2.0 and 5.0 for the mandarin bagasse [24] and almond green hull [25].

3.2. Effect of temperature on Cr (VI) removal by *L. chinensis* Sonn shell

Temperature is found to be a critical parameter in the bioadsorption of Cr (VI) by *L. chinensis* Sonn shell (Figure 2). The highest removal was observed at 40 and 50°C. At this point the total removal of the metal is carried out. The results are coincident for tamarind seeds with 95% of

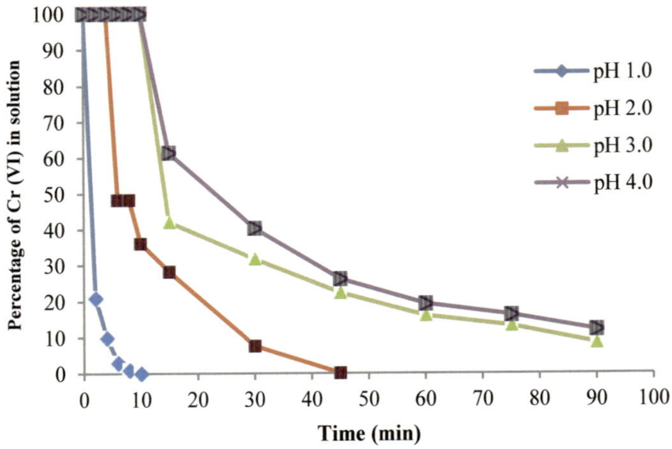

Figure 1. Effect of incubation time and pH on Chromium (VI) removal by *L. chinensis* Sonn shell. 100 mg/L Cr (VI). 28°C, 100 rpm. 1 g biomass.

removal at 58°C and 3 h [26], for the adsorption of cadmium (II) from aqueous solution on natural and oxidized corncob (40°C and 5 days) [27], but this are different for the mandarin waste [24], *Caladium bicolor* (wild cocoyam) biomass [29] and *Saccharomyces cerevisiae* [30].The increase in temperature increases the rate of removal of chromium (VI) and decreases the contact time required for complete removal of the metal, to increase the redox reaction rate [26].

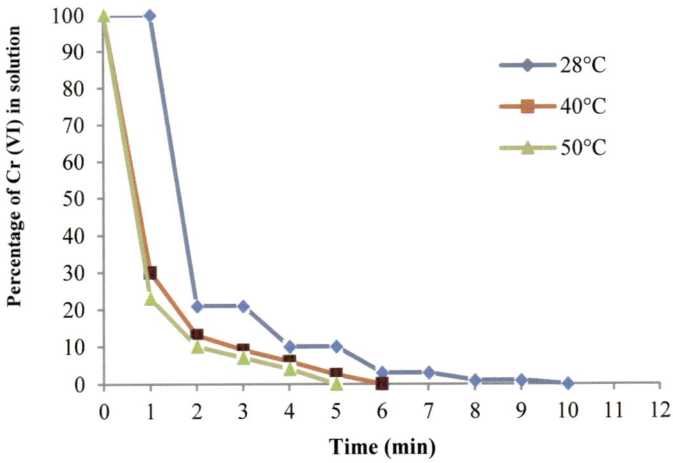

Figure 2. Effect of temperature on Chromium (VI) removal by *L. chinensis* Sonn shell. 100 mg/L Cr (VI). pH 1.0, 100 rpm. 1 g biomass.

3.3. Effect of initial metal concentration

On the other hand, at low metal concentrations (100 and 200 mg/L), biomass studied, shows the best results for removal, adsorbing 100% at 10 and 20 min. respectively, while 1000 mg/L of metal is removed 100% up to 195 min of incubation at 28°C (Figure 3). Also, we observed the development of a blue-green and a white precipitate (Cr $(OH)_3$), which changes more rapidly at higher temperatures (Figure 4). The results are coincident for tamarind peel and seeds, and *C. limonium* [10, 26, and 29]. The increase in initial concentration of Cr (VI) results in the increased uptake capacity and decreased the percentage of Cr (VI) removal. This was due to the increase in the number of ions competing for the available functions groups on the surface of biomass [26].

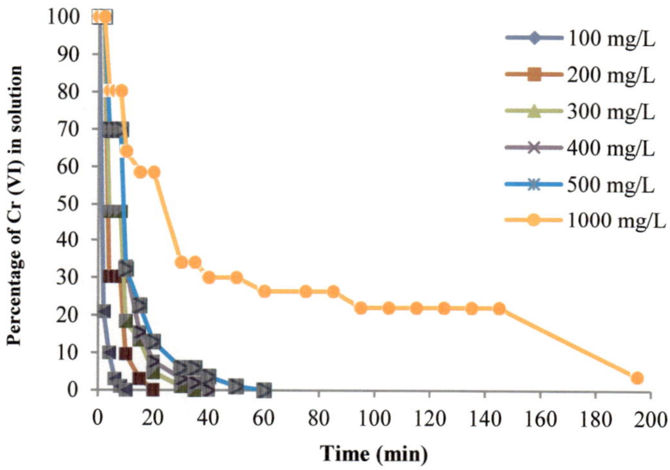

Figure 3. Effect of initial metal concentration on Chromium (VI) removal by *L. chinensis* Sonn shell. 28°C. pH 1.0, 100 rpm. 1 g biomass.

4. Time course of Cr (VI) decrease and Cr (III) production

The ability of the *L. chinensis* Sonn shell to decrease the initial Cr (VI) of 1.0 g/L and Cr (III) production in solution are analyzed. Figure 5 shows that the shell exhibited a remarkable efficiency to diminish Cr (VI) level with the concomitant production of Cr (III) as $Cr(OH)_3$ in the solution (indicated by the formation of a blue-green color and a white precipitate (Cr $(OH)_3$) and his determination for Cromazurol S, (Figures 4 and 6) [19, and 20].

Thus, after 1 h of incubation, the shell biomass caused a drop in Cr (VI) from its initial concentration of 1.0 g/L to almost undetectable levels and the decrease level occurred with no significant change in total Cr content. As expected, total Cr concentration remained constant over time, in solution control. These observations indicate that Litchi shell is able to reduce Cr (VI) to Cr (III) in solution. Furthermore, as the *L. chinensis* Sonn shell contains vitamin C and

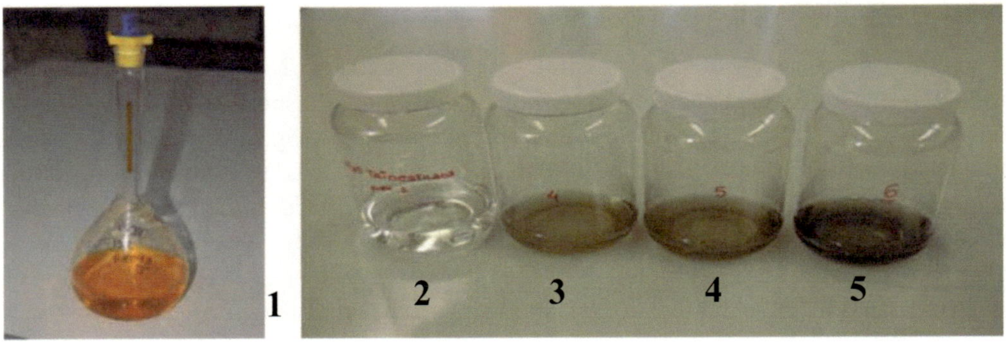

Figure 4. Formation of blue-green color by different chromium (VI) concentrations at 28 ° C and 60 ° C in the presence of Litchi shell. pH 1.0. 1 g biomass. **1- Chromium (VI) standard solution 2.- Trideionized water 3.- 200 mg/L 4.- 500 mg/L. 5.- 1 000 mg/L**

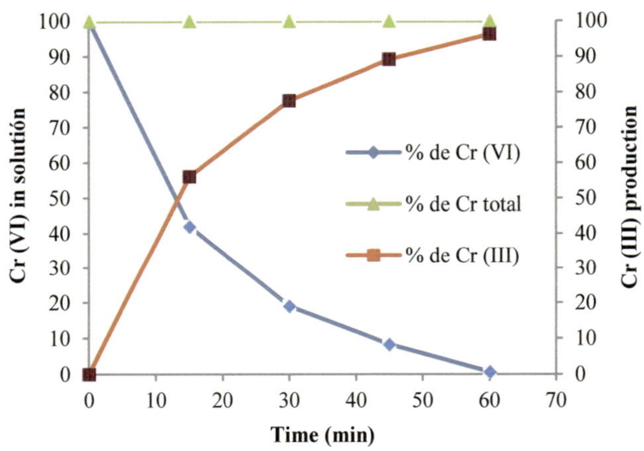

Figure 5. Time-course of Cr (VI) decrease and Cr (III) production in solution with 1.0 g/L Cr (VI). 100 rpm, 28°C, pH 1.0

some carbohydrates [32], we found that vitamin C and Cystine quickly reduce Cr (VI) to Cr (III) and could be a very important part in the metal reduction (Table 1), according to some reports in the literature [2, 12, 13, 31, 35, 36, 37, and 38]. There are two mechanisms by which chromate could be reduced to a lower toxic oxidation state by an enzymatic reaction. Currently, we do not know whether the shell biomass used in this study express Cr (VI) reducing enzyme(s). Further studies are necessary to extend our understanding of the effects of coexisting ions on the Cr (VI) reducing activity of the biomass reported in this study. Cr (VI) reducing capability has been described in some reports in the literature [2, 3, 7, 12, 13, 15, 18, 31, 39, and 40]. Biosorption is the second mechanism by which the chromate concentration could be reduced, because the biomass shell can be regarded as a mosaic of different groups that could form coordination complexes with metals and our observations are like to the most of the reports [2, 3, 7, 12, 15, 18, 39, and 40].

http://dx.doi.org/10.5772/56152

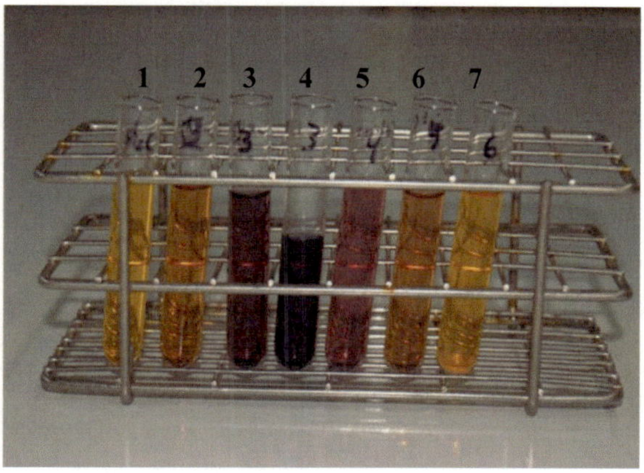

Figure 6. Cromazurol S Reaction 1.-Trideionized water pH=1.0 2.- K_2CrO_4 (50 µg/mL) 3.- $Cr(NO_3)_3$ $9H_2O$. (50 µg/mL) 4.- $Cr(NO_3)_3$ $9H_2O$. (500 µg/mL) 5.- Problem (50 µg/mL) 6.- Vitamin C (50 µg/mL) 7.- $Al_2(SO_4)_3$ (50 µg/mL)

Cr (VI) Concentration (mg/L)	Time (min)	
	60°C	28°C
0*	--------	-------
200	10	45
300	15	60
400	15	60
500	15	80
1000	20	120
Cystine	N.D. **	5
Vitamin C	N.D. **	5

*Control: 100 mL of trideionized water, pH 1.0. There were no variations in color.

**Not determinated.

Table 1. Formation of blue-green color by different chromium (VI) concentrations to 28°C and 60°C, in the presence of Litchi shell. pH 1.0. 1 g biomass.

4.1. Effect of biosorbent dose

The influence of biomass on the removal capacity of Cr (VI) was depicted in Figure 7. If the researchers increase the amount of biomass also increases the removal of Cr (VI) in solution (100% of removal, with 5 g of biomass, 20 minutes), with more biosorption sites of the same, because the amount of added biosorbent determines the number of binding sites available for metal biosorption [27]. Similar results have been reported for modified corn stalks [9], tamarind shell [12], and *Mucor hiemalis* and *Rhizopus nigricans*, although latter with 10 g of biomass [1, 28], but they are different from those reported for biomass wastes from the mandarin (bagasse), with an optimal concentration of biomass of 100 mg/L [24].

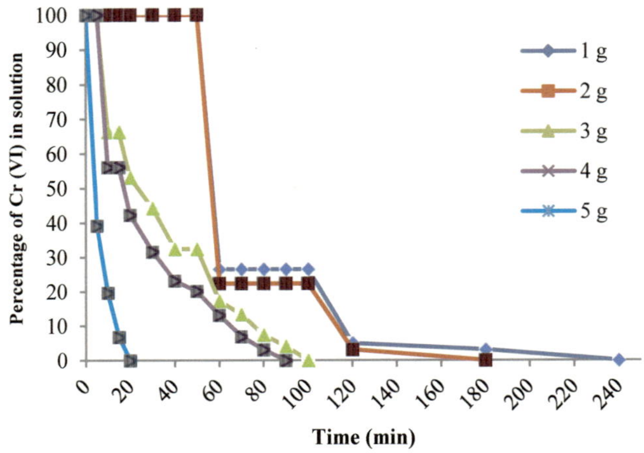

Figure 7. Effect of biomass concentration on Chromium (VI) removal by *L. chinensis* Sonn shell. 1 g/L Cr (VI). 28°C. pH 1.0, 100 rpm.

4.2. Cr (VI) Removal in the presence of different heavy metals

The researchers analyzed whether the presence of different metals interfere with the Cr (VI) removal (500 mg/L) at a pH of 1.0, with 1 g of Litchi shell, finding that none of the added metals (salts of cadmium, copper, zinc and mercury) interferes with the Cr (VI) removal, but in the presence of zinc and mercury takes 10-20 min longer to remove 100% of the metal (Figure 8). This is consistent with many reports in the literature [1, 12, 13, 16, 31, 32, and 33].

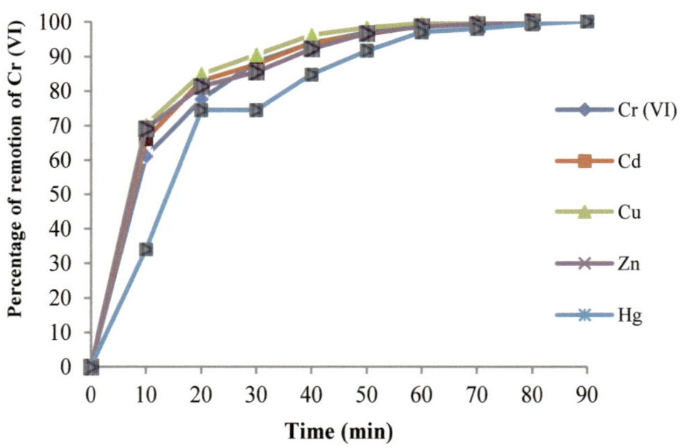

Figure 8. Effect of different metals concentration on Chromium (VI) removal by *L. chinensis* Sonn shell. 500 mg/L of metal. 1.0 g biomass. pH 1.0. 100 rpm. 28°C

4.3. Cr (VI) Removal by different biomasses

The researchers studied the Cr (VI) (100 mg/L) removal, with 1 g of different biomass. Litchi shell was the most efficient, because in 10 min at 28°C remove 100% of the metal, followed by xylan and polygalacturonic acid (150 and 300 min at 60 °C, respectively) and starch and cellulose were less efficient (43.6% at 28°C and 300 min of incubation and 21.83% at 60°C, at the same time of incubation, respectively) (Figure 9). With respect to other biomasses used, most authors report lower removal efficiencies of metal, for example: 45 mg/L for eucalyptus bark [16], 13.4 and 17.2 mg/L for bagasse and sugar cane pulp, 29 mg/L coconut fibers, 8.66 mg/L for wool [22], 25 and 250 mg/L of chitin and chitosan [41] and 1 mg/L for cellulose acetate [42].

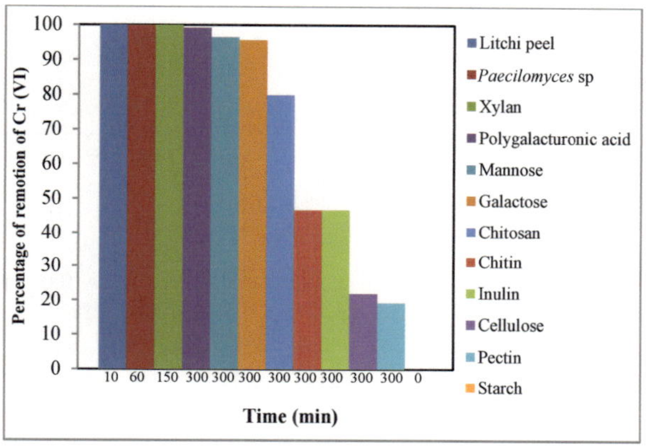

Figure 9. Chromium (VI) removal by different biomasses. 100 mg/L Cr (VI). 1.0 g biomass. pH 1.0. 100 rpm. 60°C, Litchi peel 28°C.

4.4. Removal of Cr (VI) in industrial wastes with *Litchi chinensis Sonn* shell

The researchers adapted a water-phase bioremediation assay to explore possible usefulness of *L. chinensis* Sonn shell, for eliminating Cr (VI) from industrial wastes, the biomass (5 g) was incubated with 20 and non-sterilized contaminated soil containing 297 mg Cr (VI)/g, suspended in trideionized water, and 100 mL of contaminated water with 373 mg/Cr (VI) /L. It was observed that after five and six days of incubation with the biomass, the Cr (VI) concentration of soil and water samples decrease 100% (Figure 10), and the decrease level occurred without change significant in total Cr content, during the experiments. In the experiment carried out in the absence of the biomass, the Cr (VI) concentration of the soil and water samples decreased by about of 18%, and 8%, respectively (date not shown); this might be caused by indigenous microflora and (or) reducing components present in the soil and water.

The chromium removal abilities of *L. chinensis* Sonn shell are equal or better than those of other reported biomass, for example tamarind shell [12], *M. americana* [38], *Candida maltosa* RR1 [39]. In particular, this biomass was superior to the other biomass because it has the capacity for efficient chromium reduction under acidic conditions. Many of the Cr (VI) reduction studies were carried out at neutral pH [43]. *Aspergillus niger* also has the ability to reduce and adsorb Cr (VI) [41]. When the initial concentration of Cr (VI) was 500 ppm, *A. niger* mycelium removal 8.9 mg/L of chromium/g dry weight of mycelium in 7 days.

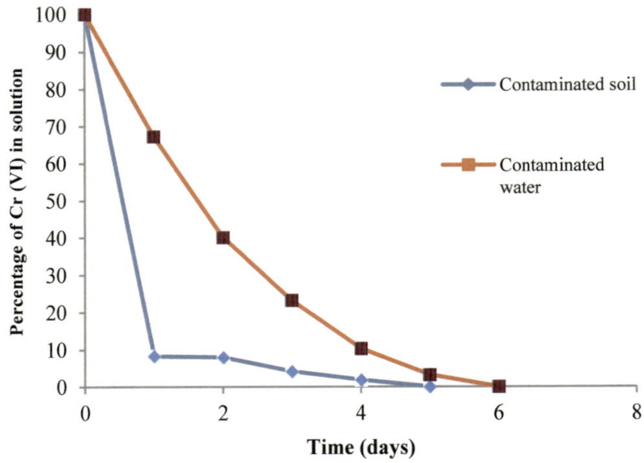

Figure 10. Chromium (VI) removal in industrial wastes incubated with the biomass. 100 rpm. 28°C. 5 g Litchi biomass. Contaminated soil and water (297 mg/Cr(VI)/g soil] and 373 mL Cr(VI)/L, respectively.

4.5. Biorremediation assay *in situ*

100 kg of contaminated soil (345 mg Cr(VI)/ g soil), with 20 g of natural biomass of *M. Americana* were incubated in a greenhouse at 28°C. After 10 weeks of incubation, the natural biomass removal 83% of the metal from contaminated soil (Figure 11), without change significant in total Cr content. In the experiment carried out in the absence of the biomass, the Cr (VI) concentration of the soil sample decreased by about of 15% (date not shown); this might be caused by indigenous microflora and (or) reducing components present in the soil, like microorganisms lactic acid producers, which reduce Cr (VI) to Cr (III) (Figure 12). Reports on applications of microorganisms for studies of bioremediation of soils contaminated with chromates are rare. Such study involved the use of *Pseudomona mendocina* for the removal of the metal from cooling tower effluent [44], and soil microcosms [45]. In the first process when carried out in a 20 liter continuous stirred tank reactor removed 25-100 mg chromate/L, in 4.5-8 hours with >99.9% efficiency in the presence of sugarcane molasses as nutrient, and in soil microcosms could immobilize 100 μg (2mM) chromate/g soil in 8 hours by converting into trivalent form, and the chromate contaminated soil, after microbiological treatment, supported growth of wheat seedlings without exerting any toxic effects. Other study involved the use of unidentified bacteria native from the contaminated site, which is used in bioreactors to treat soil contaminated with Cr (VI). It was found that the maximum reduction of Cr (VI) occurred with the use of 15 mg of bacterial biomass/g of soil (wet weight), 50 mg/g of soil molasses as carbon source, the bioreactor operated under these conditions, completely reduced 5.6 mg/Cr (VI)/g of soil at 20 days [46]. In another study using unidentified native bacteria reducing Cr

(VI) of a contaminated site, combined with *Ganoderma lucidum*, the latter used to remove by biosorption Cr (III) formed. The results showed that the reduction of 50 mg/L of Cr (VI) by bacteria was about 80%, with 10 g / L of peptone as a source of electrons and a hydraulic retention time of 8 h. The Cr (III) produced was removed using a column with the fungus *G. lucidum* as absorber. Under these conditions, the specific capacity of adsorption of Cr (III) of *G. Lucidum* in the column was 576 mg/g [47]. In other studies, has been tested the addition of carbon sources in contaminated soil analyzed in column, in one of these studies was found that the addition of tryptone soy to floor with 1000 mg/L of Cr (VI) increase reduction ion, due to the action of microorganisms presents in the soil, although such action is not observed in soil with higher concentrations (10.000 mg/L) of Cr (VI) [48]. Another study showed that the addition of nitrate and molasses accelerates the reduction of Cr (VI) to Cr (III) by a native microbial community in microcosms studied, in batch or columns of unsaturated flow, under similar conditions to those of the contaminated zone. In the case of batch microcosms, the presence of such nutrients caused reduction of 87% (67 mg/L of initial concentration) of Cr (VI) present in the beginning of the experiment, the same nutrients, added to a column of unsaturated flow of 15 cm, added with 65 mg/L of Cr (VI) caused the reduction and immobilization of 10% of metal, in a period of 45 days [49]. Finally, Cardenas-Gonzalez and Acosta-Rodríguez [40], adapted a water-phase bioremediation assay to explore possible usefulness of strain of *Paecilomyces* sp to eliminate Cr (VI) from industrial wastes, the mycelium biomass was incubated with non-sterilized contaminated soil containing 50 mg Cr (VI)/g, suspended in Lee's minimal medium [50] pH 4.0. It was observed that after eight days of incubation with the *Paecilomyces* sp biomass, the Cr (VI) concentration (50 mg/g) of soil sample decrease fully (100%).

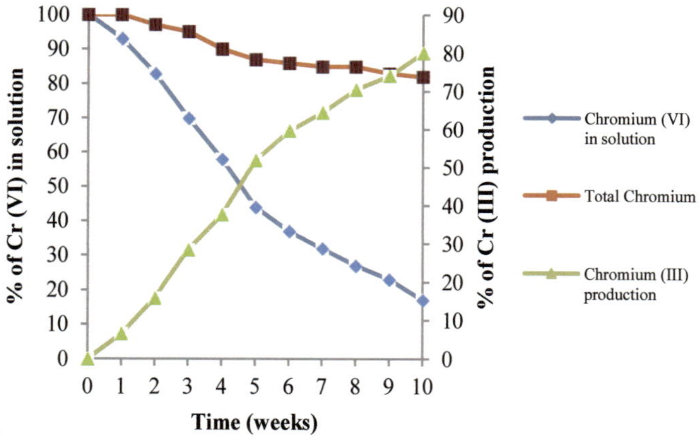

Figure 11. Bioremediation of Cr (VI) *in situ*, by *M. americana* biomass. 100 Kg of contaminated soil with 345 mg de Cr (VI)/g soil. (20 Kg biomass. 28°C).

Figure 12. Chromium (VI) Reduction by lactic acid. 1. - Lactic acid standard solution (85%) 2. - Chromium (VI) standard solution (1.0 g /L). pH= 1.0 3.-100 mg Cr (VI)/L with lactic acid (100 mL) 4. - 1000 mg Cr (VI)/L with lactic acid (100 mL)

4.6. Desorption of Cr (VI) by different solutions

Furthermore, the researchers examined the ability of different solutions to desorb the metal bioadsorbed (250 mg/L) for the Litchi biomass, obtaining high efficiency with 0.1 N NaOH and 0.5 N (80 and 61% respectively (Figure 13), which are less those reported for desorption of Chromium (VI) with alkaline solutions (100%, pH 9.5), 1.0 N NaOH (95%) and a hot solution of $NaOH/Na_2CO_3$ (90%), respectively, [21, 51], and are higher than that reported (14.2%) using 0.2 M NaOH [52]. This indicates that binding of metal to biomass is not as strong and that it can be used up to 6 desorption cycles of removal, which further lowers the metal removal process of niches contaminated with it.

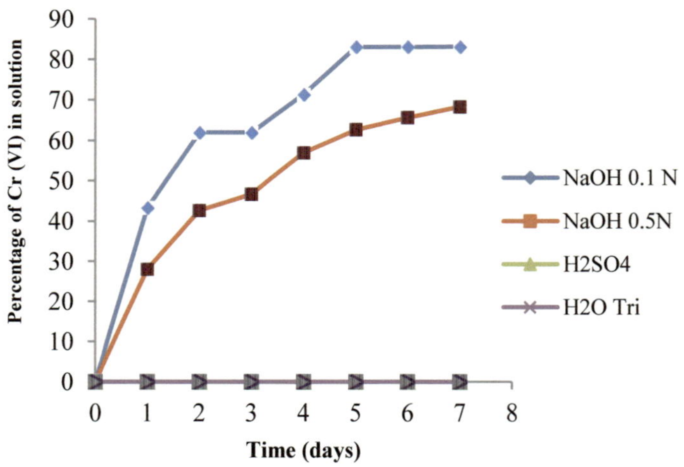

Figure 13. Desorption of Chromium (VI) (250 mg/L) by different solutions (1 g biomass. 28°C, 100 rpm)

4.7. Biosorption of Chromium (VI) in solution by different natural biomasses

In Table 2, the researchers show the biosorption of Chromium (VI) *by* the different biomasses analyzed. It was found that the biomass of L. *chinensis* Sonn, *T. indica, M. Americana,* and *C. reticulata,* shells were the most efficient at removing the metal in solution (100% at 20, 40, 50 and 60 minutes, respectively), at pH 1.0 at 50°C, the biomasses reduce the metal in solution, they can remove it from contaminated industrial wastes, and can be used six times efficiently.

Parameter	*Litchi chinensis*	*Citrus reticulata*	*Mammea americana*	*Citrus sinensis*	*Citrus limonium*	*Tamarindus indica*	*Musa cavendishii*	*Cuccumis melo* L.
pH optimum	1.0	1.0	1.0	1.0	1.0	1.0	1.0	1.0
Incubation time (100 mg/L,28°C)	10 min	40 min	50 min	120 min	80 min	60 min	60 min	230 min
Temperature (50°C, 1.0 g/L)	35 min	90 min	110 min	120 min	200 min	140 min	80 min	105 min
Biomass concentration (5 g. 1.0 g/L)	20 min	60 min	50 min	75 min	85 min	40 min	25 min	600 min
Presence of different heavy metals (500 mg/L)	Not Interfere	Not Interfere	Not Interfere	Not Interfere	Not Interfere	Not Interfere	Not Interfere	Not Interfere
Reduction of Cr VI to Cr III	Yes	Yes	Yes	Yes	Yes	Yes	Yes	Yes
Biorremediation of contaminated sites (100%)	Soil: 5 days Water: 6 days	Soil: 5 days Water: 6 days	Soil: 5 days Water: 6 days	Soil: 5 days Water: 6 days	Soil: 5 days Water: 6 days	Soil: 5 days Water: 6 days	Soil: 5 days Water: 6 days	Soil: 5 days Water: 6 days
Desorption (7 days)	81.2%	80.1%	78.3%	83%	79.3%	80%	78%	78%

Table 2. Chromium (VI) removal of 1.0 g/L with different natural biomasses.

5. Conclusion

The use of biomaterials like natural biomasses has demonstrated to be a promising alternative for removal of Chromium hexavalent from aqueous solution. The screening and selection of the most effective biomaterial (biomasses) with sufficiently high metal binding capacity and selectivity for heavy metal ions, in this case, Chromium (VI), are prerequisite for a full process.

The natural biomasses showed complete capacity of biosorption and reduction concentrations of 1.0 g/L Cr (VI) in solution after different incubation times, and *L. chinensis* Sonn, *C. reticu-*

lata, and *M. americana* shells, were the most efficient, at 28°C, 100 rpm with 1 g of biomass, and after of 10 weeks the natural biomass of *M. americana*, removed 83% of the metal in contaminated soil (100 kg), with 345 mg Cr (VI)/g of soil. These results suggest the potential applicability of these biomasses for the remediation of Cr (VI) from polluted soils and waters in the fields, and this biomasses are naturals, they can be obtained in big amount, cheaper, and could be removal selectively heavy metals from aquatic mediums.

Author details

Ismael Acosta-Rodríguez[1*], Juan F. Cárdenas-González[1],
María de Guadalupe Moctezuma-Zárate[1] and Víctor M. Martínez-Juárez[2]

*Address all correspondence to: iacosta@uaslp.mx

1 Autonomous University of San Luis Potosí, Chemical Science Faculty, Center for Research and Graduate Studies, Experimental Mycology Laboratory, Dr. Manuel Nava, University Area, San Luis Potosí, S.L.P., México

2 Autonomous University of Hidalgo's State, Institute of Agricultural Science, Academic Area of Veterinary Medicine, University University Ranch C.P., Tulancingo of Bravo Hidalgo, México

References

[1] Bai, R.S. and Abraham, T.E. Biosorption of chromium (VI) from aqueous solution by *Rhizopus nigricans*. Bioresource Technology 2001; 79(1) 73-81.

[2] Seng, H. and Wang, Y.T. Biological reduction of chromium by *E. coli*. Journal of Environmental Engineering1994; 120(4) 560-572.

[3] Marsh, T.L. and McInerney, M.J. Relationship of hydrogen bioavailability to chromate reduction in aquifer sediments. Applied and Environmental Microbiology 2001; 67, 1517-1521.

[4] Nriagu, J.O. and Nieboer. E. Chromium in the Natural and Human Environments. Wiley-Interscience, New York; 1988.

[5] Lofroth, G. and Ames, B.N. Mutagenicity of inorganic compounds in *Salmonella typhimurium*: Arsenic, chromium and selenium. Mutation Research 1978; 53(2), 65-66.

[6] Park, D., Yung, Y.S. and Park, J.M. Reduction of Hexavalent chromium with the brown seaweed *Ecklonia* Biomass. Environmental Science Technology 2004; 38(18), 4860-4864.

[7] Sahin, Y. and Öztürk, A. Biosorption of chromium (VI) ions from aqueous solution by the bacterium *Bacillus thuringiensis*. Process Biochemistry 2004; 40 (5), 1895-1901.

[8] Volesky, B. and Holan. Z.R. Biosorption of heavy metals. Biotechnology Progress 1995; 11, 235-250.

[9] Chen, S., Yue, Q., Gao, B., Li, Q. and Xu, X. Removal of Cr (VI) from aqueous solution using modified corn stalks: Characteristic, equilibrium, kinetic and thermodynamic study. Chemical Engineering Journal 2011; 168, 909-917.

[10] Cimino, G., Passerini, A. and Toscano, G. Removal of toxic cations and Cr (VI) from aqueous solution by hazelnut shell. Water Research 2000; 34, 2955-2962.

[11] Perez-Marin, A.B., Meseguer, V., Zapata, J.F., Ortuno, M. and J. Aguilar, J. Removal of cadmium from aqueous solutions by adsorption onto orange waste. Journal of Hazardous Materials 2007; 139, 122-131.

[12] Acosta, I., López, V., Coronado, E., Cárdenas, J.F. and Martínez, V.M. 2010. Remoción de Cromo (VI) por la biomasa de la cascara de Tamarindo, (*Tamarindus indica*). Revista de Biotecnología y Bioingeniería 2010; 14, 11-23.

[13] Acosta-Rodríguez, I., Martínez-Pérez, R., Cárdenas-González, J.F., Moctezuma-Zárate, M.G. and Martínez-Juárez, V.M. Hexavalent Chromium Removal by *Litchi chinensis* Sonn Peel. American Journal of Biochemistry and Biotechnology 2012; 8(1) 7-13.

[14] Razmovski, R.N. and Sciban, M.B. Effect of different conditions on Cu (II) and Cr (VI) biosorption by dried waste tea fungal biomass. APTEFF 2007; 38,149-156.

[15] Aldrich, M.V., Torresdey, J.L.G., Videa, J.R.P. and Parsons, J.G. Uptake and reduction of Cr(VI) to Cr(III) by mesquite (*Prosopis* spp.): chromate–plant interaction in hydroponics and solid media studied using XAS. Environmental Science Technology 2003; 37, 1859-1864.

[16] Sarin, V. and Pant, K.K. Removal of chromium from industrial waste by using eucalyptus bark. Bioresource Technology 2006; 97, 15-20.

[17] Shafqat, F., Bhatti, H.N., Hanif, M.A. and Zubair, A. kinetic and equilibrium studies of Cr (III) and Cr (VI) sorption from aqueous solution using rosa gruss an teplitz (red rose) waste biomass. Journal of Chile Chemical Society 2008; 53, 1667-1672.

[18] Fiol, N., C. Escudero, C. and Villaescusa, I. Chromium sorption and Cr (VI) reduction to Cr (III) by grape stalks and yohimbe bark. Bioresource Technology 2008; 99, 5030-5036.

[19] Clesceri, L.S., A.D. Eaton and A.E. Greenberg. Standard Methods for the Examination of Water and Wastewater. 20th Ed., American Public Health Association, Washington, D.C.; 1998; pp. 1220.

[20] Pantaler, R.P. and Pulyaeva, I.V. A spectrophotometric study of complexation between chromium and chromazurol S. Journal of Analytical Chemistry (Moscow) 1985; 40, 1634-1639.

[21] Gupta, S. and Babu, B.V. Removal of toxic metal Cr (VI) from aqueous solutions using sawdust as adsorbent: Equilibrium, kinetics and regeneration studies. Chemical Engineering Journal 2009; 150, 352-365.

[22] Dakiki, M., Khamis, M., Manassra, A. and Mer'eb, M. Selective adsorption of chromium (VI) in industrial wastewater using low-cost abundantly available adsorbents. Advances in Environmental Research 2002; 6, 533-540.

[23] Popuri, R.S., Jammala, A., Suresh, K.N. and Abuburi, K. Biosorption of hexavalent chromium using tamarind (*Tamarindus indica*) fruit shell-a comparative study. Journal of Biotechnology 2008; 10, 358-367.

[24] Zubair, A., Bhatti, H.N., Hanif, M.A. and Shafqat, F. Kinetic and equilibrium modeling for Cr (III) and Cr (VI) removal from aqueous solutions by *Citrus reticulata* waste biomass. Water Air Soil Pollution 2008; 191, 305-318.

[25] Sharanavard, M., Ahmadpourand, A. and Reza Doosti, M. Biosorption of hexavalent chromium ions from aqueous solutions using almond green hull as a Low-Cost biosorbent. European Journal of Science Research 2011; 58,392-400.

[26] Agarwal, G.S., Kumar, H. and Chaudari, S. Biosorption of aqueous chromium (VI) by *Tamarindus indica* seeds. Bioresource Technology 2006; 97, 949-956.

[27] Cervantes, C., Campos-García, J., Devars, S., Gutiérrez- Corona, F., Loza-Tavera, H., Torres-Gúzman, J.C. and Moreno-Sánchez, R. Interactions of Chromium with Microorganisms and Plants. FEMS Microbiology Review 2001; 25(3), 335-347.

[28] Tewari, N., Vasudevan, P. and Guha, B. Study on Biosorption of Cr (VI) by *Mucor hiemalis*, Biochemical Engineering Journal 2005; 23(2), 185-192.

[29] Jnr, M.H. and Spiff, A.I. Effects of temperature on the sorption of Pb^{2+} and Cd^{2+} from aqueous solution by *Caladium bicolor* (wild cocoyam) biomass. Electronical Journal of Biotechnology 2005; 8, 162-169.

[30] Ozer, A. and Ozer, D. Comparative study of the biosorption of Pb(II), Ni(II) and Cr(VI) ions onto *S. cerevisiae*: Determination of biosorption heats. Journal of Hazardous Materials 2003; 100, 219-229.

[31] Vargas-Morales, J.M. Bautista-Mata, D., Cárdenas-González, J.F. Martínez-Juárez, V.M. and Acosta-Rodríguez, I. Hexavalent Chromium Removal by *Citrus limonium* Peel. Open Journal of Inorganic Non-metallic Materials 2012; 8(1), 7-13.

[32] Leyva-Ramos, R., Bernal-Jacome, L.A. and Acosta- Rodriguez, I. 2005. Adsorption of cadmium (II) from aqueous solution on natural and oxidized corncob. Separation and Purification Technology 2005; 45, 41-49.

[33] Shaukat, S.S. Progress in Biomass and Bioenergy Production. 1st Ed., InTech; 2011; pp. 444.

[34] Castro López, N.M. Lichis: Delicia de Reyes. México Desconocido 1996; 235, pp. 33-48.

[35] Xu, X. R., Li, H.B., Gu, J.D. and Li, X.Y. Kinetics of the reduction of Chromium (VI) by Vitamin C. Environmental Toxicology and Chemistry 2005; 24(6), 1310-1314.

[36] Yong, L., Xin-Hua, X and Ping, H. Remediation of Cr (VI) in solution using vitamin C. Journal of Zhejiang University Science 2005; 6B(6), 540-542.

[37] Balderas-Hernández, P., Hernández-Monroy, C.A., Roa-Morales, G., Gómez-Espinosa, R.M. y Barrea-Díaz, C.E. Reducción de Cromo (VI) utilizando fructosa, glucosa y sacarosa. Memorias del VI Congreso Internacional y XII Nacional de Ciencias Ambientales 2006; 423-427.

[38] Acosta, I., Sandoval, P., Bautista, D., Hernández, N., Cárdenas, J.F. y Martínez, V.M. Bioadsorción de Cromo (VI) por la cáscara de Mamey (*Mammea americana* L.). Avances en Ciencias e Ingeniería 2012; 3(2), 1-9.

[39] Ramirez-Ramirez, R., Calvo-Mendez, C., Avila- Rodriguez, M., Lappe, P., Ulloa, M. and Gutierrez-Corona, J.F. Cr (VI) reduction in a chromate-resistant strain of *Candida maltosa* isolated from the leather industry. Antonie Van Leeuwenhoek 2004; 85, 63-68.

[40] Cárdenas-González, J.F. and Acosta-Rodríguez, I. Hexavalent chromium removal by a *Paecilomyces* sp fungal. In: Progress in Biomass and Bioenergy Production. Edited by Syed Shahid Shaukat. INTECH. Croatia, 2011 Chapter 7. pp: 133-150.

[41] Sag, Y. and Aktay, Y. Kinetic studies on sorption of Cr(VI) and Cu(II) ions by chitin, chitosan and *Rhizopus arrhizus.* Biochemical Engineering Journal 2002; 12,143-153.

[42] Arthanareeswaran, G., Thanikaivelan, P., Jaya, N., Mohan, D. and Raajenthiren, M. Removal of chromium from aqueous solution using cellulose acetate and sulfonated poly (ether ketone) blend ultrafiltration membranes. Journal of Hazardous Materials 2007; 139, 44-49.

[43] Fukuda, T., Ishino, Y., Ogawa, A., Tsutsumi, K. and Morita, H. Cr(VI) reduction from contaminated soils by *Aspergillus* sp. N2 and *Penicillium* sp. N3 isolated from chromium deposits. Journal of General and Applied Microbiology 2008; 54, 295-303.

[44] Bhide, J.V., Dhakephalkar, P.K. and Paknikar, K.M. Microbiological process for the removal of Cr(VI) from chromate-bearing cooling tower effluent. Biotechnology Letters 1996; 18(6) 667-672.

[45] Salunkhe, P.B., Dhakephalkar, P.K. and Paknikar, K.M. Bioremediation of hexavalent chromium in soil microcosms. Biotechnology Letters 1998; 20(8), 749-751.

[46] Jeyasingh, J. and Philip, L. Bioremediation of chromium contaminated soil: optimization of operating parameters under laboratory conditions. Journal of Hazardous Materials 2005; 118(1-3), 113-120.

[47] Krishna, K.R. and Philip, L. Bioremediation of Cr (VI) in contaminated soils. Journal of Hazardous Materials 2005; 121 (1-3), 109-117.

[48] Tokunaga, T.K., Wan, J., Firestone, M.K., Hazen, T.C., Olson, K.R., Donald, J. Herman, D.J. Sutton, S.R. and Lanzirotti, A. Bioremediation and Biodegradation. In situ reduction of chromium (VI) in heavily contaminated soils through organic carbon amendment. Journal of Environmental Quality 2003; 32(5), 1641-1649.

[49] Oliver, D.S., Brockman, F.J., Bowman, R.S., Thomas L. and Kieft, T.L. Microbial reduction of hexavalent chromium under vadose zone conditions. Journal of Environmental Quality 2003; 32(1), 317-324.

[50] Lee, K. L., Buckley, M. R., Campbell, C. An amino acid liquid synthetic medium for development of mycelia and yeast forms of *Candida albicans*. Sabouraudia 1975; 13, 148-153.

[51] Singh, K.K., Hasan, S.H., Talat, M., Singh, V.K. and Gangwar, S.K. Removal of Cr (VI) from aqueous solutions using wheat bran. Chemical Engineering Journal 2009; 151, 113-121.

[52] Wang, X.S., Tang, Y.P. and Tao, S.R. Kinetics, equilibrium and thermodynamic study on removal of Cr (VI) from aqueous solutions using low-cost adsorbent Alligator weed. Chemical Engineering Journal 2009; 148, 217- 225.

Bioremediation of Waters Contaminated with Heavy Metals Using *Moringa oleifera* Seeds as Biosorbent

Cleide S. T. Araújo, Dayene C. Carvalho,
Helen C. Rezende, Ione L. S. Almeida,
Luciana M. Coelho, Nívia M. M. Coelho,
Thiago L. Marques and Vanessa N. Alves

Additional information is available at the end of the chapter

http://dx.doi.org/10.5772/56157

1. Introduction

Water is not only a resource, it is a life source. It is well established that water is important for life. Water is useful for several purposes including agricultural, industrial, household, recreational and environmental activities. Despite its extensive use, in most parts of the world water is a scarce resource. Ninety percent of the water on earth is seawater in the oceans, only three percent is fresh water and just over two thirds of this is frozen in glaciers and polar ice caps. The remaining unfrozen freshwater is found mainly as groundwater, with only a small fraction present above ground or in the air. Thus, almost all of the fresh water that is available for human use is either contained in soils and rocks below the surface, called groundwater, or in rivers and lakes.

The contamination of soil and water resources with environmentally harmful chemicals represents a problem of great concern not only in relation to the biota in the receiving environment, but also to humans. The continuing growth in industrialization and urbanization has led to the natural environment being exposed to ever increasing levels of toxic elements, such as heavy metals. Approximately 10% of the wastes produced by developed countries contain heavy metals. Figure 1 gives some indication of the amounts of metal-containing waste produced in developed countries. Much of the discharge of metals to the environment comes from mining, followed by agriculture activities.

Figure 1. Waste containing heavy metals produced in developed countries [1].

Many different definitions have been proposed for heavy metals, some based on density, some on atomic number or atomic weight, and others on chemical properties or toxicity, which are not necessarily appropriate. For example, cobalt, iron, copper, manganese, molybdenium, vanadium, strontium and zinc are required to perform vital functions in the body and therefore cannot be considered as compounds with high toxicity or ecotoxic properties. Regarding the meaning of the term "heavy metal" it was found that there can be misinterpretation due to the contradictory definitions and lack of a coherent scientific basis [2].

In conventional usage "heavy" implies high density and "metal" refers to the pure element or an alloy of metallic elements. According to Duffus [2], a new classification should reflect our understanding of the chemical basis of toxicity and allow toxic effects to be predicted. Various publications have used the term "heavy metals" related to chemical hazards and this definition will also be used herein. Among the classes of contaminants, heavy metals deserve greater concern because of their high toxicity, accumulation and retention in the human body. Moreover, heavy metals do not degrade to harmless end products [3, 4]. It is well established that the presence of heavy metals in the environment, even in moderate concentrations, is responsible for producing a variety of illnesses of the central nervous system (manganese, mercury, lead, arsenic), the kidneys or liver (mercury, lead, cadmium, copper) and skin, bones, or teeth (nickel, cadmium, copper, chromium) [5].

Due to its properties, water is particularly vulnerable to contamination with heavy metals. Table 1 shows the maximum limits for some metals in drinking water, according to the US Environmental Protection Agency (US EPA) [6]. The US EPA requires that lead, cadmium and total chromium levels in drinking water do not to exceed 0.015, 0.005 and 0.1 mg L^{-1}, respectively. Corresponding values for other metals are presented in Table 1.

Within this context, and considering that heavy metals do not decay and are toxic even at low concentrations, it is necessary to remove them from various types of water samples. Of the conventional treatments used for the removal of metals from liquid waste, chemical precipitation and ion exchange are the predominant methods. However, they have some limitations since they are uneconomical and do not completely remove metal ions, and thus new removal processes are required [7-9]. Table 2 illustrates in more detail the advantages and limitations of the traditional methods applied to treat effluents.

Element	US EPA Limit (mg L^{-1})
Antimony	0.006
Arsenic	0.010
Beryllium	0.004
Chromium (total)	0.1
Cadmium	0.005
Cupper	1.3
Lead	0.015
Mercury	0.002
Selenium	0.05
Silver	0.1

Table 1. Maximum acceptable concentrations of metals in drinking water according to the US EPA [6].

Process	Disadvantages	Advantages
Precipitation and filtration	For high concentrations Separation difficult Not very effective Produces sludge	Simple Low cost
Biological oxidation and reduction	When biological systems are used the conversion rate is slow and susceptible to adverse weather conditions	Low cost
Chemical oxidation and reduction	Requires chemicals Applied to high concentrations Expensive	Mineralization Enables metal recovery
Reverse osmosis	High pressures Expensive	Pure effluent (for recycling)
Ion exchange	Responsive to the presence of particles Resins of high cost	Effective Enables metal recovery
Adsorption	Not effective for some metals	Conventional sorbents (coal)
Evaporation	Requires an energy source Expensive Produces sludge	Pure effluent obtained

Table 2. Traditional process used in wastewater treatment: advantages and disadvantages [10].

For these reasons, alternative technologies that are practical, efficient and cost effective for low metal concentrations are being investigated. Biosorption in the removal of toxic heavy metals is especially suited as a 'nonpolluted ' wastewater treatment step because it can produce close to drinking water quality from initial metal concentrations of 1-100 mg L^{-1}, providing final concentrations of < 0.01-0.1 mg L^{-1} [11]. Biosorption has been defined as the ability of certain biomolecules or types of biomass to bind and concentrate selected ions or other molecules from aqueous solutions. It should to be distinguished from bioaccumulation which is based on active metabolic transport; biosorption by dead biomass is a passive process based mainly on the affinity between the biosorbent and the sorbate [12]. The biosorption of heavy metals by non-living biomass of plant origin is an innovative and alternative technology for the removal of these pollutants from aqueous solution and offers several advantages such as low-cost biosorbents, high efficiency, minimization of chemical and/or biological sludge, and regeneration of the biosorbent [13].

Recently, natural adsorbents have been proposed for removing metal ions due to their good adsorption capacity. Technologies based on the use of such materials offer a good alternative to conventional technologies for metal recovery. In this context, *Moringa oleifera* represents an alternative material for this purpose [14-16].

2. *Moringa oleifera*

Moringa oleifera is the best known species of the *Moringaceae* family. *Moringaceae* is a family of plants belonging to the order *Brassicales*. It is represented by fourteen species and a single genus (*Moringa*), being considered an angiosperm plant. It is a shrub or small tree which is fast growing, reaching 12 meters in height. It has an open crown and usually a single trunk (Figure 2). It grows mainly in the semi-arid tropics and subtropics. Since its preferred habitat is dry sandy soil, it tolerates poor soils, such as those in coastal areas [17].

Native to northern India, it currently grows in many regions including Africa, Arabia, Southeast Asia, the Pacific and Caribbean Islands and South America [3, 16, 19]. It is cultivated for its food, medicinal and culinary value and its leaves, fruits and roots are the parts used. It is commonly known as the 'horseradish' tree arising from the taste of a condiment prepared from the roots or 'drumstick' tree due to the shape of the pods. Figures 3 and 4 show the pods and seeds of this tree. *M. oleifera* has a host of other country-specific vernacular names, an indication of the significance of the tree around the world [16, 20-23].

Research has focused on the use of *M. oleifera* seeds and fruits in water purification and the treatment of turbid water is the best-known application. The seeds of various species contain cationic polyelectrolytes which have proved to be effective in the treatment of water, as a substitute for aluminum sulfate. Interest in the study of natural coagulants for water clarification is not new. The coagulant is obtained from a byproduct of oil extraction and the residue can be used as a fertilizer or processed for animal fodder. Compared to the commonly used

Figure 2. Tree of *Moringa oleifera* species [18].

Figure 3. Pods of *Moringa oleifera* [18].

Figure 4. Seeds of *Moringa oleifera* [18].

coagulant chemicals, *Moringa oleifera* has a number of advantages including low cost, biodegradable sludge production and lower sludge volume, and also it does not affect the pH of the water. Apart from turbidity removal, *M. oleifera* seeds also possess antimicrobial properties [24, 25], although the mechanism by which seeds act upon microorganisms is not yet fully understood.

Tissues of *M. oleifera* from a wide variety of sources have been analyzed for glucosinolates and phenolics (flavonoids, anthocyanins, proanthocyanidins, and cinnamates). *M. oleifera* seeds reportedly contain 4-(α-L-rhamnopyranosyloxy)-benzylglucosinolate in high concentrations. Roots of *M. oleifera* have high concentrations of both 4-(α-L-rhamnopyranosyloxy)-benzylglucosinolate and benzyl glucosinolate. Leaves contain 4-(α-L-rhamnopyranosyloxy)-benzylglucosinolate and three monoacetyl isomers of this glucosinolate and only 4-(α-L-rhamnopyranosyloxy)-benzylglucosinolate has been detected in *M. oleifera* bark tissue [26]. Every glucosinolate contains a central carbon atom which is bonded to the thioglucose group (forming a sulfated ketoxime) via a sulfur atom and to a sulfate group via a nitrogen atom. These functional groups containing sulfur and nitrogen are good metal sequesters from aqueous solution. The leaves of *M. oleifera* reportedly contain quercetin-3-*O*-glucoside and quercetin-3-*O*-(6''-malonyl-glucoside), and lower amounts of kaempferol-3-*O*-glucoside and kaempferol-3-*O*-(6''-malonyl-glucoside), along with 3-caffeoylquinic acid and 5-caffeoylquinic acid. Neither proanthocyanidins nor anthocyanins have been detected in any of the tissues [26]. Although *M. oleifera* seeds have been most widely applied as a coagulant agent, many studies have been performed in order to explore other potential applications of this material, especially in the removal of metals from aqueous systems.

http://dx.doi.org/10.5772/56157

3. Biosorption of metals using *Moringa oleifera*

Since *Moringa oleifera* seeds have the ability to retain metals, it is necessary to define and to understand the functional groups responsible for the adsorption phenomenon. Biosorption by dead biomass or by some molecules and/or their active groups is a passive process based mainly on the affinity between the biosorbent and the sorbate. In this case, the metal is sequestered by chemical sites naturally present in the biomass. The diagram in Figure 5 illustrates the main steps in this process. In most cases, the biosorption process is rapid and takes place under normal temperature and pressure. After the process of phase separation a biomass "charged" with metal ions and an effluent free of contamination are obtained. Two paths can be followed to deal with the "contaminated" biomass, the one of greatest interest being biosorbent regeneration and metal recovery. This process is the most attractive because biomass can be used for the removal of other metal species from other contaminated effluents. The other option is the destruction of the biomass, which offers no possibility of reuse.

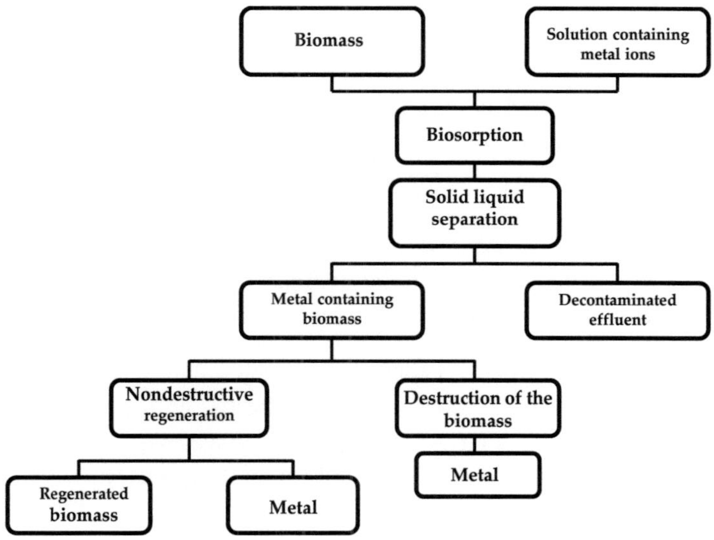

Figure 5. Main steps in biosorption process [27].

The mechanisms associated with heavy metal biosorption by biomass are still not clear; however, it is important to note that this process is not based on a single mechanism. Since metals may be present in the aquatic environment in dissolved or particulate forms, they can be dissolved as free hydrated ions or as complex ions chelated with inorganic ligands, such as hydroxide, chloride or carbonate, or they may be complexed with organic ligands such as amines, humic or fulvic acids and proteins. Metal sequestration occurs through complex mechanisms, including ion-exchange and complexation, and it is quite possible that at least some of these mechanisms act simultaneously to varying degrees depending on the biomass, the metal ion and the solution environment.

In reference [28] indicated that ion-exchange is an important concept in biosorption, because it explains many of the observations made during heavy metal uptake experiments. In this context, the term ion-exchange does not explicitly identify the mechanism of heavy metal binding to biomass, and electrostatic or London–van der Waals forces should be considered as the precise mechanism of chemical binding, i.e., ionic and covalent bonds. Figure 6 provides a schematic representation of an ion-exchange mechanism for a biosorbent material where "Me" represents a metal with valence +2.

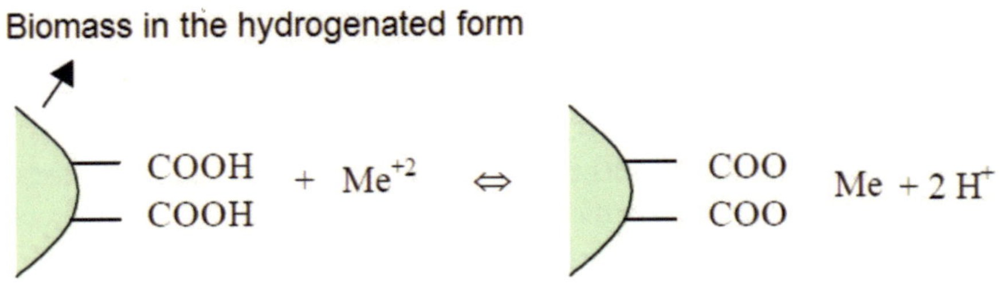

Figure 6. Schematic diagram of an ion exchange mechanism [29].

The seeds of *Moringa oleifera* and its parts can be classified as lignocellulosic adsorbents, consisting mainly of cellulose, hemicellulose and lignin. These functional groups are comprised of macromolecules that have the ability to absorb metal ions through ion exchange or complexation [30] phenomena which occur on the surface of the material through the interaction of the metal with the functional groups present. In order to understand the adsorption process it is also important to characterize the biomass material. Several techniques can be used to define the functional groups responsible for the adsorption phenomenon.

Infrared spectroscopy is an important technique in the qualitative analysis of organic compounds, widely used in the areas of natural products, organic synthesis and transformations. It is applied as a tool to elucidate the functional groups which may be present in substances [31], particularly with respect to the availability of the main groups involved in adsorption phenomena.

Figure 7 shows FT-IR spectra for *Moringa oleifera* seeds which verify the presence of many functional groups, indicating the complex nature of this material. The bandwidth centered at 3420 cm^{-1} may be attributed to the stretching of OH bonds present in proteins, fatty acids, carbohydrates and lignin units [32]. Due to the high content of protein present in the seed there is also a contribution in this region from N-H stretching of the amide bond. The peaks present at 2923 cm^{-1} and 2852 cm^{-1}, respectively, correspond to asymmetric and symmetric stretching of the C-H bond of the CH_2 group. Due to the high intensity of these bands it is possible to assign them to the predominantly lipid component of the seed, which is present in a high proportion similar to that of protein [33]. In the region of 1800-1500 cm^{-1} a number of overlapping bands are observed and between 1750 and 1630 cm^{-1} this can be attributed to C=O stretching. Due to the heterogeneous nature of the seed, the carbonyl group may be bonded to different neighborhoods as part of the fatty acids of the lipid portion or amides of the protein portion. The carbonyl component that appears due to the presence of lipids can be seen at 1740 and 1715 cm^{-1}, as can be observed in the infrared spectra as small peaks, and the shoulders forming part of the main band that appears at 1658 cm^{-1} are attributed to the carbonyl amides present in the protein portion. The peak observed at 1587 cm^{-1} may be attributed to stretching connecting CN and also the deformation of the N-H bond present in the proteins of seeds [34, 35].

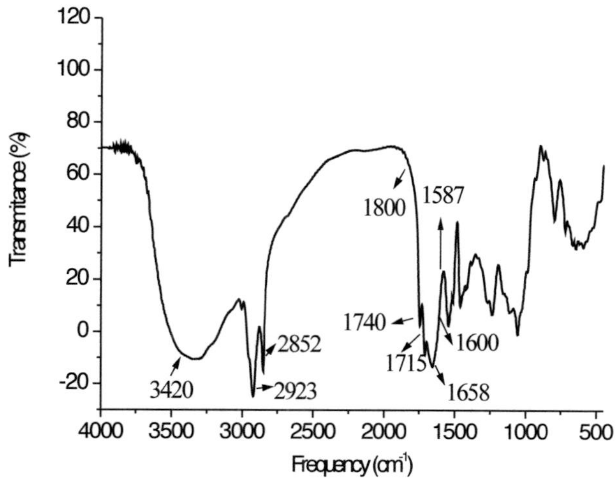

Figure 7. FT-IR spectrum of *Moringa oleifera* seeds. The arrows indicate the maximum signal obtained [36].

Among the various techniques for material characterization, the X-ray diffraction (XRD) technique is recommended for the evaluation of the presence of crystalline phases present in natural materials. In general, we can classify materials as amorphous, semicrystalline or crystalline. Figure 8 shows the XRD patterns for *M. oleifera* seeds. The XRD pattern for crushed seeds, due to the high amount of oils and proteins present in the composition of the material which represent around 69% of the total mass [36], shows unresolved signals (predominantly amorphous). For this reason intact seeds are analyzed, constituting a complex matrix comprised of a wide variation of substances including proteins, lipid structures and, to a lesser extent, carbohydrates. It was possible to separate a broad peak at around 2θ equals 10º. The presence of this peak is probably associated with the diffraction of the protein constituent surrounded by other components which have a more amorphous pattern [37]. The amorphous nature of the biosorbent suggests that the metal ion could more easily penetrate the biosorbent surface.

Thermogravimetric (TG) analysis was used to characterize the decomposition stages and thermal stability determined through the mass loss of a substance subjected to a constant heating rate for a specified time. The mass loss curve for a sample of *Moringa oleifera* seeds can be observed in Figure 9, showing a typical profile that indicates several stages of the decomposition process. This thermogravimetric curve verifies the sample heterogeneity, since the intermediates formed are a mixture of several components. The mass loss curve can be divided into three stages: i) the first step occurs from 30°C to 128°C where a mass loss in the order of 8%, associated with water desorption, was observed. The amount of water loss from seeds determined by this technique is similar to the value of 8.9% found in [38]; ii) in the second step

32% of mass loss was observed in the temperature range of 128–268°C. This stage occurs due to the decomposition of organic matter, probably the protein component, present in seeds; and iii) the third step occurs from 268°C to 541°C with decomposition of the greater part of the seed components, which probably includes fatty acids, for example, oleic acid has a boiling point of 360°C. At 950°C a total residue of around 14.6% was observed, due to the ash content and probably inorganic oxides.

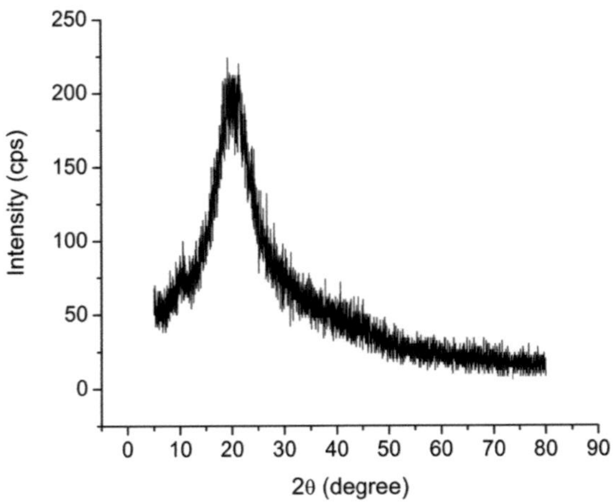

Figure 8. X–ray diffractogram for *Moringa oleifera* seeds [36].

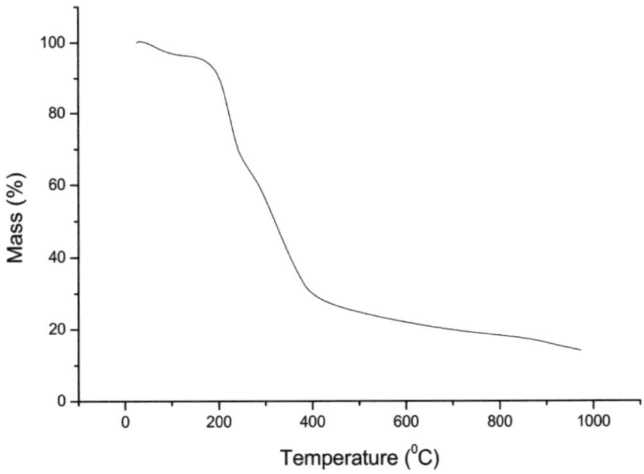

Figure 9. Thermogravimetric curve for *Moringa oleifera* seeds [36].

The morphological characteristics of the crushed seeds obtained using a scanning electron microscopy (SEM) can be seen in Figure 10. The results reveal that the material exhibits a relatively porous matrix with heterogeneous pore distribution. This feature is attributed to the fact that the whole seed comprises a wide variety of biomass components. The presence of some deformations on the surface of the plant tissue can be observed, containing available sites, from which it is possible to infer that the adsorbent provides favorable conditions for the adsorption of metal species in the interstices [35].

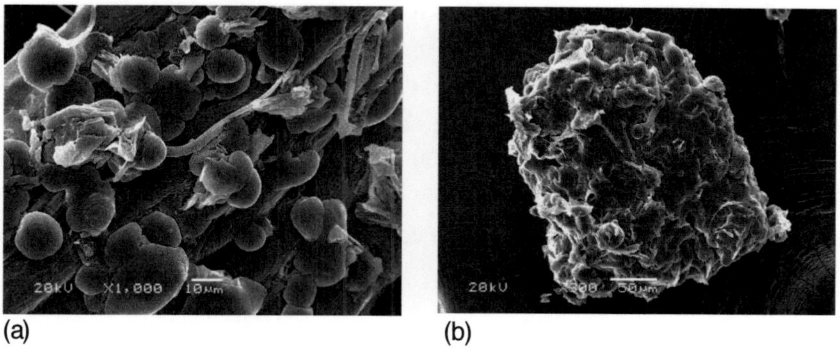

Figure 10. Scanning electron micrographs of *Moringa oleifera*. In the order of (a) 10 μm and (b) 50 μm [36].

4. Influence of parameters in biosorption process

Many variables can influence metal biosorption and experimental parameters such as temperature, stirring time, pH, particle size of the biomass, ionic strength and competition between metal ions can have a significant effect on metal binding to biomass. The biomass mass also influences the adsorption process because as the adsorbent dose increases the number of adsorbent particles also increases and there is greater availability of sites for adsorption. Some of the most important factors affecting metal binding are discussed below. In general, adsorption experiments are carried out in batch mode.

The pH is one of the most important parameters affecting any adsorption process. This dependence is closely related to the acid-base properties of various functional groups on the adsorbent surfaces [39]. The literature shows that a heterogeneous aqueous mixture of *M. oleifera* seeds contains various functional groups, mainly amino and acids groups. These groups have the ability to interact with metal ions, depending on the pH. An increase in metal adsorption with increasing pH values can be explained on the basis of competition between the proton and metal ions for the same functional groups, and a decrease in the positive surface charge, which results in a higher electrostatic attraction between the biosorbent surface and the metal [40]. Low pH conditions allow hydrogen and hydronium ions to compete with metal binding sites on the biomass, leading to poor uptake. Biosorbent materials primarily contain weak acidic and basic functional groups. It follows from the theory of acid–base equilibrium that, in the pH range of 2.5–5, the binding of heavy metal cations is determined primarily by the dissociation state of the weak acidic groups. Carboxyl groups (–COOH) are important groups for metal uptake by biological materials. At higher solution pH, the solubility of a metal complex decreases sufficiently for its precipitation, leading to a reduced sorption capacity.

Therefore, it is recommendable to study biosorption at pH values where precipitation does not occur. Biomasses are materials with an amphoteric character; thus, depending on the pH of the solution, their surfaces can be positively or negatively charged. At pH values greater than the point of zero discharge (pH_{pzc}), the biomass surface becomes negatively charged, favoring the adsorption of cationic species. However, adsorption of anionic species will be favored at pH < pH_{pzc}. The pH_{pzc} of the *M. oleifera* seeds is between 6.0 and 7.0 [41], indicating that the surface of the biosorbent presents acid characteristics. Figure 11 illustrates the surface charge or the point of zero net proton charge of *Moringa oleifera* seeds. The surface charge of the seeds is positive at pH < PZC, is neutral at pH = PZC and is negative at pH > PZC. The variation in pH caused by protonation and deprotonation of the adsorbent reflects the presence of functional groups. Table 3 shows the use of components of the *M oleifera* in the pH range of 2.5 to 8.0.

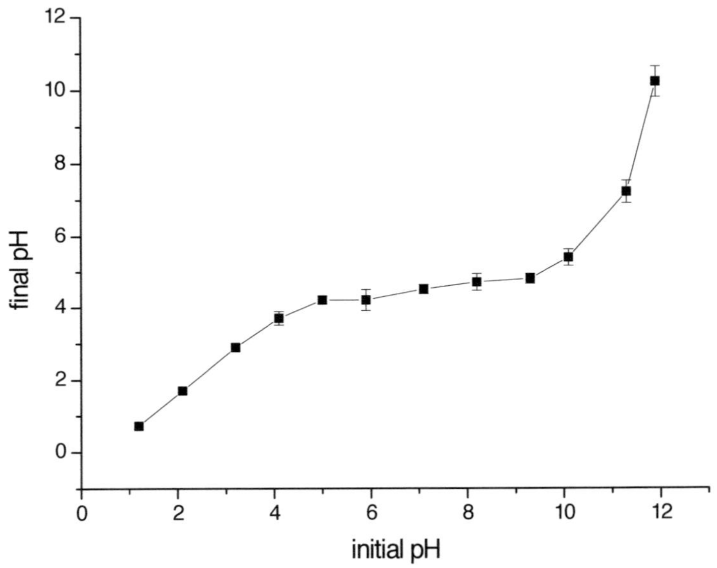

Figure 11. Point of zero net proton charge of *Moringa oleifera* seeds.

It has been noted that the temperature can influence the sorption process. Simple physical sorption processes are generally exothermic, i.e., the equilibrium constant decreases with increasing temperature. According to data reported in the literature (Table 3), the binding of the metal to different parts of the *M. oleifera* plant can be observed when the temperature is raised from 22 to 50 °C.

The contact time (or stirring time) is another important parameter that influences the efficiency of the adsorption process. As can be seen in Table 3, a period of 5 min was chosen for the nickel sorption process and good results were obtained; however, longer times (240 min) are required when using activated carbon.

Moringa oleifera is capable of directly sorbing metal ionic species from aqueous solutions. An interesting characteristic assigned to these biosorbents is the high abrasive content and the relative chemical resistance, allowing them to be subjected to different chemical treatments to increase their affinity and/or specificity for metal ions. Results previously published show the potential use of untreated seeds, although biosorbent materials are generally derived from plant biomass through different kinds of simple procedures. They may be chemically pre-treated for better performance and/or suitability for process applications. However, good results have been obtained when the seeds were treated with NaOH. This treatment can remove organic and inorganic matter from the sorbent surface. Chemical treatments are commonly performed employing alkaline solutions or with phosphoric and citric acids [42]. Recently, however, efforts have been made to remove and subsequently also recover metals. Metal-saturated biosorbent materials can be easily regenerated applying a simple (e.g. acidic) wash which then contains a very high concentration of released metals in a small volume, making the solution quite amenable to metal recovery.

Moringa Oleifera	Modifying agent(s)	Heavy metal	Temperature (°C)	pH	Contact time (min)	Ref.
Seeds	Petroleum ether	Cd (II) Cu (II) Co (II) Ni (II) Pb (II)	22	3.5 – 8.0	60	[4]
Leaves	NaOH and Citric acid	Cd (II) Cu(II) Ni(II)	40	5.0	50	[32]
Bark	Original state	Ni(II)	50	6.0	60	[35]
Wood	Activated carbon	Cu(II) Ni(II) Zn(II)	30	6.0	240	[31]
Leaves	NaOH and Citric acid	Pb(II)	40	5.0	50	[34]
Bark	Original state	Pb(II)	25	5.0	30	[19]
Pod	Original state NaOH, H_2SO_4 CTAB HCl $Ca(OH)_2$ Triton X-100 H_3PO_4 $Al(OH)_3$	Zn(II)	30	7.0	50	[16]

Moringa Oleifera	Modifying agent(s)	Heavy metal	Temperature (°C)	pH	Contact time (min)	Ref.
	SDS					
Shelled seeds	Original state	Cd(II) Cr(III) Ni(II)	-	6.5 6.5 7.5	40	[15]
Shells	Original state	As (III) As (V)	-	7.5 2.5	60	[43]
Husk and pods	Unmodified CTAB H_3PO_4 H_2SO_4 HCl	Pb(II)	30	5.8	120	[3]
Shelled seeds	Original state	Cd(II)	-	6.5	40	[14]
Seeds	Original state	Ag(I)	25	6.5	20	[36]
Seeds	NaOH	Ni(II)	25	4.0-6.0	5	[44]

CTAB: Cetyl trimethylammonium bromide, SDS: Sodium dodecyl sulfate

Table 3. Study parameters for the removal of metal ions using *Moringa oleifera*.

5. Adsorption models

An important physicochemical aspect in terms of the evaluation of sorption processes is the sorption equilibrium. Adsorption isotherms are a basic requirement in understanding how the adsorbate is distributed between the liquid and solid phases when the adsorption process reaches the equilibrium state [45, 46]. Over the years a wide variety of isotherm models have been introduced. The most commonly used isotherm models include Langmuir [47], Freundlich [48], Dubinin-Radushkevich [49] and Temkin [50].

It can be observed that in most of the cases the Langmuir adsorption model has been successfully used to predict metal adsorption processes. The Langmuir isotherm model assumes monolayer adsorption onto an adsorbent surface containing a finite number of identical sites and without interaction between adsorbed molecules. The Langmuir isotherm model assumes that: each site can accommodate only one molecule or atom; the surface is energetically homogenous; there is no interaction between neighboring adsorbed molecules or atoms; and there are no phase transitions [51]. The Langmuir equation is expressed as follows:

$$q_e = \frac{q_m K_L C_e}{1 + K_L C_e} \tag{1}$$

where q_e is the amount of metal adsorbed at equilibrium (mg g^{-1}), C_e is the concentration of metal in solution at equilibrium (mg L^{-1}), and q_m (mg g^{-1}) and K_L (L mg^{-1}) are the Langmuir constants related to the adsorption capacity (amount of adsorbate needed to form a complete monolayer) and adsorption energy, respectively. The constants q_m and K_L can be calculated from the intercepts and the slopes of the linear plots of C_e/q_e versus C_e.

The Freundlich model describes adsorption onto an energetically heterogeneous surface not limited by the monolayer capacity [48]. It can be presented in the following form:

$$q_e = K_f \; C_e^n \tag{2}$$

where q_e is the amount of metal adsorbed at equilibrium (mg g^{-1}), C_e is the concentration of metal in solution at equilibrium (mg L^{-1}), and K_f (mg g^{-1})(L mg)$^{(1/n)}$ and n (g L^{-1}) are the Freundlich constants related to the multilayer adsorption capacity and adsorption intensity, respectively. According to the theory, n values between 1 and 10 represent favorable adsorption conditions [52]. Values of K_f and n can be calculated from the slope and intercept of the plot of Log q_e versus Log C_e. Experimental adsorption results with high coefficient correlation (R^2) values obtained for Freundlich isotherms have been reported as shown in Table 4.

The Dubinin-Radushkevich model has been used to distinguish between physical and chemical adsorption [53]. The Dubinin-Radushkevich is more general than the Langmuir model because it does not assume a homogenous surface or constant sorption. The Dubinin-Radushkevich equation is given by:

$$q_e = q_m e^{(-\beta \in^2)} \tag{3}$$

where, q_m (mg g^{-1}) is the theoretical sorption capacity (mol g^{-1}), ε is the Polanyi potential which is related to the equilibrium concentration and the constant β gives the mean energy of sorption, E (KJ mol^{-1}). The constants q_m and β are obtained from the intercept and slope of ln q_e versus ε^2, respectively. If the magnitude of E is between 8 and 16 KJ mol^{-1} the adsorption process proceeds by ion-exchange or chemisorptions, while for values of $E < 8$ KJ mol^{-1} the adsorption process is of a physical nature [54]. In [31] reported that the sorption energy (E) values obtained with the Dubinin-Radushkevich model showed that the interaction between metal ions and the adsorbent proceeded principally by physical adsorption.

The Temkin isotherm model is based on the assumption that the heat of adsorption of all the molecules in the layer decreases linearly with coverage due to adsorbent-adsorbate interactions, and the adsorption is characterized by a uniform distribution of binding energies, up to a maximum binding energy [50]. The model is represented by the following equation:

$$q_e = B \; ln \left(AC_e \right) \tag{4}$$

where *A* $(L\ g^{-1})$ and *B* $(J\ mg^{-1})$ are Temkin isotherm constants relating to adsorption potential and heat of adsorption, respectively. A plot of q_e versus ln C_e gives the values of Temkin constants *A* and *B*. In the adsorption of copper, nickel and zinc onto activated carbon produced from *Moringa oleifera* wood [31] the Temkin isotherm showed a higher correlation coefficient, which may be due to the linear dependence of the heat of adsorption on low or medium coverage. The repulsive force probably occurs between the different adsorbate species or for intrinsic surface heterogeneity may be associated with the linearity.

Table 4 details some of the results for the biosorption studies using *Moringa oleifera* which have been reported in the literature from 2006 onwards. From this table it is clear that *Moringa oleifera* shows versatility, removing a variety metals under favorable conditions and is among the most promising metal biosorbents.

Comparing the Langmuir and Freundlich models, *M. oleifera* seeds demonstrated a good removal capacity for Co(II), Cu(II), Pb(II), Cd(II) and Ag(I), as compared to reports related to other parts of the plant (Table 3). The variations in the removal percentage for metal ions can be explained by the different ionic radii of chemical species. In general, for the single metal solutions, ions with larger ionic radii are preferentially adsorbed. Among the metals tested, Pb(II) has the largest ionic radius and hence shows the highest adsorption percentage, whereas Co(II) presents the lowest level of adsorption [55].

Kinetics models are important in evaluating the basic qualities of an adsorbent as well as the time required for the removal of particular metals, the effectiveness of the adsorbent and the identification of the types of mechanisms involved in an adsorption system [56-58]. In order to investigate the mechanism of biosorption and its potential rate-controlling steps, which include the mass transfer and chemical reaction processes, kinetics models are exploited to test experimental data obtained in kinetics studies. These usually show an initial period of rapid metal adsorption with a subsequent decreased until reaching equilibrium of the system. This occurs due to the rapid adsorption of metallic ions by the surface of the adsorbent followed by a step of slow diffusion of ions from the surface film to the adsorption sites in the micropores which are less accessible [59].

In practice, the kinetic studies are carried out in batch experiments, typically varying the adsorbate concentration, the adsorbent mass, the agitation time and the temperature, as well as the type of adsorbent and adsorbate. Subsequently, the data are processed and used in the linear regression to determine the kinetics model which provides the best fit. However, for the validity of the order of the adsorption process two criteria should be evaluated, the first based on the regression coefficient (R^2) and the second on the calculated q_e values, which must approach the experimental q_e [60]. The main models used to evaluate the kinetics model profile are pseudo-first-order and pseudo-second order. However, other models are also are applied, such as Bangham's model and the Weber and Morris sorption kinetic model.

Heavy metal	Langmuir model			Freundlich model			Ref.
	qm (mg g^{-1})	KL (L mg^{-1})	R^2	Kf (mg g^{-1}) (L mg)$^{(1/n)}$	n (g L^{-1})	R^2	
Cd (II)	171.37	0.037 0.029 0.023	> 0.99	-	-	-	[32]
Cu(II)	167.90		> 0.99				
Ni(II)	163.88		> 0.99				
Ni(II)	30. 38	0.31	0.9994	-	-	-	[35]
Cu(II)	11.534	0.2166	0.9979	3.8563	2.9214	0.9976	[31]
Zn(II)	17.668	0.1430	0.9528	3.7708	2.2528	0.9996	
Ni(II)	19.084	0.6165	0.9973	-	-	-	
Pb(II)	209.54	0.038	> 0.99	-	-	-	[34]
Ni(II)	29.6	-	0.9913	-	-	-	[44]
Ag(II)	23.13	0.1586	0.9935	-	-	-	[36]
Zn(II)	52.08	0.150	0.9994	50.35	-	0.9953	[16]
Cd(II)	1.06	0.51	0.94	-	-	-	[15]
Cr(III)	1.01	0.40	0.96				
Ni(II)	0.94	0.34	0.96				
As (III)	1.59	0.04	0.96	-	-	-	[43]
As (V)	2.16	0.09	0.98				
Pb(II)	-	-	0.9981	-	-	-	[3]
Pb (II)	-	-	-	8.6	2.8	0.9981	[19]
Cd(II)	-	-	-	3.04	1.37	-	[14]

Table 4. Langmuir and Freundlich isotherm parameters for *Moringa oleifera.*

The pseudo-first order equation, also known as the Lagergren equation, is expressed as follows [16, 61]:

$$\log(q_e - q_t) = \log q_e - \frac{k_1}{2.303}t \quad (5)$$

where q_t and q_e (mg g^{-1}) are the amount of metal ions adsorbed per unit weight of the adsorbent at time t and equilibrium, respectively; and k_1 (min^{-1}) is the pseudo-first order rate constant of the sorption process and t (min) is the mixing time [60]. Table 5 presents the data of calculated q_e, pseudo-first order rate (k_1) and correlation coefficient (R^2). This kinetics model is based on the assumption that the adsorption rate is proportional to the number of free sites available, occurring exclusively onto one site per ion [34, 62].

In most studies discrepancies occurred between the value of q_e calculated by the pseudo-first order model and the experimental q_e, as shown in Table 5, highlighting the inability of this model to describe the kinetics of the metal ion adsorption processes. In general, calculated q_e values are smaller than the experimental q_e, which may occur because of a time lag, probably due to the presence of the boundary layer or external resistance at the beginning of the sorption process [63]. Considering the papers detailed in Table 5, only in [43] and [14] used only the pseudo-first order kinetics model to examine the data obtained, even though in the latter case the correlation values obtained were relatively low. In [43] noted no change in the adsorption rate constant when varying the concentrations of As(III) and As(V) and therefore this model could describe the adsorption process. In [14] used this model to compare the adsorption rate constants of ternary metal ions and single metal ions and noted that these constants were lower for ternary metal ions. Their explanation for this was that metal ions compete for vacant sites and uptake by binding sites within the shortest possible time.

The pseudo-second-order kinetics model is also based on the assumption that the sorption rate is controlled by a chemical sorption mechanism involving electron sharing or electron transfer between the adsorbent and adsorbate [64]. It can be expressed as:

$$\frac{t}{q_t} = \frac{1}{k_2 q_e^2} + \frac{t}{q_e} \tag{6}$$

where q_t and q_e (mg g^{-1}) are the amount of metal ions adsorbed per unit weight of the adsorbent at time t and equilibrium, respectively; and k_2 (g mg^{-1} min^{-1}) is the pseudo-second order rate constant of the sorption process and t (min) is the mixing time.

Table 5 presents the data of calculated q_e, pseudo-second order rate (k_2) and correlation coefficient (R^2). For most of the pseudo-second order kinetics models the calculated q_e values approach the experimental q_e values and the correlation coefficients are close to 1, indicating a good ability of this model to describe the kinetics of the metal ion adsorption process. This observation indicates that the rate-limiting steps in the biosorption of metallic ions are chemisorption involving valence forces through the sharing or exchange of electrons between the sorbent and the sorbate, complexation, coordination and/or chelation, in which mass transfer in the solution was not involved.

Considering that neither the pseudo-first-order nor the pseudo-second-order model can identify the diffusion mechanism, other kinetic models are needed to study this process, such as Bangham's model and the Weber and Morris sorption kinetics model [65]. The latter model is also known as the intra-particle diffusion model, this process in many cases being the rate-limiting step, which can be determined through the following equation:

$$q_t = k_{id} t^{1/2} + c_{id} \tag{7}$$

where q_t (mg g^{-1}) is the amount of metal ions adsorbed per unit weight of the adsorbent at time t, c_{id} (mg g^{-1}) is a constant of Weber and Morris, and k_{id} (mg g^{-1} min$^{-1/2}$) is the intra-particle diffusion rate constant and t (min) is the mixing time [66]. The value of the intercept gives an idea of the thickness of the boundary layer, i.e., the larger the intercept the greater the boundary layer effect will be. When there is a complete fit of the model the value of c_{id} should be zero, and the deviation of this constant is due to differences in the mass transfer rate during the initial and final stages of adsorption. This is indicative that there is some degree of boundary layer control and shows that the intra-particle diffusion is not the only rate-limiting step, and thus several processes operating simultaneously may control the adsorption [34].

According to this model, if the plot of q_t versus $t^{1/2}$ gives a straight line, then the sorption process is controlled by intra-particle diffusion, while if the data exhibit multi-linear plots then two or more steps influence the adsorption process [67]. In two studies performed in [32, 19], multi-linear plots were observed with three distinct steps involved in the biosorption, the initial region of the curve relative relating to the adsorption on the external surface. The second region corresponds to the gradual uptake, where the intra-particle diffusion is the rate-limiting step. The final plateau region indicates the equilibrium uptake.

In reference [3] compared different types of carbon through the k_{id} values and observed that the effect of intra-particle diffusion may be significantly increased by chemically modifying the adsorbents. Although none of the data collected in the studies detailed in Table 5 were well-described by the kinetics model proposed by Weber and Moris, the intraparticle diffusion may not be the only rate-limiting step in these studies.

Bangham's model evaluates whether pore diffusion is the only rate-controlling step in the adsorption process [65] and can be represented by the following equation:

$$\log\left[log\left(\frac{c_o}{c_o - q_t m}\right)\right] = \log\left(\frac{k_o m}{2.303\, V}\right) + \sigma \log t \tag{8}$$

where q_t (mg g^{-1}) is the amount of metal ions adsorbed per unit weight of the adsorbent at time t; c_o (mg L^{-1}) is the initial metal ion concentration in liquid phase; m (g L^{-1}) is adsorbent concentration at time t (min); V (L) solution volume, t (min) is the mixing time and k_o (L g^{-1}) and σ ($\sigma < 1$) are constants of Bangham's model. Of the studies published, only in [3] used Bangham's kinetic model to compare the rate constants for the adsorption of Pb(II) onto different types of functionalized carbon prepared from the seed husks and pods of *M. oleifera* and thereby assess the efficiency of the functionalization of this material.

The temperature is reportedly an important parameter for the adsorption of metal ions. An increase or decrease in temperature can cause a change in the amount of metal removed or adsorbed by the adsorbent. A change in temperature causes a change in the thermodynamic parameters of free energy ($\Delta G°$), enthalpy ($\Delta H°$) and entropy ($\Delta S°$). These parameters are important to understand the adsorption mechanism [68]. For a given temperature, a phenomenon is considered to be spontaneous if the $\Delta G°$ has a negative value. Moreover, if $\Delta H°$ is positive the process is endothermic and if it is negative the process is exothermic [69]. Negative values of $\Delta S°$ show a decreased randomness or increased order at the metal-biomass interface. The positive value showed a change in the biomass structure during the sorption process, causing an increase in the disorder of the system [68]. The parameters $\Delta G°$ (kJ mol^{-1}), $\Delta H°$ (kJ mol^{-1}) and $\Delta S°$ (J mol^{-1} K^{-1}) can be evaluated from the following equations [70].

$$\Delta G^\circ = -RT \ln K_c \tag{9}$$

$$\ln K_c = -\frac{\Delta H^\circ}{RT} + \frac{\Delta S^\circ}{R} \tag{10}$$

where R(8.314J mol^{-1} K^{-1}) is the gas constant, T(K) the absolute temperature and K_c(mL g^{-1}) the standard thermodynamic equilibrium constant defined by q_e/C_e. ΔH° and ΔS° can be determined from the slope and the intercept of the linear plot of Ln K_c versus $1/T$.

The studies performed on *Moringa oleifera* using chemically-modified leaves for the adsorption of Pb(II) [34], bark for Ni(II) [35] and leaves for Cd(II), Cu(II) and Ni(II) [32] showed the endothermic nature and spontaneity of the adsorption process. The positive values of ΔS° suggest an increase in randomness at the solid/liquid interface with some structural changes in the sorbate.

6. Final considerations

Although the biosorption of heavy metals from aqueous solutions is a relatively new process that has proven very promising in the removal of contaminants from aqueous effluents, offering significant advantages like the low-cost, availability, profitability, easy of operation and efficiency. Other technologies have also been very attractive ensuring an appropriate process to treat industrial waste effluents [71-77]. However, biosorption is becoming a potential alternative to the existing technologies for the removal and/or recovery of toxic metals from wastewater. The major advantages of biosorption technology are its effectiveness in reducing the concentration of heavy metal ions to very low levels and the use of inexpensive biosorbent materials.

7. Conclusions

The studies described herein indicate that *Moringa oleifera* seeds are an alternative sorbent for metal ion removal from contaminated waters. This can be found in most papers which report 60 to 90% removal of metals (Cd(II), Cu(II), Ni(II), Pb(II), As(III), As(V), Cr(III) and Zn(II)). In these cases, not only the seeds were used, but also leaves, bark and pods showing the great versatility of this plant. The results show that even with the high heterogeneity of the matrix confirmed through characterization techniques there is a great potential for the application of these seeds in effluent treatment without component separation, which makes the process economically and technically attractive.

Biosorption is the most economical and eco-friendly method for the removal of heavy metals from domestic as well as industrial wastewater and it is particularly important to promote the development of biosorption for industrial processes. Notable advantages are: (a) low cost of

Metal	c_o (mg L^{-1})	qe, exp (mg g^{-1})	Pseudo-first-order			Pseudo-second-order				Ref.
			qe (mg g^{-1})	k1 (min $^{-1}$)	R^2	qe (mg g^{-1})	k2 (g mg^{-1} min^{-1})	h_0 (mg g^{-1} min^{-1})	R^2	
Zn (II)	50[a]	45.00	-	-	-	46.94	4.51 10-4	-	0.997	[16]
	50[b]	45.76	-	-	-	47.16	2.85 10-4	-	0.999	
	50[c]	42.80	-	-	-	43.47	2.04 10-5	-	0.997	
Pb(II)	30[d]	-	-	-	-	24.57	0.0085	5.131	0.999	[3]
	30[e]	-	-	-	-	27.70	0.0052	3.989	0.998	
	30[f]	-	-	-	-	28.49	0.0060	4.868	0.998	
	30[g]	-	-	-	-	29.08	0.0062	5.243	0.999	
	30[h]	-	-	-	-	29.46	0.0087	7.550	0.999	
As (III)	25	-	-	0.047	-	-	-	-	-	[43]
	50	-	-	0.049	-	-	-	-	-	
As (V)	25	-	-	0.063	-	-	-	-	-	
	50	-	-	0.065	-	-	-	-	-	
Cd(II) Cr(III) Ni(II)	25	1.06	-	0.51	-	-	-	-	-	[14]
	25	1.01	-	0.40	-	-	-	-	-	
	25	0.94	-	0.34	-	-	-	-	-	
Pb(II)	10	12.7343	-	-	-	13.26	15.35	0.20	0.9974	[34]
	25	19.8988	-	-	-	20.64	10.08	0.21	0.997	
	40	23.9233	-	-	-	25.01	9.3	0.23	0.9995	
Pb(II)	10.4	8.7	-	-	-	8.8	27.8	2.5	0.9999	[19]
	30.1	10.2	-	-	-	10.3	18.2	1.9	0.9999	
	50.4	12.5	-	-	-	12.53	12.6	1.6	0.9998	
Ni(II)	10	9.7	-	-	-	10.29	1.91	2.03	0.9971	[35]
	25	6.74	-	-	-	7.14	2.70	1.38	0.9964	
	50	3.27	-	-	-	3.43	12.74	1.05	0.996	
Cu(II)	30	8.3406	-	-	-	8.3264	0.0848	-	0.9998	[31]
Zn(II)	30	13.2537	-	-	-	13.2450	0.2457	-	1	
Ni(II)	30	9.5847	-	-	-	9.6154	0.0957	-	0.9999	
Cd (II) Cu(II) Ni(II)	10	13.54	-	-	-	10.99	1.39	2.73	0.9951	[32]
	25	13.80	-	-	-	10.50	1.46	3.09	0.9969	
	40	20.86	-	-	-	15.24	1.22	5.61	0.9981	
	10	11.92	-	-	-	11.03	3.4	5.58	0.9992	
	25	13.55	-	-	-	12.45	1.66	3.37	0.9958	
	40	16.01	-	-	-	12.86	1.56	4.31	0.9977	
	10	10.24	-	-	-	10.24	1.51	1.70	0.9951	
	25	12.49	-	-	-	12.49	1.31	2.29	0.9952	
	40	14.07	-	-	-	14.07	1.52	3.27	0.9967	

Table 5. Kinetics parameters for metal biosorption using Moringa oleifera.

the biosorbent, (b) high efficiency for metal removal at low concentration, (c) potential for biosorbent regeneration and metal valorization, (d) high sorption and desorption rates, (e) limited generation of secondary residues, and (f) relatively environmentally-friendly life cycle of the material (easy to eliminate compared to conventional resins, for example).

However, after the metal removal from aqueous solutions by the biomass, the recovery of the metal is an important issue. This can be achieved through a metal desorption process, aimed at weakening the metal-biomass linkage. Thus, studies to evaluate the reversibility of the adsorption reactions involved in the biosorption of heavy metals are of great importance. The problems associated with the disposal of exhausted adsorbent can be solved either by its activation or incineration or its disposal after proper treatment. For biosorption and desorption processes, another important aspect is the biosorbent reuse in successive biosorption-desorption cycles, the viability of which is determined by the cost-benefit relationship between the loss in biosorption capacity during the desorption steps and the operational yield in the metal recovery. Thus, further studies need to focus on the development of new clean environmentally-acceptable technologies.

Acknowledgements

The authors are grateful for financial support from the government agencies Conselho Nacional de Desenvolvimento Científico e Tecnológico (CNPq), Fundação de Amparo à Pesquisa do Estado de Minas Gerais (FAPEMIG), Fundação de Amparo à Pesquisa do Estado de Goiás (FAPEG) and Coordenação de Aperfeiçoamento de Pessoal de Nível Superior (CAPES).

Nomenclature

A	Temkin isotherm constant relating to adsorption potential, (L g^{-1})
B	Temkin isotherm constant relating to heat of adsorption, (J mg^{-1})
β	mean energy of sorption, E (KJ mol^{-1})
C_o	initial metal ion concentration in liquid phase, (mg L^{-1})
C_e	concentration of metal in solution at equilibrium, (mg L^{-1})
C_{id}	constant of the Weber and Morris model, (mg g^{-1})
CAPES	Coordination of Improvement of Higher Education Personnel
CNPq	National Council for Scientific and Technological Development
ε	Polanyi potential which is related to the equilibrium concentration
FAPEMIG	Research Foundation of the State of Minas Gerais
FAPEG	Research Foundation of the State of Goiás

http://dx.doi.org/10.5772/56157

FT-IR	Fourier transform infrared
$\Delta G°$	free energy, (kJ mol^{-1})
$\Delta H°$	enthalpy, (kJ mol^{-1})
K_c	standard thermodynamic equilibrium constant defined by q_e/C_e, (mL g^{-1})
K_L	Langmuir constant related to adsorption energy, (L mg^{-1})
K_f	Freundlich constant related to the multilayer adsorption capacity
k_1	pseudo-first order rate constant of the sorption process, (min^{-1})
k_2	pseudo-second order rate constant of the sorption process, (g mg^{-1} min^{-1})
k_{id}	intra-particle diffusion rate constant, (mg g^{-1} min$^{-1/2}$)
k_o	constant of the Bangham's model, (L g^{-1})
σ	constant of the Bangham's model
m	adsorbent concentration at time t (min), (g L^{-1})
n	Freundlich constant related to the multilayer adsorption intensity
q_e	amount of metal adsorbed per unit weight of the adsorbent at equilibrium, (mg g^{-1})
q_m	Langmuir constant related to the adsorption capacity, (mg g^{-1})
q_t	amount of metal ions adsorbed per unit weight of the adsorbent at time t, (mg g^{-1})
R	gas constant, (8.314J mol^{-1} K^{-1})
R^2	coefficient correlation
$\Delta S°$	entropy, (J mol^{-1} K^{-1})
SEM	scanning electron microscopy
t	mixing time, (min)
T	absolute temperature, (K)
TG	thermogravimetric
US EPA	United States Environmental Protection Agency
V	solution volume, (L)
XRD	X-ray diffraction

Author details

Cleide S. T. Araújo[1], Dayene C. Carvalho[2], Helen C. Rezende[2], Ione L. S. Almeida[2], Luciana M. Coelho[3], Nívia M. M. Coelho[2*], Thiago L. Marques[2] and Vanessa N. Alves[2]

*Address all correspondence to: nmmcoelho@ufu.br

1 State University of Goiás, Anápolis, GO, Brazil

2 Institute of Chemistry, Federal University of Uberlândia, Uberlândia, MG, Brazil

3 Department of Chemistry, Federal University of Goiás, Catalão, GO, Brazil

References

[1] Miliarum: Ingenieria Civil y Medio Ambiente. Cursos Edafología on line: Contaminación por metales pesados. http://www.miliarium.com/Marcos/SuelosContaminados.asp (accessed 27 September 2012).

[2] Duffus JH. "Heavy metals" - A meaningless term? (IUPAC Techinical Report). Pure and Applied Chemistry 2002; 74(5) 793-807.

[3] Nadeem M, Mahmooda A, Shahid SA, Shah SS, Khalid AM, McKaye G. Sorption of lead from aqueous solution by chemically modified carbon adsorbents. Journal of Hazardous Materials 2006; 138(3) 604-13.

[4] Obuseng V, Nareetsile F, Kwaambwa HM. A study of the removal of heavy metals from aqueous solutions by Moringa oleifera seeds and amine-based ligand 1,4-bis[N,N-bis(2-picoyl)amino]butane. Analytica Chimica Acta 2012; 730 87-92.

[5] Zevenhoven R, Kilpinen P. Control of pollutants in flue gases and fuel gases. Finland: Espoo; 2001.

[6] EPA: Environmental Protection Agency. National Primary Drinking Water Regulations. http://www.biovir.com/Images/pdf054.pdf (accessed 27 September 2012).

[7] Babel S, Kurniawan TA. Cr(VI) removal from synthetic wastewater using coconut shell charcoal and commercial activated carbon modified with oxidizing agents and/or chitosan. Chemosphere 2004; 54(7) 951-67.

[8] Bai RS, Abraham TE. Low-cost supports used to immobilize fungi and reliable technique for removal hexavalent chromium in wastewater. Bioresource Technology 2003; 87(17) 17-26.

[9] Petroni SLG, Pires MAF, Munita CS. Adsorção de zinco e cádmio em colunas de turfa. Química Nova 2000; 23(4) 477-81.

[10] Volesky B. Detoxification of metal-bearing effluents: biosorption for the next century. Hydrometallurgy 2001; 59(2) 203-16.

[11] Naja G, Murphy V, Volesky B. Biosorption, Metals - Encyclopedia of Industrial Biotechnology. United Kingdom: Wiley; 2010.

[12] Melcakova I, Horvathova H. Study of biomass of reynoutria japonica as a novel biosorbent for removal of metals from aqueous solutions. Czech Republic: GeoScience Engineering; 2010. http://gse.vsb.cz/2010/LVI-2010-1-55-70.pdf (accessed 27 September 2012).

[13] Dermirbas, A, Heavy metal adsorption onto agro-based waste materials: A review. Journal of Hazardous Materials 2008; 157 220–29

[14] Sharma P, Kumari P, Srivastava MM, Srivastava S. Removal of cadmium from aqueous system by shelled Moringa oleifera Lam seed powder. Bioresource Technology 2006; 97 299–305.

[15] Sharma P, Kumari P, Srivastava MM, Srivastava S. Ternary biosorption studies of Cd(II), Cr(III) and Ni(II) on shelled Moringa oleifera seeds. Bioresource Technology 2007; 98 474–77.

[16] Bhatti HN, Mumtaz B, Hanif MA, Nadeem R. Removal of Zn(II) ions from aqueous solution using Moringa oleifera Lam. (horseradish tree) biomass. Process Biochemistry 2007; 42 547–53.

[17] Keer WE, Silva AR. "Moringa: uma nova hortaliça para o Brasil". Uberlândia: UFU/ DIRIU, 1999.

[18] Almeida ILS. Avaliação da capacidade de adsorção da torta da Moringa oleifera para BTEX em amostras aquosas. Masters dissertation. Federal University of Uberlândia; 2010.

[19] Reddy DHK, Seshaiah K, Reddy AVR, Rao MM, Wang MC. Biosorption of Pb2+ from aqueous solutions by Moringa oleifera bark: equilibrium and kinetic studies. Journal of Hazardous Materials 2010; 174 831–838.

[20] Nautiyal BP, Venkataraman KG. Moringa (drumstick) - an ideal tree for social forestry: growing conditions and uses - part 1. Myforest 1987; 23(1) 53-8.

[21] Ndabigengesere A, Narasiah KS, Talbot BG. Active agents and mechanism of coagulation of turbid waters using Moringa oleifera. Water Research 1995; 29(2) 703-10.

[22] Jahn SAA. Proper use Moringa oleifera for food and water purification - Selection of clones and growing of annual short-stem. Pflanzenzucht 1989; 4 22-5.

[23] Oliveira JTA, Silveira SB, Vasconcelos IKM, Cavada BS, Moreira RA. Compositional and nutritional attributes of seeds from the multiple purpose tree Moringa oleifera Lamarck. Journal of the Science of Food and Agriculture 1999; 79 815-20.

[24] Olsen A. Low technology water purification by bentonite clay and Moringa oleifera seed flocculation as performed in Sudanese village: effects on Schistoma mansoni cercariae. Water Research 1987; 21(5) 517-22.

[25] Madsen M, Schumundt J, Omer EFE. Effect of watter coagulation by seeds of Moringa oleifera on bacterial concentration. Journalof tropical Medicine and Hygiene 1987; 90(3)101-9.

[26] Bennett RN, Mellon FA, Foidl N, Pratt JH, Dupont MS, Perkins L, Kroon A. Profiling Glucosinolates and Phenolics in Vegetative and Reproductive Tissues of the Multi-Purpose Trees Moringa oleifera L. (Horseradish Tree) and Moringa stenopetala L. Journal of Agricultural and Food Chemistry 2003; 51(12) 3546-53.

[27] Moreira DR. Desenvolvimento de adsorventes naturais para tratamento de efluentes de galvanoplastia. Masters dissertation. Pontifical Catholic University of Rio Grande do Sul; Porto Alegre; 2010.

[28] Davis TA, Mucci A, Volesky B. A review of the biochemistry of heavy metal biosorption by brown algae. Water Research 2003; 37(18) 4311-30.

[29] Schneider IAH, Rubio J. Plantas Aquáticas: Adsorventes Naturais para a Melhoria da Qualidade das Águas, 2003. XIX Prêmio Jovem Cientista - 2003 – Água: Fonte de Vida, Porto Alegre, Brazil, School of Engineering, Federal University of Rio Grande do Sul, 2003.

[30] Pagnanelli F, Mainelli S, Veglio F, Toro L. Heavy metal removal by olive pomace: biosorbent characterization and equilibrium modeling. Chemical Engineering Science 2003; 58 4709-17.

[31] Kalavathy MH, Miranda LR. Moringa oleifera—A solid phase extractant for the removal of copper, nickel and zinc from aqueous solutions. Chemical Engineering Journal 2012; 158 188–99.

[32] Reddy DHK, Seshaiaha K, Reddyb AVR, Leec SM. Optimization of Cd(II), Cu(II) and Ni(II) biosorption by chemically modified Moringa oleifera leaves powder. Carbohydrate Polymers 2012; 88 1077–86.

[33] Brito ES, Damasceno LF, Gallão MI. Avaliação química e estrutural da semente de moringa. Revista Ciência Agronômica 2006; 37 106-9.

[34] Reddy DHK, Harinatha Y, Seshaiaha K, Reddy AVR. Biosorption of Pb(II) from aqueous solutions using chemically modified Moringa oleifera tree leaves. Chemical Engineering Journal 2010; 162 626–34.

[35] Reddy DHK, Ramana DKV, Seshaiah K, Reddy AVR. Biosorption of Ni(II) from aqueous phase by Moringa oleifera bark, a low cost biosorbent. Desalination 2011; 268 150–57.

[36] Araújo CST, Melo EI, Alves VN, Coelho NMM. Moringa oleifera Lam. seeds as a natural solid adsorbent for removal of AgI in aqueous solutions. Journal of the Brazilian Chemical Society 2010; 21 1727-32.

[37] Fifield FW, Kealey D. Principles and Practice of Analytical Chemistry. Oxford: Blackwell Science; 2000.

[38] Anwar F, Rashid U. Physico-chemical characteristics of Moringa oleifera seeds and seeds oil from a wild provenance of Pakistan. Pakistan Journal of Botany 2007; 39(5) 1443-53.

[39] Pereira GM, Arruda MAZ. Trends in preconcentration procedures for metal determination using atomic spectrometry techniques. Microchimica Acta 2003; 141 115-31.

[40] Senthilkumaar S, Bharathi S, Nithyanandhi D, Subburam V. Biosorption of toxic heavy metals from aqueous solutions. Bioresource Technology 2000; 75 163-5.

[41] Alves VN, Mosquetta R, Coelho NMM, Bianchin JN, Roux KCP, Martendal E, Carasek E. Determination of cadmium in alcohol fuel using Moringa oleifera seeds as a biosorbent in an on-line system coupled to FAAS. Talanta 2010; 80(3) 1133-38.

[42] Marshall WE, Johns MM. Agricultural by-products as metal adsorbents: sorption properties and resistance to mechanical abrasion. Journal of Chemical Technology 1996; 66(2) 192-8.

[43] Kumari P, Sharma P, Srivastava S, Srivastava MM. Biosorption studies on shelled Moringa oleifera Lamarck seed powder: removal and recovery of arsenic from aqueous system. International Journal Mineral Process 2006; 78 131-9.

[44] Marques TL, Alves VN, Coelho LM, Coelho NMM. Removal of Ni(II) from aqueous solution using Moringa oleifera seeds as a bioadsorbent. Water Science & Technology 2012; 68 1435-40.

[45] Shen D, Fan J, Zhou W, Gao B, Yue Q, Kang Q. Adsorption kinetics and isotherm of anionic dyes onto organo-bentonite from single and multisolute systems. Journal of Hazardous Materials 2009; 172 99-107.

[46] Abdullah MA, Chiang L, Nadeem M. Comparative evaluation of adsorption kinetics and isotherms of a natural product removal by Amberlite polymeric adsorbents. Chemical Engineering Journal 2009; 146 370-6.

[47] Ma X, Li L, Yang L, Su C, Wang K, Yuan S, Zhou J. Adsorption of heavy metals ions using hierarchical CaCO3-maltose meso/macroporous hybrid materials: Adsorption isotherms and kinetic studies. Journal of Hazardous Materials 2012; 209-210 467-477.

[48] Tofighy MA, Mohammadi T. Adsorption of divalent heavy metal ions from water using carbon nanotube sheets. Journal of Hazardous Materials 2011; 185 140-7.

[49] Olu-owolabi BI, Oputu OU, Adebowale KO, Ogunsolu O, Olujimi OO. Biosorption of Cd2+ and Pb2+ ions onto mango stone and cocoa pod waste: Kinetic and equilibrium studies. Scientific Research and Essays 2012; 7(15) 1614-29.

[50] Kumar PS, Ramalingam S, Senthamarai C, Niranjanaa M, Vijayalakshmi P, Sivanesan S. Adsorption of dye from aqueous solution by cashew nut Shell: Studies on equilibrium isotherm, kinetics and thermodynamics of interactions. Desalination 2010; 261 52-60.

[51] Mahmoud DK, Salleh MAM, Karim WAA. Langmuir model application on solid-liquid adsorption using agricultural wastes: Environmental application review. Journal of Purity, Utility Reaction and Environment 2012; 1(4) 170-99.

[52] Nwabanne JT, Igbokwe PK. Mechanism of cooper (II) removal from aqueous solution using activated carbon prepared from different agricultural materials. International Journal of Multidisciplinary Sciences and Engineering 2012; 3(7) 46-52.

[53] Kumar PS, Ramalingam S, Kirupha SD, Murugesan A, Vidhyadevi T, Sivanesan S. Adsorption behavior of nickel (II) onto cashew nut Shell: equilibrium. Thermodynamics, kinetics, mechanism and process design. Chemical Engineering Journal 2011; 167 122-31.

[54] Kumar PS, Ramalingan S, Sathyaselvabala V, Kirupha SD, Murugesan A, Sivanesan S. Removal of cadmium(II) from aqueous solution by agricultural waste cashew nut shell. Korean Journal of Chemical Engineering 2012; 29(6) 756-68.

[55] Matos GD, Arruda MAZ. Online preconcentration/determination of cadmium using grape bagasse in a flow system coupled to thermospray flame furnace atomic absorption spectrometry. Spectroscopy Letters 2006; 39 1-14.

[56] Febrianto J., Kosasih A. N., Sunarso J., Ju. Y., Indraswati N., Ismadji S. Equilibrium and kinetic studies in adsorption of heavy metals using biosorbent: A summary of recent studies. Journal of Hazardous Materials 2009; 162, 616-45.

[57] Kumar KV, Sivanesan S. Pseudo-second-order kinetic models for safranin onto rice husk: comparison of linear and non-linear regression analysis. Process Biochemistry 2006; 41 1198–1202.

[58] Savic JZ, Vasic VM. Thermodynamics and kinetics of 1,8-dihydroxy-2-(imidazol-5-yl-azo)-naphthalene-3,6-disulphonic acid immobilization on Dowex resin, Colloids and Surfaces A. Physicochemical and Engineering Aspects 2006; 278 197–203.

[59] Zhao G, Wu X, Tan X, Wang X. Sorption of Heavy Metal Ions from Aqueous Solutions: A Review. The Open Colloid Science Journal 2011; 4 19–31.

[60] Li X, Zheng W, Wang D, Yang Q, Cao J, Yue X, Shen T, Zeng G. Removal of Pb (II) from aqueous solutions by adsorption onto modified areca waste: Kinetic and thermodynamic studies. Desalination 2010; 258 148–53.

[61] Bark N, Abdennouri M, Boussaoud A, Makhfouk MEL. Biosorption characteristics of Cadmium (II) onto Scolymus hispanicus L. as low-cost natural biosorbent. Desalination 2010; 258(1-3) 66–71.

[62] Lalhruaitluanga H, Jayaram K, Prasad MNV, Kumar KK. Lead(II) adsorption from aqueous solutions by raw and activated charcoals of Melocanna baccifera Roxburgh (bamboo)-A comparative study. Journal of Hazardous Materials 2010; 75 311–8.

[63] Vijayaraghavan K, Palanivelu K, Velan M. Biosorption of copper(II) and cobalt(II) from aqueous solutions by crab shell particles, Bioresource Technology 2006; 97 1411–19.

[64] Ding Y, Jing D, Gong H, Zhou L, Yang X. Biosorption of aquatic cadmium(II) by unmodified rice straw. Bioresource Technology 2012; 114 20–5.

[65] Yaneva Z, Koumanova B, Georgieva N. Study of the Mechanism of nitrophenols sorptions on expanded perlite – Equilibrium and Kinetics Modeling. Macedonian Journal of Chemistry and Chemical Engineering 2012; 31(1) 101–14.

[66] Baidas S, Gao B, Meng X. Perchlorate removal by quaternary amine modified reed. Journal of Hazardous Materials 2011; 189 54–61.

[67] Bilgili MS. Adsorption of 4-chlorophenol from aqueous solutions by xad-4 resin: isotherm, kinetic, and thermodynamic analysis. Journal of Hazardous Materials 2006; 137 157–64.

[68] Farooq U, Kozinski JA, Khan MA, Athar M. Biosorption of heavy metals ions using wheat based biosorbents - A review of the recent literature. Bioresource Technology 2010; 101 5043-53.

[69] Gueu S, Yao B, Adouby K, Ado G. Kinetics and thermodynamics study of lead adsorption on to activated carbons from coconut and seed hull of the palm tree. International Journal Science Technology 2007; 4(1) 11-17.

[70] Zhu HY, Fu YQ, Jiang R, Jiang JH, Xiao L, Zeng GM, Zhao SL, Wang Y. Adsorption removal of congo red on magnetic cellulose/Fe3O4/activated carbon composite: Equilibrium, kinetic and thermodynamic studies. Chemical Engineering Journal 2011; 173 494-502.

[71] Bhat V, Rao P, Patil Y. Development of an integrated model to recover precious metals from electronic scrap - A novel strategy for e-waste management. Procedia - Social and Behavioral Sciences 2012; 37: 397-406.

[72] Patil YB. Development of an innovative low-cost industrial waste treatment technology for resource conservation - A case study with gold-cyanide emanated from SMEs. Procedia- Social and Behavioral Sciences 2012; 37: 379-388.

[73] Gaddi SS, Patil YB. Screening of some low-cost waste biomaterials for the sorption of silver-cyanide [Ag(CN)2-] from aqueous solutions. International Journal of Chemical Sciences 2011; 9: 1063-1072.

[74] Patil YB, Paknikar KM. Biological detoxification of nickel-cyanide from industrial effluents. Process Metallurgy 2001; 11: 391-400.

[75] Patil YB, Paknikar KM. Development of a process for biodetoxification of metal cyanides from wastewaters. Process Biochemistry 2000; 35: 1139-1151.

[76] Patil YB, Paknikar KM. Biodetoxification of silver-cyanide from electroplating industry wastewater. Letters in Applied Microbiology 2000; 30: 33-37.

[77] Patil YB, Paknikar KM. Removal and recovery of metal cyanides using a combination of biosorption and biodegradation processes. Biotechnology Letters 1999; 21: 913-919.

Chapter 11

Resource Recovery from Industrial Effluents Containing Precious Metal Species Using Low-Cost Biomaterials — An Approach of Passive Bioremediation and Its Newer Applications

Nilisha Itankar, Viraja Bhat, Jayati Chourey,
Ketaki Barve, Shilpa Kulkarni, Prakash Rao and
Yogesh Patil

Additional information is available at the end of the chapter

http://dx.doi.org/10.5772/56965

1. Introduction

Industrial wastes can generally be classified as wastes rich in organic matter on one hand and wastes rich in inorganic matter on the other. Cyanide (CN^-) and heavy metals (viz. copper, nickel, iron, zinc, cadmium, chromium, silver, gold, etc.) form a significant part of the latter type of wastes. Free cyanide (CN^-) is industrially important chemical because of its some unique properties of binding various transition metals to form metal-cyanides (M_xCN) complexes of variable stability and toxicity (Sharpe 1976). Therefore, cyanide finds enormous applications in variety of industrial processes. Industries like gold and silver mining, electroplating, printed circuit board manufacturing and jewellery units emanate large-volume low-tenor effluents containing anionic M_xCN complexes like gold-cyanide i.e. $[Au(CN)_2]^-$ and silver-cyanide i.e. $[Ag(CN)_2]^-$ (Vapur et al 2005). The total cyanide, gold (Au) and silver (Ag) content in these effluents ranges from 5-25, 1-2 and 5-10 mg/L, respectively (Patil 1999). The discharge limits for total cyanide is 0.2 mg/L, while for Au and Ag the standards are yet to be set and currently not available. Apart from Au and Ag many other heavy metals normally occur in the effluents in low quantity and concentrations. If inappropriately managed, cyanide and metals or their complexes can be mobilised and carried into the food web as a result of leaching from waste dumps, contaminated soils and waters. At each level of food chains, the concentration of metals increases which results into a phenomenon called biomagnification.

Since cyanide is toxic and Au and Ag being precious metals, non-renewable and finite resource; their complete removal from effluents is the key.

The conventional methods adopted for the treatment of M_xCN contaminated effluents is alkaline chlorination oxidation process (Ganczarczyk et al 1985). Although this method of treatment can be very efficient in detoxifying free cyanide bearing wastes, it is not effective when challenged with anionic metal species such as M_xCN (Eckenfelder 1989). Other methods, such as copper catalyzed hydrogen peroxide oxidation, ozonation, electrolytic decomposition, etc. requires large inputs of energy, cost intensive materials and are rarely used for treatment of M_xCN containing wastes. Furthermore, at low concentration, metal-cyanide recovery by conventional means is either not possible and/or very expensive. Thus, there is a big technological breach, which needs to be bridged immediately.

Biological treatment systems (i.e. bioremediation) for the detoxification of toxic and hazardous wastes has immense potential of becoming and effective alternative because of their several advantages over conventional methods; and therefore being explored by the researchers all over the world (Patil et al., 2012; Patil and Paknikar, 1999b). However, biological methods like biodegradation / biodetoxification using live microorganisms are subject to toxicity of cyanide and metals. Therefore, removal of precious metal-cyanides species from wastes requires immediate attention of scientists and technologists. The challenge is not limited only to their removal, but also extends to finding competent and inexpensive ways of possible recovery and recycling. It was assumed that if a competent process for removal/recovery could be established, M_xCN could be conserved, which in the authors opinion would be an innovative strategy of resource recovery. Since M_xCN are anionic chemical species, therefore in principle, a well established physico-chemical methods can be used for removal and recovery of precious metal species. A few physico-chemical methods have been tried for adsorption of Au- and Ag-cyanide (Niu and Volesky 2001) and M_xCN using activated carbon or inorganic chemically active adsorbents (Lee et al 1998). However, the practical utility and cost-effectiveness of these processes are not yet established. Biosorption of metal "cations" have been studied extensively (Paknikar et al 2003). However, very few attempts have been made to adopt this technology for possible removal and recovery of "anions" such as M_xCN (Patil 2012; Patil and Paknikar 1999); especially Au- and Ag-cyanide (Niu & Volesky 2001). Literature clearly shows the paucity of references on the removal/recovery of precious Au- and Ag-cyanide using low-cost waste biomass.

It is known that biomass like bacteria, fungi, algae, plants, agricultural biomass and different agro-based industrial waste and byproducts have the ability to bind metals, in some cases selectively, from aqueous solutions (Paknikar *et al.*, 2003). This phenomenon is named as 'metal biosorption' and the biomass responsible for the process are known as 'biosorbents'. Biosorption is combination of the processes such as electrostatic interactions, ion exchange, complexation, formation of ionic bonds, precipitation, nucleation, etc. Biomass surfaces are usually charged. The functional groups like phosphoryl, carboxyl, sulphahydryl and hydroxyl of membrane proteins, lipids and of other cell wall components are responsible for adsorption of metal (both cationic and anionic species). The overall interfaces are a result of complexity of biomass surfaces and chemical/ physical properties of metal ions (Modak and Natarajan,

1995). The advantages of passive bioremediation (i.e. biosorption) are: (i) non-living biomass is not subject to toxicity limitations; (ii) costly nutrients and aseptic conditions not required; (iii) the process is very rapid; (iv) waste biomass from fermentation and many other natural and industrial sources could be a cheap source of biomass; (v) biosorbent could be operated at wider range of pH, temperature and metal concentration; (vi) established theories, conventions and formulae could be applied to the adsorbent and many others (Paknikar et al. 2003). Although biosorption of metal cations have been studied extensively, however, very little is known about the removal and recovery of anionic metal species (viz. M_xCN) from aqueous wastes.

With the above background in mind, investigations were aimed at screening diverse range of low-cost biomass obtained from different sources for the removal and recovery of precious anionic metal species like gold- and silver-cyanide from aqueous wastes with the emphasis on the development of a laboratory / pilot scale technology.

2. Materials and methods

2.1. Low-cost biosorbents

The low-cost biosorbents in the present study were collected from diverse sources (as given below). Some of these biosorbents are reported for the removal of diverse metal species from acqueous solutions (Mohan & Pittman, 2006), while some of them have been employed for the first time.

i. **Agricultural wastes/by-products:** Coconut fibres, Cow dung cakes, Groundnut shells, Rice husk and Rice straw

ii. **Industrial wastes/by-products:** Dairy waste sludge, Saw dust, Sugarcane bagasse and Tea powder waste.

iii. **Municipal solid waste components:** Nirmalya (waste flowers), Compost and Vegetable waste.

iv. **Fungal and Bacterial (waste) biomass:***Ganoderma lucidum*, Yeast waste biomass, *Mucor heimalis*, *Penicillium* sp., *Streptomyces* waste biomass, *Streptoverticillium* waste biomass, Wood rotting fungal waste and Bacterial consortium (capable of degrading cyanide and thiocyanate). Biomass of the fungal cultures viz. *Mucor heimalis* and *Penicillium* sp. available in our laboratory were obtained by cultivating them aseptically in Sabouraud's medium (Composition: Glucose 20 g; Peptone 10 g; Distilled water 1000 ml; pH 4.5-5.0) for 4-5 days at 30°C on rotary shaker incubator (150 rpm). Biomass was harvested after growth by filtering through muslin cloth and washed with distilled water 3-4 times in order to remove the organic traces. After washing, the biomass was subjected to drying in oven at 50-60°C for 2-3 days till the constant weight was obtained. Biomass was then ground using electric mixer in order to obtain particle size of ≤ 500 μm (0.5 mm). Later the biomass were stored in glass bottles with suitable air tight caps for further use.

v. **Algae:** Mixed algae biomass obtained from lake Rankala, Kolhapur

vi. **Terrestrial and aquatic plant species:***Parthenium* sp., *Eichhornia* root biomass, *Eichhornia* stem biomass, *Eichhornia* leaf biomass, Runners, *Tectona grandis* waste leaves and *Lantana camara.*

vii. **Reference material:** Activated charcoal was employed as a reference material in order to obtain comparative data.

Biomass samples were collected in polythene bags and transported to laboratory. The samples were washed several times with tap water to remove the dirt and other contaminants, if any, and was then finally washed with deionised water (< 5 μS). Biomass were then subjected for drying at 50°C for 48-72 h to a constant weight and powdered. Dried biomass was pulverized employing electric mixer and sieved; so as to get uniform particle size of ≤ 500 μm (0.5 mm).

2.2. Synthesis of stock anionic M_xCN solutions

The stock solutions of Au-cyanide i.e. $[Au(CN)_2]^-$ (Dicyanoaurate-DCAU) and Ag-cyanide i.e. $[Ag(CN)_2]^-$ (Dicyanoargentate - DCAG) were prepared stoichiometrically by combining their respective salts with sodium cyanide in the molar proportion of 1:2 (Patil & Paknikar, 2000a; Patil and Paknikar,2000b; Rollinson et al., 1987). Spectral properties were checked and confirmed periodically using UV spectrophotometer. The synthesized DCAU and DCAG solutions were refrigerated at 8-10°C.

2.3. Chemicals and analyses

Chemicals used for all experiments were of analytical grade (AR). Glassware used were made of borosilicate material. Stock solutions and reagents were prepared in deionized water (< 5 μS) and stored in refrigerator (8-10°C). Gold (Au), silver (Ag), nickel (Ni), copper (Cu), zinc (Zn) and iron (Fe) in the experimental solutions and effluents were analysed using Atomic Absorption Spectrophotometer (Elico, India SL-173). Total cyanide and chemical oxygen demand (COD) content in the solutions were estimated by pyridine-barbituric acid and reflux method, respectively as described in Standard Methods (APHA-AWWA-WEF, 1998). Phosphates (PO_4^{-3}) from effluents were analysed by phenol- disulphonic acid method; sulphates (SO_4^{-2}) were determined by barium chloride method while chlorides (Cl^-) were determined by argentometric method, as per the methods prescribed in Standard Methods (APHA-AWWA-WEF, 1998). Colour and turbidity were recorded by visual observations. pH and electrical conductivity from solutions was measured by their respective meters.

In order to determine the inherent/actual pH of each powdered unconditioned biomass, the biomass sample and RO water were mixed and serially diluted in the ratio of 1:20, 1:30, 1:40 and 1:50 (w/v) in conical flasks. The contents were stirred vigourously and kept for one hour in stationary conditions and was followed by determining the pH of each dilution ratio. pH value obtained for each dilution was then plotted on graph of "pH against water-to-biomass ratio (v/w)". The straight line obtained after joining all the points was extrapolated backwards so as to intersect with Y-axis (i.e. the pH scale).

2.4. Gold-cyanide (DCAU) and silver-cyanide (DCAG) biosorption studies

A batch equilibration method was used to determine the sorption of DCAU (0.02 mM i.e. 3.94 mg/l Au and 2.08 mg/l CN^-) and DCAG (0.1 mM i.e. 10.78 mg/l Ag and 5.2 mg/l CN^-). Biosorbent (0.05 to 0.2 g) was contacted with 10 ml solution of DCAU or DCAG of desired pH in a set of 50 ml capacity conical flasks. The flasks were incubated on rotary shaker incubator adjusted to a speed of 150 rpm at 30°C for 1 h. Contents of flasks were filtered using ordinary filter paper and then analysed for residual Au, Ag and cyanide. All experiments were performed in duplicates and repeated twice to confirm the results. Appropriate controls were run simultaneously to detect the air stripping of cyanide, if any, to confirm biosorption.

Influence of pH on biosorption of DCAU/DCAG was checked in the range of 4-10 with preconditioned biomass. On the basis of maximum DCAU/DCAG uptake values obtained under optimum pH conditions, efficient biosorbents were selected for further studies. DCAU/DCAG loading capacity (μmol DCAU/DCAG bound per gram weight of biosorbent) of each biosorbent was determined by contacting 0.1 g powdered biomass several times with fresh batches of 10 ml DCAU/DCAG solution till saturation was achieved. To determine optimum biosorbent amount, DCAU/DACG was contacted with varying amounts of biomass powder, ranging from 0.1 to 5% (w/v). Rate of DCAU/DCAG uptake was studied by contacting the biosorbent for a period ranging between 0 to 5 h. Under optimised conditions, effect of various competing cations viz. Cu^{2+}, Ni^{2+}, Zn^{2+}, Cd^{2+}, Pb^{2+}, Fe^{2+}, Ag^+, etc. (0.01-0.1 mM) and anions viz. SO_4^{2-}, NO_3^-, Cl^-, PO_4^{3-}, etc. (0.1-1 mM) on biosorption of DCAU/DCAG was also checked. In order to test the effect of pre-treated biomass on uptake of DCAU/DCAG, the biosorbent were treated for one hour using L-cysteine, boiling water, sodium hydroxide, formaldehyde, acetone, acetate, methanol and acetic anhydride prior to sorption.

2.5. Adsorption isotherms

To study the impact of initial concentration on adsorption, varying concentration of DCAU/DCAG was used in the range of 0.01 to 1 mM. In order to obtain sorption data, uptake value (Q) was calculated using the following equation:

$$Q = V\left(C_i - C_f\right) / 1000\ m \quad (1)$$

Where, Q is DCAU/DCAG uptake (mmol per gram biomass); V is the volume of DCAU/DCAG solution (ml); C_i is the initial concentration (μmol); C_f is the final concentration (μmol); m is mass of sorbent (g). Based on the 'Q' value obtained adsorption isotherms were plotted according to Freundlich and Langmuir equations (Freundlich, 1926; Langmuir, 1918):

$$\ln Q = \ln K + \left(1/n\right) C_{eq} \text{ Freundlich equation} \quad (2)$$

$$C_{eq} / Q = \left(1 / b\, Q_{max}\right) + \left(C_{eq} / Q_{max}\right) \text{ Langmuir equation} \quad (3)$$

Where, C_{eq} is the liquid phase concentration of DCAU/DCAG; b is Langmuir constant; Q_{max} is maximum DCAU/DCAG uptake; K is constant; n is the number of metal reactive sites and Q is the specific metal uptake.

2.6. Adsorption/desorption of DCAU and DCAG

Samples of 1 g biosorbent loaded with target M_xCN was eluted using desorbing agent (1-3 N NaOH) in concentrated form and analysed. Following the elution, biosorbent was washed with DW and then again conditioned to appropriate optimum pH to use in next adsorption/desorption cycle.

2.7. Biosorption of Au- and Ag-cyanide from industrial wastewaters

Two types of effluents were procured from silver and gold plating industry. Both effluents were subjected to characterization using Standard methods (APHA-AWWA-WEF, 1998). The proximate analysis of the samples is shown in Table 8 and 9. Batch equilibration method was followed as mentioned earlier. Rice husk (0.1 g) and *Eichhornia* root (0.1 g) biomass was contacted with 10 ml of gold-cyanide and silver-cyanide effluents, respectively. Prior to sorption, the gold- and silver-cyanide effluents were adjusted to desired optimum biosorption pH. All the batch sorption experiments were carried out under optimum conditions as given in Table 1. After contact, the contents of the flasks were filtered and then analysed for residual metal (i.e. gold and silver) and cyanide. Appropriate controls were run simultaneously.

Parameters	For DCAU experiments	For DCAG experiments
Biomass	**Rice husk (L-cysteine treated)**	***Eicchornia* root biomass (L-cysteine treated)**
pH	4.0	6.0
Temperature (°C)	30	30
Biomass quantity (w/v)	1.0	2.0
Contact time (min)	60	60
Rotation speed (rpm)	150	150

Table 1. Optimum conditions used for biosorption experiments

2.8. Continuous biosorption studies using fixed bed column at laboratory level

Scale-up studies in fixed bed continuous mode at laboratory level for biosorption of gold- and silver-cyanide was carried out in two separate fabricated glass columns of height 44 cm, internal diameter 1.3 cm and filter media height being 30 cm. The total volume of the column was 58.37 cm^3, while the working volume was 39.80 cm^3 (figure not shown). Glass column no. 1 was filled with 21 g rick husk biomass pretreated with L-cysteine, while the glass column no. 2 was filled with 24 g *Eicchornia* root biomass also pretreated with L-cysteine. The target effluents were passed through the columns in upward direction in continuous mode at a flow rate of 40 ml/h using programmable peristaltic pump (Enertech-Victor, India). All connecting silicon tubings used in the experiments were of 0.5 cm outer diameter and 0.3 cm inner diameter. Gold-cyanide effluent was passed through the column no. 1 upto 50 bed volumes, while silver-cyanide effluent was passed upto 34 bed volumes till the breakthrough curve (S-shaped) was obtained. Samples were collected periodically after every two hours and analysed for Au, Ag and total cyanide content.

2.9. Biodegradation of residual M_xCNs

Unrecoverable (residual) gold-cyanide and silver-cyanide in the solutions after biosorption treatment (in batch studies) were subjected to biodegradation using "live chemoheterotrophic bacterial consortium". The consortium (comprising of three *Pseudomonas* sp. in a proportion of 1:1:1) capable of degrading free cyanide and thiocyanate as the source of nitrogen was isolated by enrichment culture technique (Patil, 2008) and was available in my laboratory. Biodegradation experiment were conducted under aerobic and optimum conditions of pH (7.0), temperature (30°C), inoculum size (10^7 cells/ml) and glucose concentration (1 mM). The biodegradation process was used as a polishing step to clean the effluent containing traces of cyanide in order to meet the requirements of statutory agencies.

3. Results

3.1. Screening of low-cost waste biomass for DCAU and DCAG sorption

Data in Table 2 summarizes the results obtained for DCAU and DCAG sorption under optimal pH conditions. The results showed that optimum sorption in terms of Q (i.e. μmol M_xCN sorbed per gram biomass) of 0.02 mM DCAU and 0.1 mM DCAG for most of the waste biomass/ sorbents tested were at pH 4.0 and 6.0, respectively. It was observed that biosorption of DCAU and DCAG was less above pH 7.0 for all the biomass tested. In acidic pH conditions, sorption of DCAU and DCAG increased significantly. The table also shows that other than activated charcoal (chosen as reference material) which showed highest biosorption capacity, biomass of Rice husk (3.65 μmol/g) and *Eichhornia* roots (3.56 μmol/g) were efficient biosorbents for DCAU sorption; while *Eichhornia* roots (4.76 μmol/g) and Tea powder waste (4.73 μmol/g) were efficient biosorbents for DCAG. The overall Q values observed for all the waste sorbents tested for DCAU and DCAG were in the range of 2.69 - 3.65 μmol/g and 2.74 - 4.76 μmol/g, respectively. The observed Q values for efficient biomass were found to be marginally below the Q values obtained for activated charcoal (3.80 - 5.00 μmol/g). As far the optimum pH for sorption was concerned, DCAU uptake was maximal at pH 4.0 for all the biomass tested, while DCAG uptake for majority of the biomass was at pH 5.0 to 6.0. There was no loss of DCAU or DCAG in the control flasks without sorbent during the tested time period.

Table 2 also shows the data on pH values of all unconditioned biomass. Other than the reference materials, the lowest pH observed was that of coconut fibers (pH 4.24), while the highest pH was of mixed algae biomass (pH 7.61). pH of unconditioned Rice husk, Tea powder waste and *Eichhornia* root biomass observed were 5.94, 4.94 and 7.01, respectively, while their optimum pH of biosorption was 4.0 (for DCAU biosorption), 5.0 - 9.0 (for DCAG biosorption) and 4.0 (for DCAU biosorption) and 7.0 – 9.0 (for DCAG biosorption).

On the basis of maximum DCAU/DCAG uptake values obtained under optimum pH conditions, Rice husk and *Eichhornia* root biomass were selected for DCAU sorption, while *Eichhornia* root and Tea powder waste biomass were selected for DCAG sorption for further experiments. Activated charcoal acted as a reference material.

3.2. Influence of temperature on biosorption of DCAU and DCAG

It was observed that biosorption of DCAU and DCAG by the selected biomass did not had any significant impact with the change in temperature of the system from 5-45°C.

3.3. DCAU and DCAG loading capacity

Table 3 and 4 depicts the data on DCAU and DCAG loading capacity of pre-conditioned (at pH 4.0 for DCAU and pH 5.0-7.0 for DCAG) biosorbents selected on the basis of maximum sorption under optimum pH, as described earlier. Also the results were compared with the unconditioned biomass (i.e. the original pH of the biomass itself). It could be seen that Rice husk biomass had the maximum loading capacity for DCAU (7.63 μmol/g) sorption among the two tested biomass; and was followed by *Eichhornia* root biomass (7.04 μmol/g). It was also observed that the loading capacity of activated charcoal was found relatively lower when compared with the Rice husk. While the loading capacity of unconditioned biomass dropped by 7.6% and 43% for Rice husk and *Eichhornia* root biomass, respectively.

In case of DCAG, *Eichhornia* root biomass showed highest loading capacity (9.74 μmol/g) followed by Tea powder waste (9.41 μmol/g). Loading capacity values of *Eichhornia* root biomass was highly competitive and comparable with activated charcoal (9.95 μmol/g), which was used as reference material. Furthermore, the loading capacity of unconditioned biomass was not affected when compared with the conditioned biomass (Table 4).

Sr. No.	Biosorbent	pH of unconditioned biomass	Q (μmol M_xCN sorbed per gram biomass)	
			DCAU	DCAG
(A)	**Agricultural waste/by-products**			
1.	Coconut fibers	4.24	3.32 (4.0)*	4.62 (5.0)*
2.	Cow dung cakes	7.73	3.07 (4.0)	4.64 (6.0)
3.	Groundnut shells	5.49	3.22 (4.0)	4.62 (6.0)
4.	Rice husk	5.94	3.65 (4.0)	4.68 (6.0)
5.	Rice straw	6.13	3.25 (4.0)	4.69 (6.0)
(B)	**Industrial waste/by-products**			

Sr. No.	Biosorbent	pH of unconditioned biomass	Q (μmol M_xCN sorbed per gram biomass)	
			DCAU	DCAG
6.	Dairy waste sludge	6.88	2.91 (4.0)	4.71 (6.0)
7.	Saw dust	5.59	3.52 (4.0)	4.64 (6.0)
8.	Sugarcane Bagasse	5.92	3.16 (4.0)	4.03 (6.0)
9.	Tea powder waste	4.94	2.94 (4.0)	4.73 (5.0-9.0)
(C)	**Municipal solid waste components**			
10.	Nirmalya (Waste flowers)	6.20	3.43 (4.0)	4.60 (6.0)
11.	Compost	7.28	3.09 (4.0)	3.22 (6.0)
12.	Vegetable waste	6.77	2.91 (4.0)	3.88 (6.0)
(D)	**Fungal and Bacterial waste/biomass**			
13.	*Ganoderma* sp.	6.04	3.01 (4.0)	4.07 (6.0)
14.	Yeast biomass	4.39	2.69 (4.0)	3.06 (5.0)
15.	*Mucor heimalis*	4.45	1.97 (4.0)	2.29 (6.0)
16.	*Penicillium* waste	4.26	3.08 (4.0)	3.99 (6.0)
17.	*Streptomyces* waste	4.86	3.00 (4.0)	2.78 (6.0)
18.	*Streptoverticillium* waste	4.67	2.77 (4.0)	3.52 (6.0)
19.	Wood rotting fungi	6.04	3.18 (4.0)	4.21 (6.0)
20.	Bacterial consortium	6.83	3.12 (4.0)	4.00 (6.0)
(E)	**Algae biomass**			
21.	Mixed algae biomass	7.61	3.29 (4.0)	4.16 (6.0)
(F)	**Photosynthetic trees/plants waste**			
22.	*Parthenium* sp.	6.69	3.07 (4.0)	4.22 (6.0)
23.	*Eichhornia* leaves	5.57	3.20 (4.0)	4.62 (6.0)
24.	*Eichhornia* roots	7.01	3.56 (4.0)	4.76 (7.0-9.0)
25.	*Eichhornia* stem	5.58	3.38 (4.0)	4.66 (6.0)
26.	Runners	6.52	3.06 (4.0)	4.67 (6.0)
27.	*Tectona grandis* leaves	5.40	3.42 (4.0)	4.63 (6.0)
28.	*Lantana camara* leaves	6.59	2.98 (4.0)	2.74 (5.0)
(G)	**Reference materials**			
29.	Activated charcoal	5.59	3.80 (4.0)	5.00 (6.0)

Sr. No.	Biosorbent	pH of unconditioned biomass	Q (μmol M_xCN sorbed per gram biomass)	
			DCAU	DCAG
30.	Bagasse Fly ash	8.75	3.40 (4.0)	4.70 (6.0)
	Control (without biomass)	-	0 (2.0)	0 (6.0)

All the values in table are average of two readings; *Values in parentheses indicates optimum pH (Gaddi and Patil, 2011; Patil, 2012)

Table 2. Biosorption of DCAG and DCAU at optimum pH

Sorbent / Biosorbent	Loading capacity (μmol/g of biomass)		
	Conditioned biomass (at optimal pH)	Unconditioned biomass (at original biomass pH)	Remarks
Rice husk	7.63 (4.0)	7.05 (5.94)	Moderately affected
Eichhornia roots	7.04 (4.0)	4.38 (7.01)	Significantly affected
Activated charcoal	7.61 (4.0)	7.58 (5.59)	Not affected

All the values presented in table are average of two readings

Table 3. DCAU loading capacity of selected biosorbents

Sorbent / Biosorbent	Loading capacity (μmol/g of biomass)		
	Conditioned biomass (at optimal pH)	Unconditioned biomass (at original biomass pH)	Remarks
Eichhornia roots	9.74 (7.0)	9.77 (7.01)	Not affected
Tea powder waste	9.41 (5.0)	9.40 (4.94)	Not affected
Activated charcoal	9.95 (6.0)	9.94 (5.59)	Not affected

All the values presented in table are average of two readings

Table 4. DCAG loading capacity of selected biosorbents

Considering the above results, selection of the biosorbent was further narrowed down to Rice husk and *Eichhornia* root biomass for DCAU and DCAG biosorption, respectively, (using conditioned biomass) for further experiments.

3.4. Influence of biosorbent quantity

The effect of biomass quantity (% w/v) on DCAU and DCAG biosorption was studied at optimal pH values. Varying amount of biomass ranging from 0.1 to 5.0 g were used keeping the volume of both the metal-cyanides (M_xCNs) solution constant (10 ml); thereby giving the solid-to-liquid ratio in the range of 0.01 to 0.5. The results showed that the biomass quantity increased the % biosorption of both DCAU and DCAG also increased. Maximum uptake in terms of Q (3.84 μmol/g) was observed at 1% (w/v) of Rice husk biomass for DCAU sorption. However, from 1 to 5 % (w/v) there was no significant increase. In case of DCAG sorption, *Eichhornia* root biomass showed highest Q for biomass quantity from 2.0 to 5.0% (w/v).

3.5. Rate of DCAU and DCAG uptake

The effect of contact time on DCAU and DCAG biosorption was studied at their optimum pH (pH 4.0 and pH 6.0 for DCAU and DCAG sorption, respectively), temperature (30°C) and biomass quantity of 1% (w/v) and 2% (w/v) for DCAU and DCAG, respectively. 10 ml of precious M_xCN solution having concentration 0.02 mM (in case of DCAU) and 0.1 mM (in case of DCAG) was contacted with respective biomass (Rice husk and *Eichhornia* root biomass for DCAU and DCAG, respectively) for the period upto 180 min. The time intervals chosen for study were 0 to 180 minutes. Periodically the flask contents were removed by filtration and the filtrates were analyzed for Au, Ag and cyanide concentration.

It was observed that rate of both the M_xCN uptake was maximum in the 15-20 minutes, with over 80% of biosorption. Later, the sorption rate slowed down until it reached a plateau after 30-40 min, indicating equilibration of the system. Maximum sorption of both the precious M_xCNs was observed at 40 min (88.2% for DCAU and 94.3% in case of DCAG).

3.6. Adsorption isotherm models

The effect of initial concentration provides an important driving force to overcome all mass transfer resistance of target inorganic ion between the aqueous and solid phases. The biosorption of both DCAU and DCAG were carried out at different initial concentrations ranging from 0.01 mM to 1.0 mM (corresponding to approximately 10 to 1000 μmol) at pH 4.0 and 6.0 using 1 and 2% (w/v) of Rice husk and *Eichhornia* root biomass, respectively. It was found that the equilibrium sorption capacity of the sorbent increased with increasing initial concentration of M_xCNs from 0.01 mM to 1.0 mM, due to the increase in the number of ions competing for the available binding sites in the biomass. The uptake of M_xCNs approached towards plateau above 0.5 mM. There was a significant increase in the specific uptake of both M_xCNs

3.7. Equilibrium models to fit experimental data

To examine the relationship between sorption isotherm models are widely employed for fitting the data. Langmuir and Freundlich were used to describe the equilibrium between the two M_xCNs sorbed on Rice husk and *Eichhornia* root biomass and M_xCNs in solution. Data obtained show that M_xCNs uptake values could be well fitted to the Langmuir and Freundlich isotherm models with the regression value >0.98.

3.8. Influence of cationic and anionic moieties on DCAU and DCAG sorption

It was observed that biosorption of both the metal-cyanides were not significantly affected by the presence of various metal cations and anions in majority of the cases. Biosorption of DCAU was affected by zinc, chromium and cadmium up to certain extent (33-40%). In case of DCAG, biosorption was affected significantly by the presence of cadmium, zinc, iron and chromium (37-67%). Biosorption in the presence of other metals cations (copper, nickel and silver) and anions (phosphates, sulphates and chlorides) was consistently above 80%.

When the low-cost biomass was pre-treated with different chemicals, it was found that (Table 5 and 6) there was greater degree of variation in the biosorption of DCAU and DCAG using Rice husk and *Eichhornia* root biomass, respectively. Rice husk and *Eichhornia* root biomass treated with 1% L-cysteine enhanced the biosorption capacity of both the M_xCNs. In contrast, the NaOH pretreated biomass significantly hampered the biosorption process. It was 0% in case of DCAU sorption and 2.5% in case of DCAG biosorption.

DCAU + Chemicals used for pre-treatment	% DCAU biosorption	Relative % biosorption
DCAU without sorbent (control)	0	0
DCAU + Rice husk without pretreatment	90.1	100
DCAU + Rice husk (treated with boiled water)	86.5	96.0
DCAU + Rice husk (treated with 1% L-cysteine)	100	110.9
DCAU + Rice husk (treated with 1 N NaOH)	0	0
DCAU + Rice husk (treated with 1 N Formaldehyde)	87.2	96.7
DCAU + Rice husk (treated with acid i.e. HCl)	79.0	87.7
DCAU + Rice husk (treated with acetate)	73.6	81.7
DCAU + Rice husk (treated with methanol)	27.1	30.1
DCAU + Rice husk (treated with acetic anhydride)	49.2	54.6
DCAU + Rice husk (treated with acetone)	53.7	59.6

Table 5. Impact of pretreatment on DCAU biosorption by Rice husk

The experiment on pretreatment of biomass with L-cysteine clearly showed enhanced biosorption of DCAU and DCAG from solutions. It was therefore thought worthwhile to find out the loading capacity of both the biomass pretreated with L-cysteine. Experiment on loading capacity of Rice husk and *Eichhornia* root biomass was performed as mentioned earlier (section 2.4). It could be seen from Table 7 that the loading capacity of Rice husk and *Eichhornia* root biomass enhanced the biosorption of DCAU and DCAG up to 175% and 140%, respectively compared to untreated biomass (i.e. in absence of L-cysteine loaded biomass).

DCAU + Chemicals used for pre-treatment	% DCAG biosorption	Relative % biosorption
DCAU without sorbent (control)	0	0
DCAU + *Eichhornia* root without pretreatment	94.0	100
DCAU + *Eichhornia* root (treated with boiled water)	83.0	88.3
DCAU + *Eichhornia* root (treated with 1% L-cysteine)	100	106.4
DCAU + *Eichhornia* root (treated with 1 N NaOH)	2.5	0.5
DCAU + *Eichhornia* root (treated with 1 N Formaldehyde)	89.4	95.1
DCAU + *Eichhornia* root (treated with acid i.e. HCl)	81.7	86.9
DCAU + *Eichhornia* root (treated with acetate)	64.9	69.0
DCAU + *Eichhornia* root (treated with methanol)	34.5	36.7
DCAU + *Eichhornia* root (treated with acetic anhydride)	56.6	60.2
DCAU + *Eichhornia* root (treated with acetone)	59.0	62.7

Table 6. Impact of pretreatment on DCAG biosorption by *Eichhornia* root biomass

Biosorbent	Loading capacity (µmol/g of biomass)	
	Biomass without pretreatment	L-cysteine treated biomass
Rice husk (for DCAU biosorption)	7.60 (100%)	13.34 (175%)
Eichhornia roots (for DCAG biosorption)	9.72 (100%)	13.62 (140%)

All the values presented in table are average of two readings

Table 7. Loading capacity of untreated and L-cysteine treated biomass

3.9. Adsorption-desorption of DCAU and DCAG

The loaded DCAU and DCAG on Rice husk and *Eichhornia* root biomass, respectively, could be desorbed with more than 97% efficiency using 1 N sodium hydroxide solution. Final concentrations of M_xCNs in the eluent were 28-30 folds of initial concentration of DCAU and 22-25 fold of the initial concentration of DCAG. However, during the second cycle of M_xCN adsorption, the loading capacity of the biosorbent decreased by 10-15%.

3.10. Biosorption of DCAU and DCAG from industrial wastewaters

The gold-cyanide and silver-cyanide from the effluents procured from the industries could be effectively biosorbed/treated by Rice husk and *Eichhornia* root biomass which were pretreated with L-cysteine. Table 8 and 9 depicts the data on gold-cyanide and silver-cyanide before and after biosorption along with their percentage removal. Gold and cyanide removal efficiency from gold-cyanide effluent was 91.53% and 82.69%, respectively. However, the cyanide content in the treated effluent after biosorption although very less (0.59 and 0.78 mg/l for Au- and Ag-cyanide effluents, respectively) but was not complying with the prescribed Indian

Sstandards (0.2 mg/l). Overall, the results indicated that both Rice husk and *Eichhornia* root biomass were very effective in treating the effluents by biosorption process.

Physicochemical parameters	Before biosorption	After biosorption	% Removal efficiency
Color	Colorless	Colorless	-
Turbidity	Clear	Clear	-
pH	6.87	4.12	-
Total cyanide	3.41	0.59	82.69
Gold	1.30	0.11	91.53
Silver	0.48	0.03	93.75
Copper	0.95	0.18	81.05
Nickel	BDL	-	-
Zinc	0.50	0.10	80.00
Iron	0.11	BDL	100.00
Phosphates	97.9	76.1	22.26
Sulfates	63.5	61.3	3.46
Chlorides	173.0	155.2	10.28
Chemical oxygen demand (COD)	42	31	26.19

All the figures given in the table are in mg/l, except pH; BDL: Below detectable limits

Table 8. Biosorption of gold-cyanide from industrial effluent in batch mode using rice husk pretreated with L-cysteine

In order to treat the residual (unrecoverable) cyanide remaining in the solutions after biosorption were subjected to biodegradation process using bacterial consortium under optimized conditions in further experiments.

3.11. Biodegradation of unrecoverable (residual) metal-cyanides

Typical residual concentrations of gold and cyanide in gold-cyanide effluent after biosorption were 0.11 and 0.59 mg/l, respectively. Similarly, the residual silver and cyanide concentration in the silver-cyanide effluent were 0.98 and 0.78 mg/l, respectively. When these solutions were subjected to biodegradation using bacterial consortium under optimum conditions as mentioned earlier, it was observed that the consortium could degrade the said cyanide from both effluents with an efficiency exceeding 90% within a period of 3-4 h. pH, cyanide and chemical and oxygen demand (COD) of the treated effluent were within the permissible limits prescribed by statutory agencies in India (Table 10). Percent cyanide removal efficiency was >90% for both types of effluents. Gold and silver metals were not detected in bacterial free treated solutions. Findings indicated that biodegradation could be used as a polishing step in the treatment of M_xCNs containing wastewaters.

http://dx.doi.org/10.5772/56965

Physicochemical parameters	Before biosorption	After biosorption	% Removal efficiency
Color	Colorless	Colorless	-
Turbidity	slightly turbid	Clear	-
pH	7.16	6.44	-
Total cyanide	5.02	0.78	85.0
Gold	-	-	-
Silver	7.29	0.98	86.55
Copper	1.56	0.31	80.12
Nickel	BDL	-	-
Zinc	0.92	0.27	70.65
Iron	0.18	BDL	100.0
Phosphates	117.5	115.0	2.12
Sulfates	94.1	95.2	0
Chlorides	199.6	193.3	3.15
Chemical oxygen demand (COD)	47	29	38.29

All the figures given in the table are in mg/l, except pH; BDL: Below detectable limits

Table 9. Biosorption of silver-cyanide from industrial effluent in batch mode using *Eichhornia* root biomass pretreated with L-cysteine

Parameter	Biodegradation		*BIS Standards	% Removal Efficiency
	Before	After		
Gold-cyanide				
pH	6.99-7.04	7.05-7.12	5.5 – 9.0	-
Total cyanide	0.59 mg/l	0.04	0.2	93.22%
Chemical oxygen demand (COD)	102 mg/l	23 mg/l	250	77.45%
Gold	0.11	BDL	NA	-
Silver-cyanide				
pH	6.95-7.03	7.07-7-11	5.5 – 9.0	-
Total cyanide	0.78 mg/l	0.05	0.2	93.58%
Chemical oxygen demand (COD)	98 mg/l	19 mg/l	250	80.61%
Silver	0.98	BDL	NA	-

*BIS- Bureau of Indian Standards; BDL-Below Detectable Limits; NA-Not Available; All the values presented in table are average of two readings

Table 10. Treatment of residual gold-cyanide and silver-cyanide by a cyanide and thiocyanate degrading heterotrophic bacterial consortiumg

3.12. Biosorption of gold- and silver-cyanide effluent in packed bed column

Biosorption studies on gold-cyanide and silver-cyanide effluents were performed in continuous mode in two separate packed bed glass columns consisting of Rice husk (column 1) and *Eichhornia* root biomass (column 2), respectively. Biosorption results showed that breakthrough point observed for gold and cyanide in column 1 was 60 h, while the breakthrough time observed for silver and cyanide in column 2 was 40 h (figures not shown). The total effluent passed through the column 1 and 2 was equivalent to 50 and 34 bed volumes, respectively. Column 1 and 2 got completely saturated after 90 and 70 h, respectively.

4. Discussion

Review of literature show that biosorption of heavy metal cations from aqueous solutions have been studied widely (Paknikar *et al.*, 2003). Studies have also been carried out on biosorption of anionic metal species like chromates (Basha *et al.*, 2008; Itankar and Patil, 2012), free cyanide (Azab *et al.*, 1995) and metal-cyanides (Patil and Paknikar, 1999a) using microbial biomass, especially the waste fungal biomass obtained from fermentation industry and laboratory cultivated biomass. In contrast, biodetoxification of metal-cyanides and thiocyanate using live bacterial consortium was also studied (Patil, 2006; Patil, 2008a; Patil, 2008b; Patil, 2011; Patil and Paknikar, 2000; Patil and Paknikar, 2001). Safety aspects of cyanide use in mining industries have been well emphasized by Patil and Kulkarni (2008). Prashanth and Patil (2007) have also studied the impact of free cyanide on edible fish *Catla catla*.

Another important and precious chemical species that are normally encountered in the industrial effluents emanating from mining, electroplating, printing circuit board manufacturing, photography units, etc. are gold-cyanide and silver-cyanide. These species are active and important members of cyano-group chemicals that occur in water environment. Some research has been carried out on the removal of gold-cyanide and silver-cyanide species by biodegradation/biodetoxification method (Karavaiko et al. 2000; Kiruthika and Shrinithya, 2008). However, very little information is available on the removal and recovery of gold-cyanide and silver-cyanide from high volume low tenor effluents (Gaddi and Patil, 2011). Much work has been restricted to the removal of metal-cyanides and thiocyanates using anion exchange resins and activated charcoal (Kononova *et al.*, 2007) and polyurethane foam (Hasany, 2001). Some papers on removal of free thiocyanate and metal-thiocyanate have also been published using low-cost materials (Namasivayam, 2007; Thakur and Patil, 2009). Overall literature survey shows that very little work has been carried out on the removal of anionic species like Au-cyanide i.e. $[Au(CN)_2]^-$ (Dicyanoaurte-DCAU) (Niu and Volesky, 2000) and Ag-cyanide i.e. $[Ag(CN)_2]^-$ (Dicyanoargentate-DCAG) (Gaddi and Patil, 2011) from waste solutions using low-cost materials emanated either by natural or manmade activities. Since all the cyano-group chemicals like free cyanide and thiocyanate, metal cyanides and metal thiocyanate are toxic to all classes of living cells their removal and recovery from waste prior to discharge in environment is the key.

In the light of above, the present research work was focused to study removal and recovery of Au- and Ag-cyanide from effluents using low-cost biosorbents (using waste biomass from various sources); followed by biodegradation (using active bacterial consortium). It was contemplated that if an efficient process for removal and recovery could be developed, then precious M_xCN or metals could be conserved, which according to the project investigators opinion would be an innovative approach of resource recovery.

It is well-known that certain type of microbial or waste biomass has high degree of competency to adsorb heavy metals. This sorption is solely due to the chemical composition of biomass (Volesky, 2003). With biosorption applications in mind it makes sense to screen variety of biomass types that are readily available in large quantities. There are basically two types of biomass sources that can practically be considered with low costs and availability in mind. First, the industrial waste biomass generated as a by-product of large scale (for example fermentation industry) with virtually no uses for it and disposal is a problem. Secondly, the biomass generated in large quantities from water environment (for examples unwanted plants like *Eichhornia* sp. and algae). It can be easily collected or harvested as raw material for biosorbents. Also, there are many other sources from where low-cost biomass could be procured especially in a developing country like India. These include the vegetable waste, yard wastes, waste flowers and coconut fibres from temples, etc. Energy generation potential from biomass and MSW have been reported by Saini et al (2012). In order to find the right biosorbent candidate, it is imperative to screen variety of biomass occurring in human environment.

Free cyanide and thiocyanate, metal-cyanides and metal-thiocyanates can occur in the wastewaters in various forms depending upon the chemical nature of the compounds and the concentration of metal, cyanide and thiocyanate, provided if metal moiety is bound to cyanide and thiocyanate. For example, free cyanide and thiocyanate can occur in the waters in its anionic form like CN^- and SCN^-, respectively. While metal-cyanide like for example - copper-cyanide occur in water in various forms such as $Cu(CN)_2^-$, $Cu(CN)_3^{2-}$, $Cu(CN)_4^{2-}$, etc. Similarly, copper-thiocyanate complex occur in waste waters in various forms like $Cu(SCN)_2^-$, $Cu(SCN)_3^{2-}$, $Cu(SCN)_4^{2-}$, etc. Therefore, it was thought worthwhile to explore the possibility of their recovery by adsorption on low-cost biomass procured from various places. Activated charcoal, a conventional material, was used for obtaining comparative data.

It is well known that the process of biosorption is regulated by aqueous solution pH (Puranik and Paknikar, 1997). The first step in present study was therefore determination of optimum solution pH for biosorption of gold-cyanide and silver-cyanide. It was found that biosorption of both the M_xCNs (for all the low-cost biomass) increased with pH and then declined rapidly with further increase in pH. As seen from the Table 2 that maximum sorption of DCAU was at pH 4.0 while DCAG sorption by most of the biomass was at pH 6.0. Sorption decreased in the alkaline pH. It was found that other than activated charcoal (which was used as reference material) biomass like Rice husk & *Eichhornia* roots and *Eichhornia* roots and Tea powder waste were efficient biosorbents for DCAU and DCAG sorption. There was no auto-oxidation loss of both the M_xCNs in controls without sorbent confirming that (bio)sorption is the only mechanism by which M_xCNs are being removed from solution. In the previous study carried out by Niu and Volesky (2000) found that the maximum adsorption of DCAU by biomass was in the acidic pH ranging from 2.0 to 4.0. These results corroborates with the results obtained in our studies.

Increased sorption under acidic conditions may be due to the protonation of the functional groups acquiring net positive charges. Probably therefore, the formation of species such as H^+-$AuCN_2^-$ and H^+-$AgCN_2^-$ on the biomass might have taken place thereby accommodating more number of M_xCN species on the biosorbent sites. Waste biomass from natural origin contains large number of surface functional groups like hydroxyl, carbonyl, carboxyl, sulphydryl, amine, imine, amide, phosphonate, phosphodiester, etc. Probably some of these functional groups might have played the crucial role in the sorption of DCAU and DCAG from aqueous solution.

Matheickal and Yu (1996) have reported that pH dependence of cationic and anionic adsorption can largely be related to type and ionic state of these functional groups and the chemistry of target compound in solution. DCAU and DCAG in our studies could be compared with anionic metal species like hexavalent chromium (an oxyanion) and arsenic. At low pH values, cell wall ligands are protonated and compete significantly with metal binding. With increasing pH, more ligands such as amino and carboxyl groups, would be exposed leading to attraction between these negative charges and the metals and hence increases in biosorption on to cell surface (Aksu, 2001). As the pH increased further, the overall surface charge on the cells could become negative and biosorption decreased (Aksu, 2001). Patil and Paknikar (1999) have reported the optimum pH of 4.0 for the sorption of copper- and nickel-cyanide from aqueous solutions using *Cladosporium cladospoiroides* biomass.

Free cyanide (CN^-) bearing effluents are highly alkaline in nature and have pH ranging from 9.5 to 12.5, whilst M_xCN effluents have pH in range of 6.0 to 10.0. Obviously, suitable pH alterations of the effluents would be required before biosorption. Unlike free cyanide, M_xCNs does not evolve potent hydrogen cyanide (HCN) gas because of their high stability constants (APHA-AWWA-WEF, 1998; Sharpe, 1976). Therefore, biosorption under acidic conditions would be a safe procedure. On the basis of screening studies under optimum pH conditions, Rice husk and *Eichhornia* root biomass were selected for DCAU sorption, while *Eichhornia* roots and Tea powder waste biomass were selected for DCAG sorption for further experiments.

The DCAU and DCAG loading capacity of the biosorbent could be taken as an equivalent measure of binding sites present. It was found that Rice husk biomass had the maximum loading capacity for DCAU (7.63 μmol/g) sorption among the two tested biomass; and was followed by *Eichhornia* root biomass (7.04 μmol/g). Loading capacity of Activated charcoal was less (7.61 μmol/g) when compared with Rice husk. In case of DCAG biosorption, *Eichhornia* root biomass showed highest loading capacity (9.74 μmol/g) followed by Tea powder waste (9.41 μmol/g). Loading capacity of *Eichhornia* root biomass though marginally less, but was highly competitive and comparable with that of activated charcoal (9.95 μmol/g). This unlocks newer opportunities of developing an efficient biosorption process for the removal and recovery of anionic species like gold-cyanide and silver-cyanide from low tenor waste solutions. In the study carried out by Patil (1999) it was found that the biomass of *C. cladosporoides* had higher loading capacity (34-40 μmol/g) than activated charcoal (27.5-30 μmol/g) for the sorption of metal-cyanides viz. copper- and nickel-cyanide. These results also indicate that more such biomass screening programmes are needed in search of right candidate for efficient sorption.

In the present study loading capacity of conditioned biomass was also compared with that of unconditioned biomass (Table 3 and 4). For DCAU sorption, the unconditioned biomass showed lowered loading capacity compared to conditioned biomass. This reduction in loading capacity might be due to pH at which the loading capacity was determined. For conditioned biomass, the optimum pH for sorption was 4.0 as against the pH of sorption of unconditioned biomass i.e. the pH of original biomass (pH of Rice husk 5.94; pH of *Eichhornia* root biomass 7.01). In case of DCAG biosorption, it was observed that pH of unconditioned and conditioned biomass did not have any effect on the loading capacity of *Eichhornia* root and Tea powder waste biomass. This could be illustrated by the fact that original pH (unconditioned) of both *Eichhornia* root (pH 7.01) and Tea powder waste (pH 5.94) were similar to the obtained optimum pH values of our experiments. This result is very important from the view point of actual use of the biosorption process at commercial scale is concerned. Use of unconditioned biomass at commercial scale will save both time and money thereby making the cost of treatment economical which otherwise would have required for conditioning the biomass. Considering these results, selection of biosorbent was further narrowed down to Rice husk and *Eichhornia* root biomass for DCAU and DCAG biosorption, respectively.

For cost effective treatment of industrial effluents, it is imperative to discern the biomass quantity (i.e. solid-to-liquid ratio) required. In our experiments, it was found that as the biomass quantity increased the % biosorption of both the M_xCNs also increased. Maximum uptake in terms of Q (3.84 μmol/g) was observed at 3% (w/v) of Rice husk biomass for DCAU sorption. However, from 1 to 5 % (w/v) there was no significant increase. In case of DCAG sorption, *Eichhornia* root biomass showed highest Q value for the biomass-to-sorbent quantity from 2.0 to 5.0% (w/v). However, as the concentration of biomass was further increased the M_xCN uptake did not increase the biomass loading which is attributable to the interference between binding sites at higher quantities (de Rome and Gadd, 1987).

Process of biosorption is fundamentally a surface interaction and is characterized by rapid uptake of ions by biomass surfaces. Rapidity of the process makes it a worthy candidate for use in effluent treatment on a commercial scale. Kinetics showed that rate of uptake of both the M_xCN was maximum in first 15-20 minutes with over 80% of biosorption. Later, the sorption rate slowed down until it reached a plateau after 35-40 min, indicating the equilibration of system. Maximum sorption of DCAU and DCAG was 88% and 94% in 40 min. The quick equilibrium time may be attributed to the particle size. The effective surface area is high for small particles. Such type of result is typical for biosorption of metals involving no energy-mediated reactions, where metal removed from solution is due to purely physico-chemical interactions between the biomass and metal in solution. Basha *et al.* (2008) observed similar results in case of biosorption of oxyanion species viz. chromium using seaweed *Cystoseira indica*. The rapid kinetics has significant practical importance as it will facilitate smaller reactor/column volumes ensuring efficiency and economy.

The influence of starting DCAU and DCAG concentration on biosorption by Rice husk and *Eichhornia* roots biomass showed that equilibrium sorption capacity of the sorbent increased with increasing starting concentration of M_xCNs from 0.01 to 1 mM (10 to 1000 μmol). This might be due to the increase in number of ions competing for available binding sites in the biomass. Uptake of M_xCNs at various concentrations reached a plateau when the concentration was in the range of 0.5 mM (500 μmol). This might be due to the saturation of binding sites, which clearly showed that M_xCN uptake by Rice husk and *Eichhornia* root biomass was a chemically equilibrated and saturable phenomenon. The higher starting concentration of target compound offers increased driving force to overcome all mass transfer resistance of target chemical ions between the aqueous and solid phases resulting in higher probability of collision between M_xCN ions and the biosorbent. This results in higher uptake of the target compound. Moreover, the biomass cell membrane comprises host of functional groups made of polysaccharides, proteins, lipids that have the potential of binding to M_xCN ions.

It is well known that biosorption resembles physical adsorption process and follows an adsorption type isotherm (Tsezos, 1990). Adsorption isotherms are the plots of solute concentration in the adsorbed state as a function of its concentration in the solution at constant temperature. Equilibrium sorption isotherms give useful evidence for selection of an adsorbent and facilitate evaluation of adsorption process for a given application (Weber, 1985). Isotherm indicates the relative affinity of biosorbent for target ions and the adsorption capacity of biosorbent. Also, the sensitivity of biosorption to changes in target compound concentration can be determined by the relative steepness of the isotherm line. Some of the important equilibrium models developed to describe adsorption isotherm relationships include single layer adsorption (Langmuir, 1918; Freundlich, 1926) and multilayer adsorption (Branauer *et al.*, 1938).

Adsorption isotherms are known to have been largely used for projected industrial applications (Tsezos and Volesky, 1981). In the present study, it was decided to fit the DCAU and DCAG sorption data with two most widely accepted adsorption models viz. Freundlich and Langmuir. Linear transformation of the adsorption data using Freundlich and Langmuir models (R^2 = >0.96) allowed computation of the M_xCN adsorption capacities. Experimental data was found to obey the basic principles underlying these models, that is, heterogeneous surface adsorption and monolayer adsorption at constant adsorption energy, respectively (Langmuir 1918; Freundlich 1926).

Other than the M_xCN species many additional cations and anions are normally encountered in the effluents emanated from industries like metal mining, electroplating, photofinishing units, printed circuit board manufacturing, etc. These species might inhibit the removal of DCAU and DCAG from aqueous solutions. The impact of commonly occurring cations and anions was therefore studied on biosorption of DCAU and DCAG by Rice husk and *Eichhornia* root, respectively. It was observed that M_xCNs were not significantly affected in most of the cases. However, biosorption of DCAU reduced by 33-40% in the presence of zinc, chromium and cadmium. In case of DCAG, sorption reduced by 37-67% by the presence of cadmium, zinc, iron and chromium. Biosorption in the presence of other metals cations (copper, nickel and silver) and anions (phosphates, sulphates and chlorides) was consistently above 80%.

Pretreated Rice husk and *Eichhornia* biomass with variety of chemicals showed greater degree of variation in the biosorption of DCAU and DCAG. Pretreatment of Rice husk and *Eichhor-*

nia root biomass with 1% L-cysteine enhanced the biosorption capacity of both the M_xCNs, while the NaOH pretreated biomass significantly hampered the biosorption process. Based upon the results obtained, it was thought worthwhile to determine the loading capacity of L-cysteine pretreated biomass as well. It was observed that the loading capacity of Rice husk and *Eichhornia* root biomass enhanced the biosorption of DCAU and DCAG upto 175% and 140%, respectively compared to untreated biomass (i.e. in absence of cysteine loaded biomass). These result corroborated with the findings obtained by Niu and Volesky (2000). This could be explained by the fact that in the acidic pH (pH 4.0 to 6.0 in our study), weak base groups either on cysteine or on the biomass becomes protonated and acquires a net positive charge. Roberts (1992) had reported the p*K* ranging from 3.5 to 6.0 of the positively charged weak base amine groups. Carboxyl group on the biomass could be protonated in their neutral for as the p*Ka* is in the range of 3 to 5 (Buffle, 1988). In acidic pH range of 2.0 to 6.0, some of the carboxyl groups on cysteine may still be dissociated since the dissociated constant of carboxyl group on cysteine is 1.90, whereas the amino group is protonated and with a positive charge. This allows the cysteine binding to biomass through the integration/combination of negative cysteine carboxyl groups with some of the positively charged biomass functional groups. Thus, the positively charged cysteine amino group were available for binding of anionic M_xCN species like $[Au(CN)_2^-]$ and $[Ag(CN)_2^-]$ which are the target compounds in our studies. In other words, the anionic species $[Au(CN)_2^-]$ and $[Ag(CN)_2^-]$ adsorbed by ionizable functional groups on cysteine loaded biomass carrying a positive when protonated.

(Waste Biomass --- Cysteine --- H^+) --- $Au(CN)_2^-$

(Waste Biomass --- Cysteine --- H^+) --- $Ag(CN)_2^-$

When the target compound is rare and costly, it is always desirable to recover the target compound from industrial effluents having low concentration and high volumes. For an effective and viable biosorption technology, elution methods for the recovery of target compound should be highly efficient, economical and should not cause damage to the biomass. Several eluting agents have been reported in the literature which includes mainly mineral acids, alkalis, organic acids, etc. In the present study, the loaded DCAU and DCAG on Rice husk and *Eichhornia* root biomass, respectively, could be desorbed with more than 95% efficiency using 1 N sodium hydroxide solution. Final concentration of DCAU and DCAG in the concentrated eluent was 28-30 and 22-25 folds, respectively, of the starting concentration. Such high tenor solution of recovered gold-cyanide and silver-cyanide may be recycled back to the user industry.

The next major task in the study was to test the selected biomass viz. Rice husk and *Eichhornia* root biomass for the removal of gold-cyanide and silver-cyanide from their respective industrial effluents in batch mode. As mentioned earlier that the project investigator encountered great difficulty in procuring the effluent samples from industries. In the end, third party intervention helped the investigator to get the sample. In developing country like India, most of industrial personnel are reluctant to give any information regarding toxic chemical waste like cyanide. Moreover, they don't allow the outsider to invade into their industry mainly due to the risk and threat that is associated with cyanide disposal. With the stricter statutory limits imposed by statutory agencies, the conventional physic-chemical methods for the treatment of metal-cyanide bearing effluents are proving to be expensive and also inadequate to meet the required standards. This techno-economic impasse has led to closure of several industries especially the plating industries.

The gold-cyanide and silver-cyanide from the effluents procured from the industries could be effectively biosorbed by Rice husk and *Eichhornia* root biomass which were pretreated with L-cysteine. Gold and cyanide removal efficiency from gold-cyanide industrial effluent was 91.53% and 82.69%, respectively. However, the cyanide level in the biosorbed treated effluent although very less (0.59 mg/l) but was not below the standard limits prescribed by Indian statutory agencies, which is 0.2 mg/l. Similarly, the cyanide concentration after biosorption treatment to silver-cyanide effluent was also not complying with the standards prescribed by Indian statutory agencies. Overall, the studies on industrial effluents indicated that both the biomass viz. Rice husk and *Eichhornia* root biomass were very effective in treating both the effluents by biosorption process. Therefore, it is possible to employ Rice husk and *Eichhornia* root biomass for the treatment of industrial effluents on commercial scale. The residual (unrecoverable) cyanide remaining in the solutions after biosorption were subjected to biodegradation process using bacterial consortium under optimized conditions.

When the residual gold-cyanide and silver-cyanide biodegradation experiment was run under optimized conditions in batch mode, it was found that the live bacterial consortium previously isolated by Patil (2008) could degrade the cyanide present in the solution within a period of 5 h with an efficiency of >90% for both types of effluents. The resulting treated solution could comply with the disposal standards prescribed by statutory agencies in India. These findings indicated that biodegradation could be used as a polishing step in the treatment of precious M_xCNs containing industrial waste waters.

In process applications, the most effective apparatus for sorption/desorption and making the most effective use of the reactor volume, is a fixed-bed column. The column makes optimum use of the concentration gradient between the solute sorbed by the solids and that remaining in the liquid phase thereby providing the driving force for the biosorption process. The process of biosorption (metals and their related species) is governed by three key regimes: (i) the sorption equilibrium, (ii) the sorption particle mass transfer and (iii) the flow pattern through the packed bed. These three regimes determine the overall performance of the sorption column which is judged by its 'service time'. Service time is the length of time until the sorbed species breaks through the bed to be detected at a given concentration in the column effluent. The breakthrough point indicates that the column is saturated practically and could be taken out of operation for some kind of its regeneration (Volesky, 2003).

The column bed is being saturated at inflow concentration which represents equilibrium concentration for the part of the bed upstream from the transfer zone. The saturation of the bed/column varies from zero to the full saturation. This zone of partial saturation moves the column in the direction of the flow at a certain velocity which is predominantly determined by the biomass loading, sorbent capacity and the feed rate to column. The column is operational until this zone reaches the end of the column. Until that time the effluent leaving the column has no trace of the sorbate in it. When the transfer zone reaches the column end, the sorbate concentration in the effluent starts to gradually increase and for all practical purposes, the working life of the column is over and the "breakthrough point" occurs marking the usable column "service time". These two parameters are very important from the process design point of view because they directly affect the feasibility and economics of the sorption process (Volesky, 2003).

After successfully treating both the industrial effluents in batch mode using Rice husk and *Eichhornia* root biomass, further biosorption studies were carried out in continuous mode using packed bed column. It was found that the service time offered by the column beds for gold-cyanide (from column 1) and silver-cyanide (from column 2) effluents were 60 h and 40 h, respectively. In other words, these itself were the breakthrough points. For both the columns the transfer zone observed was of 30 h each. The total effluent passed through the column 1 and 2 was equivalent to 50 and 34 bed volumes, respectively, while the complete saturation occurred after 90 and 70 h, respectively. Continuous study clearly showed that both the effluents were biosorbed and treated successfully in the packed bed columns for the removal of both precious and toxic species. Further, in these studies the project investigator did not immobilized any of the biomass primarily because the present work was focused on low tenor effluents containing precious gold and silver and toxic chemical species like cyanide (all below 10 mg/l). Secondly, the results obtained through batch and continuous studies showed that both the biomass were efficient enough to sorb and treat the effluents and therefore the project investigator felt that immobilization of the biomass probably is not required in this case.

Thus, it could be concluded that the waste biomass used in the present study has immense potential "as biosorbents" for the removal/management of low tenor precious and toxic pollutants, as evident from the example of gold-cyanide and silver-cyanide management in the present study. Further, biosorption technology used could also become an economical, non-destructive and reliable alternative to the conventional processes for the management of industrial effluents employed on the commercial scale.

5. Application of biosorption to some newer wastes and products

Apart from the removal and recovery of precious heavy metal species from industrial effluents, passive bioremediation technology (PBT) can also be employed for some newer type of wastes and products that have emerged in the recent times.

A novel approach of combined biosorption-biodegradation processes was used by Patil and Paknikar (1999) for the removal and recovery of copper- and nickel-cyanide from electroplating effluents. *Cladosporium cladosporioides* biomass was found to be highly efficient sorbent in this case. The unrecoverable (residual) metal-cyanides after biosorption was subjected to biodegradation process using bacterial consortium. The treated effluent was free of cyanide and metals and complied with the statutory limits (Patil and Paknikar, 1999).

The problem of waste photovoltaic cells was addressed by Paknikar et al (1997) by way of recovering and recycling of expensive metals like silver, cadmium and tellurium. In this study, the researchers used scrapings of waste photovoltaic cells, which were dissolved in suitable mineral acid and was diluted to obtain desired metal concentration. The metal solution was then subjected to biosorption column consisting of inactive granulated biomass of *C. cladosporioides* #1 for selective removal of silver. Similarly, in the next step, cadmium was removed by biosorption process by passing the solution through biosorbent column. The silver and cadmium free solution after treatment was then fed to tellurium reducing bioreactor consisting *P. mendocina*. The overall removal and recovery efficiency of these metals was >90%. With the rising demand and shift towards renewable energy sources, the number of photovoltaic cells producing units/industries will increase in the years to come; and so the use of non-renewable resources like metals and it wastages in the form of rejections. Although the economic feasibility of the process was not studied by the researchers) but the study certainly add to the advancement of knowledge by employing combined passive and active bioremediaton (Paknikar et al 1997).

Pethkar et al. (2001) reported an interesting study on the removal of toxic metals like lead and cadmium from fruit juices of carrot, grapes and oranges, and extracts of Jatamansi herb and raisin by passive bioremediation using the biomass of *C. cladosporioides* #2. With a growth rate of 15%, the annual turnover of herbal medicinal industry in India is Rs. 75,000 million. As per ASSOCHAM (Associated Chamber of Commerce and Industry), the turnover of herbal industry is projected to double to Rs.1,50,000 million (USD 3 billion) by 2015. However, the business is getting severely affected by the presence of toxic heavy metals into food and herbal products thereby making them unacceptable in foreign markets because of their stringent statutory norms. Therefore, removal of these toxic metals from such products using biosorption process is crucial and has great prospectus. Sun et al (2007) had reported sorption of heavy metal ions by polyaspartyl polymers from Chinese herbal medicines. However, there is paucity of literature on biosorption of toxic metals from herbal medicines and food products.

Bhat et al (2012) had proposed a novel integrated model for the recovery of gold/silver from e-waste using an integrated hydrometallurgical (chemical) and biometallurgical (low cost biomass) processes. Feasibility study was conducted to explore the possibility of removal/ recovery of silver-cyanide using low-cost biosorbents. *Eicchornia* root biomass and Waste tea powder were found to be an efficient low-cost biosorbents for leached silver-cyanide from electronic scrap. The concentrated silver-cyanide recovered had the potential for its further use as input material for electroplating industry (Bhat et al, 2012). Awareness among the urban population regarding disposal and management e-waste has also been studied by Bhat and Patil (2012).

In the twenty-first century, entire world is witnessing a paradigm shift in the overall waste management practices, which is rapidly changing its face and orientation. Waste is no more considered as waste but is recognized as a 'Resource'. This lost resource could potentially be recovered from the wastes using suitable strategies and technologies. Therefore, in a real sense, model like recovery and recycling of waste resource is gaining remarkable importance in today's so called 'Technological Society'. Application of concepts similar to this work will ultimately reduce the demand for natural resources thereby extending its sustenance. In view of this, the present chapter on passive bioremediation will certainly add to the advancement of knowledge in the field of resource recovery and industrial pollution management, waste

http://dx.doi.org/10.5772/56965

minimization and will help profitability of business community at large. It has not escaped through authors mind that the recovered resource from the waste of one industry has all the potential for its use as an input material for other industry thereby strengthening the emerging discipline of 'Industrial Ecology'.

Acknowledgements

Dr. Yogesh Patil gratefully acknowledges the International Foundation for Science (IFS), Stockholm, Sweden, in cooperation with The Organization for the Prohibition of Chemical Weapons (OPCW), The Hague, The Netherlands, for providing the research grant.

Author details

Nilisha Itankar[1], Viraja Bhat[2], Jayati Chourey[2], Ketaki Barve[2], Shilpa Kulkarni[2], Prakash Rao[2] and Yogesh Patil[3*]

*Address all correspondence to: head_respub@siu.edu.in

1 Symbiosis Institute of Technology (SIT), Symbiosis International University (SIU), Near Lupin Research Park, Lavale, Pune, India

2 Symbiosis Institute of International Business (SIIB), Symbiosis International University (SIU), Rajiv Gandhi Infotech Park, Hinjewadi, Pune, Maharashtra, India

3 Symbiosis Institute of Research and Innovation (SIRI), Symbiosis International University (SIU), Lavale, Pune, Maharashtra, India

References

[1] Aksu Z (2001) Equilibrium and kinetics modeling of cadmium (II) biosorption by *C. vulgaris* in a batch system: effect of temperature. *Separation and Purification Technology* 21:285-294.

[2] APHA, AWWA, WEF (1998) Standard Methods for the Examination of Water and Wastewater Analysis, 20th Ed. American Public Health Association, Washington, DC.

[3] Azab HM, El-Shora HM and Mohammed HA (1995) Biosorption of cyanide from industrial waste water. *Al-Azhar Bulletin of Science* 6:311-323.

[4] Basha S, Murthy ZVP and Jha B (2008) Biosorption of hexavalent chromium by chemically modified seaweed, *Cystoseira indica*. *Chemical Engineering Journal* 137:480-488.

[5] Bhat Viraja and Patil Y.B. (2012) Mobile user's perspective towards e-waste: A case study of Pune city. *International Journal of Academic Conference Proceedings* 1(2).

[6] Bhat Viraja, Rao Prakash and Patil Yogesh (2012) Development of an integrated model to recover precious metals from electronic scrap - A novel strategy for e-waste management. *Procedia - Social and Behavioral Sciences* 37: 397-406.

[7] Braunauer S *et al* (1938) Adsorption of gases in multimolecular layers. *Journal of American Chemical Society* 60: 309-319.

[8] DeRome L and Gadd GM (1987), Copper adsorption by Rhizopus arrhizus, Cladosporium resinae and Penicillium italicum. *Applied Microbiology and Biotechnology* 26: 84-90.

[9] Eckenfelder WW (1989) Industiral Water Pollution Control. New York, McGraw Hill.

[10] Freundlich H (1926) *Colloid and Capillary Chemistry,* London: Methuen.

[11] Gaddi Shivanand S and Patil Yogesh B (2011) Screening of some low-cost waste biomaterials for the sorption of silver-cyanide [Ag(CN)2-] from aqueous solutions. *International Journal of Chemical Sciences* 9: 1063-1072.

[12] Ganczarczyk JJ, Takoaka PT, Ohashi DA (1985) Application of polysulfide for pretratment of spent cyanide liquors. *Journal of Water Pollution Control Federation,* Cont. Fed., 57: 1089-93.

[13] Hasany SM, Saeed MM and Ahmed M (2001) Sorption of traces of silver ions onto polyurethane foam from acidic solution. *Talanta* 30(1): 89-98.

[14] Itankar, Nilisha and Patil, Yogesh (2012) Sorption of toxic hexavalent chromium (Cr^{6+}) from aqueous solutions by some (bio)materials. In: *Proceedings of International Conference on Functional Materials for Defence (ICFMD)* jointly organized by Deference Institute of Advanced Technology, Pune, Naval Postgraduate School, USA & Office of Naval Research (Global), USA, during 18-20 May 2012, p. 137.

[15] Karavaiko GI, Kondrat'eva TF, Savari EE, Grigor'eva NV and Avakyan ZA (2000) Microbial degradation of cyanide and thiocyanate *Microbiology* 69(2): 167-173.

[16] Kiruthika AJ and Shrinithya (2008) Cyanide detoxification and recovery of gold from gold effluent. *Advanced Biotech* pp. 20-26.

[17] Kononova ON, Kholmogorov AG, Danilenko SV, Kachin SV, Kononov YS and Dmitrieva, ZV (2007) Sorption of gold and silver on carbon adsorbents from thiocyanate solutions. *Carbon* 43(1): 17-22.

[18] Langmuir J (1918) The adsorption of gases on plane surfaces of glass, mica and platinum. *Journal of American Chemical Society* 40: 1361-1403.

[19] Lee JS, Deorkar NV & Tavlarides LL (1998) Adsorption of copper cyanide on chemically active adsorbents. *Industrial and Engineering Chemical Research.,* 37: 2812-2820.

[20] Matheickal JT and Yu Q (1996) Biosorption of lead from aqueous solutions by marine alga *Ecklonia radiate. Water Science and Technology* 34: 1-7.

[21] Modak JM and Natarajan KA (1995) Biosorption of metals using nonliving biomass – A review.*Mineral and Metallurgy Process*12: 189-196.

[22] Mohan D and Pittman CU Jr. (2006) Activated carbons and low cost adsorbents for remediation of tri- and hexavalent chromium from water. *Journal of Hazardous Materials,* B137: 762-811.

[23] Namasivayam C and Sangeetha D (2007) Kinetic studies of adsorption of thiocyanate onto zinc chloride activated carbon from coir pith, an agricultural solid waste. *Chemosphere* 60(11): 1616-1623.

[24] Niu H & Volesky B (2001) Gold adsorption from cyanide solution by chitinous materials. *Journal of Chemical Technology and Biotechnology,* 2001, 76(3), 291-297.

[25] Niu H and Volesky B (2000) Gold-cyanide biosorption with L-cysteine. *Journal of Chemical Technology and Biotechnology* 75: 436-442.

[26] Paknikar et al (1997) An integrated chemical-microbiological approach for the disposal of waste thin film cadmium telluride photovoltaic modules. In: *Environmental, Safety and Health Issues in IC Production,* edited by R Reif et al. Materials Research Society, Pittsburgh, Pennsylvania, USA, pp. 133-138.

[27] Paknikar KM, Pethkar AV & Puranik PR (2003) Bioremediation of metalliferous wastes and products using inactivated microbial biomass. *Indian Journal of Biotechnology.* 2: 426-443.

[28] Patil Yogesh B (1999) Studies on biological detoxification of metal-cyanides containing industrial effluents. Ph.D. Thesis, University of Pune, Pune, India.

[29] Patil YB and Paknikar KM (1999a) Removal and recovery of metal cyanides using a combination of biosorption and biodegradation processes. *Biotechnology Letters* 21: 913-919.

[30] Patil YB and Paknikar KM (1999b) Removal and recovery of metal cyanides from industrial effluents. *Process Metallurgy* 9: 707-716.

[31] Patil YB and Paknikar KM (2000a) Biodetoxification of silver-cyanide from electroplating industry wastewater. *Letters in Applied Microbiology* 30: 33-37.

[32] Patil YB and Paknikar KM (2000b) Development of a process for biodetoxification of metal cyanides from wastewaters. *Process Biochemistry* 35: 1139-1151.

[33] Patil YB and Paknikar KM (2001) Biological detoxification of nickel-cyanide from industrial effluents. *Process Metallurgy* 11: 391-400.

[34] Patil Yogesh B (2006) Isolation of thiocyanate degrading chemoheterotrophic bacterial consortium. *Nature Environment and Pollution Technology* 5(1): 135-138.

[35] Patil Yogesh B (2008a) Biodegradation of thiocyanate from aqueous waste by a mixed bacterial community. *Research Journal of Chemistry and Environment* 12(1): 69-75.

[36] Patil Yogesh B (2008b) Thiocyanate degradation by pure and mixed bacterial cultures. *Bioinfolet* 5(3): 308-309.

[37] Patil Yogesh B (2011) Utilization of thiocyanate (SCN-) by a metabolically active bacterial consortium as the sole source of nitrogen. *International Journal of Chemical, Environmental & Pharmaceutical Research*, 2: 44-48.

[38] Patil Yogesh B (2012) Development of a low-cost industrial waste treatment technology for resource conservation – An urban case study with gold-cyanide emanated from SMEs. *Procedia – Social and Behavioural Sciences* 37: 379-388.

[39] Patil Yogesh B and Kulkarni Anil R (2008) Environmental sensitivity and management of toxic chemical waste in mining industry with special reference to cyanide. In: *High Performing Organizations: Needs and Challenges*, Tata McGraw Hill Publications, Part D: pp. 183-196.

[40] Patil Yogesh, ChoureyJayati and RaoPrakash (2012) Biotechnological strategy for the management of industrial waste with concurrent mitigation of global warming – A feasibility study using microalgae. In: *Inclusiveness and Innovation – Challenges for Sustainable Growth of Emerging Economies* (Edited by Rajiv Divekar and BR Londhe), Excel India Publishers, New Delhi, India, pp. 249-255.

[41] Pethkar AV et al. (2001) Biosorptive removal of contaminating heavy metals from plant extracts of medicinal value. *Current Science* 80, 1216-1219.

[42] Prashanth MS and Patil Yogesh B (2007) Behavioural surveillance of Indian major carp *Catla catla* (Hamilton) exposed to free cyanide. In: *Disaster Ecology and Environment*, Book Edited by Arvind Kumar, Daya Publishing House, New Delhi, pp. 210-215.

[43] Roberts GAF (1992) Chitin Chemistry, Macmillan, London, UK, pp. 204-206.

[44] Rollinson G, Jones R, Meadows MP, Harris RE and Knowles CJ (1987) The growth of a cyanide-utilizing strain of *Pseudomonas fluorescens* in liquid culture on nickel cyanide as a source of nitrogen *FEMS Microbiology Letters* 40: 199-205.

[45] Saini Samir, RaoPrakash and Patil Yogesh (2012) City based analysis of MSW to energy generation in India, calculation of state-wise potential and tariff comparison with EU. *Procedia - Social and Behavioral Sciences* 37: 407-416.

[46] Sharpe AG (1976) The chemistry of cyano complexes and the transition metals. London: Academic Press, 1976.

[47] Sun Bo, Mi Zhen-Tao, An Gang and Liu Guozhu (2007) Binding of several heavy metal ions by polyaspartyl polymers and their application to some Chinese herbal medicines. Journal of Applied Polymer Science. 106(4): 2736-2745.

[48] Thakur Ravindra Y and Patil Yogesh B (2009) Management of thiocyanate pollution using a novel low cost natural waste biomass. *South Asian Journal of Management Research* 1(2): 85-96.

[49] Tsezos M (1990) Engineering aspects of metal bonding by biomass. In: *Microbial Mineral Recovery*, edited by HL Ehrlich and CL Brierley, McGraw-Hill Publishing Co., New York, USA, pp. 325-339.

[50] Tsezos M and Volesky B (1981) Biosorption of uranium and thorium. *Biotechnology and Bioengineering* 23: 583-604.

[51] Vapur H, Bayat O, Mordogan H and Poole C (2005) Effects of stripping parameters on cyanide recovery in silver leaching operations. *Hydrometallurgy* 77(3-4): 279-286.

[52] Volesky B (2003) *Sorption and Biosorption*, BV Sorbex, Inc, Montreal – St. Lambert, Quebec, Canada.

[53] Weber WJ Jr. (1985) Adsorption theory, concepts and models. In: *Adsorption Technology: A Step-By-Step Approach to Process Evaluation and Application*, Edited by FL Slejko, Marcel Dekker, New York, pp. 1-35.

Chapter 12

A Novel Bioremediation Method for Shallow Layers of Soil Polluted by Pesticides

Naofumi Shiomi

Additional information is available at the end of the chapter

http://dx.doi.org/10.5772/56153

1. Introduction

Excess amounts of pesticide have often been used to enhance the yield of agricultural products, and simultanously causing serious problems by polluting the soil and groundwater all over the world. For example, in the United States, large amounts of atrazine were used as a weed killer. The excessive atrazine sprayed by planes polluted rivers and soil near the farms, and this eventually was transferred to urban areas. The soil pollution in China is even more serious. Many agricultural fields and groundwater supplies are no longer used because of their contamination with pesticides and chemical fertilizers. An investigation by the FAO/ UNEP/WHO suggested that 1-5 million patients develop pesticide poisoning every year, and several thousand patients die [1].

As a result, many countries have passed laws about the remediation and conservation of soil. In the United States, the remediation and monitoring of polluted soil must be done based on the federal Superfund Act. In Germany, a law about the preservation of water and soil was passed in 1997. The remediation of soil is carried out according to these laws in these countries.

Among the various processes used for the remediation of soil, bioremediation has been remarkable because it is high safety and inexpensive running cost compared to the physical and chemical methods, such as burning and adsorption. Bioremediation methods can be classified into two broad classes (*in situ* and *ex situ* bioremediation), and *in situ* bioremediation processes employs the treatment method without moving the polluted soil. Some pilot studies have applied *in situ* bioremediation methods, such as bioventing and oxygen release compounds (ORC) methods, to remediate soil polluted by pesticides [2]. Such projects have succeeded in decreasing the herbicide concentration to a safe level, but large scale equipment and high expenses were major drawback associated with these methods. As most of the agriculture-related soil pollution occurs over a large area, bioremediation processes with high costs cannot be applied for this pollution. Therefore, the development of a new bioremediation method, which is applicable to huge areas and is low cost, has been desired.

In this chapter, the author proposed a novel bioremediation method called "Bioremediation with the self-immobilized system (BSIS)". Using this method, degrading microorganism can be self-immobilized by the help of *Bacillus subtilis,* and the immobilized cells rapidly degrade the pollutant in the shallow layer of the soil. The equipment and running cost associated with BSIS are low, and therefore, I think that this method is a superior method for remediating soil polluted by pesticides.

2. Necessity of remediation of soil and groundwater polluted by pesticides

2.1. Harmful influences of pesticides on living organisms

Typical pesticides employed in the agriculture sector are shown in Figure 1. Until the 1980s, organochlorine insecticides, such as DDT and BHC, were the most commonly used agents. However, Rachel L. Carson warned in her book "Silent Spring," which was published in 1962, that these insecticides were hard to degrade and subject to accumulation in the environment. In fact, many deformed birds and fish that had accumulated the organochlorine compounds were found. Based on this book and the findings of numerous other scientists, the production and utilization of DDT is now forbidden in many countries. Moreover, the discontinuation of persistent organic pollutants (POPs) such as DDT and BHC was decided to be a worldwide goal at the Stockholm Convention on Persistent Organic Pollutants in 2001. As a result, DDT cannot be used except for eliminating mosquitoes carrying malaria.

Following the banning of POPs, organophosphorous insecticides, such as parathion, dichlorvos, penitrothione and diazinon, were gradually applied to patties and fields instead of DDT and BHC. The persistence of organophosphorous compounds is much lower than that of DDT or BHC. These compounds, however, strongly inhibit the activity of acethylcholine esterase in nerve cells, and some of them showed strong toxicity to humans [3]. The use of many organophosphorous compounds was prohibited by the European Union (EU) in the 2000s, but they are still used in other countries. The derivatives of pyrethroid, such as chrysanthemic acid and pyrethrolone, and the derivatives of nicotine, such as imidacloprid, were similarly developed to decrease the toxicity of these herbicides. Imidacloprid is currently speculated to be the cause of colony collapse disorder, but remains widely use all over the world.

Another compound, 2,4-dichlorophenoxyacetic acid (2,4-D), was first developed in 1944 as a weed killer for treating wheat and corn fields. The compound kills dicotyledons, but functions as a phytohormone for monocotyledons. It is still used today. Triazine herbicides, such as atrazine and simazine, were developed as weed killers in the 1970s. These compounds kill weeds after absorption from the leaves and roots. They were used in Europe and the United States with great success, but atrazine was subsequently found to act as an endocrine disruptor [4,5]. According to Hayes *et al.* [4], when 40 male frogs (*Xenopus laevis*) were exposed to water containing 2.5 ppm of atrazine for three years, 30 frogs became infertile and four frogs became females. The four female frogs could be fertilized by a male frog. These results strongly suggested that atrazine was harmful. The EU decided to prohibit the use of atrazine after noting that concentrations often exceeded the upper limit in groundwater. The US Environmental Protection Agency (EPA) set the upper limit of atrazine as 3 ppm in the drinking water, which was much higher than that in the EU. The EPA does not currently prohibit the utilization of atrazine, because the proof of adverse effects was not considered to be reliable. Therefore, atrazine is still used in the United States. Some specialists also warn that the sharp decrease of frogs in all over the world might be related to the excess use of triazine herbicides.

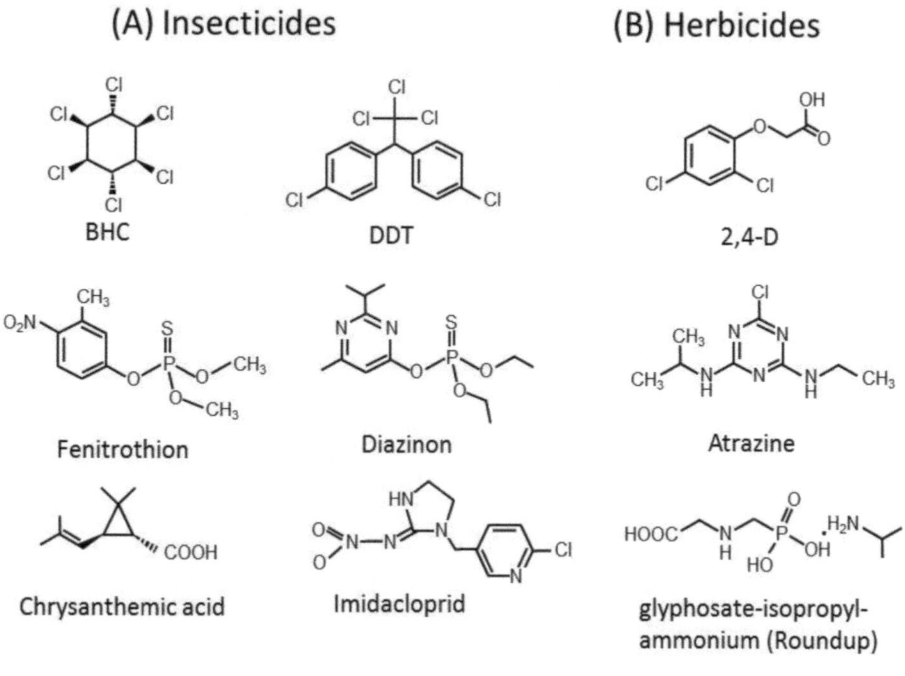

Figure 1. Pesticides commonly used for agriculture.

In the 1980s, Roundup was developed by Monsant Company. Glyphosate-isopropyl ammonium is a main component of the herbicide. It inhibits the synthesis of an amino acid, and works against all kinds of plants. The company also developed genetically modified organisms (GMO) which were resistant to Roundup. Based on these successes, Roundup is widely used to grow GMO in the United States. However, excess utilization of the non-selective herbicides (2, 4-D and Roundup) causes a decrease in soil insects and bacteria, which supply the nitrogen, carbon and phosphate sources to plants.

As described above, the use of many harmful herbicides has been prohibited or regulated since the 2000s, and the pesticides showing low toxicity and a short retention time are now being used. However, dangerous pesticides continue to be used in many countries, especially developing countries, and the acreage where serious damage has been inflicted have been increasing all over the world.

2.2. Soil and groundwater pollution by pesticides

Atrazine and simazine have been used extensively in the United States and Europe. In the United States, 36,000 tons/year of atrazine are still being used in the patties and fields. The fields and rivers in the northwest part of the United States have been monitored for several years, and many kinds of pesticides were detected there [6-9]. In fact, 75% and 40% of samples collected in rivers and groundwater contained atrazine and other harmful pesticides. The concentrations of harmful pesticides were gradually decreased, but the concentrations of diazinon, chlorpyrifos and malathion often exceeded the benchmark value [6].

Atrazine sprayed from the airplanes is often carried by wind to other locations at a distance of over 1000 km from the spraying place, and these fall with rain. As noted above, the EU forbade the use of triazine herbicides, because their concentrations in groundwater exceeded the limit they set for safety standards. However, because the yield of crops can be enhanced by the use of atrazine, the farmers in the EU do not necessarily obey this decision, and harmful pesticides are still used and detected from the river water and groundwater in some countries in Europe [10-12].

Soil pollution is much more serious in the countries in Asia. Although the current amount of pesticide consumption in developing countries is only 25% of the total amount produced, 99% of the people killed by pesticide poisoning are from developing countries. Soil pollution also induces groundwater pollution. As groundwater is used for drinking water without any treatment in many countries in Asia due to the water shortage, this may lead to many human cases of poisoning. Additionally, many kinds of POPs which were produced before 2001 are still used in some countries, even though their use and production is prohibited by the Stockholm Convention. For example, in Vietnam, the soil is polluted by 4 tons of harmful pesticides that were buried in soil to discard them, and 108 tons of toxic POPs are still being stored in a warehouse. Such POPs can be easily obtained at markets in Nepal and India. BHC is still used for the production of crops and cotton cultivation in those countries, because it is cheap and effective [13].

In China, the pollution of soil and groundwater are extremely serious [14-16]. Approximately one-third of the wastewater from industries, and 90% of the drainage from households is directly discharged to the river. It has been estimated that 40% of river water (95% in urban areas) is already impossible to utilize as drinking water due to pollution, and thus, it has the potential to harm the 160 million people who are known to use this water.

The amount of pesticide consumption in China has increased ten times compared to that in the 1990s. Because of excess use (13.4 kg/ha) for agricultural fields, 60% of the pesticides used remain in the soil without degradation. Therefore one-sixth of the agricultural fields need to be remediated. Chemical fertilizers are also excessively used, and approximately 60% of the fertilizer also remains in the soil. Excrement (3 billion tons) discarded in pastures is also present in the soil. These contaminants cause pollution due to nitrate nitrogen. The use of plastic sheeting buried in soil, which help to maintain nutrients and moisture, exacerbates these problems. In addition, discarded plastic sheeting (500,000 tons/year) spoils the soil by interrupting the exchange of air and water.

Soil pollution in China is caused by not only pesticides, but also wastewater that is insufficiently treated by industry. For example, melamine, which is used for the production of melamine resin, contains a very high level of nitrogen. In 2004-2007, melamine was intentionally added to dairy products by some food manufacturers in China to increase the apparent content of protein. Consequently, serious food-related toxicity occurred in the people, dogs and cats that consumed the dairy products. Wastewater containing melamine was discarded into the environment without any treatment until the incidents occurred. Therefore, the soil, groundwater and crops near the manufacturing facilities were polluted with melamine [17]. A similar incident occurred related to the production of rice in China. Methamidophos, an organophosphorous compound, was mistakenly included in rice, although the precise reason for this accident was not clarified. One million tons/year of food is affected by soil pollution in China according to an investigation by a public organization.

The shortage of drinking water caused by groundwater pollution is also serious. To avoid groundwater polluted by toxic compounds, wells were dug to much deeper layers in the aquifer [18-19], but the groundwater at these layers contains arsenic and fluoride. Around 25 million people who drank this water showed arsenic toxicity and dental fluorosis. Since the 1980s, poisoning by arsenic has been gradually increased in many countries, such as India, Thailand and Bangladesh [20].

3. A novel bioremediation process targeting the shallow layers of soil

3.1. Problems with conventional bioremediation processes

As described in section 2, excess use of pesticides in agricultural fields causes serious pollution of the soil and groundwater. The characteristics of soil pollution by pesticides are that the area of pollution is huge, and the pollution reaches to deep layers. *Ex situ* bioremediation, which is a process performed after movement of the polluted soil, is not adequate to control such widespread pollution. Therefore, *in situ* bioremediation, which involves treating the soil in place, had been applied for pollution with pesticides.

Several *in situ* bioremediation processes have been proposed [2], and the ORC method developed by the REGENESIS Corp and bioventing are often used. Figure 2 shows a schematic diagram of these processes. In the bioventing method, air is supplied from the upstream side of the pollutant at deeper layers of soil by using high-pressure pumps. In the ORC method, oxygen release compounds (ORC) are injected into the deeper layer by using high-pressure pumps. The supply of oxygen activates the microorganisms present in the polluted soil, enhancing the degradation rate.

Bioventing is suitable for the remediation of contaminants at deeper layer and over wide areas of soil, and some projects using bioventing have been performed. For example, bioremediation of organochloride compounds was examined using a pilot plant [21]. Anaerobic bacteria could be activated by injecting vegetable oil as a carbon source, and the concentration of tetrachloroethylene was decreased until the level was lower than the limit of the benchmark. The ORC method also succeeded in the remediation of oil pollution in practical cases.

However, there are several conditions that must be overcome in cases of soil pollution by pesticides: (1) The polluted area is huge and requires the use of large-scale equipment; (2) the cost for the remediation should be low; (3) the operation for remediation must be continued permanently, because pesticides are sprayed periodically every year; (4) the polluted area cannot be dug up, because grass, crops and other plants are already present there. The bioventing and ORC methods don't conform to these conditions and are therefore not applicable. A novel bioremediation method for soil pollution due to pesticides is strongly needed.

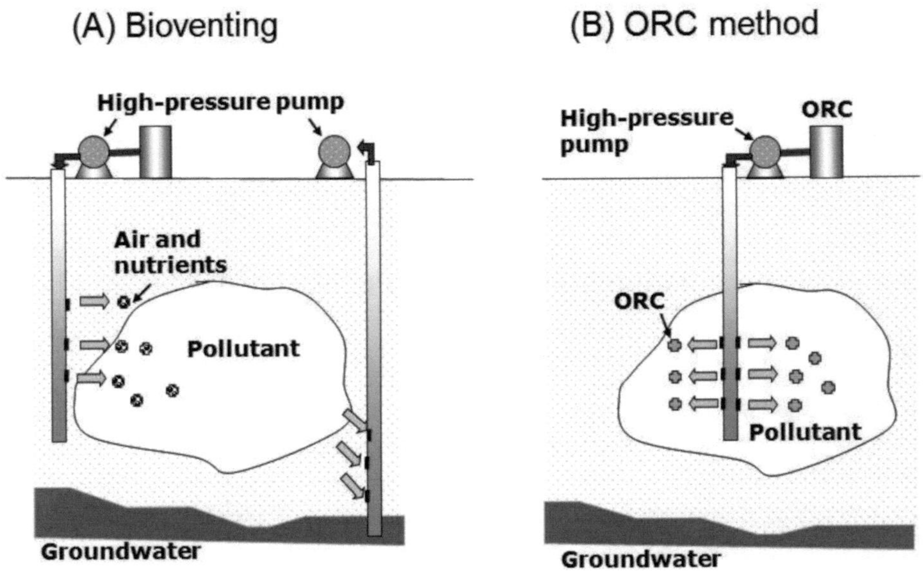

Figure 2. *In situ* bioremediation methods: Bioventing (A) and the ORC method (B).

3.2. Advantages of targeting the shallow layer of soil as the degrading zone

The author previously proposed a new concept for bioremediation [22]: the shallow layer of soil should be targeted for the bioremediation of pesticides. Targeting the shallow layer (A0 and A1 horizons) of soil has many advantages compared to the deeper layers (B and C horizons). Figure 3 shows a schematic diagram of the advantages of targeting the shallow layer, and can be explained as follows:

The first advantage is that the degradation rate of the pollutant at shallow layers is much higher than that in the deeper layers. Pesticide is sprayed in the air, and drops on the soil surface with rain or sprayed water. Because the pesticide diffuses from the soil surface to the deeper layer, the concentration is the highest at the soil surface. The degradation rate of a pollutant is generally proportional to the concentration, and thus, the highest degradation rate can be obtained at the shallow layer.

The second advantage is that oxygen is sufficiently present at the shallow layer. Many microorganisms require oxygen for their activation. In the shallow layer, water in soil contains sufficient oxygen, because air passes through the cavities in soil. Therefore, the exogenous supply of oxygen is not necessary at the shallow layer, making large-scale equipment unnecessary.

The third advantage is that it is easy to supply a degrading bacterium and nutrients to the shallow layer. Pollution is partially caused by the shortage of degradation by microorganisms. Therefore, microorganisms showing high degradation potential should be supplied to compensate for the shortage. In the deeper layer of the soil, it is very hard to supply microorganisms, and huge amounts of cells must be supplied. In the shallow layer, however, microorganisms can be easily supplied by spraying them with a sprayer, and the amount of cells required is much lower than that required for the deeper layer.

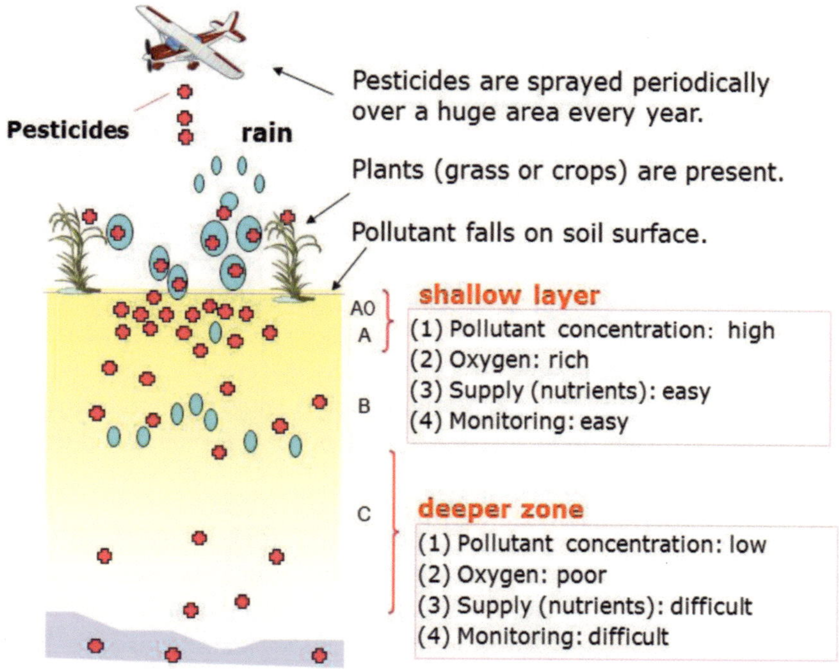

Figure 3. A schematic diagram of the advantages of targeting the shallow layer of soil for bioremediation.

The fourth advantage is that the maintenance and monitoring of the degradation are very easy. A pesticide is periodically sprayed every year, and there is always some contamination. Therefore, the ability to degrade the compounds must be permanently maintained. It is important to provide a supply of nutrients for the microorganisms to maintain the degradation activity. In the bioventing method, nutrients are often unable to reach the polluted areas. In contrast, the supply of nutrients in the shallow layers can be easily provided and assessed. Furthermore, in the agricultural fields in China or other countries in Asia, it is unnecessary to provide carbon and nitrogen sources or phosphate because chemical fertilizers are excessively used, and acid rain containing SOx and NOx often falls.

3.3. Bioremediation in the shallow layer using cells immobilized by a carrier

Although there are many advantages to bioremediation targeting the shallow layer, the passage time of pesticides in the shallow layer is short, and the degradation must be completed during this short period so that it does not reach the deeper layers. This can be realized by covering the soil surface or the shallow layer with immobilized cells. By using immobilized cells, the degradation rate and activity can be maximized.

Many studies have reported the immobilization of microorganisms [23] and its application for the treatment of waste water [24,25]. To use immobilized cells for bioremediation, it is necessary to ensure that the techniques are inexpensive and that the pore size is large enough to provide a high flow rate. Immobilization by biomass-supported particles (BSPs) [26], as shown in Figure 4A, may be the best applicable method. Yeast and bacteria showing cohesiveness can be immobilized by BSPs [27]. A high flux and low cost can be obtained because of large pore size for the immobilization, as shown in Figure 4B. The excessive growth of cells in BSPs can be easily remedied by their flowing out from the carrier, and the high degrading activity can be maintained by these released cells.

Charcoal also may be suitable for this purpose. It is inexpensive and can be used in the immobilization of bacteria. Immobilization with charcoal has been examined previously in the treatment of waste water. For example, Takagi *et al.* [28] examined the degradation of simazine in an experimental plot of a golf course. The *Arthrobactor sp.* cells that had the capacity to degrader simazine were immobilized with charcoal particles (1 cm^3 average size), and covered the soil surface. Simazine (25 g) was sprayed twice every 10 days. Consequently, around 90% of the simazine was degraded, and the degradation capacity was maintained for at least two months.

These results suggest that the degradation in the shallow layer is very useful for the remediation of the pollution by pesticides. However, two problems must be solved. One is the cost. When the immobilization method must be applied to a large area, the immobilization operation will likely be costly, even though the running cost is low. The other issue is digging. Most fields cannot be dug to cover them with the immobilization particles, because grass or crops are present. If the cover by these particles is inadequate, an insufficient degradation rate will be obtained.

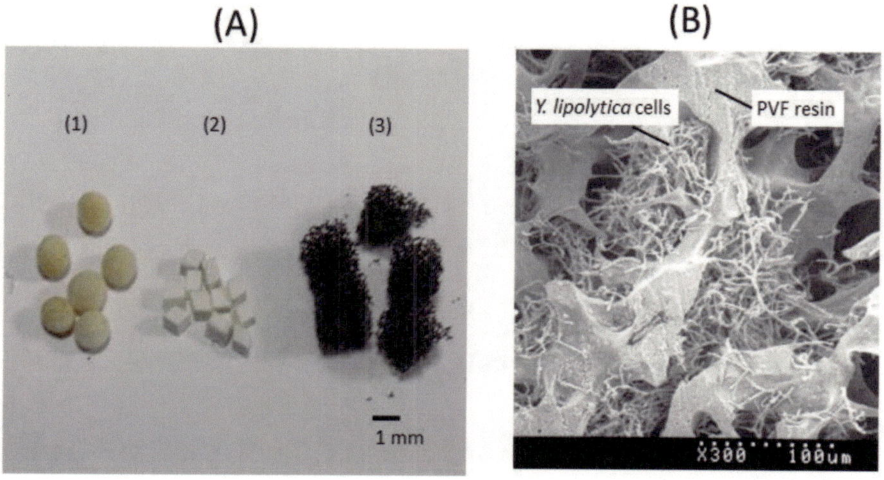

Figure 4. Immobilization with BSPs. Various kinds of BSPs (A) and a photograph of *Yarrowia lipolytica* cells immobilized by polyvinyl formal (PVF) resin (B). (1) The cellulose form; (2) the polyvinyl formal resin; (3) the urethane form

3.4. Bioremediation with the Self-Immobilized System (BSIS)

To compensate for the defects of the immobilized method, the author proposed a novel bioremediation method, which the author named "bioremediation with a self-immobilized system (BSIS)". BSIS is based on the idea that a polymer secreted by a microorganism can be utilized instead of an immobilization carrier, and the microorganism can immobilize itself in the shallow layer of soil without any other immobilization carrier (BSPs or charcoal). Below, BSIS is described in detail.

Figure 5 shows a schematic diagram of the protocol for BSIS. The *Bacillus natto* (*Bacillus subtilis*) strain, which is used for the production of "Natto" (a traditional fermented food in Japan), secretes a poly-glutamate (PGA) polymer. The PGA polymer shows high viscosity and can be used as a humectant, and was used in an attempted greening of a desert. The author used the PGA polymer as an immobilization carrier. As the first and second steps in Figure 5, a cell suspension of *B. natto* (*Bacillus subtilis*) is sprinkled on the soil surface, and medium is supplied for several hours. The *Bacillus* cells grow and form cell aggregates by secreting the PGA polymer in a shallow layer. As the third and fourth steps, the cells showing the ability to degrade pollutant (degrading cells) are sprinkled on the soil surface, and the medium is supplied. The degrading cells are trapped by the PGA polymer previously secreted by *B. subtilis* and grow. The medium is supplied for another several hours until cell aggregation at high-density constructs in the shallow layer of soil is achieved. By these steps, *B. subtilis* and degrading cells are co-immobilized by attaching to the PGA polymer, and the zone for degradation is constructed in the shallow layer of soil.

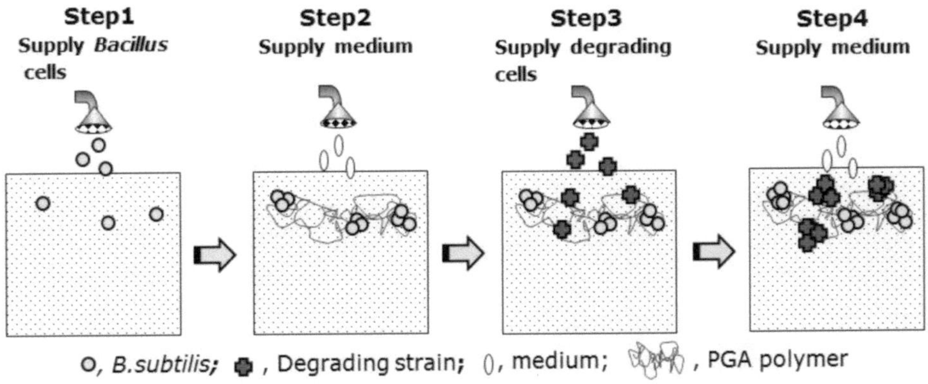

Figure 5. The protocol for the BSIS. Step 1, Supply *B. subtilis*; Step 2; Supply the medium; Step 3, Supply the degrading strain; Step 4, Supply the medium.

BSIS has many advantages compared to the other bioremediation methods. For example, BSIS does not require a supply of exogenous oxygen or nutrients with a high pressure pump, because BSIS treats pollutants at a shallow layer in the soil, where oxygen is rich and nutrients are easily supplied by an inexpensive sprinkler. BSIS therefore does not require expensive equipment. The cost for the equipment and operation of BSIS are much lower than those of the other methods. BSIS can be applied even in cases where many crops and grass are present, because it can be performed without having to dig up the soil. These advantages suggest that BSIS can be applied for the bioremediation of huge areas, and that continuous operation can be easily performed for many years. Thus, the BSIS method is an optimal remediation technique for soil polluted by pesticides.

3.5. Bioremediation of acid rain and pesticides by using BSIS

Finally, this section precisely explains how the idea of BSIS can be realized by introducing the author's studies. In a previous investigation [22], the author selected the *B. subtilis* NBRC3335 strain, which showed the highest cohesion among the stock bacteria at Kobe College. Figure 6 shows the cohesion of the strain. The *B. subtilis* NBRC3335 or *Eshcerichia coli* (control) bacteria were poured and penetrated into soil or sand packed in a column (5 cm at height), and 160 ml of water was poured into the column. A large amount of *E. coli* cells were passed through the column, especially in the case of the column packed in sand, which had a larger particle diameter (average diameter 400 μm) than that of the soil (average diameter 40 μm). On the contrary, *B. subtilis* cells were retained even in the column packed with sand. Therefore, *B. subtilis* cells have ability to attach to the shallow layer of various types of soil, and can be used as the first step of BSIS, as shown in Figure 5 (although *B. natto* which is used in the actual production of natto is superior to the NBRC3335 strain).

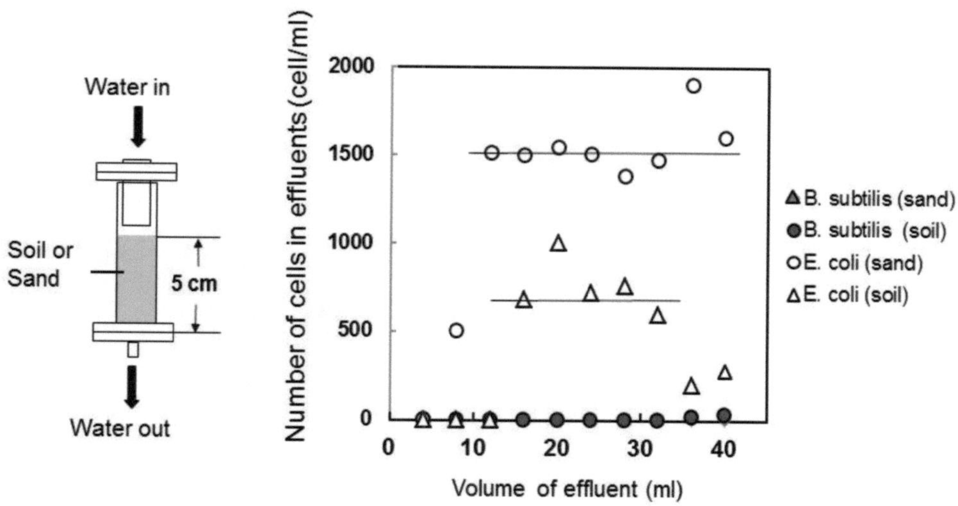

Figure 6. The self-immobilization capacity of *B. subtilus* NBRC3335 in the shallow layer. Four milliliters of *B. subtilis* or *E.coli* cell suspensions (0.01 OD) were supplied to the soil or sand in a column (5 cm height), and water was added. The numbers of cells in effluent samples (4 ml) were respectively counted.

Moreover, the author examined whether BSIS can be applied for the remediation of soil acidified by acid rain [29,30]. In recent years, acid rain has damaged many kinds of plants through the leaching of calcium and magnesium ions from the soil, and it has become a serious problem worldwide [31]. Preservation of the pH value of soils is important to avoid this problem. To protect the soil from acidification by acid rain, calcium carbonate powder or stone (marble) has often been used for neutralization of acids in Europe.

The author tried to use microorganisms showing a high neutralization potential instead of using calcium carbonate powder. The *Aureobasidium pullulans* NPH6 strain was isolated [32], and showed the high neutralization by degrading nitrate and secreting ammonia, as shown in Figures 7A and B. A neutralization test for BSIS was performed using the NPH6 strain and *B. subtilis* NBRC3335. Figure 7C shows the apparatus used for the experiment. An acrylic column (5 cm in inside diameter and 57 cm in height) was packed with sterilized silica sand (600-800 μm of particle sizes), which was used as a model soil. A mixture of *A. pullulans* NPH6 and *B. subtilis* IFO3335 cells (1.0 OD) was applied to the top of the column at a flow rate of 16.6 ml/h, then the modified Luria-Bertani medium (LB medium; 10 g/L bactopeptone, 10 g/L yeast extract, 5 g/L NaCl, 1g/L glucose, pH7.2) was supplied at a rate of 5.6 ml/h for 18 h to grow and co-immobilize these cells (mixed culture: BSIS). For comparison, immobilization was carried out by application of only NPH6 cells (pure culture: control). Figure 8A shows the distribution of the cell number of the NPH6 strain, which is expressed as the total cell number in each section of the column, after supplying the LB medium for 18 h. By supplying of LB medium, the cell number in the column increased about 17 times the number initially supplied in the mixed culture (BSIS). This increase in the cell number is almost the same as that in the batch culture. Furthermore, in the mixed culture, over 80% of the total cells existed in the layer

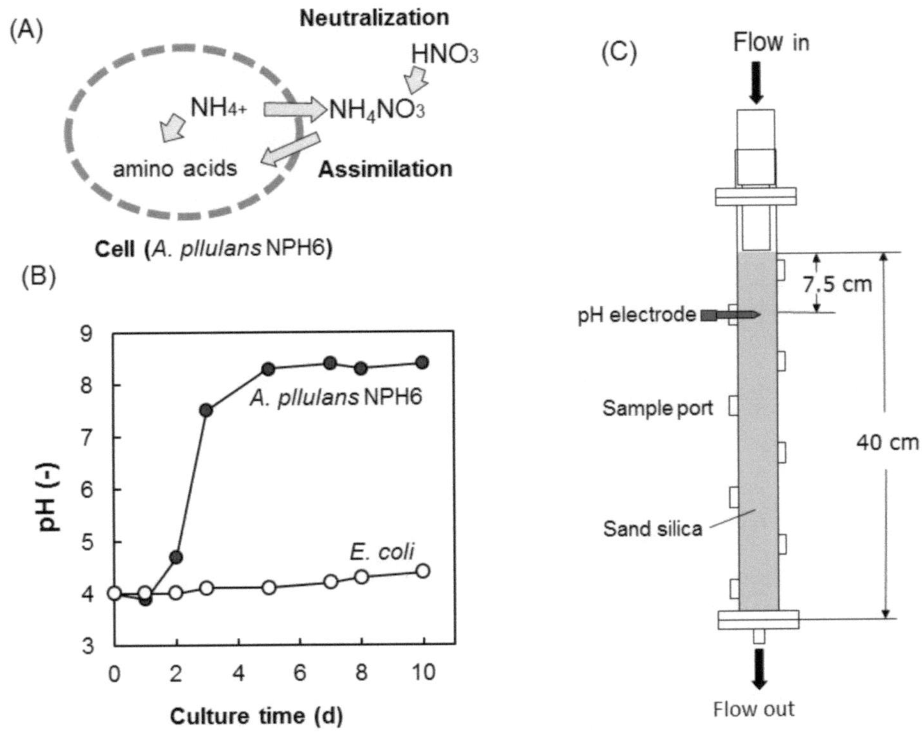

Figure 7. The strain and apparatus used for the BSIS experiment. A schematic diagram of the neutralization of nitric acid by *A. pllulans* NPH6 (A) and the time courses of the pH changes (B) when the *A. pllulans* NPH6 or *E. coli* cells were cultured in the LB medium. (C), A schematic diagram of the column used for the experiment. Sand silica was packed in the column, and the cells and medium were supplied at the top of the column with a micro-pump.

0-5 cm from the surface of the packed silica. The DO value measured by a DO electrode inserted to a depth of 7.5 cm was about 50% of air saturation after supplying the LB medium for 18 h.

Figure 8B shows the neutralization characteristics of the sulfuric acid solution (pH 4.9) by BSIS after the cells were immobilized (as shown in Figure 8A), when the sulfuric acid solution was supplied at the top of the column at a flow rate of 24.9 ml/ h. The pH value in the mixed culture was kept above 6.5 for 96 hours, while it decreased below 5 in the pure culture. The total amount of solution corresponded to a rain fall of 1200 mm, which is the average value of rain fall during a one year period in Japan. This means that BSIS is applicable for the neutralization of a large amount of acid rain.

The author also tried to apply the BSIS for the remediation of soil polluted by pesticides [33]. Figure 9 shows the metabolic pathway of atrazine. In this pathway, cyanuric acid is a key intermediate, and the cleavage of its ring is the rate limiting step in its degradation [34]. Thus, the author examined the bioremediation of cyanuric acid instead of atrazine by using the *Pseudomonas sp*. NRRL B-2227, which showed a high degradation capacity for cyanuric acid. The same apparatus and model soil (Figure. 8A) were used for the experiment. To co-immobilize the *Pseudomonas sp.* NRRLB-12227 with *B. subtilis,* 20 ml of a cell suspension of *B. subtilis* NBRC3335 at 1.0 OD was supplied to the top of the column at a flow rate of 16.6 ml/h. Next, LB medium was supplied at a rate of 5.6 ml/h for 24 h to grow *B. subtilis* cells. A 20 mL suspension of *Pseudomonas sp.* at 1.0 OD was supplied at a flow rate of 16.6 ml/h. The R medium was supplied at a rate of 10 ml/h for 18 h to co-immobilize the *Pseudomonas* cells with *Bacillus* cells (mixed culture: BSIS). For comparison with the mixed culture, *Pseudomonas sp.* cells were immobilized without *Bacillus* cells (pure culture: control).

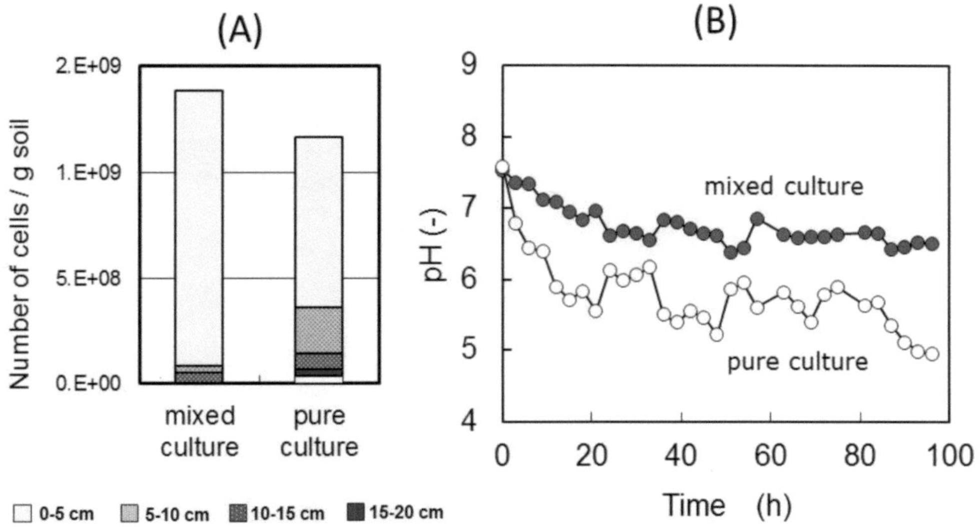

Figure 8. Neutralization of sulfuric acid with *A. pllulans* NPH6 in a mixed culture (BSIS) and a pure culture (control). (A), The distribution of the *A. pllulans* NPH6 cells along the column axis after the cells were respectively immobilized by mixed culture (BSIS) and pure culture (Control). (B), the pH value was measured 7.5 cm from the top, when sulfuric acid solution (pH 4.7) was supplied at a flow rate of 24.9 ml/h for 96 h.

After immobilization on the packed soil, the R medium containing 1 mM cyanuric acid was supplied to the column at a flow rate of 24.9 ml/h. Figure 10A shows the viable cell number in each layer of soil in the column after supplying the R medium containing 1 mM cyanuric acid. In both cases, 60-80% of the cells were immobilized in a shallow layer (0-7.5 cm) of packed soil, and the cell numbers in the mixed culture were much higher than those in the pure culture. Figure 10B shows the time course for the cyanuric acid concentrations for liquid flowing 7.5 cm from the surface of the packed bed and in the effluent solution (35 cm), when the medium containing 1 mM cyanuric acid was supplied for 72 h to the mixed culture. The rate of degradation of cyanuric acid in the BSIS gradually increased after 24 h. This increase may have been caused by acclimation of the cells to cyanuric acid and an increase in cell activity during cultivation in the medium containing 1 mM cyanuric acid. The rate of degradation in the BSIS was higher than that in the control, and the cyanuric acid concentration was decreased to 0.1-0.2 mM at the column exit. The degradation rate could be maintained for a long period of time (two weeks) by supplying suitable nutrients in another test. Therefore, we concluded that BSIS is an excellent method for the bioremediation of soil polluted by pesticides.

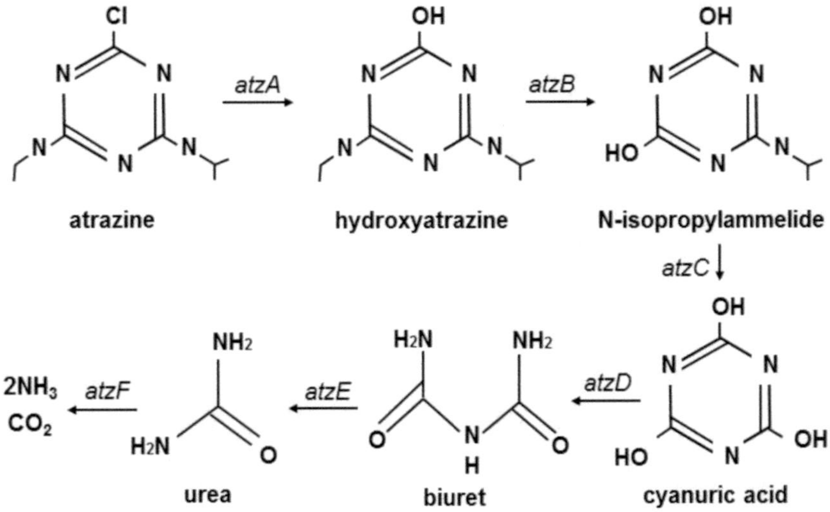

Figure 9. The biological degradation pathway of atrazine, and the genes corresponding to the enzymes.

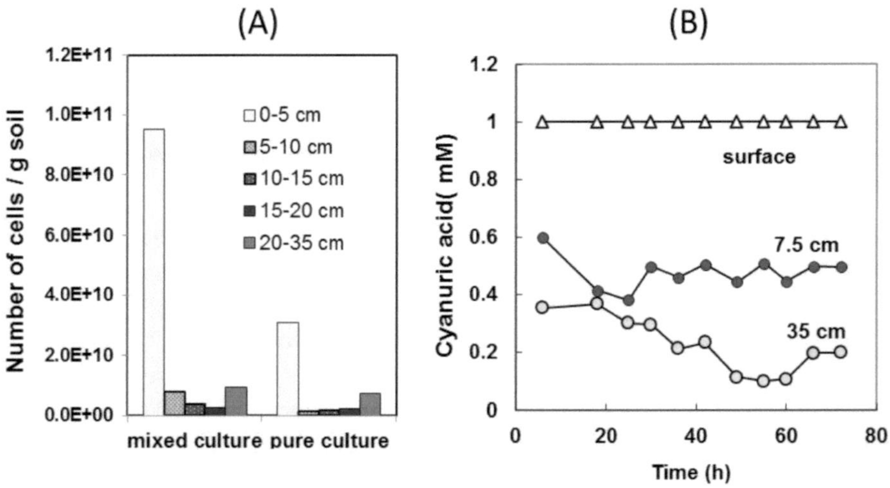

Figure 10. Bioremediation of soil polluted with cyanuric acid by BSIS. (A), The distribution of the *Pseudomonas sp.* cells along the column axis after the cells were respectively immobilized by a mixed culture (BSIS) and mono culture (Control). (B), the cyanuric acid concentration was measured 0, 7.5 and 35 cm from the top, when 1 mM cyanuric acid was supplied at a flow rate of 25 ml/h for 72 h.

3.6. Enhancement the performance of degradation and stability using the optimal microorganisms in the BSIS

Despite these successful experiments, there were some points that need to be improved in further studies. Enhancement of the performance in the BSIS is necessary for its practical use, and the following two improvements in the degrading microorganisms are important.

One is the activity for degrading pesticides. A very high degradation rate is necessary to complete the degradation in the shallow layer. A higher degradation rate can be obtained by a higher density of degrading cells, but there is an upper limit for the rate. Additionally, the extremely high-density operation can cause a shortage of nutrients and oxygen, and it decreases the levels of other important natural soil microorganisms. Therefore, during the operation of BSIS, the increase of degrading activity per cell is more effective and important than providing a higher density operation.

Gene manipulation is one of the most effective methods to enhance the activity of the microorganisms. Many microorganisms showing the ability to degrade triazine and organophosphorous herbicides have been isolated [35-39], and the metabolic pathways and their genes are known. For example, Figure 9 shows the genes encoding the enzymes involved in the metabolism of atrazine. The atzA-atzD genes shown in Figure 9 were cloned from bacteria, and their activity has been examined [36]. The performance of BSIS will be increase if recombinant bacteria showing a higher degrading capacity can be used.

Other interesting studies were also reported on other aspects that increased the activity. The rate limiting step of degradation is the permeability of the cell membrane. To decrease the time required for the agent to cross the membrane, recombinant strains that can express the organophosphorous hydrolase on the surface of the cells [40,41] and secrete it into the periplasmic space [42] were constructed. As another method, a bacterium carrying a high functionality catabolic plasmid was also constructed, which showed the ability to perform conjugal transfers [43]. When this bacterium is present in the soil, the degradation ability can be increased by conjugal transfer. A recombinant strain carrying a special sensor to search for pollution has also been sought [44]. These results suggest that the performance of BSIS will be able to approach to an ideal value by using these recombinant strains if they are successfully constructed and their use is permitted.

The other point that could use improvement is the synergy between microorganisms. Many microorganisms play important roles in maintaining the healthy condition of the soil. For example, a nitrogen-fixing bacterium provides nitrogen sources to plants, and photo-chemical bacteria help degrade chemical compounds. Degradation of pollutants by multiple strains is much more effective than that by a single strain. Some studies have reported on the optimal combinations that can be used to enhance the synergy between strains [45-49]. As many different strains can be simultaneously immobilized in the BSIS, the optimal combination will be help to increase the performance and stability of the operation. Recently, microorganisms producing electricity have been reported. Those strains are present even in the agricultural field, and can produce electricity in the field [50]. If the electrical power generated by these bacteria can reach the level of a solar battery, a permanent operation in the BSIS using the battery may be possible, further minimizing the cost of operation.

4. Conclusion

The pollution of soil and groundwater by the use of pesticides has become a serious condition all over the world. The soil pollution is not only present in huge areas, but also is repeated continuously, leading to permanent contamination, because the pesticide is used many times each year. Therefore, the conventional methods require high budgets, making their set-up and maintenance are impractical. The author proposed the use of a novel bioremediation method, namely BSIS. Using BSIS, degrading microorganisms can be self-immobilized at high density in the shallow layer with the help of a PGA polymer secreted by the *Bacillus* strain, and can rapidly degrade a pollutant *in situ*. The author applied the BSIS for acid rain and triazine herbicides by using model soil packed in a column. BSIS was effective to remediate soil acidified by the acid rain and polluted by the pesticide. Moreover, this BSIS has advantages in that it can be applied without any expensive apparatus and can operate easily and permanently. These results suggest that BSIS may be the best method for the bioremediation of soil pollution by pesticides, although further improvements in the microorganisms might be necessary for the practical application of the technique.

Everyone can freely study and use our BSIS method without any restrictions by a patent. Therefore, the author expects that many researchers will recognize the advantages of BSIS, and by further improvements, it can applied for the pollution by pesticides, which is a serious worldwide problem.

Acknowledgements

I am grateful to Drs. Shigeo Katoh, Tomohisa Katsuda and Takako Yasuda. The study on BSIS was cooperatively performed with them, when they were a Professor, Assistant Professor and doctoral student, respectively, at Kobe University. I also thank the graduate students in Prof. Katoh's laboratory at Kobe University and the students in my laboratory at Kobe College. The research was supported by grants-in-aids from the Central Laboratory and the School of Human Sciences at Kobe College.

Author details

Naofumi Shiomi*

Address all correspondence to: n-shiomi@mail.kobe-c.ac.jp

Department of Human Sciences, Kobe College, Hyogo, Japan

References

[1] World Health Organization: Children Face High Risks from Pesticide PoisoningJoint note for the media WHO/FAO/UNEP. www.who.int/mediacentre/news/notes/2004/np19/en/accessed 10 February (2013).

[2] Cookson-jr, J. T. Bioremediation Engineering Design and Application. New York: McGraw-Hill; (1995).

[3] Cardon, N, Vaillant, C, Cren, P, Gruffat, B, Rappold, J. P, & Corbé, H. Acute Organophosphorus Pesticide Poisoning and Cholinesterases Activities. Annales Biologie Clinique (Paris) (2005). , 63(3), 329-34.

[4] Hayes, T, Haston, K, Tsui, M, Hoang, A, Haeffele, C, & Vonk, A. Atrazine Induced Hermaphroditism at 0.1 ppb in American Leopard Frog (*Rana pipiens*): Laboratory and Field Evidence. Environmental Health Perspectives (2003). , 111(4), 568-575.

[5] Tarbah, F. A, Kardel, B, Pier, S, Temme, O, & Daldrup, T. Acute Poisoning with Hhosphamidon: Determination of Dimethyl Phosphate (DMP) as a Stable Metabolite in a Case of Organophosphate Insecticide Intoxication. Journal of Analytical Toxicology (2004). , 28(3), 198-203.

[6] Johnson, H. M, Domagalski, J. L, & Saleh, D. K. Trends in Pesticide Concentrations in Streams of the Western United States, 1993-2005. Journal of the American Water Resources Association (2011). , 47(2), 265-286.

[7] Vogel, J. R, Majewski, M. S, & Capel, P. D. Pesticides in Rain in Four Agricultural Watersheds in the United States. Journal of environmental Quality (2008). , 37(3), 1101-1115.

[8] Dörfler, U, & Feicht, E. A. Scheunert I. S-triazine Residues in Groundwater. Chemosphere (1997).

[9] Rebich, R. A, Coupe, R. H, & Thurman, E. M. Herbicide Concentrations in the Mississippi River Basin-the Importance of Chloroacetanilide Herbicides degradates. Science of the Total Environment (2004).

[10] Konstantinou, I. K, Hela, D. G, & Albanis, T. A. The Status of Pesticide Pollution in Surface Waters (River and Lakes) of Greece. Part I. Review on occurrence and levels. Environmental Pollution (2006). , 141(3), 555-570.

[11] Tappe, W, Groeneweg, J, & Jantsch, B. Diffuse atrazine pollution in German aquifers. Biodegradation (2002). , 13(1), 2-10.

[12] Webster, L, Russell, M, Walsham, P, Phillips, L. A, Hussy, I, Packer, G, Dalgarno, E. J, & Moffat, C. F. An Assessment of Persistent Organic Pollutants in Scottish Coastal and Offshore Marine Environments. Journal of Environmental Monitoring (2011). , 13(5), 1288-1307.

[13] NowPublic News Coverage: Growing cotton in India's killing fieldswww.nri.org/projects/inthefield/india_pests.htm (accessed 10 February (2013).

[14] Li, X, Zhu, Y, Fu, S, Xu, X, & Cheng, H. Distribution of HCHs and DDT in Soils from Beijing City, China, Archives of Environmental Contamination and Toxicology (2006). , 51(3), 329-336.

[15] Li, X. H, Wang, J, Cao, X. L, Wang, X. F, Liu, J. C, Liu, X. F, Xu, X. B, & Jiang, X. N. Contamination of Soils with Organochlorine Pesticides in Urban Parks in Beijing China. Chemosphere (2008).

[16] Bao, L. J, Maruya, K. A, Snyder, S. A, & Zeng, E. Y. China's Water Pollution by Persistent Organic Pollutants. Environmental Pollution (2012). , 163, 100-108.

[17] Qin, Y, Lv, X, Qi, G, Diao, Q, Liu, G, & Xue, M. Wang J Tong J, Zhang L, Zang K. Assessment of Melamine Contamination in Crop, Soil and Water in China and Risks of Melamine Accumulation in Animal Tissues and Products. Environmental International (2010). , 36(5), 446-452.

[18] Mukherjee, A, Sengupta, M. K, Hossain, M. A, Ahamed, S, Das, B, Nayak, B, Lodh, D, Rahman, M. M, & Chakraborti, D. Arsenic Contamination in Groundwater: A Global Perspective with Emphasis on the Asian Scenario. Journal of Health, population, and nutrition (2006). , 24(2), 142-163.

[19] Zhang, L, & Chen, C. Geographic Distribution and Exposure Population of Drinking Water with High Concentration of Arsenic in China. Journal of hygiene research (1997). , 26(5), 310-313.

[20] Majumdar, K. K. Guha Mazumder DN. Effect of Drinking Arsenic-Contaminated Water in Children. Indian Journal of Public Health (2012). , 56(3), 223-6.

[21] Homma, N. *In-situ* Bioremediation of Soil and Groundwater Contaminated by Chlorinated Organic Compounds. Technical Report of Mitui Engineering & Shipbuilding (Japanese) (2008). , 193, 17-22.

[22] Yasuda, T, Katoh, S, Inoue, Y, & Shiomi, N. A Bioremediation Method Based on Self-Immobilization of Cells in Shallow Layer of Soil. Journal of Chemical Engineering of Japan (2003). , 36(2), 216-219.

[23] Zhou, L, Li, G, An, T, Fu, J, & Sheng, G. Recent Patents on Immobilization Microorganism Technology and Its Engineering Application in Wastewater Treatment. Recent Patents on Engineering (2008). , 2, 28-35.

[24] Patil, Y. B. Biodegradation of Thiocyanate from Aqueous Waste by a Mixed Bacterial Comunity. Research Journal of Chemistry and Environment (2008). , 12(1), 69-75.

[25] Patil, Y. B, & Paknikar, K. M. Development of a Process for Biodetoxification of Metal Cyanides from Wastwaters. Process Biochemistry (2000). , 35, 1139-1151.

[26] Albrechtsen, H. J, Smith, P. M, Nielsen, P, & Christensen, T. H. Significance of Biomass Support Partecles in Laboratory Studies on Microbial Degradation of Organic Chemicals in Aquifers. Water Research (1996). , 30(12), 2977-2984.

[27] Shiomi, N, Inoue, Y, Tomioka, K, & Yasuda, T. Immobilization of *Candida* Cells Showing Mycelium-Like Shapes in Porous Polyvinyl Formal Resin and Their Applications. Journal of Chemical Engineering of Japan (2003). , 36(2), 161-165.

[28] Takagi, K, Kataoka, R, & Yamazaki, K. Review. Recent Technology on Bio-remediation of POPs and Persistent Pesticides. Japan Agricultural Research Quaterly (2011). , 45(2), 129-136.

[29] Yasuda, T, Shiomi, N, Iwasaki, S, Yamaguchi, Y, Katsuda, T, & Katoh, S. Neutralization of Acids by Microorganisms Co-Immobilized with *Bacillus subtilis* in a Shallow Layer of Model Soil. Journal of Chemical Engineering of Japan (2004). , 37(12), 1445-1451.

[30] Shiomi, N, & Katoh, S. Bioremediation of Soil and Water Acidified by Acid Rain with Fungi Cells. In: Roglesfield LG. (ed.) Acid Rain Research Focus, Nova Scientific Piblishers. Inc.; (2008). , 104-124.

[31] Evans, J. S. Biological Effects of Acidity in Precipitation on Vegtation, A Review. Environmental and Experimental Botany (1982). , 22, 155-169.

[32] Shiomi, N, Yasuda, T, Inoue, T, Kusumoto, Y, Iwasaki, N, Katsuda, S, & Katoh, T. S. Characteristics of Neutralization of Acids by Newly Isolated Fungal Cells. Journal of Bioscience and Bioengineering (2004). , 97(1), 54-58.

[33] Shiomi, N, Yamaguchi, Y, Nakai, H, Fujita, T, Katsuda, T, & Katoh, S. Degradation of Cyanuric Acid in Soil by *Pseudomonas* sp. NRRLB-12227 Using Bioremediation with Self-Immobilization System. Journal of Bioscience and Bioengineering (2006). , 102(3), 206-209.

[34] Cook, A. M, Beilstein, P, Grossenbacher, H, & Hütter, R. Ring Cleavage and Degradative Pathway of Cyanuric Acid in Bacteria. The Biochemical Journal (1985). , 231(1), 25-30.

[35] Wackett, K. P, Sadowsky, M. J, & Martinez, B. Shiodegradation of Atrazine and Related s-Triazine Compounds: from Enzyme to Field Studies. Applied Microbiology and Biotechnology (2002). , 58, 39-45.

[36] Martinez, B, Tomkins, J, Wackett, L. P, Wing, R, & Sadowsky, M. J. Complete Nucleotide Sequence and Organization of the Atrazine Catabolic Plasmid pADP from *Pseudomonas* sp. Stain ADP. Journal of Bacteriology (2001). , 183(19), 5684-5697.

[37] Singh, B. K, & Walker, A. Microbial Degradation of Organophosphorous Compounds. FEMS Microbiology Reviews (2006). , 30(3), 428-471.

[38] Theriot, C. M, & Grunden, A. M. Hydrolysis of Organophosphorous Compounds by Microbial Enzyme. Applied Microbiology and Biotechnology (2011)., 89, 35-43.

[39] Bhushan, B, Chauhan, A, & Saman, R. K. Kinetics of Biodegradation of p-Ntrophenol by Dfferent Bacteria. Biochemical and Biophysical Research Communications (2000)., 270, 733-740.

[40] Wernérus, H, & Ståhl, S. Biotechnological Applications for Surface-engineered Bacteria. Biotechnolgy and Applied Biochemistry (2004)., 40(3), 209-228.

[41] Zang, J, Lan, W, Qiao, C, Jiang, H, Muchndani, A, & Chen, W. Bioremediation of Organophosphorous Pesticides by Surface-Expressed Carboxylesterase from Mosquito on *Escherichia coli*. Biotechnology Progress (2004)., 20(5), 1567-1571.

[42] Kang, D. G, Nim, G. B, & Cha, H. J. Functional Periplasmic Secretion of Organophosphorous Hydrase using the Twin-Arginine Translocation Pathway in *Escherichia coli*. Journal of Biotechnology (2005)., 118(4), 379-385.

[43] Ikuma, K, & Gunsch, C. Genetic Bioaugmentation as an Effective Method for *in situ* Bioremediation: Functionality of Catabolic Plasmids Following Conjugal Transfers. Bioengineerd. (2012)., 3(4), 236-241.

[44] Ishida, T, Ikeda, T, Takiguchi, N, Kuroda, A, Ohtake, H, & Kato, J. Inhibition of Quorum Sensing in *Pseudomonas aeruginosa* by N-Acetyl Cyclopentylamides. Applied Environmental Microbiology (2007)., 73(10), 3183-3188.

[45] Kleinsteuber, S, Schleinitz, K. M, & Vogt, C. Key Players and Team Play: Anaerobic Microbial Communities in Hydrocarbon-Contaminated Aquifers. Applied Microbiology and Biotechnology (2012)., 94(4), 851-873.

[46] Zhuang, K, Ma, E, Lovley, D. R, & Mahadevan, R. The Design of Long-Term Effective Uranium Bioremediation Strategy Using a Community Metabolic Model. Biotechnology and Bioengineering (2012)., 109(10), 2475-83.

[47] Dugat-bony, E, Biderre-petit, C, Jaziri, F, David, M. M, Denonfoux, J, Lyon, D. Y, Richard, J. Y, Curvers, C, Boucher, D, Vogel, T. M, Peyretaillade, E, & Peyret, P. In situ TCE Degradation Mediated by Complex Dehalorespiring Communities during Biostimulation Processes. Microbial Biotechnology (2012)., 5(5), 642-653.

[48] Zhang, C, & Katayama, A. Humin as an Electron Mediator for Microbial Reductive Dehalogenation. Environmental Science & Technology (2012)., 46(12), 6575-83.

[49] Zhang, J, Zhang, Y, & Quan, X. Electricity Assisted Anaerobic Treatment of Salinity Wastewater and Its Effects on Microbial Communities. Water Research (2012)., 46(11), 3535-3543.

[50] Zhao, Y, Watanabe, K, & Hashimoto, K. Poly(bis-2,6-diaminopyridinesulfoxide) as an Active and Stable Electrocatalyst for Oxygen Reduction Reaction. Journal of Materials Chemistry (2012)., 22, 12263-12267.

Chapter 13

Persistence and Bioaccumulation of Persistent Organic Pollutants (POPs)

Tomaz Langenbach

Additional information is available at the end of the chapter

http://dx.doi.org/10.5772/56418

1. Introduction

Some organic pesticides are the most widespread POPs (Persistent Organic Pollutants) used in agriculture worldwide. Other POPs such as volatilized industrial pollutants as well as pollutants from fuel combustion can also contaminate large areas due to dispersion by wind and rainfall. Non biodegradable molecules persist in the environment and the soil serves as a large sink for them. These pollutants are able to move in the environment and in low concentrations over a long period of time, but through bioaccumulation they can have hazardous effects on the biota of the region. This article focuses mainly on pesticides and some other POPs that contaminate large areas. The chemical characteristics of persistent molecules that are resistant to biodegradation will be discussed in the context of the environmental conditions that enhance persistence.

Adsorption or not of a molecule in the soil is a key step to define bioavailability. The concept of "preferential partition" helps us understand the competitive mechanisms between the soil and biosphere. Bioaccumulation occur when concentrations in the biosphere are much higher than those found in the surrounding environment. The uptake of persistent molecules to levels of high concentrations can have hazardous effects on flora and, through the food chain, on fauna and human health.

Xenobiotics not adsorbed by the soil can become mobile and when the molecule is persistent it may be dispersed by water or air through the biosphere, polluting large geographical areas and affecting the biodiversity of the flora and fauna. A brief overview of this dispersion will be necessary to show how recalcitrant molecules in the environment make remediation inefficient. The application of pesticides over large areas can only be sustainable if their molecules biodegrade naturally. Remediation can be applied usefully to accidents in restricted areas but it is not feasible over large areas due to the high costs. The only efficient policy for recalcitrant molecules that can not be controlled by remediation is the restriction of their use or banning them altogether.

2. How molecules can escape degradation in the environment

The adsorption of molecules, due to their affinity to other chemical components of the soil matrix or soil biosphere, protects them from chemical degradation as well biodegradation. Non adsorbed molecules are bioavailable and therefore are exposed to degradation. Experiments with soils weathered for long time with ethylene bromide contamination in which was added ^{14}C-ethylene bromide, shows strong biodegradation of the new applied radioactive substance after sixty days at the same time that weathered contamination substance remains in the same concentration [1]. Microorganisms involved in biodegradation are active against a freshly applied radioactive molecule but are not efficient for the same substance entrapped in the soil. Weathered molecules by soil friction process in which the larger soil particles are broken into smaller ones results in the gradually desorption of the molecules with enhanced bioavalability [1]. Therefore it is a pitfall to conclude that the hazardous effects are only caused by pesticide concentrations in the soil [2].

The fate of pollutants in the soil depends on the soil properties and the physical chemical characteristics of these molecules. Essentially the uptake of pollutants in the soil can be understood to be the result of chemical attraction and bond strength. These processes in the soil are governed by a number of well described phenomena including: Van der Waals forces, hydrogen bonding, ion exchange, charge transfer mechanisms, lipophilic affinity, entrapment and covalent reactions to humic acids [3, 4]. In essence, pollutant molecules will move to the most attractive sites in the soil environment. Competition between plant roots, soil mesofauna, microbes and soil organic and inorganic components makes residue to bind in the most attractive site where residues will accumulate preferentially. Pollutants, such as pesticides, which are not held or bound to the living and non-living parts of the soil, will be leached through the soil, resulting in pollution of ground water, rivers and reservoirs or they will be volatilized.

2.1. Where molecules adsorb

In this section the factors that regulate the distribution of pollutants between the soil matrix and the biosphere including microorganisms, mesofauna and bioaccumulation in plants will be discussed. In aquatic environments, where polar water molecules predominate, lipophilic pollutants have affinity for organic matter and therefore always move toward the biosphere [5] (Figure 1). In this case the bioaccumulation factor depends on molecular size and correlates positively with a lipophilic character measured by the K_{ow} value [6]. However, polarity on its own cannot explain to which compartment POP's move to in the soil.

Soils are a highly variable mixture of mineral and organic materials with living, dead and decaying biologic components. There is a lipophilic fraction as well that can adsorb lipophilic pollutants. The binding process is complex considering the diversity of compartments such as microorganisms, mesofauna and plants that compete for the uptake of lipophilic pollutants. Despite difficulties to standardize a methodology for lipid determination [7], this parameter is used to calculate the bioaccumulation factor [8, 9]. Soils can function as a filter when they adsorb the remaining residues, which then become unavailable for the biosphere. On the other hand low adsorption capacity pollutants remain bioavailable and can contaminate water, air, fodder plants, livestock and moving along the food chain up to humans. The soils may function as a filter or as a source of pollutants and this depends mainly on the kind of soil.

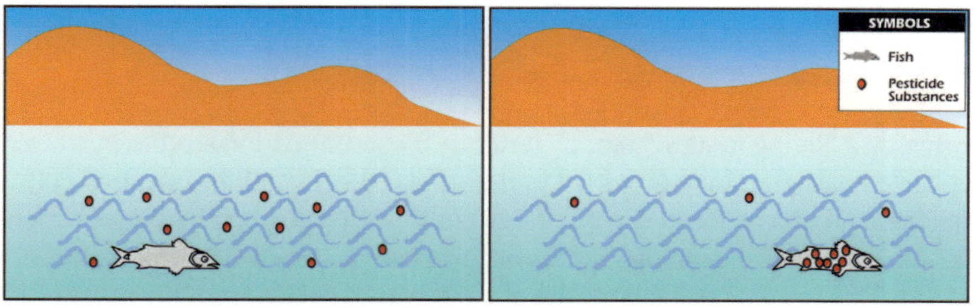

Figure 1. In aquatic systems the movement of lipophilic substance is unidirectional towards living forms, resulting in bioaccumulation

Despite enormous amounts of published scientific literature about bioaccumulation, the distribution of pollutants in the soil or biosphere is not well understood. Here the mechanisms of bioaccumulation in the soil environment will be discussed including the development of the proposed "preferential partition" concept.

2.1.1. Lipophilic molecule uptake by microorganisms

Soil microorganisms represent a large part of the living biomass but in general are not used for bioaccumulation studies since they cannot be separated from the soil to measure the pollutants. Nevertheless important information could be obtained from an experiment in which antibiotic resistant bacteria with bioaccumulated difocol were introduced into the soil [10]. Radioactive ^{14}C-dicofol was bioaccumulated during the incubation of *Pseudomonas fluorescens* (soil bacteria) resistant to kanamicyn and rifampycin. These bacteria can be measured in the soil using CFU counts with petri dishes containing the above mentioned antibiotics able to suppress all other soil microorganisms. ^{14}C-dicofol was incubated and bioaccumulated in Pseudomonas fluorescens strains after which it was poured on top of a soil and subjected to a succession of simulated rainfall events. At the end of the simulated rainy period of 24h, 95% of the radioactive labeled insecticide remained in the upper 1cm of the soil, whereas 60% of the microorganisms had been transported 10 cm through the soil and were recovered in the leachate. Only 7% of the bacteria poured onto the soil remained within the first 1 cm of soil. In this experiment the acaricide dicofol moved away from the bacteria toward the soil particles [10] showing that it was more strongly attracted to the soil matrix than to the living cells. Therefore in this case the soil acted as a filter/sponge, protecting microorganisms, mesofauna, plant roots and preventing ground water pollution. Molecules will move to the sites they find most attractive and in this case the soil matrix showed a higher uptake of the pesticide, thus reducing the dicofol content in the bacterial cell envelope. In organic matter rich soils, it is not uncommon for many POP's to show **preferential partition** towards soil organic matter adsorption rather than to plants or other living forms [11]. A good parameter to evaluate **preferential partition** of pollutants between soil and biosphere is to compare octanol/water (K_{ow}) ratios of each compartment. A higher octanol/water ratio (K_{ow}) for organic matter than for the soil bacteria explains why the dicofol moves out of the cell and into the organic component of the soil [12, 13].

2.1.2. Lipophilic molecule uptake by earthworms

Earthworms, as a "living system" model, facilitate bioaccumulation studies in soil since they can be collected easily and analyzed for pollutant uptake. Papini and Andrea [14] working with simazine, a relatively non-polar (K_{ow} 2-2.3) herbicide, [15, 16] and Paraquat, a highly polar herbicide, found that simazine did not bioaccumulate in the earthworm *Eisenia foetida* but Paraquat did. This result was the opposite of what was expected from the point of view that non polar substances bioaccumulate in the biosphere and polar substances do not. In a separate but similar study using an Argisol soil, the herbicide atrazine did not bioaccumulate in the earthworms *Pontoscolex corethrurus* either [17]. However, these results are not fully explained by the polarity of the pesticides. To interpret these results more precisely requires an understanding of K_{ow} as well as knowing the importance of organic matter in the soil. Soil organic carbon content (K_{oc}) correlates positively with soil K_{ow} and is an important factor to attract lipophilic substances [13, 18]. Given that K_{oc} and K_{ow} in general correlate positively, soil organic matter (OM) content can be used to select soils for study. Andréa and Papini used this method to compare how simazine and paraquat [19, 20] would behave in soil in the presence of the earthworm *Eisenia foetida* with different amounts of soil organic matter.

Two soils, one with 12 g. L^{-1} organic matter (pH 5.7) and the other with 93 g. L-1 organic matter (pH 6.4), were treated with simazine and paraquat. Soils were incubated for 90 days with *Eisenia foetida* [19, 20] and the bioaccumulation coefficient factor (Bcf.) was determined. Simazine did not bioaccumulate (Bcf. 0.9) in the earthworms in the high organic matter soil [19]; however, in the low organic matter soil it did (Bcf. 6.9). With simazine, it seems likely that molecular polarity controlled the different distribution of pesticides between soil organic matter and biota. In the soil with low organic matter, **preferential partition** was toward earthworms that provided more attractive sites for the pesticide than the soil matrix. In the high organic matter soil, **preferential partition** was toward the soil organic matter, less polar than the earthworms and therefore chemically more attractive to lipophilic pesticides (Figure 2). The results at first appear to be at odds with one another, in terms of where the pesticide accumulated, until we recognize that pesticide movement is not necessarily unidirectional.

2.1.3. Lipophilic molecule uptake between earthworms and plants

Experiments with p,p`-DDE in soils with the earthworms *Ersenia fetida, Lumbricus terrestris* or *Apporectodea caliginosa* in the presence of different plant subspecies showed that *Curcurbita pepo*

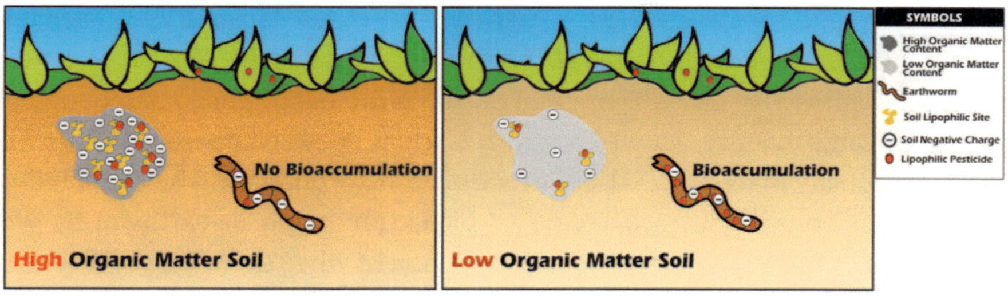

Figure 2. The movement of lipophilic pesticides in soil is influenced by the relative abundance of lipophilic sites in the living and non living soil components.

ssp. reduces bioaccumulation in the three species of earthworms. Similar experiments in the presence of *Curcurbita ovifera ssp.* showed a reduction as well as an increase of bioaccumulation in the different earthworm specie [21]. The authors observed that bioaccumulation in the plant C. pepo were enhanced with the three earthworm species which did not occur with *C. Ovifera* where only a slight increase was observed. In this set of experiments **preferential partition** occurs mainly toward plants and less to earthworms.

An increase in the bioaccumulation in plants indicates higher bioavailability of this residue due to presence of the earthworm [22] which through chemolysis is able to change humic acid and increase the protein and carbohydrate moieties and degrade the carboxylic and aliphatic groups. In consequence the hydrophobic index HI = 0.0433 – 0.0811 in the soil decreases in the presence of earthworms and by *C. pepo* to 0.0231- 0.0286, a condition that reduces soil adsorption and increases bioavailability and therefore bioaccumulation is enhanced in the plant [21]. Other data showed that the phytoextraction capacity of plants is related to the capacity of inorganic uptake from soil. Fertilizer amendment with N and P enhances phytoextraction and increases bioaccumulation in *C. Pepo* [23].

Bioaccumulation of lipophilic substances such as chlordane is quite different between plant species, as observed by zucchini with a low and pumpkin with a high uptake [24]. These lipophilic substances are normally bioaccumulated in roots and only a small amount is translocated in a decreasing sequence to stems, leaves and fruits [21, 24, 25]. When K_{ow} is higher than 5, plant uptake is considered to occur mainly via the air-to-plant route [26]. These data agree with the observations of Schnoor et al. [27] that plant uptake is very efficient for moderate hydrophobic organic chemicals with a K_{ow} of 0.5 to 3. For a K_{ow} higher than 3 these chemicals bind more and more strongly to the surface roots with decreasing translocation within the plant. However, translocation of chemicals such as terbuthylazine, with a K_{ow} of 3, and atrazine can occur in high amounts [28, 29, 30]. Fairly soluble chemicals with a K_{ow} lower than 0.5 are not sufficiently sorbed to roots and are not actively transported through plant membranes.

Soil amendment with manure compost may reduce bioavailability by retaining the toxic organic chemicals in the organic matter and therefore reduce the hazardous effects [31] but the literature shows controversy data in which organic amendment can reduce adsorption of pesticides by increasing the desorption effects [32]. The increasing addition of sludge as the final disposal on soil introduces POPs in much higher amounts than air deposition [33].

Nevertheless lipophilic substances with a high octanol-water partition coefficient (log K_{ow}) remain preferentially in soils and with little bioavailability they have low bioaccumulation in earthworms [8]. Radioactive atrazine applied on soil with low organic matter content previously covered by cattle manure, showed a slower leachate speed compared to control but with a low retention capacity in the soil [34]. Soils modulate adsorptions and bioavailability and an inverse correlation occurs with a decrease of bioaccumulation in earthworms when K_{ow} increases, which is different from an aqueous environment when there is a positive correlation between K_{ow} and bioaccumulation [6].

2.1.4. Cation bioaccumulation

Many evidences indicate that the lipophilic character of soil organic matter is one of the most important factors for **preferential partition** of lipophilic substances toward soil; nevertheless this process can be carried out by chemical bonds such as ionic charged bonds of organic toxic chemicals. Below is a description of how **preferential partition** works between the soil and the living biosphere with cationic charged molecules. As far as the author know bioaccumulation in terms of cationic charged molecules between the soil and biosphere has not been reported in the literature before.

Based on the hypothesis that polarity is the main factor controlling bioaccumulation, one would expect that a strongly polar pollutant, like paraquat, would be accumulated in the most polar parts of the soil. Thus in soils with low organic matter, one would expect no bioaccumulation in earthworms and paraquat would be bound within the soil matrix. Nevertheless, Papini and Andrea[14] found the opposite. **Preferential partition** moved this compound towards earthworms, depending on the amount of ^{14}C-paraquat (1.2; 12 and 120 μg. a.i.g^{-1}) applied to the soil [20]. To understand these results we have to note the importance of the relative abundance of charge-binding sites (attractive/exchange sites) in the biosphere and soil. In soils with low OM and low ion exchange capacities, the exchange sites, which sequester positive charged pesticides that gradually become saturated and consequently makes possible simultaneous available pesticides to bind anionic sites on or in earthworms [35]. This is a competition between the earthworms and attractive soil sites for paraquat. In this low SOM soil, paraquat bioaccumulation did not surpass Bcf 5 probably because the anionic sites on the surface of the earthworm cells were limited and already saturated (Figure 3).

In a high OM soil and with a low application of paraquat (1.2 μg a.i.g^{-1}), the bioaccumulation factor (Bcf.) was 1.1 and increased with higher concentrations of applied paraquat up to Bcf. 3.8. With increasing paraquat concentrations (12 to 120 μg a.i.g^{-1}) one would predict that as the soil charge sites gradually became saturated then gradually more paraquat would become attracted to the earthworms. From these experiments, we noted the predominance of electrostatic binding in the soil and the importance of an abundance of exchange sites. In comparison, lipophilic attraction is driven by affinity without limits of concentrations which are different from the electrostatic bonds were the charged sites involve higher bond energy and therefore are predominant but have quantitative limitations with pesticides up to saturation with consequences in bioaccumulation. The correlation between K_{oc} (soil organic matter) with K_{ow}

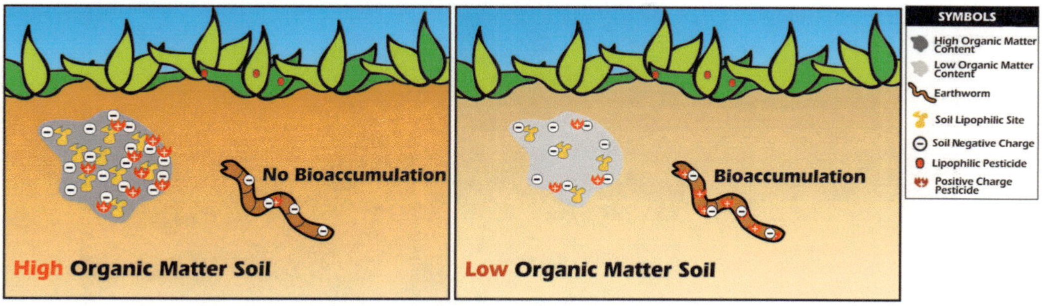

Figure 3. In preferential partition charged pesticides are attracted to anions between the abundance in soil and biosphere.

(molecular polarity) and availability of cation exchange sites (electrostatic bonds) in the soil shows that the **Preferential Partition** concept can explain pesticide bioaccumulation.

Ecotoxicology depends on soil organic matter. When the SOM is high the ecotoxicological effects are low and when the SOM is low the effects are high. In this latter case the residues remain bio-available and are intensely absorbed in the biosphere and therefore are hazardous to flora and fauna. Polarity and/or anion charge capacity of living organisms compete with the parameters in the soil organic matter. The highest attraction capacity of these different compartments defines the way in which molecules move in the environment. Molecules inserted in the soil matrix by "**Preferential Partition**" can be protected from hydrolysis, oxidation/reduction reactions as well escape enzymatic action.

2.2. Microbial constraints of biodegradation

All natural molecules can be biodegraded in the environment. An important constraint of xenobiotic biodegradation is the absence of microorganisms with efficient biodegradation capacity in a specific environment. Recently there has been an intense research effort to develop or transfer microbial efficient biodegradation genes to a microorganism that is adapted to a specific environment but lacks biodegradation capacity. Other constraints can occur when microorganisms do not enter in small soil microspores and therefore could not be present to promote biodegradation (Figure 4).

3. Movement of persistent molecules in the environment

The link between bioavailable persistent substances and the spread of these substances in the environment, causing dissemination of hazardous effects, will be discussed in this section. Desorbed or non sorbed molecules are bio-available and can move into the food chain and can also get into ground/surface water or reach the atmosphere through volatilization and are thus be randomly disseminated in the biosphere (Figure 5) [36]. When persistent molecules are adsorbed in the soil they cause less hazardous effects than when they are bioavailable.

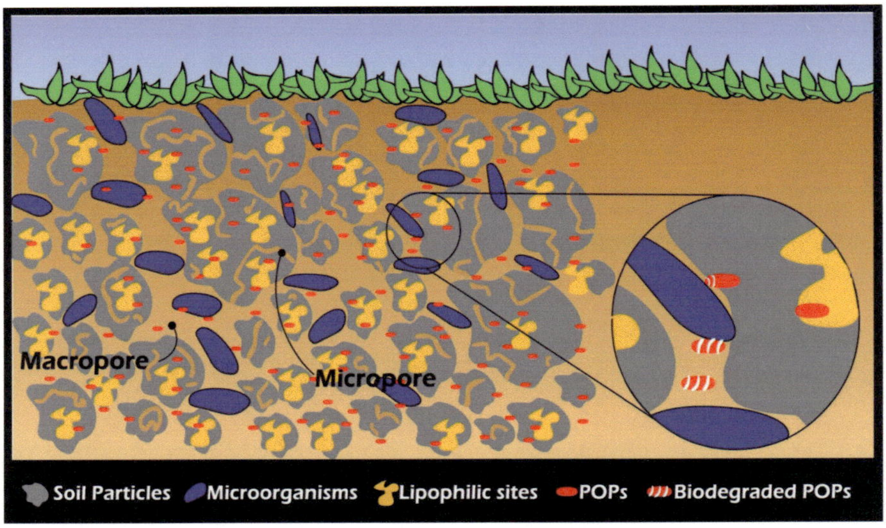

Figure 4. Restriction access of microorganism in the soil by micropores.

Biodegradable substances, differently from persistent molecules, cannot survive for a long time in the environment. The problem is not only the actual pollution of pesticide applications, which is strongly dependent on the climatic conditions [37, 38] but also the composts with the organic wastes containing POPs that are added to soils [39] as well the inadequate forms in which POPs used to be stored [40, 41].

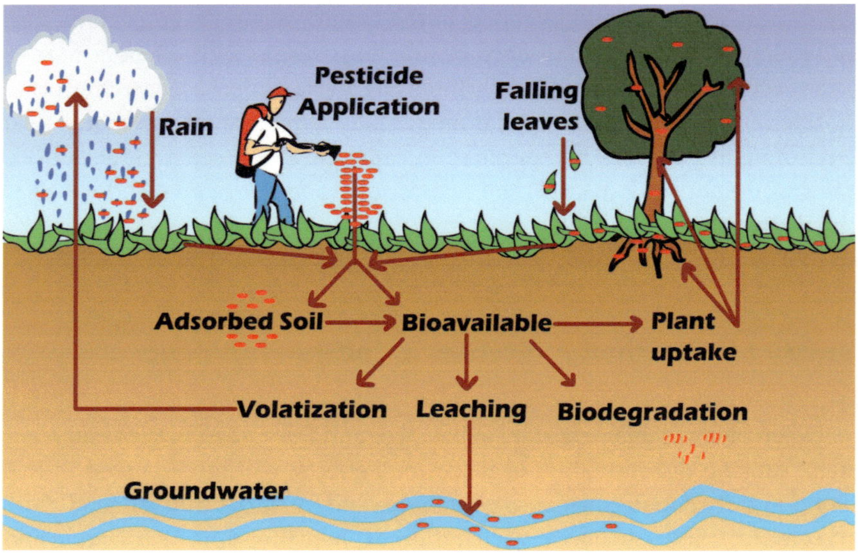

Figure 5. Overview of POP dispersion in the environment of air, water and biosphere.

3.1. Movement from soil to water

The vertical displacement of substances in the soil can leach from the surface through the unsaturated zone into deeper layers down to ground water [42]. It is well-known that the soil infiltration capacity depends on soil texture characteristics, porosity, and humidity [43]. Soils with a sandy texture are more susceptible to the process of leaching while clay soils have greater pesticide adsorption potential. Leaching can occur in an irregular manner known as preferential flow [44]. This is due to irregularities in the soil that facilitate an easy path for water and pollutants at specific points. These paths can be formed by soil cracks, worm trails, rotten roots, termites, and other biological activities [45]. Soil in which surface layer rich in organic matter was removed by flattened to built roads or courtyard enhance strongly the leaching process [34]. Water consumption from wells exposed to pesticide applications can be very dangerous for humans and cattle. Generally, ground water in the dry season drains to rivers and in the raining period receives water from rivers and pollutants follow these processes [46, 47].

Surface water contamination with persistent molecules occurs not only by the drainage processes from ground water but also strongly increases with runoff [48]. When the quantity of rainfall exceeds the soil infiltration capacity, superficial runoffs occur [49]. Runoffs begin with small lamina flows that build up until they form turbulent flows which cause pesticide losses and even gully erosion. This process has been shown to be an important source of pollution in surface waters [50]. Another source of pollution is the discharge of effluents from waste water treatment plants to surface waters [51]. Pollutant molecules can be distributed by rivers and affect different living forms in the water as well as the fauna that use it for drinking. Persistent molecules contaminating these waters are also spread to agriculture products through irrigation [51].

3.2. Movement in the air

The sources of POP in the air are pesticide applications, oil combustion, industries and indoors at home. During the application of pesticides a part remains in the air due to drift. Another very important source is volatilization of pesticides from soil or plant surfaces [52]. Pesticide dissipation occurs when the product is being sprayed. Droplets are aerially dispersed and reach non-target areas [53]. The spray drift of pesticides is influenced by the size of the pulverized droplets, application pressure, distance of the pulverizer nozzle to the target and the velocity of the wind during application [54]. By strong volatilization the relative loss of the applied amount can be very high [55].

These molecules contaminating the air are transported by air currents and redistributed globally, polluting all kinds of ecosystems [56]. Today we may find many such molecules in the Antarctic and in obscure ocean islands [57, 36]. This pollution affects man as well as many animal species [58]. Besides the adsorption capacity that is able to immobilize these molecules, another important variable is the molecular vapor pressure which is influenced by temperature – an increase of 10°C makes these molecules about four times more volatile [59]. This process is much greater in the tropics with higher temperatures than in temperate regions of the world.

Also these losses can affect people, domestic animals, pollinating insects as well as contaminating lakes, rivers, and fish or other aquatic plants and animals. Air pollution is quite difficult to control. Trees and forests can filter the pollution that is spread by horizontal wind but not by rainfall [60]. Air pollutants can move from rural areas to towns and can be present everywhere. This distribution is determined by the wind and physical barriers such as mountains etc. The challenge is how to reduce this process as much as possible considering that these molecules have been lost for the purpose that they were intended and non target flora and fauna are not able to escape from their hazardous effects efficiently.

Air pollution can also begin indoors and then move outdoors. Polybrominated diphenyl ethers (PBDEs) and phthalates among others are some examples. Brominated compounds make up an important group of flame retardants. Some of these products, such as Penta-BDE, seem to show toxicological effects at very low concentrations [61]. PBDEs are structurally similar to thyroid hormones acting as endocrine disruptors via alterations in thyroid hormone homeostasis [62] and demonstrate neurodevelopmental effects [64]. The most sensitive populations are likely to be pregnant women, developing fetuses, and infants [65].

In recent years, PBDEs have been recognized as significant pollutants of the indoor environment [66]. These additives are mixed into plastics and foams but do not form chemical bonds. This makes them much more likely to leach out of goods and products by volatilization. PBDEs are lipophilic compounds [67], and when released into the environment bioaccumulation can occur in living organisms and this is followed by biomagnification in the food chain. Bioaccumulation in wildlife has been reported in numerous studies, even in places with no local sources or industrial production [68] like the Arctic. Due to their high production volume, widespread usage, and environmental persistence, PBDEs have become ubiquitous contaminants in environmental media, biota and humans [69]. As their levels are rapidly increasing in the environment, these chemicals have evolved from 'emerging contaminants' to globally-distributed organic pollutants [70].

3.3. Movement in the food chain

Movement of persistent molecules in the biosphere toward soil microorganisms, mesofauna and plant uptake will be discussed in this section. Here some aspects of the way in which persistent molecules can be transferred from one life form to other will be emphasized, considering that the overall food chain sequence processes are well known [71].

3.3.1. Cell bioaccumulation

How do persistent molecules move into cells in higher concentrations than the surrounding environment [72]? "**Preferential partition**" describes the "choice" of the substance, due to affinity, between cell membranes or components of the environment. As mentioned earlier affinity is characterized mainly by lipophilicity and charge binding. With the photo-binding technique it is possible to identify in which cell component the molecule will preferentially bind. This technique preserves the binding when only weak bonds between the pollutant and cell are available. These weak bonds are not resistant to the harsh conditions of cell fractionation. The method establishes covalent bonds by UV irradiation only between pollutants containing aromatic structures and the cell component where weak bonding occurs and this covalent form of bonding can resist cell fractionation. Studies with the gram negative soil bacteria *Azospirillum lipoferum*, have shown that dicofol, which is easily hydrolyzed [73] turns persistence by becoming imbedded in the membrane [74]. In this case dicofol was found in the neutral lipids of membranes [75] where it influences the microdomain of membrane bound enzymes, as also occurs with other molecules [76]. This characterizes a non specific mode of action as seen with other molecules that interfere with different membrane bound enzymes such as ATPase [77] and nitrogenase [78].

3.3.2. Ecotoxicological effects in soil living organisms

The key message is that ecotoxicological effects are not only defined by the intrinsic chemical nature of each compound but also by the interaction of their effects properties within a given environment. Applying the concept of **"Preferential Partition"** helps us to understand conflicting reports about bioaccumulation and the ecotoxicological effects in soils. Bioaccumulation is the natural process that gradually concentrates non toxic levels of pollutants into toxic levels within a biota causing unpleasant side effects [72].

Different side effects have been described for the same pesticide by distinct authors. For example in the nitrogen cycle, some authors reported an inhibition effect of a pesticide while others reported, for the same molecule, an increasing or no effect on the ammonification, nitrification and nitrogen fixation processes. This depends mainly on the differences in the soil used and concentrations applied [79, 80, 77]. **"Preferential Partition"** explains why substances remain in the soil matrix, when they are likely to be bioaccumulated in living forms or move through the soil to pollute water resources or are suitable for biodegradation.

Non soil bound residues are an essential condition for biodegradation. Persistence occurs mainly when low numbers of or no biodegrading microorganisms are present or most commonly when despite the presence of biodegradation microorganisms, the residues are entrapped and not bioavailable.

Hazardous organic substances that are strongly adsorbed in soil organic matter pose less risk than those with low adsorbance. Sufficient soil organic matter reduces bioavailability and decreases bioaccumulation in the living biosphere but in turn results in persistence with long term pollution in soils [81, 74]. The balance between adsorbed and bioavailable molecules determines the ecotoxicity levels [2] and this is determined by the physic-chemical characteristics of the soil as well as the pollutant molecule. This balance is changed by the weathered pollutants in the soil that have a lower bioavailability than the recently applied products [1]. Therefore it is a pitfall to conclude that the hazardous effects are only caused by pesticide concentrations in the soil.

3.3.3. Plant uptake and food chain up to fauna

One of the most important sources of pollution in the food chain is soil contaminated with pesticides and other POPs mainly introduced by air deposition. Another way is the direct surface uptake mainly in leaves during application but also by plant uptake powered by plant evapotranspiration in which persistent molecules dissolved in soil water are moved from roots to shoots. From the roots the pesticides move by translocation to stems and then often a strong bioaccumulation occurs in the leaves [29, 30] or fruits. Crops where pesticides are used intensively are consumed by cattle, humans or wild life. A strong increase the of concentrations of these molecules can occur in this process called biomagnifications as described above (Figure 6).

Figure 6. Human contamination with pesticide moved by food chain.

In order to minimize ecotoxicity we need to restrict the inappropriate use of pesticides and thereby remove them from the food chain and water reserves. **Preferential Partition** presents a concept - tool which is able to estimate what happens to pesticides in crop and fodder plants. In soils with high organic matter content, lipophilic and or charged pesticides are retained in the soil organic matter and the uptake into plants decreases [11, 82, 83]. The reverse occurs in low organic matter sandy soils with low cation exchange capacity CEC. Light textured sandy soils do not adsorb and retain in high amount hazardous products and will both bioaccumulate in living tissues and pollute water resources. In soils of this type, plants strongly adsorb pesticides resulting in enhanced contamination with subsequent phytotoxicity and toxicological effects on fauna [84]. This soil type would be a worst case scenario adequate to be the bench mark for pesticide registration. The pesticides that could be used under these conditions can be used generally in all soils. Registration based on strongly adsorbing soils needs to be avoided because it camouflages the eco-toxicological effects that occur when the pesticide is applied on less adsorbent soils.

3.3.4. Food chain up to fauna

The food chain sequence is where persistent molecules are first adsorbed by plants which are then consumed by animals that are in turn consumed by predatory species. In the first scenario, the animal serves as a filter with systemic eliminations and the concentration of a xenobiotic that remains present in the consumable product is much lower than the original concentration

http://dx.doi.org/10.5772/56418

in a feed material. For meat (including adjunctive fatty tissue), this would mean that in many cases human exposure is very low or even negligible for almost all compounds. The second scenario addresses the accumulation of chemical residues in animal tissues such as liver and kidney with notorious examples including the accumulation of heavy metals, melamine and other nephrotoxins in the kidney, and dioxins and polyhalogenated persistent environmental pollutants (POPs) in the liver and fatty tissues [71]. Persistent molecules increase in each ingestion with the uptake of higher concentrations along the food chain that can result in very poisonous and hazardous concentrations for humans (Figure 6), cattle and wild life. The worst bioaccumulation seems to be salmon produced on fish farms [85].

The metabolic pathways of digestion in diverse animal species are different. Strict herbivorous species digest soluble fibers in their large caeca (the equivalent of the rumen) with the help of complex microbiological flora. This microbial fermentation may result in the release of bound (plant) toxic products which are often conjugated to sugars (glucosides) in plants and in this form are biologically inactive (or less active) [71]. The morphological and physiological characteristics of the gastro-intestinal tract in farm animal species largely determine the rate of absorption of a contaminant. The internal dose of a toxicant that reaches the post-hepatic circulation, the rate of excretion and subsequently the potential carry over and cumulative potential in tissues need to be considered. The most significant diversity in species is observed at the level of hepatic biotransformation expressed by cytochrome P450 enzymes and their polymorphisms [86]. Diversity can also be observed in all phase of enzymes such as glutathione-s-transferases, glycine conjugates, UDP-glucuronosyltransferases that are expressed in the intestinal wall and in the liver [87].

The choice of toxicity endpoints may differ between animal health and human risk assessments. Hence, clinical reports from intoxications in different animals when available may partly provide a basis for hazard characterization whereas other important endpoints maybe applied to human risk assessment as an extrapolation of the dose response from laboratory animal species to humans or using the dose response provided by human epidemiological data [88]. This is typically illustrated with genotoxicity and carcinogenicity data, which are prominent endpoints in human hazard characterization, but which are not commonly used in farm animal health risk assessment considering their relatively short lifespan [71]. The quantitative transfer of a substance from feed to an animal-derived product is commonly expressed as the carry-over rate and serves as a basis to establish maximum tolerable limits (MLs) in animal feed and animal-derived foods.

3.3.5. Chronic toxicology for animals and humans

Bioaccumulation is the natural process that gradually concentrates non toxic levels of pollutants into toxic levels within a biota causing unpleasant side effects [72, 5]. Bioaccumulation involves silent natural processes which stealthily and inevitably affect all biota. Today hazardous substances are ubiquitous, albeit mainly in low concentrations, and for the majority of us they are an unconscious part of our daily diet. The regular intake of sub toxic levels of persistent pollutants can gradually bioaccumulate up to toxic levels and after time produce chronic effects which today are recognized and understood as such.

More recently low concentrations have been shown to cause effects such as endocrine disruptors (EDs) quite different from those observed by high concentrations [89]. EDs are recognized as being partly responsible for a global reduction in male fertility [90], sexual abnormalities [91]), adrenal function disorders [92], human obesity [93] and other metabolic disturbances not to mention similar and worse effects on other living organisms [94, 95]. Recently new evidences have shown the effects of POPs (mainly pesticides) on metabolic disturbances related to obesity [96], insulin resistance promoting diabetes 2 [97] and the association of organochlorine with vitamin D deficiency [98].

The bioaccumulation of POPs by mothers, over many years, exposures the fetus to these pollutants during pregnancy [65]. Despite the difficulty of experimentation with POPs in humans, follicles could be analyzed when "induced reproduction" occurs and results have shown that contamination in the oocyte follicle can reduce human embryo quality with consequences for future generations caused by the harmful effects of these hazardous chemicals [89, 99].

Clearly, the threat to the environment and human health is reduced when pesticides are not bioavailable and cannot concentrate in the food chain. Food grown in soils where pesticides are strongly adsorbed has lower health risks.

3.3.6. Cell mortality and redistribution of persistent molecules

The presence of persistent molecules in different forms of life is temporary limited by plant senescence and/or microbial/fauna death. Two possible routes can occur; one following the food chain as described above and the other by a rotting process in which recalcitrant molecules can be released. In plants the rotting process of wood and straw [100] is a complex process in which the twisted lignin/hemicellulose/cellulose fibers follow a cascade where the first step is the "loosing" of the cell wall. This occurs by non enzymatic peroxidation (Fenton reaction) [101] that changes the lignin structure and reduces the barrier effect of the cell wall allowing the second step of enzyme diffusion of ligninase peroxidase, Mn peroxidase and laccase into areas where polysaccharides can be hydrolyzed [102]. Laccase is a phenol oxidase that has the ability to degrade many persistent xenobiotics [103, 104]. Other lignolytic enzymes could be released and are also able to biodegrade many persistent molecules [105, 106].

Nevertheless in this rotting process many recalcitrant substances are not biodegraded or only partially and therefore these molecules follow the normal distribution: adsorption in the soil, free moving molecules that can be removed by runoff, leaching, volatilization and uptake in plants, mesofauna or soil microorganisms [48, 37].

4. How to control hazardous effects of persistent molecules

The rules that enhance molecular persistence and the parameters that make the wide spread distribution of POPs possible are driven by natural factors, out of human control. Nevertheless recently remediation technology has been developed and in many cases has shown good results to remove pollutants. The approach in this text is to point out different possibilities with reference to some excellent reviews. The technology used was based on physical, chemical and biological methods.

The first step of POP remediation is to promote desorption [107]. Physical methods need to remove the pollutants from soils or water in order to destroy the molecules completely by burning [107]. This has a high cost as the incinerator plant needs filter systems to eliminate other pollutants such as dioxins. Physical methods such as electro kinetic (EK) remediation technology use a low-level direct current as the "cleaning agent", inducing several transport mechanisms (electro-osmosis, electro-migration and electro-phoresis) and electrochemical reactions (electrolysis and electrodeposition). This technology has already proven its value, especially in contaminated fine-grain soils [108].

Ozon can be used for chemical remediation of pollutants and there are other methods currently being researched [109]. Chemical degradation reactions in the environment depend on water content, pH, temperature and oxygen reduction potential. These conditions are normally not optimum in nature and therefore degradation is frequently low. Physical and chemical remediations are powerful technologies to remediate small areas that have been polluted by industry or transport accidents. For large areas these process are generally considered too expensive.

Biological degradation processes are promoted by living organisms such as animals, plants but mainly microorganisms. The advantages of microbes are the ubiquitous distribution in normal and extreme environments, fast biomass growth, easy manipulation and high diversity of catabolic enzymes. Bioremediation using microorganisms compared to the other methodologies can be used in polluted soils and waters and is less expensive than all other methods. Nevertheless the high costs are limitations for large polluted agricultural areas. One of the most powerful approaches is phytoremediation as some plants have a strong capacity to extract pollutants from the soil and incorporate these molecules into their own biomass [29, 30]. Nevertheless in phytoremediation of hydrocarbon pollutants (HCs) the accumulation between roots and shoots is quite low compared to plant-promoted biodegradation in the rhizosphere [110]. Root exudation stimulates microorganisms in the rhizosphere resulting in enhanced mineralization [111].

Today efforts are being made to identify the most adequate alternative remediation in a framework of possibilities considering parameters that characterize cost-risk tradeoffs and uncertainty impacts [112].

5. Conclusions

Two points related to persistent molecules need to be emphasized. The first is the limits of biodegradation of persistent molecules as reported here in which the best microorganism to degrade a given molecule can not be efficient if the molecule is entrapped in the soil. The second point is that there is no technology available to clean the environment considering the large dispersion of persistent substances.

The relationship between molecular characteristics and environmental conditions that determine persistence as described here can be helpful to understand what happens to other polluting molecules like non biodegradable plastic materials, bioactive products discharged by sewage treatment plants and nano-molecules. Many of these molecules get into the environment on a global scale and in fact this is due to a lack of regulatory rules such as those used for the registration process of pesticides. Little is known about what happens to many of these new molecules in the environment.

As shown here, to clean the environment of persistent products that were applied over large areas has limited possibilities. Prevention by only using biodegradable products is the best approach. However, the solution for the sustainable use of chemicals in the environment is to ban persistent molecules and substitute them for others with a high biodegradation capacity and less non-target effects.

Acknowledgements

The author thanks Prof. Andrew Macrae for his valuable contributions to the text. Financial support was provided by PRONEX – FAPERJ/CNPq and fellowship CNPq.

Author details

Tomaz Langenbach*

Address all correspondence to: tomazlange@yahoo.com.br

Department of Civil Engineering, PUC – Rio, Rio de Janeiro, Brazil

References

[1] Pignatello J.J., Frin, R., Mari, P.A. Drost, E.X. Field-observed ethylene dibromide in an aquifer after two decades. Journal of Contaminat Hydrology 1990;5 195-214.

[2] Vasseur P., Bonnard M., Palais F., Eom I. C., Morel J. L. Bioavailability of Chemical Pollutants in Contaminated Soils and Pitfalls of Chemical Analyses in Hazard Assessment. Environmental Toxicology 2008;23 652-656.

[3] Wang X., Guo X., Yang Y., Tao S., Xing B. Sorption mechanisms of phenanthrene, lindane, and atrazine with various humic acid fractions from a single soil sample. Environ Sci Technol. 2011;45(6) 2124-2130.

[4] Dörfler U., Scheunert I., Viswanathan R., Zsolnay A. Physikalisch- Chemisches Verhalten von s-Triazinen im Boden. In: Dörfler U., Scheunert I., Viswanathan R., Zsol-

nay A. (eds.) Standortgerechte Bewertung von s-Triazinen in Böden. GSF: Bericht 22/97; 1997 p99-123.

[5] Connell D.W. General Characteristics of Organic Compounds which Exhibit Bioaccumulation. In: Connell, D.W. (ed.) Bioaccumulation of xenobiontic Compounds. CRC Press; 1990. p47-57.

[6] Saarikoski J., Viluksela M. Relations between physichochemical properties of phenols and their toxicity and accumulation in fish. Ecotoxicology Environmental Safety 1983;6 501.

[7] Randall R.C., Lee I., Ozretich R.J., Lake J.L., Pruel RJ. Evaluation of selected lipid methods for normalizing pollutant bioaccumulation. Environmental Toxicology Chemistry 1991;19 1431-1436.

[8] Matscheko N., Tysklind M., De Wit C., Bergek S., Anderson R., Sellström U. Application of sewage sludge to arable land-soil concentrations of polybrominated diphenyl ether and polychlorinated dibenzo-o-dioxin, dibenzofurans, and biphenyls and their accumulation in earthworms. Environmental Toxicology Chemistry 2002;21 2515-2525.

[9] Cornelissen G., Gustavson O., Bucheli T.D., Jonker M.T.O., Koelmans A.A. Van Noort P.C.M. Extensive sorption of organic compounds to black carbon, coal, and Kerogen in sediments and soils: mechanisms and consequences for distribution, bioaccumulation, and biodegradation. Environmental Science and Technology 2005;39 6881-6895.

[10] Brunninger B.M., Mano D.M.S., Scheunert I. Langenbach T. Mobility of the Organochlorine Compound Dicofol in Soil Promoted by Pseudomonas fluorescens. Ecotoxicology and Environmental Safety 1999;44 154-159.

[11] Scheunert I. Aufnahme von s-triazinen in Planzen. In Dörfler U., Scheunert I., Viswanathan R., Zsolnay A. (eds.) Standortgerechte Bewertung von s-triazinen in Böden. GSF: Bericht; 1997(222) p173-183.

[12] Steen W.C., Karickhoff S.W. Biosorption of hydrophobic organic pollutants by mixed microbial populations. Chemosphere 1981;10 27-32.

[13] Karickoff S.W. Semi-empirical estimation of sorption of hydrophobic pollutants on a natural sediments and soils. Chemosphere 1981;10 833-846.

[14] Papini S., Andrea M.M. Ação de minhocas Eisinia foetida sobre a dissipação dos herbicidas simazina e paraquat aplicados no solo. Revista Brasileira de Ciências dos Solos 2004;28 67-73.

[15] Dörfler U. Anwendung, Eigenshaften und Wirkungen der s-Triazinen. In: Dörfler U., Scheunert I., Viswanathan R., Zsolnay A. (eds.) Standortgerechte Bewertung von s-triazinen in Böden. GSF: Bericht 22/97; 1997 p2-19.

[16] Perkow W. Wirksubstanzen der Pflanzenschutz und Schädlings-bekämpfungsmittel 2. In: Perkow (ed.) Verlag Paul Parey. 2 Auflage; 1988.

[17] Langenbach T., Inácio M.V.S., Aquino A.M., Brunninger B. Influência da minhoca Pontoscolex corethrurus na distribuição do acaricida dicofol em um Argissolo. Pesquisa Agropecuária Brasileira 2002; 37 1663-1668.

[18] Baker J.R., Mihelcic J.R., Shea, E. Estimating K_{oc} for persistent organic pollutants: limitations of correlations with K_{oc}. Chemosphere 2000;41 813-817.

[19] Andréa M.M., Papini S. Influence of soil properties on bioaccumulation of ^{14}C-simazine in earthworms Eisenia foetida. Journal of Environmental Science and Health Part B 2005;40 55-58

[20] Papini S., Langenbach T., Luchini L.C., Andréa M.M. Influence of substrate on bioaccumulation of ^{14}C-paraquat in compost earthworms Eisenia foetida. Journal of Environmental Science and Health 2006; 41(B) 523-530.

[21] Kelsey J. W., Slizovskiy I. B., Petriello M. C., Butler K. L. Influence of plant–earthworm interactions on SOM chemistry and p,p0-DDE bioaccumulation. Chemosphere 2011;83 897–902.

[22] Mangrich A.S., Lobo M.A., Tank C.B., Wypch F., Toledo E.B.S., Guimarães E. Criterious preparation and characterization of earthworms-compost in view of animal waste recycling. Part 1. Correlation between chemical, thermal and FTIR spectroscopic analysis of four humic acids from earthworm-composted animal manure. Journal of Brazilian Chemistry Society 2000;11 164-169.

[23] White J.C., Parrish D., Isleyen M., Gent M.P., Ianucci-Berger W., Eitzer B.D., Mattina M. I. Influence of nutrient amendments on the phytoextraction of weathered 2, 2-Bis(p-chlorophenyl)-1,1-dichloroethylene by cucurbits. Environmental Toxicology and Chemistry 2005;24 987–994.

[24] Mattina M. I., White J. C., Eitzer B. D., Iannucci-Berger W. Cycling of weathered chlordane residues in the environment: Compositional and chiral profiles in contiguous soil, vegetation, and air compartments. Environmental Toxicology and Chemistry 2002;21 281–288.

[25] Mattina M. I., Eitzer B. D., Iannucci-Berger W., Lee W. Y., White J. C. Plant uptake and translocation of higly wethered, soil-bound technical Chlordane residues: data from field and rhizotron studies. Environmental Toxicology and Chemistry 2004;23 2756-2762.

[26] Fries G. F. Transport of organic environmental contaminants to animal products. In: Ware G. W. (ed.). Reviews of environmental contamination and toxicology. Springer Verlag 1995; 41 p71-109.

[27] Schnoor J. L., Licht L. A., McCutcheon S.C., Wolfe N.L., Carreira L. H. Phytoremediation of organic and Nutrient contaminants. Environmental Science and Technology 1995;29 318-323.

[28] Schroll R., Bierling B., Cao G., Dörfler U., Lahaniati M., Langenbach T., Scheunert I., Winkler. Uptake pathways of organic chemicals from soil by agricultural plants. Chemosphere 1994;28 297-303.

[29] Marinho D.A., Bicalho S.T., Ferreira E.M., Langenbach T. Distribution and Mineralization of 14C-Hexazinone in Soil-Plant Microcosm with the Riparian Forest Specie Cecropia Hololeuca. Special Issue Article J Bioremed Biodegrad 2011; S1-002

[30] Bicalho S.T.T., Langenbach T. Distribution of the herbicide atrazine in a microcosm with riparian forest plants. Journal of Environmntal Science and Health Part. B 2012;47(6) 505-511.

[31] Jiang L., Ma L., Sui Y., Han S. Q., Wu Z. Y., Feng Y. X., Yang H. Effect of manure compost on the herbicide prometryne bioavailability to wheat plants. Journal of Hazardous Materials 2010;184 337-344.

[32] Celis R., Barriuso E., Houot S. Sorption and desorption of atrazine by sludge-amended soil: dissolved organic matter effects. Journal of Environmental Quality 1998;27 1348-1356.

[33] Sellström U., De Witt C. A., Lundgren N., Tysklind M. Effect of sewage –sludge applications on concentrations of higher-brominated diphenyl ethers in soils and earthworms. Environmental Science and Technology 2005;39 9064-9070.

[34] Langenbach T., Correia F.V., Macrae A., Vargas E.A., Campos T.M.P. Atrazine leaching through surface and subsurface of a tropical Oxisol. Journal of Environment Science and Health 2008;(B) 43 214-218.

[35] Rytwo G., Nir S., Margulies L.. A Model for Adsorption of Divalent organic Cations to Montmorillonite. Journal of Colloidal and Interface Science 1996;181 551-560.

[36] Weber J., Halsall C.J., Muir D., Teixeira C., Small J., Solomon K., Hermanson M., Hung H., Bidleman T. Endosulfan, a global pesticide: a review of its fate in the environment and occurrence in the Arctic. Sci Total Environ. 2010;408(15) 2966-84.

[37] Langenbach R., Schroll I., Sheunert. Fate of herbicide 14C-terbuthylazine in Braxzilian soils under various climatic conditions. Chemosphere 2001; 45 387-398

[38] Kookana R., Holz G., Barnes C., Bubb K., Fremlin R., Boardman B. Impact of climatic and soil conditions on environmental fate of atrazine used under plantation forestry in Australia. J Environ Manage Dec 2010;91(12) 2649-56

[39] Cañero A.I., Becerra D., Cornejo J., Hermosín M., Albarrán A., Piñeiro A., Cox, L. Transformation of organic wastes in soil: Effect on bentazone behavior. Science of the Total Environment 2012; 433 198–205

[40] Östereicher-Cunha, P., Langenbach, T., Torres, J.P., Lima, A.L.C., Campos, T.M.P., Vargas E.A.Jr. and Wagener, A.R.. HCH distribution and microbial parameters after liming of a heavily contaminated soil in Rio de Janeiro. Environment. Res 2003;93 316-327.

[41] Dvorská M. Šír, Honzajková Z., Komprda J., Čupr P., Petrlík J., Anakhasyan E., L. Simonyan, Kubal M. Obsolete pesticide storage sites and their POP release into the environment—an Armenian case study. Environ Sci Pollut Res. 2012;19 1944–1952.

[42] Jørgensen L.F., Kjær J. a, Olsen P. b, Rosenbom A. Leaching of azoxystrobin and its degradation product R234886 from Danish agricultural field sites. Chemosphere 2012;88 554–562.

[43] Correia F.V., Langenbach T., Campos T.M. Avaliação do transporte de atrazina em solos sob diferentes condições de manejo agrícola. R. Bras. Ci. Solo 2010;34 525-534.

[44] Vryzas Z., Papadakis E. K., Papadopoulou-Mourkidou E. Leaching of BrL, metolachlor, alachlor, atrazine, deethylatrazine and deisopropylatrazine in clayey vadoze zone: A field scale experiment in north-east Greece. Water Res. 2012;46 1979-1989.

[45] Reichenberger S.; Amelung W.; Laabs V.; Pinto A.; Totsche K.U.; Zech W. Pesticide displacement along preferential flow pathways in a Brazilian Oxisol. Geoderma 2002;110 63-86.

[46] Winter, T. C., Harvey, J.W., Franke O.L.Alley W.M... Ground Water and Surface Water. A Single Resource. U.S. Geological Survey. 1998 Circular 1139. Ed. U.S. Governmment Printing Office.

[47] Belamie R., Gouy V. Introduction des pollutants dans le milieu fluvial. Influence du ruissellement des sols. Oceanis 1992;18 505-521.

[48] Carluer N.; Gouy V.E., Grill J.N. Contamination des eaux de surface par dês produits phyosanitaires et modelisation. Ingenieries-EAT 1996;6 19-30.

[49] Correia F.V., Macrae A., Guilherme L.R.G., Langenbach, T. Atrazine sorption and fate in a Ultisol from humic tropical Brazil. Chemosphere. 2007; 67 847-854.

[50] Sangchan W., Hugenschmidt C., Ingwersen J., Schwadorf K., Thavornyutikarn P., Pansombat K., Streck T. Short-term dynamics of pesticide concentrations and loads in a river of an agricultural watershed in the outer tropics. Agriculture, Ecosystems and Environment 2012;158 1– 14.

[51] Fenet H., Mathieu O., Mahjoub O., Zhi L., Buys D., Casellas C., Gomez E. Carbamazepine, carbamazepine epoxide and hydroxycarbamazepine sorption to soil and occurrence in a wastewater reuse site in Tunisia. Chemosphere 2012;88 49–54.

[52] Rudel H. Volatilization of pesticides from soil and plant surfaces. Chemosphere 1997;35 143–152.

[53] Ganzelmeier H., Rautmann D., Spangenberg R., Streloke M., Herrmann M., Wenzelburger H.J. Studies on the spray drift of plant protection products. Mitteilungen aus

der Biologischen Bundesanstalt fur Land-und Forstwirtschaft. Berlin-Dahlem: Heft; 1995; 305 111.

[54] Arvidsson T. Spray drift as influenced by meteorological and technical factors. A methodological study. Swedish University of Agricultural. Sciences, Acta Universitatis.Agriculturae Sueciae; 1997. 71. 144.

[55] Willis G., Mcdowell L., Harper L., Southwick L., Smith S. Seasonal disappearance and volatilization of toxaphene and DDT from a cotton field. J. Environ. Qual. 1983;12 80–85.

[56] Lohmann R., Breivik K., Dachs J., Muir D. Global fate of POPs: Current and future research directions. Environ. Pollut. 2007;150(1) 150-165.

[57] Taylor A. Post-application volatilization of pesticides under field conditions. J. Air Pollut. Con. Ass. 1978;28 922–927.

[58] Jasma J., Linders J. Volatilization of pesticides from soil and plants after spraying, Report no. 679102030. National Institute of Public Health and Environmental Protection Bilthoven: The Netherlands; 1995.

[59] Lyman W.J., Reehl W.F., Rosenblatt D.H. Handbook of Chemical Property Estimation Methods. American Chemical Society: Washington, DC. 1990.

[60] Mast MA, Foreman WT, Skaates SV. Current-use pesticides and organochlorine compounds in precipitation and lake sediment from two high-elevation national parks in the Western United States. Arch Environ Contam Toxicol. 2007;52 294-305.

[61] Darnerud, P.O., Eriksen, G.S., Jóhannesson, T., Larsen, P.B., Viluksela, M. Polybrominated diphenyl ethers: occurrence, dietary exposure, and toxicology. Environmental Health Perspectives 2001;109 49-68.

[62] Tseng, L.H., Li, M.H., Tsai, S.S., Lee, C.W., Pan, M.H., Yao, W.J., Hsu, P.C. Developmental exposure to decabromodiphenyl ether (PBDE 209): effects on thyroid hormone and hepatic enzyme activity in male mouse offspring. Chemosphere 2008;70 640-647.

[63] Meeker J.D., Johnson P.I., Camann D., Hauser R. Polybrominated diphenylether (PBDE) concentrations in house dust are related to hormone levels in men. Science of the Total Environment 2009;407 3425-3429.

[64] Caserta D., Mantovani A., Marci R., Fazi A., Ciardo F., La Rocca C., Maranghi F., Moscarini M. Environment and women's reproductive health. Human Reproduction Update, 2011;17 418–433.

[65] McDonald, T.A. A perspective on the potential health risks of PBDEs. Chemosphere 2002;46 745e755.

[66] Besis A., Samara C. Polybrominated diphenyl ethers (PBDEs) in the indoor and outdoor environments and a review on occurrence and human exposure. Env. Pol. 2012;169 217-229.

[67] Li L., Xie S., Cai H., Bai X., Xue Z. Quantitative structure property relationships for octanol water partition coefficients of polybrominated diphenyl ethers. Chemosphere 2008;72 1602-1606.

[68] Beyer W.N., Heinz G.H. Redmon-Nordwood, editor. Environmental Contaminants in Wildlife. In CRC Lewis Publisher; 1996.

[69] Wang Y.W., Jiang G.B., Lam P.K.S., Li A. Polybrominated diphenyl ether in the East Asian environment: a critical review. Environment International 2007;33 963-973.

[70] Ikonomou M.G., Rayne S., Addison R.F. Exponential increases of brominated flame retardants, polybrominated diphenyl ethers, in the Canadian Arctic from 1981 to 2000. Environ. Sc. and Technology 2002;36 1886-1892.

[71] Dorne J.L.C.M., Fink-Gremmels J. Human and animal health risk assessments of chemicals in the food chain: Comparative aspects and future perspectives. Toxicology and Applied Pharmacology (in press) 2012.

[72] Langenbach T. Bioaccumulation. In: Pimentel D. (ed.). Encyclopedia of Pest Management. Marcel Dekker Inc.; 2002. p50-53.

[73] Walsh P.R., Hites R.A. Dicofol solubility and hydrolysis in water. Bull. Environ. Contam. Toxicol. 1979;22 305-315.

[74] Mano D.M.S., Buff K., Clausen E., Langenbach T. Bioaccumulation and enhanced persistence of the acaricide dicofol by azospirillum lipoferum. Chemosphere 1996;33 1609-1619.

[75] Mano D.M.S., Langenbach T. [14C]dicofol Association to cellular components of Azospirillum lipoferum. Pest. Sci. 1998;53 91-95.

[76] Buff K. & Bernd J. Interaction of DDT (1,1,1-trichloro2,2-bisp-chlorophenyl)-ethane) with liposomnal lipids. Bioch. Biop. Actga (BBA) 1980;643 205-12.

[77] Lal R., Shivaji S. Effects of pesticides on asymbiotic nitrogen fixation. In: Lal R., Lal S. (eds). Pesticide and Nitrogen Cycle III. CRC Press; 1988. p1-46.

[78] Subramanian G. The effect of pesticides on nitrogen fixation and ammonia excretion bu Anabaena. In: national symposium biology of nitrogen fixation. Iari. New Delhi: 1982. p567.

[79] Rao V.V.S.N. Effects of pesticides on ammonification. In: Lal R., Lal S. (eds.) Pesticide and Nitrogen Cycle II. CRC Press; 1988. p1-41.

[80] Dhanaraj P.S. Effects of pesticides on nitrification and denitrification. In: Lal R., Lal S. (eds). Pesticide and Nitrogen Cycle II. CRC Press. 1988; p43-118.

[81] Afferden M.V., Beyer M., Klein J. Significance of bioavailability for the microbial remediation of PAH-contaminated soils. Preprints. International Symposium Soil Decontamination Using Biological Processes 1992 605-610.

[82] Bell R.M., Sferra P.R., Ryan J.R., Vitello, M.P. Studies of organic Pollutant uptake by plants. In: Wolf W., Van den Brink J., Colon F.J. (Eds.) Contaminated soil '88. K. Kluwer Academic Publisher; 1988 1. p451-458.

[83] Trapp S., Mathies M., Scheunert I., Topp E. M. Modelling the bioconcentration of organic chemicals in plants. Environmental Science and Technology 1990;24 1246-1252.

[84] Covaci F. V., D'Havé A., Van den Brink H., Blust N. W., De Coen R. W., Bervoets L. Accumulation of background levels of persistent organochlorine and organobromine pollutants through the earthworm-hedgehog food chain. Environment International 2010;36, 721-727.

[85] EFSA (European Food Safety Authority). Opinion of the Scientific Committee on arequest from EFSA related to a harmonised approach for risk assessment ofsubstances which are both genotoxic and carcinogenic. The EFSA Journal 2005; 282(a) 1–31. (http://www.efsa.europa.eu/EFSA/Scientific_Opinion/sc_op_ej282_gentox_en3.pdf).

[86] Fink-Gremmels J. Implications of hepatic cytochrome P450-related biotransformation processes in veterinary sciences. Eur. J. Pharmacol 2008;585 502–509.

[87] Walton K., Dorne J.L., Renwick A.G., Uncertainty factors for chemical risk assessment: interspecies differences in glucuronidation. Food Chem. Toxicol. 2001;39 1175–1190.

[88] Schlatter J. Toxicity data relevant for hazard characterization. Toxicol. Lett. 2004;153 83–89.

[89] Petro Evi M.L., Leroy, J. L.M.R., Covaci, A., Fransen, E., Neubourg, D., Dirtu, A. C., Pauw, I. and Bols, P.E.J. Endocrine-disrupting chemicals in human follicular fluid impair in vitro oocyte developmental competence. Human Reproduction 1988;27 1025-1033.

[90] Saradha B., Mathur P.P. Effect of environmental contaminants on male reproduction. Enviromental Toxicology and Pharmacology 2006;21, 34-41.

[91] Whitehead S.A., Rice S. Endocrine-disrupting chemicals as modulators of sex steroid synthesis. Best Practice and Research: Clinical Endocrinology and Metabolism 2006; 20 45-61.

[92] Hinson J.P., Raven P.W. Effects of endocrine-disrupting chemicals on adrenal function. Best Practice and Research Clinical Endocrinology and Metabolism 2006; 20 111-120.

[93] Baillie-Hamilton P.M.B. Chemical Toxins: A hypothesis to explain the global Obesity. The J. Alternative and Complementary Medicine. 2002;82 185-192.

[94] Elango A., Shepeherd B., Chen T.C. Effects of endocrine disrupters on the expression of growth hormone and prolactin mRNA in the rainbow trout pituitary. General and Comparative Endocrinology 2006;145 116-127.

[95] 95] Walker A. N., Bush P., Puritz J., Wilson T., Chang E.S., Miller T., Holloway K., Horst M.N. Bioaccumulation and metabolic effects of the endocrine disruptor methoprene in the Lobster, Homarus americanus. Integrative and Comparative Biology 2005;45 118–126.

[96] Arsenescu V., Arsenescu R.I., King V., Swanson H., Cassis L.A. Polychlorinated biphenyl-77 induces adipocyte differentiation and proinflammatory adipokines and promotes obesity and therosclerosis. Environ. Health Perspect. 2008;116 761–768.

[97] Ruzzin J. Public health concern behind the exposure to persistent organic pollutants and the risk of metabolic diseases. Public Health 2012;12 298-306.

[98] Yang J.H., Lee Y.M., Bae S.G., Jacobs D.R. Jr., Lee D.H. Associations between Organochlorine Pesticides and Vitamin D Deficiency in the U.S. Population. Plos ONE 2012;7 30093

[99] Anway M.D., Cupp A.S., Uzumcu M., Skinner M.K. Epigenetic transgenerational actions of endocrine disruptors and mate fertility. Science 2005;308 1466–1469.

[100] Gutierrez A., Reconret, Cadena E. M., Rico A., Barth D., Río J. C., Martínez A.T. Demonstration of laccase-based removal of lignin from wood and non-wood; plant feedstocks. Bioresource Technology 2012;119 114–122.

[101] Hammel K.E., Kapich A.N., Jensen K.A. Jr., Ryan Z.C. Reactive oxygen species as agents of wood decay by fungi. Enzyme Microb Technol 2002;30 445–453.

[102] Arantes V., Jellison J., B. Goodell. MINI-REVIEW. Peculiarities of brown-rot fungi and biochemical Fenton reaction with regard to their potential as a model for bioprocessing biomass. Appl Microbiol Biotechnol 2012; 94:323–338.

[103] Thurston C.F.The structure and function of fungal laccases. Microbiology 1994;140 19–26.

[104] Baldrian P. Wood-inhabiting ligninolytic basidiomycetes in soils: ecology and constraints for applicability in bioremediation. Fungal Ecol 2008;1 4–12.

[105] Austin S.D., Benson J.T. The fungus among us: use of white rot fungus to biodegrade environmental pollutants. Environmental health perspectives 1993;101 232-233.

[106] Sinsabaugh R.L. Phenol oxidase, peroxidase and organic matter dynamics of soil. Soil Biol. Biochem. 2010;42 391–404.

[107] Tong M., Yuan S. Physiochemical technologies for HCB remediation and disposal: A review. J. of Hazardous Materials 2012;229–230 1– 14.

[108] Gomes H. I., Dias-Ferreira C., Ribeiro A. B. Review Electrokinetic remediation of organochlorines in soil: Enhancement techniques and integration with other remediation technologies. Chemosphere 2012;87 1077-1090.

[109] Ma J.W., Wang F.Y., Huang Z.H., Wang H. Simultaneous removal of 2,4- dichlorophenol and Cd from soils by electrokinetic remediation combined with activated bamboo charcoal. J. Hazard. Mater. 2010;176 715–720.

[110] Khan S., Afzal M., Iqbal S., Khan Q. M. Review: Plant–bacteria partnerships for the remediation of hydrocarbon contaminated soils. Chemosphere 2012; in press.

[111] Martínez Bueno M.J., Gomez, M.J, Herrera, S., Hernando, M.D., Agüera, A., Fernández-Alba, A.R. Occurrence and persistence of organic emerging contaminants and priority pollutants in five sewage treatment plants of Spain: Two years pilot survey monitoring. Env. Pol. 2012;164 267-273.

[112] Yang A.L., Huang G.H., Qin X.S., Fan Y.R. Evaluation of remedial options for a benzene-contaminated site through a simulation-based fuzzy-MCDA approach. J. Hazardous Material 2012;213-214 421-433.

Chapter 14

Rhizoremediation: A Promising Rhizosphere Technology

Keshav Prasad Shukla, Shivesh Sharma,
Nand Kumar Singh, Vasudha Singh,
Sandeep Bisht and Vivek Kumar

Additional information is available at the end of the chapter

http://dx.doi.org/10.5772/56905

1. Introduction

An increasingly urban population and industrialized global economy over the last century have serious consequences on the environment. Understanding the sources, pathways and contaminants in the urban environment is essential for making informed management decisions. Urban areas are major concentrators, repositories and emitters of a myriad of chemicals because of the wide range and intensity of human and anthropogenic activities. Common contaminants include petroleum hydrocarbons (PHCs), polycyclic aromatic hydrocarbons (PAHs), halogenated hydrocarbons, pesticides, solvents, metals, salt and the resulting stresses on human and ecosystem health are well documented [1]. Polycyclic aromatic hydrocarbons are a class of complex organic chemicals consisting of over hundred different organic compounds. PAHs are unique contaminants in the environment because they are generated continuously by incomplete combustion of organic matter, for instance in forest fires, home heating, traffic, and waste incineration [2]. PAHs are hydrophobic compounds and their persistence in the environment is chiefly due to their low water solubility [3]. Generally, solubility of PAHs decreases and hydrophobicity increases with an increase in number of fused benzene rings. In addition, volatility decreases with an increasing number of fused rings [4]. The major source of PAHs is from the combustion of organic material [5]. PAHs are formed naturally during thermal geologic production and during burning of vegetation in forest and bush fires [6]. PAHs and their alkyl homologous may also be derived from biogenic precursors during early diagnosis [7]. However, anthropogenic sources, particularly from fuel combustion, pyrolytic processes, spillage of petroleum products, waste incinerators and domestic heaters [8] are significant sources of PAHs in the environment. At depth 90-135 cm, only phenanthrene (1.4 mg/kg), pyrene (4.0 mg/kg), chrysene (0.9 mg/kg) and dibenzoanthracene (0.8 mg/kg) were found [9]. The concentration of PAHs in the environment varies widely, depending on the level of industrial development, proximity of the contaminated sites to the production source and the mode of PAH transport. Kanaly and Harayama [10] reported that in soil and sediment, PAHs concentrations vary from 1μg/kg to over 300 g/kg.

PAHs have been detected in a wide variety of environmental samples, including air [8], soil [11], sediments [19], water [12], oils, tars [13] and foodstuffs [14]. PAHs contamination on industrial sites is commonly associated with spills and leaks from storage tanks and with the conveyance, processing, use and disposal of these fuel/oil products [4]. PAHs are also a major constituent of creosote (approximately 85% PAH by weight) and anthracene oil, which are commonly used pesticides for wood treatment [15]. The main route for PAH transport is through the atmosphere. Results from ambient air monitoring programs have shown that PAH concentrations are usually of the order of a few nano-grams per cubic metre of air [16]. However, PAH concentrations may vary from season to season depending on emissions arising from the combustion of home heating products. Motor vehicles, including spark emission and diesel automobiles, trucks and buses, also contribute to atmospheric PAHs pollution through exhaust condensate and particulates, tyre particles and lubricating oils and greases [17]. During the combustion of fossil fuels, diesel powered vehicles are the major sources of lighter PAHs to the atmosphere, whereas gasoline vehicles are the dominant source of higher molecular weight PAHs [18]. The persistence of PAHs in the environment depends on the physical and chemical characteristics of the PAHs. PAHs are degraded by photo-oxidation and chemical oxidation [19], but biological transformation is probably the prevailing route of PAH loss [20]. The microbial metabolism of PAHs containing up to three rings (naphthalene, phenanthrene, anthracene, fluorene) has been studied extensively.

Heavy metals such as lead, mercury, and cadmium are ranked second, third, and seventh, respectively, on the 2003 Comprehensive Environmental Response, Compensation, and Liability Act (CERCLA, commonly known as Superfund) priority list for hazardous substances because they are toxic widespread pollutants. Soil contamination is a particularly serious environmental concern, as the majority of superfund sites are highly contaminated with heavy metals [21].

Therefore, for remediation of contaminated environmental soil, the traditional technologies routinely used are as excavation, transport to specialized landfills, incineration, stabilization and followed by coagulation filtration or ion exchange are expensive and disruptive to the sites [22]. However, there has been much interest in bioremediation technologies which use plants and microorganisms (including bacteria) to degrade toxic contaminants in environmental soil into less toxic and/or non-toxic substances [23]. With *in situ* techniques, the soil and associated ground water is treated in place without excavation, while it is excavated prior to treatment with *ex situ* applications [24].

Biosorption (using microbially produced metallothioneins (MTs) and phytochelatins (PCs) having heavy metal binding affinities) and immobilization are major mechanisms utilized by animals and plants to limit the concentrations of internal reactive metal species [25].

Phytoremediation, the use of plants to degrade toxic contaminants in the environment involves a number of processes including phytoextraction, phytotransformation, phytostabilization, phytovolatilization and rhizofiltration [26]. Phytoextraction (or phytoaccumulation) involves the uptake and concentration of pollutants into harvestable biomass for sequestration or incineration. Phytotransformation involves enzymatic modification resulting in inactivation, degradation (phytodegradation), or immobilization (phytostabilization) of pollutants. Phytovolatilization involves the removal of pollutants from soil and their release through leaves via evapotranspiration processes and rhizofiltration involves the filtering of water through a mass of roots to remove pollutants. While some success has been reported using plants alone in bioremediation, the use of plants in conjunction with plant associated bacteria offers much potential for rhizoremediation [27].

The importance of plant microbe partnerships in the remediation of organic contaminants was confirmed in studies at the level of rhizosphere [28, 29], the phyllosphere and inside the plant [30, 31]. Rhizoremediation is considered as the most potential approach for PAHS remediation in soil [37]. Soil microflora play vitally important role during rhizoremediation of xenobiotics [32]. The interaction among microbial degrader, plant and PAHs in soil might be regulated through rhizosphere processes [39].

Rhizoremediation systems for PAHs rely on a synergistic relationship between suitable plants and their root associated microbial communities [32]. Degradation is facilitated through a rhizosphere effect where plants exude organic compounds through their roots and thereby increase the density and activity of potential hydrocarbon degrading microorganisms in the zone, surrounding the roots [35]. The biodegradation abilities of bacteria and the expression and maintenance of bacteria in the rhizosphere are extremely important for the effective removal of contaminants in rhizoremediation [36]. Thus, bioremediation, phytoremediation and rhizoremediation contribute significantly to the fate of hazardous waste and can be used to remove these unwanted compounds from the biosphere [37, 38].

Rhizoremediation is proposed as the most potential approach for PAHs remediation in soil [39]. Plant associated bacteria, such as rhizospheric bacteria have been shown to contribute to biodegradation of toxic organic compounds in contaminated soil and could have potential for improving phytoremediation [40]. The amalgamation of the activity of plant roots and rhizospheric microbial communities like secretion of root exudates (various organic acids and amino acids etc.), production of siderophores, HCN, phytoharmone and phosphatases by plant growth promoting rhizobacteria (PGPR) are also effective for ecorestoration of polluted sites [41]. The plant growth-promoting capability of *B. aryabhattai* strains may be utilized as an environmentally friendly means of revegetating barren lands [41]. The valuable effects of some rhizobacteria on plant growth are well known, and the so called PGPR have been utilized for several decades, although their mechanisms of plant growth promotion have not been completely elucidated [42]. Some of the important mechanisms include direct phytohormonal action, increase of plant nutrient availability and the enhancement of other plant beneficial microorganisms [43]. When a suitable rhizospheric isolated strain is introduced together with a suitable plant, it inhabits on the root along with indigenous population, thereby enhancing the bioremediation process [44]. In addition, such capability for root colonizing, pollutant degrading bacteria utilize the growing root system and hence this acts as an injection system to spread the bacteria through soil [45].Plant root performs certain specialized roles, including the ability to synthesize, accumulate and secrete a diverse array of nutrient compound consequently no requirement of exogenous carbon source, roots may regulate the soil microbial community in their immediate vicinity, cope with herbivores, encourage beneficial symbioses, change the chemical and physical properties of the soil and inhibit the growth of competing plant species [46]. PGP bacteria may facilitate plant growth either directly or indirectly [47]. Though the rhizoremediation process takes place naturally, but it can be modified by premeditated exploitation of the well-equipped rhizospheric microorganisms whereas it can be proficient by using suitable plant microbe pairs.

Incorporation of plant and PGPR having the pollutant degrading activity may be performed. Similarly, Kuiper *et al.* [48] described the pair of a grass species with a naphthalene degrading microbe which protected the grass seed from the toxic effects of naphthalene and the growing roots exploited with the naphthalene degrading bacteria into soil.

Previously, several researchers have also used this symbiotic relationship of plant and microbes for degradation of hazardous and xenobiotic compounds like PCBs, PAHs and TCE [49]. Mechanical injection of contaminated sites with pollutant degrading bacteria has been used to clean polluted sites in an inexpensive and less labor intensive way than the removal and/or combustion of polluted soils [50].

2. Bioremediation

Polycyclic aromatic hydrocarbons (PAHs) are of particular concern because of their toxic, mutagenic and carcinogenic properties [51]. There is thus a chief interest in studying microorganisms present in contaminated environments as a means for bioremediation. The fate of PAHs and other organic contaminants in the environment is associated with both abiotic and biotic processes, including volatilization, photooxidation, chemical oxidation, bioaccumulation and microbial transformation. Microbial activity has been deemed the most influential and significant cause of PAHs removal [3, 12]. PAHs may also be degraded by some microorganisms in the soil [52]. The term bioremediation refers to the use of living organisms to degrade environmental pollutants [53]. Bioremediation is generally considered to include natural attenuation, biostimulation or bioaugmentation, the deliberate addition of natural or engineered microorganisms to accelerate the desired catalytic capabilities. According to the Environmental Protection Agency in the United States [54], natural attenuation processes may reduce contaminant mass (through destructive processes such as biodegradation and chemical transformations), reduce contaminant concentrations (through simple dilution or dispersion). Eventually, even the contaminants bound to the soil particles gets biodegraded by the bacterial species present in the environment.

3. Bioaugmentation

Bioaugmentation is the introduction of microorganisms with specific catabolic abilities into the contaminated environment in order to supplement the indigenous population and to speed up or enable the degradation of pollutants [48, 55]

Bioaugmentation has proven successful for remediation of PAHs in sediments with poor or lacking intrinsic degradation potential [17], while other studies demonstrated that bioaugmentation did not enhance biodegradation significantly compared to natural attenuation [56]. One of the main problems in applying bioaugmentation is to ensure the survival and activity of the introduced organisms in the environment [55]. Bioaugmentation can be inhibited by a variety of factors including pH and redox, the presence of toxic contaminants, concentration and bioavailability of contaminants or the absence of key co-substrates [48]. However, the key factor for the success of bioaugmentation process is the selection of the appropriate bacterial strain. When selecting the strain for augmentation purposes, the kind of microbial communities present in the source habitat should be considered [57]. Bioaugmentation strategies may prove successful especially in the remediation of manmade contaminants, where specialized bacteria with the appropriate catabolic pathways may not be present in the contaminated habitat [55].

4. Phytoremediation

Some workers quantified and compared the responses of soil microbial communities during the phytoremediation of PAHs in a laboratory trial [15]. A recent publication of some workers describes the development of transgenic poplars (*Populus* sp.) over expressing a mammalian cytochrome P450, a family of enzymes commonly involved in the metabolism of toxic compounds. The engineered plants showed enhanced performance about the metabolism of trichloroethylene and the removal of a range of other toxic volatile organic pollutants, including vinyl chloride, carbon tetrachloride, chloroform and benzene. Some workers suggested that transgenic plants might be able to contribute to the wider and safer application of phytoremediation [58]. Widespread phytoremediation field trials research was performed *in vitro* condition and many of the works explored the effects of plants on removal of contaminants from spiked soil and soil excavated from contaminated sites [7] and most of these experiments provided valuable insights into the specific mechanisms of phytoremediation of organic contaminants [29]. Previously, numerous organic pollutants such as TCE (trichloroethylene), herbicides such as atrazine, explosives such as TNT (trinitrotoluene), PHC, BTEX (mono aromatic hydrocarbons) and PAHs, the fuel additive MTBE (methyl tertiary butyl ether), and PCBs (polychlorinated biphenyls) have been successfully phytoremediated. [59] Major advantages of phytoremediation *viz.*, cost of the phytoremediation is lower than that of traditional processes both *in situ* and *ex situ*, plants can be easily monitored, possibility of the recovery and re-use of valuable products, use of naturally occurring organisms and preservation the natural state of the environment, low cost of phytoremediation (up to 1000 times cheaper than excavation and reburial) [60].

5. Rhizoremediation

Rhizospheric microbes can degrade the majority of environmental pollutants and degradation process stops when the microbe is deprived of food. These microbes have access to the best food source available in soil, namely root exudates [61]. Researchers have described an enrichment method for the isolation of microbes [62], which combine the properties ofdegradation of a selected pollutant and excellent root colonization. They have termed this process 'rhizoremediation' instead of phytoremediation to emphasize the roles of the root exudates and the rhizosphere competent microbe. The high concentration of metals in soils and their uptake by plants harmfully influence the growth, symbiosis and consequently the yields of crops [63] by disintegrating cell organelles and disrupting the membranes, acting as genotoxic substance [64] disrupting the physiological process, such as, photosynthesis or by inactivating the respiration, protein synthesis and carbohydrate metabolism. *Pseudomonas putida* is a root colonizer of potential interest for the rhizoremediation of pollutants and the biological control of pests [65]. According to hypothesis when a suitable rhizosphere strain is inoculated together with a suitable plant (e.g., coating bacteria on plant seed), these well-equipped bacteria might settle on the root together with the normal indigenous population, thereby enhancing the bioremediation process. Pioneer work about degradation of compounds in the rhizosphere was mainly focused for herbicides and pesticides [66]. In the past two decades, a large number of publications on rhizodegradation of various organic toxicants using different plants and/or microbial inoculants have been published [37, 67-71].

Field contaminated soils that have undergone prolonged periods of ageing generally appear to be much less responsive to rhizodegradation than freshly spiked soil [72-74]. Wenzel [67] concluded that low bioavailability is a main cause of failure of rhizodegradation in field contaminated and aged spiked soils. This has important implications for the applicability of rhizodegradation as well as for the evaluation of data obtained on freshly or only shortly aged, spiked soil material. Other strategies to enhance rhizodegradation (e.g. inoculation of degrader strains) are likely to fail where low bioavailability is the main constraint.

Interestingly, microbial treatments appeared to be successful at the laboratory experiments [75] but failed when applied to long term contaminated soil on field experiments [76]. This indicates again the importance of the experimental scale and of bioavailability. In view of the still disappointing and controversial results of traditional inoculation [77], enhanced rhizodegradation requires more sophisticated approaches. Enhanced degradation capabilities of inoculated microorganisms may be obtained by induction of a nutritional bias towards the inoculated strains. Only recently, Narasimhan *et al.* [78] identified root exudate compounds (phenylpropanoids) that created a nutritional bias in favour of enhanced PCB degradation. A successful rhizoremediation process could depend on the highly branched root system of the plant species where a large number of bacteria harbor, establishment of primary and secondary metabolism, survival and ecological interactions with other organisms [48]. Plant roots can act as a substitute for the tilling of soil to incorporate additives (nutrients) and to improve aeration in soil [48]. Plants also release a variety of photosynthesis derived organic compounds (root exudates), which might help in degradation of pollutants [59]. The root exudates consists of water soluble, insoluble, and volatile compounds including sugars, alcohols, amino acids, proteins, organic acids, nucleotides, flavonones, phenolic compounds and certain enzymes [68].

Normally a symbiotic relationship develops between plant and soil microbes in the rhizosphere, where plants provide nutrients necessary for the microbes to flourish, while the microbes provide a healthier soil environment where plant roots can grow. Specifically, plants loosen soil and transport oxygen and water into the rhizosphere [26]. In addition, plants exude specific phytochemicals (sugars, alcohols, carbohydrates, etc.) that are primary sources of food (carbon) for the specific soil organisms that aid in providing the healthier soil environment [79]. Alternatively, the exuded phytochemical may be an allelopathic agent meant to suppress other plants from growing in the same soil [26]. In return for exporting these phytochemicals, plants are protected from competition, soil pathogens, toxins and other chemicals that are naturally present or would otherwise be growing in the soil environment [26]. Microbial populations can be several orders of magnitude higher in a vegetated soil compared to an un-vegetated soil. Rhizodegradation, sometimes called phytostimulation, rhizosphere biodegradation or plant assisted bioremediation/degradation, is the enhanced breakdown of a contaminant by increasing the bioactivity using the plant rhizosphere environment to stimulate the microbial populations [26].

6. Factor influencing PAH degradation

Several factors that influence the rate of rhizoremediation of PAHs in soil e.g. soil type, texture, particle size, nutrients and organic matter content which can limit the bioavailability of pollutants [67]. The conditions that increase the possibility of degradation include the presence of low molecular weight PAHs, relatively recent PAHs emission or deposition, moderate soil pH, the presence of appropriate PAHs degrading bacteria and plants to facilitate decomposition by virtue of large root surface area or uptake affinity [80]. Root microbe interactions are considered the primary process of PAHs phytoremediation [81]. Natural attenuation in vegetated settings is thought to degrade one, two and three chain PAHs in periods ranging from16 to 126 days [82]. By-products of degradation are thought to be less toxic and may serve as an energy source for other soil organisms. Research suggests that PAHs with fewer benzene rings are more easily digested by soil microbes. Johnson *et al.* [32] suggests that microbial degradation of PAHs and other hydrophobic substrates is believed to be limited by the amounts dissolved in the water phase, with sorbed, crystalline and non-aqueous phase liquid dissolved PAHs being unavailable to PAH degrading organisms.

The main problem for soil bioremediation is the bioavailability of the pollutant. Most of organic pollutants are highly hydrophobic compounds that dissolve poorly in water and many of them can form complexes with soil particle, this lack of bioavailability often lowers removal efficiencies [26]. Bioavailability is a dynamic process, determined by the rate of substrate mass transfer to microbial cells relative to their intrinsic catabolic activity [26].

7. Bioavailability

Bioavailability refers to the fraction of a chemical that can be taken up or transformed by living organisms from the surrounding bio-influenced zone where organism mediated biochemical changes occur [83, 84]. The success of any rhizoremediation process depends on the bioavailability of the specific pollutant and root microbial modifications of their solubility, physio-chemical properties of the pollutant, soil properties, environmental conditions, biological activity and chemical speciation in the rhizosphere [59, 67]. Bio surfactants increase the bioavailability of hydrocarbons resulting in enhanced growth and degradation of contaminants by hydrocarbon degrading bacteria present in polluted soil [85].

The important pollutant properties controlling their fate in the environment include the vapour pressure and the Henry's constant [67]. The vapour pressure indicates whether or not a pollutant is easily volatilised in dry soil conditions, the Henry's constant provides a better measure of the volatilisation potential in wet and flooded soil. As the residence time in soil of highly volatile compounds such as chloroethene will be short, they are not a primary target of rhizodegradation. The solubility of a pollutant is further modified by soil properties. Organic matter quality and content, clay content, mineral composition, type of mineral surface, pH and redox potential are known as important controls of organic pollutant solubility, with hydrophobic, nonpolar organic matter being of particular importance for binding organic pollutants [86]. Binding of organic pollutants to the soil matrix is known to progress as the contact time increases, rendering pollutants less bioavailable [67]. This phenomenon is known as "ageing" and is attributed to sorption onto minerals and organic matter in soil, and subsequent interparticle diffusion in minerals and entrapment within humic complexes, nano- and microspores [83, 86].

Apart from the absorption capability of the organisms (biology), the bioavailability of a pollutant in soil not only depends on its solubility (chemistry), but also its diffusion and mass transport (physics) towards the sites and niches where degrader populations are abundant [83]. It is well established that bioavailability is one of the most limiting factors in bioremediation of persistent organic pollutants in soil [37, 86]. In bioreactor systems this problem is often addressed by agitation and mixing and addition of surfactants [34]. In recent past several microbes have been reported to be chemotactic towards different organic pollutants, for example toluene acting as chemoattractant to *Pseudomonas putida* [87]. Chemotactic bacteria might be more competent for bioremediation than their non-chemotactic counterparts [87].

8. Biodegradation of PAH

Many bacterial, fungal and algal strains have been shown to degrade a wide variety of PAHs [88]. The most commonly reported bacterial species include *Acinetobacter calcoaceticus, Alcaligens denitrificans, Mycobacterium* sp., *Pseudomonas putida, Pseudomonas fluorescens, Pseudomonas vesicularis, Pseudomonas cepacia, Rhodococcus* sp., *Corynebacterium renale, Moraxella* sp., *Bacillus cereus, Beijerinckia* sp., *Micrococcus* sp., *Pseudomonas paucimobilis* and *Sphingomonas* sp. [89].

Pseudomonas putida is a good candidate for metabolic engineering and genetic manipulation applications for expression of genes encoding several degradative enzymes [88]. Therefore, a *P. putida* strain was engineered to increase the efficiency of degradation of naphthalene and salicylate [88]. Similarly, another study demonstrated the efficiency of the naphthalene degradation process performed by different microbial strains of the genera *Pseudomonas* and *Burkholderia* in soil model systems [90]. Previous, studies have shown that bacteria can degrade BaP when grown on an alternative carbon source in liquid culture experiments [91]. Sorkhoh*et al.* [93] isolated spore forming PAHs degrading bacteria and reported their subsequent genetic studies of their degradation pathways which may lead to the discovery of novel genes involved. Various studies showed the significance of the rhizospheric effect on degradation of organic pollutant molecules [94].

In recent past, Bisht *et al.* [44] reported four bacteria from non-contaminated rhizospere of *P. deltoides* which were able to degrade 80-90% degradation of anthracene and naphthalene within 6 days. The maximum degradation pathways were reported from *Mycobacterium vanbaalenii* because of its exceptional ability to degrade a great variety of low and high molecular weight PAHs oxidatively in soil versatility of this species makes it probable inoculants in the remediation of PAHs contaminated sites [95].

Calotropis sp. as a dominant and common desert plants that grows widely in warm and urbanizing regions and has a high capacity for taking heavy metals into its tissues due to their abilities to absorb and tolerate heavy metals [96]. The use of the leaf biomass of *Calotropis procera* can be employed as good bio-sorbent for the removal of Cr (III) from aqueous solutions and as an alternative method of their removal from industrial effluent [97].

9. Rate of PAH Biodegradation

The rate and degree of PCB degradation decreases with the increase of chlorination degree. For example, 62% 2-Cl-PCBs, 28% 3-Cl-PCBs, 24% 4-Cl-PCBs, and 18% 5-Cl-PCBs were degraded during the two-month active treatment phase [98]. The reversibly sorbed PCBs will be bio stabilized within 5, 6, 12 and 15 years, respectively, during the passive phase. The rate of biodegradation of PAHs is highly erratic and is dependent not only on PAH structure, but also on the physicochemical parameters of the site as well as the number and types of microorganism present. The rate and degree of PAH degradation decreases with the increase of number of benzene ring. PAHs sorb to organic matter in solid and sediments, and the rate of their sorption strongly controls the rate of which microorganism can degrade the pollutant. Much of the current PAH research focuses on techniques to enhance the bioability and therefore, the degradation raters of PAHs at polluted site. The sequential active passive biotreatment approach is an effective scheme for degradation of both PAHs and PCBs in the land treatment systems. The quantitative model, together with laboratory and field testing, can be a useful tool for the plan, design and operation of similar land treatment systems.

10. Microbial enzymes involved in PAHs degradation process

As bacteria initiate PAHs degradation by the action of intracellular dioxygenases, oxygenase, dehydrogenase, phosphatases, dehalogenases, nitrilases, nitroreductases and lignolytic enzymes [33] (Table 1). Aromatic ring dioxygenases are multicomponent enzymes which consist of an electron transport chain containing a ferredoxin and a reductase and a terminal dioxygenase [99]. The best studied PAHs dioxygenase is naphthalene dioxygenase from *Pseudomonas putida* encoded by the NAH plasmid pDTG1 [100].

Enzyme	Target pollutant
Aromatic dehalogenase	Chlorinated aromatics (DDT, PCBs etc.)
Carboxyl esterases	Xenobiotics
Cytochrome P450	Xenobiotics (PCBs)
Dehalogenase	Chlorinated solvents and Ethylene
Glutathione s-transferase	Xenobiotics
Peroxygenases	Xenobiotics
Peroxidases	Xenobiotics
Laccase	Oxidative step in degradation of explosives
N-glucosyl transferases	Xenobiotics
Nitrilase	Herbicides
Nitroreductase	Explosives (RDX and TNT)
N-malonyl transferases	Xenobiotics
O-demethylase	Alachlor, metalachor
O-glucosyl transferases	Xenobiotics
O-malonyl transferases	Xenobiotics
Peroxdase	Phenols
Phosphatase	Organophosphates

Table 1. Some important enzymes associated with bioremediation [26].

11. Improvement in rhizoremediation

Rhizoremediation process can be designed to improve in several aspects like bioavailability of contaminant molecules, expression and maintenance of genetically engineered plantmicrobial systems and root exudates for the effectiveness of the process.

Selection of bacteria, which are able to produce biosurfactants in the rhizosphere of the plants, is an interesting alternative to improve the removal efficiency [85]. In this context Kuiper *et al.* [48] identified bacteria growing in a PAHs contaminated area that produces biosurfactants that facilitate the solubilisation of PAHs and hence biodegradation by microbes. This property is also of interest because a number of biodegradative microbes exhibit positive chemotaxis towards the pollutants [44]. Therefore, the combined action of biosurfactant and chemotaxis can contribute to bacterial proliferation and to microbial spread in polluted soils, in order that more ample zones can be cleaned [28].

Microbial degradation of contaminants in the rhizosphere provides a positive effect for the plant; the pollutant concentration is decreased in the area near the roots and the plant can grow better than those in contaminated areas [101]. Because of this mutual benefit it has been proposed that plants can select specific genotypes to be present in their roots. Experiments performed by Siciliano *et al.* [102] demonstrated that the presence of the alkane monooxygenase gene was more prevalent in endophytic and rhizosphere microbial communities than in those present in bulk soil contaminated with hydrocarbons. However, the results obtained when they studied the prevalence of the xylene monoxygenase or naphthalene dioxygenase genes were just the opposite, their presence was higher in bulk soil microbial communities than those near or on the plant. This suggests that if plants are influencing the rhizosphere, this effect is dependent on the contaminant. Some researchers also concluded that the effect depended on the type of the plant. This has led to the hypothesis that the effectiveness of rhizoremediation strategies is related to the selection of the best plant bacterium pair in each case.

In a case study it was found that rhizospheric *Pseudomonas* sp. of *Calotropis* plant a good degrader for naphthalene (78.44 %) and anthracene (63.53 %) as determined by HPLC analysis. Thus, it can be concluded that rhizosphere of *Calotropis* sp. is a source of *Pseudomonas* sp. possessing potent PGP attributes, PAH degradation and biocontrol activities against phyto-pathogenic fungi. Further studies are under- way to confirm their effectiveness in field conditions [103].

A number of scientists established chemotaxis of PAHs degrading rhizosphere bacteria (*P. alcaligenes, P. stutzeri and P. putida*) to naphthalene, phenanthrene and root exudates [104]. Fascinatingly, the bacteria were repelled by anthracene and pyrene. The attraction of competent bacteria to the root zone may improve bioavailability and increase PAHs degradation in the rhizosphere. Subjugation of the phenanthrene degrading activity of *P. putida* following exposure to root extracts and exudates recommended that enzyme induction may not occur during rhizodegradation of PAHs [92]. Genetically engineered plant microbial systems to improve the rhizoremediation techniques, in which the gene cloning of plants containing bacterial gene for the degradation of organic pollutants and of recombinant, root-colonizing bacteria (e.g. *P. fluorescens*) expressing degradative enzymes e.g. orthomonooxygenase for toluene degradation [25].

The studies pertaining to the rhizospheric microorganism associated with specific plant are still missing in the available literature even though a lot of work has been reported on bioremediation. Many researchers have carried out work on plant growth promoting (PGP) activity of rhizosphere of different plants, but no information about rhizosphere community of specific plant, its molecular characterization and utilization in sustainable agriculture, biofertilization and ecorestoration is reported in the literature. Rhizoremediation can be successfully used for restoration of contaminated sites by choosing right type of plant cultivar with right rhizobacteria or by inoculating efficient rhizobacterial strains on plant seeds [34]. Bacteria inhabiting the rhizosphere of a suitable plant may be used as 'bacterial injection system' in soils for effective growth promotion and rhizoremediation.

Author details

Keshav Prasad Shukla[1], Shivesh Sharma[2], Nand Kumar Singh[2], Vasudha Singh[2], Sandeep Bisht[3] and Vivek Kumar[4]

*Address all correspondence to: shiveshs@mnnit.ac.in; ssnvsharma@gmail.com

1 Department of Biotechnology, Hindustan College of science and Technology, Mathura (UP), India

2 Department of Biotechnology, Motilal Nehru National Institute of Technology, Allahabad (UP), India

3 Department of Microbiology, UUHF, College of Horticulture, Bharsar, Pauri. (UK), India

4 IndAmity Institute of Microbial Technology, Amity University, Noida, India

References

[1] Canadian Council of Ministers of the Environment (CCME), Canada wide standards for petroleum hydrocarbons (PHC) in soil, 2001.

[2] Johnsen, A. R., Wickb L. Y. and Harmsb H. Principles of microbial PAH degradation in soil. Environmental Pollution 133, (2005), 71-84.

[3] Cerniglia, C. E. Biodegradation of polycyclic aromatic hydrocarbons. Current Opinon in Biotechnology 4, (1993), 331-338.

[4] Wilson, S.C. and Jones, K.C. Bioremediation of soils contaminated with polynuclear aromatic hydrocarbons (PAHs): A review. Environmental Pollution 88, (1993), 229-249.

[5] Guerin, W.F. and Jones, G.E. Mineralisation of phenanthrene by a Mycobacterium sp. Applied and Environmental Microbiology 54, (1988), 937-944.

[6] Blumer, M. Polycyclic aromatic compounds in nature. Scientific American 3, (1976), 35-45.

[7] Wakeham, S.G., Schaffner, C. and Giger, W. Polycyclic aromatic hydrocarbons in recent lake sediments II compound having anthropogenetic origins. Geochimica et Cosmochimica Acta 44, (1980), 403-413.

[8] Freeman, D. J. and Cattell, F. C. R. Wood burning as a source of atmospheric polycyclic aromatic hydrocarbons. Environmental Science and Technology 24, (1990), 1581-1585.

[9] Oviasogie, P. O., Ukpebor, E. E. and Omoti, U. Distribution of polycyclic aromatic hydrocarbons in rural agricultural wetland soils of the Niger Delta Region. African Journal of Biotechnology 5, (2006), 1415-1421.

[10] Kanaly, R. A. and Harayama,S. Biodegradation of high molecular weight polycyclic aromatic hydrocarbons by bacteria. Journal of Bacteriology 182, (2000), 2059-2067.

[11] Jones, K. C., Stratford, J. A., Waterhouse, K. S., Furlong, E. T., Giger, W., Hites, R. A., Schaffner, C. and Johbston, A. E. Increases in the polynuclear aromatic hydrocarbon content of an agricultural soil over the last century. Environmental Science and Technology 23, (1989), 95-101.

[12] Cerniglia, C. E. and Heitkamp, M. A. Microbial degradation ofpolycyclic aromatic hydrocarbons in the aquatic environment. In: Varanasi, U. (Ed.), metabolism of polycyclic aromatic hydrocarbons in the aquatic environment. CRC Press, Boca Raton, (1989), pp. 42-64.

[13] Nishioka, M., Chang, H.C., Lee, V. Structural characteristics of polycyclic aromatic hydrocarbon isomers in coal tars and combustion products. Environmental Science and Technology 20, (1986), 1023-1027.

[14] Lijinsky, W. The formation and occurence of polynuclear aromatic hydrocarbons associated with food. Mutation Research 259, (1991), 251-262.

[15] Walter, U., Beyer, M., Klein, J. and Rehm, H. J. Degradation of pyrene by Rhodococcus sp. Applied Microbiology and Biotechnology 34, (1991), 671-676.

[16] Ramadah, T., Alfheim, I., Rustad, S. and Olsen, T. Chemical and biological characterisation of emissions from small residential stoves burning wood and charcoal. Chemosphere 11, (1982), 601-611.

[17] Juhasz A. L. and Naidu R. Bioremediation of high molecular weight polycyclic aromatic hydrocarbons: A review of the microbial degradation of benzo[a]pyrene. International Biodeterioration & Biodegradation 45, (2000), 57-88.

[18] Guoa, H., Leea, S.C., Hoa, K.F., Wangb, X.M. and Zou, S.C. Particle associated polycyclic aromatic hydrocarbons in urban air of Hong Kong. Atmospheric Environment 37, (2003), 5307-5317.

[19] Shiaris, M. P. and Jambard, S. D. Polycyclic aromatic hydrocarbons in surficial sediments of Boston Harbour, MA, USA.Marine Pollution Bulletin 17, (1986), 469-472.

[20] Mueller, J.G., Chapman, P. J. and Pritchard, P. H. Action of a fluoranthene utilising bacterial community on polycyclic aromatic hydrocarbon components of creosote. Applied and Environmental Microbiology 55, (1990), 3085-3090.

[21] Peters, R. W. 1999. Chelant extraction of heavy metals from contaminated soils. Journal of Hazardous Materials 66:151–210.

[22] Haritash, A. K. and Kaushik, C. P. Biodegradation aspects of polycyclic aromatic hydrocarbons (PAHs): A review. Journal of Hazardous Material 169, (2009), 1-15.

[23] Shukla, K. P., Singh, N. K. and Sharma, S. Bioremediation: developments, current practices and perspectives. Genetic Engineering and Biotechnology Journal. 3, (2010), 1-20.

[24] McGuinness, M. and Dowling, D. Plant associated bacterial degradation of toxic organic compounds in soil. International Journal of Environmental Research and Public Health 6, (2009), 2226-2247.

[25] Francova, K., Sura, M., Macek, T., Szekeres, M., Bancos, S. and Demnerova, K. Preparation of plants containing bacterial enzyme for the degradation of polychlorinated biphenyls. Fresenius Environmental Bulletin 12, (2003), 309-313.

[26] Prasad M. N. V. A State of the art report on bioremediation, its applications to contaminated sites in India. Ministry of Environment & Forests, Government of India, 2011.

[27] Mejare, M., and L. Bulow. 2001. Metal-binding proteins and peptides in bioremediation and phytoremediation of heavy metals. Trends Biotechnol. 19:67–73.

[28] Gerhardt, K. E., Huang, X. D., Glick, B. R. and Greenberg, B. M. Phytoremediation and rhizoremediation of organic soil contaminants: Potential and challenges. Plant Science 176,(2009), 20-30.

[29] Ho, C.H., Applegate, B. and Banks, M.K. Impact of microbialplant interactions on the transformation of polycyclic aromatic hydrocarbons in rhizosphere of Festuca arundinacea. International Journal of Phytoremediation 9, (2007), 107-114.

[30] Kidd, P., Prieto, F. A., Monterroso, C. and Acea, M. Rhizosphere microbial community and hexachlorocyclohexane degradative potential in contrasting plant species. Plant and Soil 302, (2008), 233-247.

[31] Sandhu, A., Halverson, L.J. and Beattie, G.A. Bacterial degradation of airborne phenol in the phyllosphere. Environmental Microbiology 9, (2007), 383-392.

[32] Barac, T., Weyens, N., Oeyen, L., Taghavi, S., van der Lelie, D., Dubin, D., Spliet, M. and Vangronsveld, J. Fiels note: Hydraulic containment of a BTEX plum using poplar trees. International Journal of Phytoremediation 11, (2009), 416-424.

[33] Johnsen, A. R., Wickb L. Y. and Harmsb H. Principles of microbial PAH degradation in soil. Environmental Pollution 133, (2005), 71-84.

[34] Kamaludeen, P. B., Ramasamy, K., Rhizoremediation of metals: harnessing microbial communities. Indian Journal of Microbiology 48, (2008), 80-88.

[35] de Carcer, D. A., Martin, M., Mackova, M., Macek, T., Karlson, U. and Rivilla, R. The introduction of genetically modified microorganisms designed for rhizoremediation induces changes on native bacteria in the rhizosphere but not in the surrounding soil. The ISME Journal 1, (2007), 215-223.

[36] Phillips L. A., Greer C. W., Richard E. F. and James J. G. Plant root exudates impact the hydrocarbon degradation potential of a weathered hydrocarbon contaminated soil. Applied Soil Ecology52, (2012), 56-64.

[37] Mohan, S. V., Kisa, T., Ohkuma, T., Kanaly, R. A. and Shimizu, Y.Bioremediation technologies for treatment of PAH contaminated soil and strategies to enhance process efficiency. Review on Environmental Science and Biotechnology 5, (2006), 347-374.

[38] Ma, Y., Prasad, M. N. V., Rajkumar, M. and Freitas, H. Plant growth promoting rhizobacteria and endophytes accelerate phytoremediation of metalliferous soils. Biotechnology Advances 29, (2011), 248-258.

[39] Ma, B., He, Y., Chen, H. H. and Xu, J. M. Dissipation of polycyclic aromatic hydrocarbons (PAHs) in the rhizosphere: Synthesis through metaanalysis. Environmental Pollution 159, (2009), 855-861.

[40] David, N. D. and Sharon L. D. Improving phytoremediation through biotechnology. Current Opinion in Biotechnology20, (2009), 1-3.

[41] Lee, S., Ka, J. O. and Gyu, S. H. Growth promotion of Xanthium italicum by application of rhizobacterial isolates of Bacillus aryabhattai in microcosm soil. The Journal of Microbiology50, (2012), 45-49.

[42] Babalola, O.O. Beneficial bacteria of agricultural importance.Biotechnology Letters 32, (2010), 1559-1570.

[43] Dodd, I., Zinovkina, N., Safronova, V. and Belimov, A.Rhizobacterial mediation of plant hormone status. Annals of Applied Biology157, (2010), 361-379.

[44] Bisht, S., Pandey, P., Sood, A., Sharma, S. and Bisht, N. S.biodegradation of naphthalene and anthracene by chemotactically active rhizobacteria of Populus deltoides. Brazilian Journal of Microbiology 41, (2010), 922-930.

[45] Harms, H. and Wick, L. Y. Dispersing pollutant degrading bacteria in contaminated soil without touching it. Engineering in Life Sciences 6, (2006), 252-260.

[46] Walker, T. S., Bais, H. P., Grotewold, E. and Vivanco, J. M. Root exudation and rhizosphere biology. Plant Physiology132, (2003), 44-51.

[47] Glick, B. R. The enhancement of plant growth by free living bacteria. Canadian Journal of Microbiology 41, (1995), 109-114.

[48] Kuiper, I., Lagendijk, E. L., Bloemberg, G. V. and Lugtenberg Rhizoremediation: A beneficial plant microbe interaction.Molecular Plant Microbe Interactions 17, (2004), 6-15.

[49] Kamaludeen, P. B. and Ramasamy, K. Rhizoremediation of metals: harnessing microbial communities. Indian Journal of Microbiology 48, (2008), 80-88.

[50] Timmis, K. N. and Pieper, D. H. Bacteria designed for bioremediation.Tibtech Journal 17, (1999), 201-204.

[51] Guijian, L., Zhiyuan, N., Daniel, V. N., Jian, X. And Liugen, Z. Polycyclic aromatic hydrocarbons (PAHs) from coal combustion: Emissions, analysis, and toxicology. Reviews of Environmental Contamination and Toxicology 192, (2008), 1-28.

[52] Abd, E. H. E., Hafez, E. E., Hussain, A. A., Ali, A. G. and Hanafy, A. A. Isolation and identification of three rings polyaromatic hydrocarbons (anthracene and phenanthrene) degrading bacteria. American Eurasian Journal of Agriculture and Environmental Science 5, (2009), 31-38.

[53] Barea, J. M., Pozo, M. J., Azcon, R.and Azcon, A. C. Microbial cooperation in the rhizosphere. Journal of Experimental Botany 56, (2005), 1761-1778.

[54] U.S. Environmental Protection Agency, Use of monitored natural attenuation at Superfund, RCRA corrective action, and underground storage tank sites (1999), pp 41.

[55] Louisa, W. P. Review: In situ and bioremediation of organic pollutants in aquatic sediments. Journal of Hazardous Materials 177, (2010), 81-89.

[56] Cosgrove, L., McGeechan, P. L., Handley, P. S. And Robson, G. D. Effect of biostimulation and bioaugmentation on degradation of polyurethane buried in soil. Applied and Environmental Microbiology 76, (2010), 810-819.

[57] Tyagi, M., da Fonseca, M. R., de Carvalho, C. R.Bioaugmentation and biostimulation strategies to improvethe effectiveness of bioremediation processes. Biodegradation22, (2011), 231-241.

[58] Frerot, H., Lefebvre, C., Gruber, W., Collin, C., Dos, S. A. and Escarre, J. Specific interactions between local metallicolous plants improve the phytostabilization of mine soils. Plant and Soil 282, (2006), 53-65.

[59] Pilon, S. E. A. H. Phytoremediation. Annual Review of Plant Biology 56, (2005), 15-39.

[60] Shukla, K. P., Singh, N. K. and Sharma, S. Bioremediation: developments, current practices and perspectives. Genetic Engineering and Biotechnology Journal 3, (2010), 1-20.

[61] Bais, H.T., Perry, L.G., Simon, G. and Vivanco J.M.The role of root exudates in rhizosphere interactions with plants and other organisms. Plant Biology57, (2006), 233-266.

[62] Kuiper, I., Bloemberg, G. V. and Lugtenberg, B. J. J. Selection of a plantbacterium pair as a novel tool for rhizostimulation of polycyclic aromatic hydrocarbon degrading bacteria. Molecular Plant Microbe Interaction 14, (2001), 1197-1205.

[63] Wani, P. A., Khan, M. S. and Zaidi, A. Cadmium, chromium and copper in green gram plants. Agronomy Sustainable Development 27, (2007), 145-153.

[64] Sharma, A. and Talukdar, G. Effects of metals on chromosomes of higher organisms. Environmental Mutagenesis 9, (1987), 191-226.

[65] Lazaro, M., Cayo, R., Estrella, D., Ronchel, M. C., Garcia, J. M., Lene, W. and Ramos, J. L. Survival of Pseudomonas putida KT2440 in soil and in the rhizosphere of plants under greenhouse and environmental conditions. Soil Biology and Biochemistry 32, (2000), 315-321.

[66] Eerd, L. L. V., Hoagland, R. E., Zablotowicz, R. M. and Hall, J. C. Pesticide metabolism in plants and microorganisms. Weed Science, 51, (2003), 472-495.

[67] Wenzel, W. W. Rhizosphere processes and management in plant assisted bioremediation (phytoremediation) of soils. Plant Soil321, (2009), 385-408.

[68] Anderson, T.A., Guthrie, E.A. and Walton, B. T. Bioremediation in the rhizosphere, plant roots and associated microbes clean contaminated soil. Environmental Science and Technology 27, (1993), 2630-2636.

[69] Susarla, S.,Medina, V. F. and McCutcheon, S. C. Phytoremediation: An ecological solution to organic chemical contamination. Ecological Engineering 18, (2002), 647-658.

[70] Newman, L. A., Reynolds, C. M. Phytodegradation of organic compounds. Current Opinion in Biotechnology 15, (2004), 225-230.

[71] Dzantor, E. K. Phytoremediation: the state of rhizosphere "engineering" for accelerated rhizodegradation of xenobiotic contaminants. Journal of Chemical Technology & Biotechnology 82, (2007), 228-232.

[72] Phillips, L. A., Greer, C. W. and Germida, J. J. Culture based and culture independent assessment of the impact of mixed and single plant treatments on rhizosphere microbial communities in hydrocarbon contaminated flare pit soil. Soil Biology and Biochemistry 38, (2006), 2823-2833.

[73] Olson, P. E., Castro, A., Joern, M., DuTeau, N. M., Pilon, S. E. A. H. and Reardon, K. F. Comparison of plant families in a greenhouse phytoremediation study on an aged polycyclic aromatic hydrocarbon contaminated soil. Journal of Environmental Quality 36, (2007), 1461-1469.

[74] Dams, R. I., Paton, G. I. and Killham, K. Rhizoremediation of pentachlorophenol by Sphinobium chlorophenolicum. Chemosphere 68, (2007), 864-870.

[75] Child, R., Miller, C. D., Liang, Y., Sims, R. C. and Anderson, A. J. Pyrene mineralization by Mycobacterium sp. strain in a barley rhizosphere. Journal of Environmental Quality 36, (2007), 1260-1265.

[76] Van Dillewijn, P., Caballero, A., Paz, J. A., Gonzales, M. M., Oliva, J. M. and Ramos, J. L. Bioremediation of 2,4,6-trinitrotoluene under field conditions. Environmental Science and Technology 41, (2007), 1378-1383.

[77] Dzantor, E. K. Phytoremediation: the state of rhizosphere "engineering" for accelerated rhizodegradation of xenobiotic contaminants. Journal of Chemical Technology & Biotechnology 82, (2007), 228-232.

[78] Narasimhan, K., Basheer, C., Bajic, V. B. and Swarup, S. Enhancement of plant microbe interactions using a rhizosphere metabolomicsdriven approach and its application in the removal of polychlorinated biphenyls. Plant Physiology 132, (2003), 146-153.

[79] Shukla, K. P., Sharma, S., Singh, N. K., Singh, V., Tiwari, K. and Singh, S. Nature and role of root exudates: Efficacy in bioremediation. African Journal of Biotechnology 10,(2011), 9717-9724.

[80] US Environmental Protection Agency (EPA) Integrated risk information, (2008), website. http//:www.epa.gov/iris.

[81] Rugh, C., Susilawati, E., Kravchenko, A. and Thomas, J. Biodegrader metabolic expansion during polyaromatic hydrocarbons rhizoremediation. Zeitschrift für Naturforschung 60, (2005). 331-339.

[82] Parrish, Z., White, J., Isleyen, M., Gent, M., Iannucci, B. W., Eitzer, B., Kelsey, J. and Mattina, M. Accumulation of weathered polycyclic aromatic hydrocarbons (PAHs) by plant and earthworm species. Chemosphere 64, (2006), 609-618.

[83] Semple, K. T., Morriss, A. W. J. and Paton, G. I. Bioavailability of hydrophobic contaminants in soils: fundamental concepts and techniques for analysis. European Journal of Soil Science 54, (2003), 809-818.

[84] Harmsen, J., Rulkens, W. and Eijsackers, H. Bioavailability: concept for understanding or tool for predicting. Land Contamination&Reclamation 13, (2005), 161-171.

[85] Płociniczak, M. P., Płaza, G. A., Seget, Z. P. and Cameotra, S. S. Environmental applications of biosurfactants: Recent advances. International Journal of Molecular Science 12, (2011), 633-654.

[86] Reid, B. J., Jones, K. C. and Semple, K. T. Bioavailability of persistent pollutants in soils and sediments a perspective on mechanisms, consequences and assessment. Environmental Pollution 108, (2000), 103-112.

[87] Paul, D., Singh, R. and Jain, R. K. Chemotaxis of Ralstonia sp. towards p-nitrophenol in soil. Environmental Microbiology8, (2006), 1797-1804.

[88] Jain, R. K., Kapur, M., Labana, S., Lal, B., Sarma, P. M., Bhattacharya, D. and Thakur, I. S. Microbial diversity: Application of microorganisms for the biodegradation of xenobiotics. Current Science 89, (2005), 101-112.

[89] Cao, Bin, Nagarajan, K. and Kai, C. L. Biodegradation of aromatic compounds: Current status and opportunities for biomolecular approaches. Applied Microbiology and Biotechnology 85, (2009), 207-228.

[90] Filonov, A. E., Puntus, I. F., Karpov, A. V., Kosheleva, I. A., Akhmetov, L. I., Yonge, D. R., Petersen, J. N. and Boronin, A. M. Assessment of naphthalene biodegradation efficiency of Pseudomonas and Burkholderia strains tested in soil model systems. Journal of Chemical Technology and Biotechnology 18, (2006), 216-224.

[91] Ye, D., Siddiqi, M. A., Maccubbin, A. E., Kumar, S. and Sikka, H. C. Degradation of polynuclear aromatic hydrocarbons by Sphingomonas paucimobilis. Environmental Science and Technology 30, (1996), 136-142.

[92] Rentz, J. A., Alvarez, P. J. J. and Schnoor, J. L. Repression of Pseudomonas putida phenanthrene degrading activity by plant root extracts and exudates. Environmental Microbiology 6, (2004), 574-583.

[93] Sorkhoh, N.A., Al-Mailem D.M., Ali, N., Al-Awadhi, H., Salamah, S., Eliyas, M. and Radwan, S. S. Bioremediation of volatile oil hydrocarbons by epiphytic bacteria associated with American grass (Cynodon sp.) and broad bean (Vicia faba) leaves. International Biodeterioration & Biodegradation 65, (2011), 797-802.

[94] Chaudhry, Q., Blom-Zandstra, M., Gupta, S. and Joner, E. J. Utilising the synergy between plants and rhizosphere microorganisms to enhance breakdown of organic pollutants in the environment. Environmental Science and Pollution Research 12, (2005), 34-48.

[95] Kim, Y.H., Freeman, J.P., Moody, J.D., Engesser, K. H. and Cerniglia, C.E. Effects of pH on the degradation of phenanthrene and pyrene by Mycobacterium vanbaalenii. Applied Microbiology and Biotechnology 67, (2005), 275-285.

[96] Al-Yemni, M. N., Sher, H., Sheikh, M. A. and Eid, E. M. Bioaccumulation of nutrient and heavy metals by Calotropis procera and Citrullus colocynthis and their potential use as contamination indicators. Scientific Research and Essays6, (2011), 966-976.

[97] Overah, L. C. Biosorption of Cr (III) from aqueous solution by the leaf biomass of Calotropis procera-'bombom'.Journal of Applied Science and Environment Management 15,1 (2011), 87-95.

[98] Liu, L., Tindall, J. A., Friedel, M. J., Biodegradation of PAHs and PCBs in Soils and Sludges. Water Air Soil Pollution. DOI 10, (2007), 11270-006-9299-3.

[99] Chadhain, S. M. N., Norman, R. S., Pesce, K. V., Kukor, J. J. and Zylstra, G. J. Microbial dioxygenase gene population shifts during polycyclicaromatic hydrocarbon biodegradation. Applied and Environmental Microbiology 72, (2006), 4078-4087.

[100] Dennis, J. J., and Zylstra, G. J. Complete sequence and genetic organization of pDTG1, the 83 kilobase naphthalene degradation plasmid from Pseudomonas putida strain. Journal of Molecular Biology 341, (2004), 753-768.

[101] Natsch, A.C., Keel, J., Troxler, M., Zala, N., Albertini, V. and Defago, G. Importance of preferential flow and soil management in vertical transport of a biocontrol strain of Pseudomonas fluorescens in structured field soil. Applied Environmental Microbiology 62, (1996), 33-40.

[102] Siciliano, S. D., Germida, J. J., Banks, K. andGreer, C. W. Changes in microbial community composition and function during a polyaromatic hydrocarbon phytoremediation field trial. Applied and Environmental Microbiology 69, (2003), 483-489.

[103] Shukla, K. P., Sharma, S., Singh, N. K., Singh, V., Deciphering rhizosphere soil system for strains having plant growth promoting and bioremediation traits. Agriculture Research 1(3), (2012), 251-257.

[104] Shukla, K. P., Sharma, S., Singh, N. K., Singh, V., Tiwari, K. and Singh, S. Nature and role of root exudates: Efficacy in bioremediation. African Journal of Biotechnology 10,(2011), 9717-9724.

Chapter 15

Purification and Partial Characterization of a Thermostable Laccase from *Pycnoporus sanguineus* CS-2 with Ability to Oxidize High Redox Potential Substrates and Recalcitrant Dyes

Sergio M. Salcedo Martínez, Guadalupe Gutiérrez-Soto,
Carlos F. Rodríguez Garza, Tania J. Villarreal Galván,
Juan F. Contreras Cordero and Carlos E. Hernández Luna

Additional information is available at the end of the chapter

http://dx.doi.org/10.5772/56374

1. Introduction

Laccases (benzenediol: oxygen oxidoreductases, EC 1.10.3.2) are enzymes that catalyze the oxidation of phenolic compounds and aromatic amines with the simultaneous reduction of molecular oxygen to water [1]. They are widely distributed in many plants and fungi, some insects and bacteria, being particularly abundant in white-rot basidiomycetes [2]. Typical fungal laccases are described as glycosylated multicopper proteins, which are produced as extracellular monomeric forms of around 60-80 kDa, containing four copper atoms and 15-20% carbohydrates. Operatively, they are moderately thermotolerant, showing optima activity at 50-55 °C, and under acidic conditions (pH 3-5); although their maxima stability occurs in the alkaline zone (pH 8-9) [3]. Their copper atoms are distributed in three different sites bringing unique spectroscopic properties: The type 1 copper (CuT1) atom, is responsible of the intense blue color of enzymes by light absorption around 610 nm; The type 2 copper (CuT2) atom exhibits a weak absorption in the visible region; and the two type 3 copper (CuT3) atoms are present as a binuclear center, which has an absorption maximum about 330 nm. Moreover, CuT2 and CuT3 copper atoms are structural and functionally arranged as a trinuclear cluster. The four copper atoms form part of the active site of enzyme contributing directly to reaction. CuT1 is involved in the initial electron subtraction from reducer substrates, while trinuclear

CuT2 and CuT3 cluster is responsible of the electron transference, from CuT1 to diatomic oxygen [4].

According to their redox potential, most of blue laccases belong to class II (-500 to -600 mV) or class III (-700 to -800 mV) laccases [5]. This is a disadvantage when compared to the ability of lignin peroxidase (LiP) and manganese peroxidase (MnP) to attack compounds with higher redox potential, including non-phenolic lignin units. To overcome this limitation laccases have evolutively developed a synergistic catalytic strategy, which combines a flexible ability to recognize a great variety of chemical compounds, with an extended capability to act at the distance through the activation of diffusible low molecular substances which serve as redox mediators. From a biological stand point this strategy let laccases to become one of the most versatile enzymes in nature, adaptable to multiple functions in plants, insects, fungi and bacteria. Another interesting possibility arises from the properties of atypical "yellow" and "white" laccases, which have shown the ability to catalyze the direct oxidation of high redox potential non-phenolic lignin model substrates or polyaromatic hydrocarbons [6, 7]. It has been proposed that the improved redox capabilities of these laccases come up either, by substituting some copper atoms for zinc, iron, or manganese in the metal clusters or by a change of the redox state of the CuT1(due to the interaction with a lignin-derived ligand) at the active site of, otherwise normal laccase protein structures. So, evolution and prevalence of laccases as a part of the lignin modifying enzyme (LME) system in white-rot basidiomycetes could also be the result of a "biochemical spring-up" mechanism acting under a short term ecophysiological selective pressure.

Whether directly or by mediation, laccases are able to oxidize a broad range of natural or xenobiotic compounds, including: mono, poly or methoxy- amine- and chloro-substituted phenols as well as aromatic heterocyclic and inorganic/organometallic substances; some of them recognized among the most recalcitrant industrial pollutants, for example; polycyclic aromatic hydrocarbons (PAH), pentachlorophenols (PCP), polychlorinated biphenyls (PCB), 1,1,1-trichloro-2,2-bis(4-chlorophenyl)ethane (DDT), trinitrotoluene (TNT), and many azo, triarylmethane, anthraquinonic, indigoid and heterocyclic textile dyes [8,9]. Therefore, laccases are considered enzymes with a great potential for the development of environmental and industrial applications. Current and potential laccase applications include biobleaching of pulp and bioremediation of pulp and paper industries, bioremediation of olive mill wastewater, bioremediation of effluents of the textile and dye manufacturing industries, biocatalytic synthesis of antibiotics and novel polymeric materials, development of biosensors, clarification and stabilization of beer, juices and wines, and panification [1, 10, 11]. Some laccase formulations have already reached a commercial significance, but general thought is that their biotechnological applications and performances could be greatly improved or expanded with the development and finding of new enzyme variants with desirable functional properties, such as higher redox potential, optimum activity at neutral or alkaline pH and thermal stability [12, 13, 14]. It has been proposed that these new laccases could be obtained by protein engineering or through the exploration of the natural biodiversity. The importance of prospective studies in natural biodiversity applying an ecophysiological approach is illustrated by reports about isolation of new thermostable laccases from fungi, either from thermophilic compost [15] or tropical environments [13,16], or by the finding of novel laccases with improved ability to oxidize substrates with a higher than normal redox potential culturing under solid phase conditions [6, 7,17]. Northeast Mexico shelters a high diversity of white-rot basidiomycetes as a result of its particular combination of physiography and climate, including species associated to pine, oak and mixed forests, sub-mountain and semi-desert scrublands, and grass-land. In this work we first present information on the isolation, identification and selection of a northeast Mexico native strain of *Pycnoporus sanguineus* CS2, as a potential producer of thermostable laccases. Results on the purification and partial characterization of its laccase are then exposed, stressing on its thermal stability and ability to attack high redox potential substrates and recalcitrant dyes without the participation of redox mediators.

2. Materials and methods

2.1. Chemicals

All chemicals used as buffers, enzyme substrates, culture media ingredients and electrophoresis reagents, were reactive grade and commercially available through local distributors of Difco, Sigma-Aldrich and Fluka, or BioRad products: PDB (potato dextrose broth), bacteriological agar, yeast extract, malt extract, peptone and dextrose, were from Difco. Acrylamide, bis-acrylamide, TEMED (N,N,N',N'- tetramethylethylenediamine), 2-mercaptoethanol, SDS (sodium dodecyl sulfate), trizma-base, glycine, Coomassie blue, and low range markers kit, were from Bio Rad. Enzyme substrates and dyes: 2,6-dimethoxyphenol (2,6-DMP); *o*-dianisidine (3,3′-dimethoxybenzidine); ABTS (2,2′–azino-bis(3-ethylbenzthiazolin-6-sulphonic acid); syringaldazine (4-hydroxy-3,5-dimethoxybenzaldehyde azine); DMAB (3-dimethylaminobenzoic acid), MBTH (3-methyl-2-benzothiazolinone hydrazone); Methyl Red (Acid Red 2; CI 13020); Reactive Black 5 (RB 5; CI 20505), were from Sigma, Fluka or Aldrich. Chromatographic matrices; DEAE-Sepharose and Q-Sepharose from Sigma-Aldrich, and Biogel P-100 from BioRad. All other chemicals, including solvents, inorganic salts, acids and bases, were from Reactivos Químicos Monterrey, S.A. or CTR-Scientific S.A. de C.V. Solutions and culture media were prepared with bidistilled water from Laboratorios Monterrey, S. A.

2.2. Isolation and identification of fungal strain

The *Pycnoporus* strain used in our experiments was isolated from fruit bodies developing on decayed logs that were gathered in a man-disturbed sub-mountain scrubland around Monterrey, N.L. (Northeast México). Mycelia cultures were obtained by standard mycological techniques, according to the procedure previously described [18]. Briefly, small flesh sections were aseptically removed from inside the carpophores, and transferred to YMGA (0.4% Yeast extract, 1.0% Malt extract, 0.4% Glucose, 1.5% agar) plates, supplemented with 10% Tartaric Acid and 0.004% Benomyl. Plates were incubated at 28 °C, and those with extensive mycelia growth were analyzed under the microscope to confirm a successful isolation. Stock cultures were maintained by periodic transfers every two or three months on YMGA plates and kept refrigerated at 4 °C. Carpophore morphologic features, measurements and photographs were registered previous to dissections for microscopic examination, and identification was done by following the taxonomical keys in reference [19], and in [20] for genera of polypores and the most common macromycetes from Mexico, respectively.

2.3. Enzyme and protein assays

Laccase activity was determined by triplicate at 25 °C in 3 ml cuvettes, monitoring the increase in absorbance at A_{468} (ε=49,600 $M^{-1}cm^{-1}$), using a Shimadzu UV-VIS mini 1240 spectrophotometer and 2,6-DMP as substrate. The assay mixture contained 0.01ml enzymatic extract, 0.1 ml of 60 mM 2.6-DMP in 2.89 ml of 200 mM citrate-phosphate buffer at pH 4.0 [21]. One unit of laccase activity was defined as the amount of enzyme required to oxidize 1 µmol of 2, 6-DMP per minute at 25 °C. In some cases, it was necessary to assess the presence of lignin peroxidase (LiP) and manganese peroxidase (MnP). LiP activity was estimated by the H_2O_2-dependant veratryl alcohol oxidation to veratraldehyde as in reference [22] MnP activity was measured by the formation of Mn^{3+}-tartrate complex during the oxidation of $MnSO_4$ in tartrate buffer as in [23]. The protein concentration was estimated by the Bradford assay (Protein Assay Bradford of BioRad) with bovine serum albumin as standard.

2.4. Strain selection and enzyme production

Isolated *Pycnoporus* strain was selected as a potential source of thermostable laccases in a preliminary screening with crude enzyme preparations. 250 ml Erlenmeyer flask, containing 50 ml of natural LME inducers containing Bran Flakes (BF) media (2% Bran Flakes® in 60 mM potassium phosphate pH 6.0) [24], were inoculated with three 0.5 cm diameter cylinders of mycelia taken from the border of a YMGA growing colony and incubated at 28 °C under agitation at 150 rpm. Aliquots (200 µl) were removed from cultures and the extracellular fluid was separated by centrifugation at 14 K (Eppendorf 5415 C). Enzyme activity in supernatants was determined with 2 mM 2, 6-DMP final concentration, as described above, after sample incubation at 60 °C during different times in a four hour period. This phase of study included four different *Pycnoporus* sp. (CS 2, CS 20, CS 43 and LE 90) strains from the native basidiomycete collection of our laboratory. Among them, *Pycnoporus sanguineus* CS 2 was selected on the basis of its ability to produce a thermostable 2,6-DMP oxidizing activity, and for showing apparently a single band of activity when incubated with 2,6-DMP and SGZ in a parallel native PAGE analysis of crude supernatants.

Enzyme production was evaluated in submerged liquid cultures on the natural containing laccase-inductors BF or a modified Kirk medium (MK) [25], with the following composition: 10 g l^{-1} dextrose, 1.0 g l^{-1} yeast extract, 5.0 g l^{-1} peptone, 2.0 g l^{-1} ammonium tartrate, 1.0 g l^{-1} KH_2PO_4, 0.5 g l^{-1} $MgSO_4$, 0.5 g l^{-1} KCl, and 1.0 ml of 100 X trace element solution (0.5 g EDTA, 0.2 g $FeSO_4$, 0.01 g $ZnSO_4$, 0.003 g $MnCl_2$, 0.03 g H_3BO_4, 0.02 g $CoCl_2$, 0.001 g $CuCl_2$, and 0.003 g $NaMoO_4$ in 100 ml); amended with 350 µM $CuSO_4$ and 3% ethanol [26]. Cultures were performed at 28 °C and 150 rpm for 14 days. 50 µL aliquots were taken every two days to determine the laccase activity.

2.5. Laccase purification

All the procedures were performed at 4 °C, unless otherwise stated. Extracellular liquid from 14 day-old submerged cultures was separated from mycelium by filtration through a cotton-polyester 50:50% cloth. Then, water-soluble polysaccharides were removed from sample solution by freezing (- 20 °C for 24 h), thawing and filtration (Whatman # 1). Culture filtrate was concentrated to approximately 200 ml by 10 kDa ultrafiltration (Millipore prep/scale TFF cartridge). The obtained fluid was further reduced to 20 ml by using a stirred ultrafiltration system equipped with an YM10 membrane (Amicon, Millipore). The reddish-brown enzyme concentrate was equilibrated by diafiltration with 20 mM potassium phosphate, pH 6.0, and applied to a pre-equilibrated anion-exchange DEAE-Sepharose column (2.5 × 17 cm). Once on the column, unadsorbed protein and most of the pigment were removed by washing with two volumes of equilibrium buffer. Retained proteins were eluted with a linear gradient of potassium phosphate pH 6.0 from 20 to 300 mM, and the eluted fractions were assayed for laccase activity and the A_{280} nm monitored. Fractions with laccase activity were pooled, concentrated, equilibrated by diafiltration with 100 mM potassium phosphate, and applied on a pre-equilibrated Biogel P-100 column (2.6 x 65 cm). The loaded proteins were eluted with the same buffer. Active fractions were pooled, concentrated and diafiltrated against 20 mM potassium phosphate buffer pH 6.0. Enzyme was further purified by anion-exchange on a pre-equilibrated Q-Sepharose column (2.5 x 17 cm). Once set the sample, active fractions were eluted with a lineal gradient of potassium phosphate from 20 to 300 mM. These fractions were pooled, concentrated and diafiltrated against water, and stored at - 20 °C.

2.6. Electrophoresis analysis

Protein purity and molecular mass were evaluated by sodium dodecyl sulfate-polyacrylamide gel electrophoresis (SDS-PAGE) as in reference [27] with 4% stacking gel and 12% resolving gel. Protein bands were stained with Coomassie brilliant blue and the molecular mass (M_r) of purified laccase was determined by calculating the relative mobility of standard protein markers: *phosphorylase b* 97.4 kDa; *serum albumin* 66.2 kDa; *ovalbumin* 45 kDa; *carbonic anhydrase* 31 kDa; *lysozyme* 14.4 kDa; *aprotinin* 6.5 kDa (SDS-PAGE molecular weight standards low range, BioRad). Native PAGE was carried out as described at reference [28]. Activity staining of laccase was performed by incubating with 2, 6-DMP or the pair MBTH + DMAB in 200 mM sodium acetate buffer, pH 4.5. For identification and comparing proposes, the corresponding laccase band was removed from a parallel gel and submitted to the Proteomic Unit of IBT-UNAM at Cuernavaca, Morelos for aminoacid sequencing; resulting in the sequencing of six internal peptides.

2.7. UV-Vis absorbance spectra

As a part of the characterization of the physicochemical properties of the laccase, its absorbance spectrum from 200 to 800 nm was obtained in a UV-Vis Shimadzu-Mini 1240 spectrophotometer. The assay was performed with 25 μM of protein diluted in 1 ml of bidistilled water.

2.8. Effect of pH on enzymatic activity

Optimum pH of activity was determined in McIlvine buffer (consisting of a combination of 100 mM citrate/50 mM potassium phosphate) adjusted in a range from 3.0 to 7.0. Activity determination was made according to described method with 2,6-DMP using 0.2 M citrate phosphate buffer, at pH 4.0.

2.9. Effect of temperature on enzyme activity and stability

The effect of temperature on reaction rate was determined using 2, 6-DMP in 0.2 M citrate/phosphate buffer, at pH 4.0. The temperature of reaction mixture was adjusted to indicate value and then the reaction was started by the addition of enzyme. The assays were done by triplicate and data in graphics appear as relative activity as a function of temperature, considering as 100% the average of maxima obtained. The activation energy of the system was calculated by the Arrhenius model, according to the expression: $\text{Log } k = [-E_a / 2.303R\,(1/T)] + \text{Log A}$, where: k is the rate constant (it depends of temperature); A is preexponential factor or frecuency factor. E_a the activation energy (expressed in J/mol); R is the gas universal constant (8.314 J K^{-1} mol^{-1}), and T the absolute temperature (°K). In the thermostability assays, the enzyme was pre-incubated at 50, 60 and 70 °C for the indicated periods of time and activity was measured at 25 °C on 2, 6-DMP in 0.2 M citrate/phosphate buffer, at pH 4. The assays were carried out by triplicate and data are expressed as percent of remaining activity as a function of incubation time; taking as 100% the average value of activity at time zero for each temperature. Inactivation process was adjusted to an exponential decay model, from which the constants of heath inactivation (k) and the half-life times were calculated according to the expression: $\ln (N_o/N) = kt$, where; N_o is the activity at the starting of incubation; N is remaining activity after a certain incubation time (t); k correspond to the first-order inactivation constant, and $t_{1/2}$ is the half-life time, calculated as $\ln 2/k = 0.693/k$

2.10. Determination of kinetic parameters

Kinetic analysis was performed on some common substrates of laccase: 2, 6-DMP, ABTS, *o*-dianisidine and SGZ. Reaction mixtures were prepared in 0.2 M citrate/phosphate buffer at pH 4.0 and the change in optical density by minute was measured by triplicate at different substrate concentrations (0.05, 0.1, 0.5, 1.0, 5.0, 10.0 mM) or those indicated for each assay. The assays were performed at 468 nm for 2, 6-DMP ($\varepsilon = 49{,}600\ M^{-1}\ cm^{-1}$), 436 nm for ABTS (29,400 $M^{-1}cm^{-1}$), 460 nm for *o*-dianisidine (11,000 $M^{-1}\ cm^{-1}$) and 525 nm for SGZ ($\varepsilon = 65\ 000\ M^{-1}\ cm^{-1}$). The values of the Michaelis constant (K_m), maximum velocity (V_{max}), turnover number (K_{cat}) and specificity constant K_{cat}/K_m were estimated according to the Lineweaver-Burk method.

2.11. Decolorization assays

The decolorizing ability of laccase was evaluated with two recalcitrant dyes, the non-phenolic azo Methyl Red (MR), and the diazo reactive black 5 (RB 5). The reaction mixture consisted of 0.890 ml of 0.2 M citrate/phosphate buffer, pH 4.0, 0.1 ml of 250 μM MR or RB5 (final concentration 25 μM), and 0.01 ml of pure laccase (final concentration 5 U/ml). Assays were performed at 25 °C and reaction was initiated with the addition of enzyme. Decolorization was estimated by the decreasing of absorbance at 530 nm for MR or 597 nm for RB5. The results are expressed as the percent of remaining color as a function of incubation time according to the relationship: remaining color (%) = [(Abs_{final}/ $Abs_{initial}$)]*100, where: Abs_{final} correspond to the absorbance value at the indicated incubation times, and $Abs_{initial}$: is the initial (t = 0) absorbance value.

3. Results and discussion

3.1. Strain identification

In this study an autochthonous strain of *Pycnoporus* sp (CS 2) was selected as a potential source of thermostable laccases for its ability to produce a thermotolerant 2, 6-DMP oxidizing activity in preliminary assays with crude filtrates from submerged cultures. This basidiomycete was initially isolated from fruit bodies, growing on decayed logs in a man disturbed sub-mountain scrubland around Monterrey, N.L. México (Figure 1), and identified by its morphological and microscopic traits. According to their morphological and microscopic characteristics, the carpophores corresponded to the species *Pycnoporus sanguineus* (L.) Murrill, for their bright orange to orange-red, red or cinnabar-red shelf-like basidiomes, which are nearly round to elongated or fan-shaped in outline, have a dry surface, smooth or finely hairy, wrinkled or warty and attain 2-12 cm diameter and 0.2 to 0.5 cm thick. Their margins are thin and the under surfaces are covered by small pores (3-4 per mm), bright orange to orange red or red ranging from 0.5-1.5 mm long. Their white spores are smooth and oblong-elliptical in shape and range from 4.2 to 5.2 microns long by 2 to 3.5 microns width and the flesh is tough, red to yellowish red, staining black with KOH. On the bases of these features, we assigned the strain under study as *Pycnoporus sanguineus* CS 2.

3.2. Production and purification of laccase

Guzmán (2003), considers that *P. sanguineus* is a tropical variant of the temperate zone species *P. cinnabarinus,* adapted to man disturbed sites, where it is common in fallen logs and fences, always in sunny places [29]. As its closely related species, *P. cinnabarinus* and *P. coccineus, P. sanguineus* is recognized as an efficient lignin decomposer, in spite of its relatively simple LME system composed of laccases [5]. These features make *Pycnoporus* species an attractive group of white-rot basidiomycetes for the production and purification of unusual laccases [16, 30]. In this study, laccase production was carried out in submerged liquid cultures on a modified Kirk basal medium (MK), amended with 3.5 mM $CuSO_4$ and 3 % ethanol, as chemical laccase inducers and on Bran-Flakes medium (BF), containing natural LME inducers. Under these conditions maxima volumetric productions were reached in both media after 14-16 days (Figure 2). As laccase titers on MK media were about thrice higher than that on BF media (7.5 U ml^{-1} vs 2.3 U min^{-1}), it was selected for enzyme production in purification assays. Consistently with other reports on LME production by *Pycnoporus* species, LiP and MnP were not detected [5, 16, 30, 31, 32].

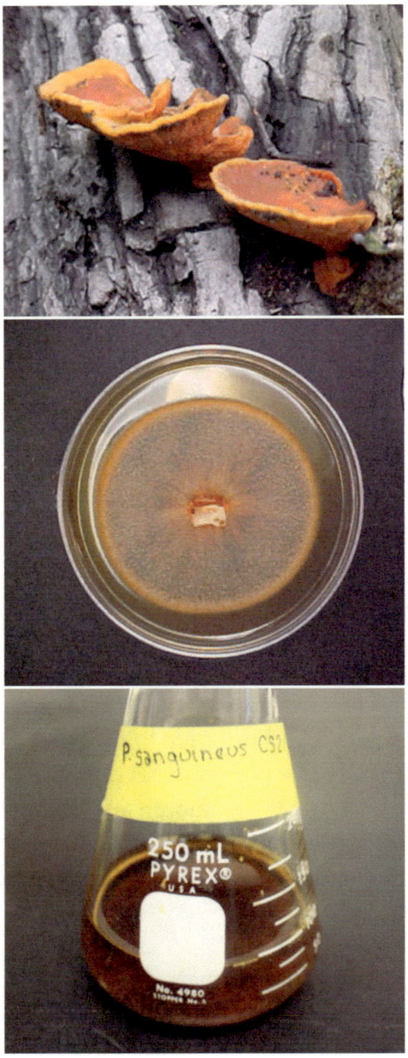

Figure 1. Fruit bodies (carpophores), mycelium colony and submerged culture of *Pycnoporus sanguineus* CS 2. Fungus identification was performed according to macroscopic and microscopic features. Strain isolation was done by tissue transference from the inner flesh of carpophores using mycological standard methodologies. Develop of orange-red pigmented mycelium on the edge of the solid plate colonies, and extracellular production of a reddish pigment under submerged conditions, were indicative of a successful isolation. Production of extracellular mucilage was also observed on submerged cultures. Isolation and identification details are given in text.

Laccase purification was started from about 1850 ml of mycelium-free filtrates from 14 day-old submerged cultures. After 10K ultraconcentration and sequential steps of anionic exchange chromatography on DEAE- Sepharose, gel filtration on Biogel P-100, and anionic exchange on Q-Sepahrose, laccase activity eluted as an apparently single protein peak with 100-140 mM phosphate (Figure 3). When aliquots of pooled laccase from this last chromatographic step were analyzed by denaturing SDS-PAGE, multiple protein bands were detected by Coomassie

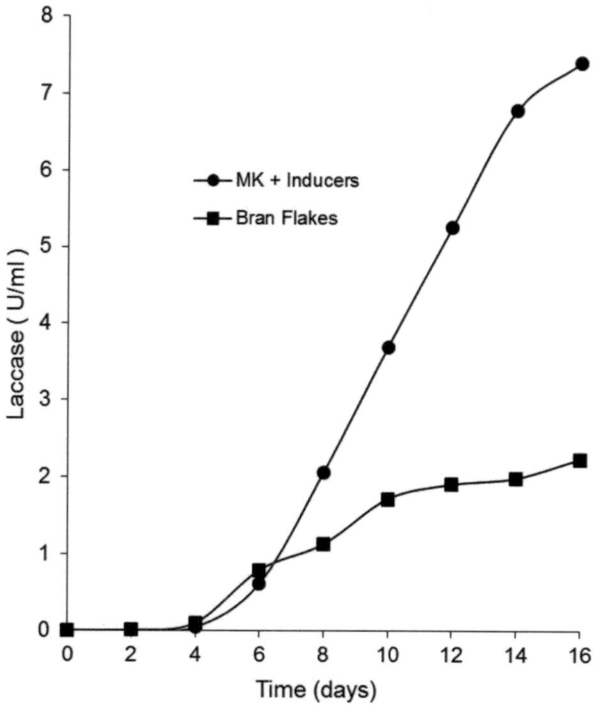

Figure 2. Time course of laccase production by *Pycnoporus sanguineus* CS2. Cultures were carried out on a modified Kirk basal medium, amended with 3.5 mM $CuSO_4$ (MK+ Inducers) and on Bran Flakes medium (BF) at 28 °C under 150 rpm agitation. Data represent the average of a representative assay in triplicate. Ethanol was added to MK medium at the third day of culture. Activity was determined with 2 mM 2, 6-DMP in 200 mM citrate-phosphate buffer pH 4.0

staining (not shown). A similar effect was reported in a work with a *Fusarium proliferum* laccase. As the multiband effect persisted after SDS substitution by other detergents, but disappeared in the absence of SDS, this phenomenon was associated to the presence of detergent on denaturing PAGE [33]. However, when we applied both a heat denatured sample and a non-boiled sample in parallel using the same SDS gel, a protein multiband and a single band were detected by Coomassie staining, respectively. Furthermore, the simple band pattern was also obtained in duplicates by activity staining using laccase substrates. This indicates that thermal treatment could be responsible of the observed multiband effect on denaturing conditions, instead of the SDS by itself. A summary of purification data is shown in Table 1. By this procedure a 16.7-fold purification and activity recovery of 25.5%, with specific activity of 69 U mg^{-1} protein was achieved. Concentrated purified enzyme showed the blue color characteristic of multi-cupper oxidases.

3.2. Biochemical properties

Electrophoresis analysis indicated that *Pycnoporus sanguineus* CS 2 produced only one laccase under the conditions used in this study. According to non-denaturing SDS-PAGE this laccase is a monomeric protein with a molecular weight of 64.4 kDa, and activity staining of native

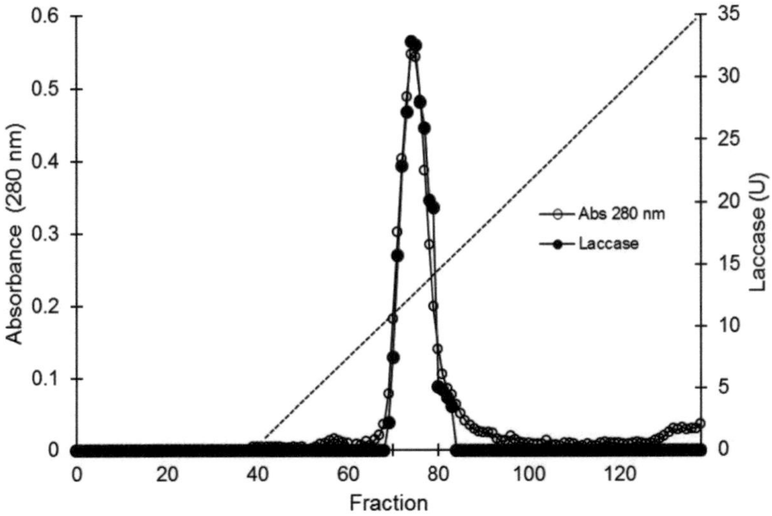

Figure 3. Elution profile for laccase from *Pycnoporus sanguineus* CS2 on anion-exchange column chromatography with Q-Sepharose (2.5 x 17 cm). The enzyme was eluted with a potassium phosphate (pH 6.0) linear gradient from 20-300 mM (dashed line) at a flow rate of 1.0 ml/min.

Purification step	Protein (mg)	Enzyme Activity (IU)	Specific Activity (U/mg)	Recovery (%)	Purification (fold)
Culture filtrate	1551	6477	4.18	100.0	1.0
Ultraconcentration 10 K	422	3724	8.8	57.4	2.1
DEAE-Sepharose FF	64	2308	35.8	35.6	8.5
Biogel P-100	32	1924	59.2	29.7	14.1
Q-Sepharose	22	1522	69.8	23.5	16.7

Table 1. Purification of *Pycnoporus sanguineus* CS 2 laccase

gels exhibited a single broad band when incubated with both, 2,6-DMP and the pair MBTH + DMAB, showing the same migration as the Coomassie blue stained band (Figure 4). Molecular mass of purified laccase was very similar to those reported for different *Pycnoporus sanguineus* strains [30, 31, 32], and it was consistent with the reported for most of basidiomycetes laccases [2, 3]. As expected for its visual appearance described above, the UV-Vis spectrum of purified enzyme was characteristic of the typical blue laccases, displaying the absorbance peak near to 600 nm related to the Cu-T1 centers, and the shoulder at 330 nm of Cu-T3 binuclear centers (Figure 5). Nonetheless, the oxidative coupling of MBTH and DMAB in the absence of mediators was indicative that *Pycnoporus sanguineus* CS2 laccase has the capability to catalyze reactions requiring a higher than usual redox potential for typical laccases [15].

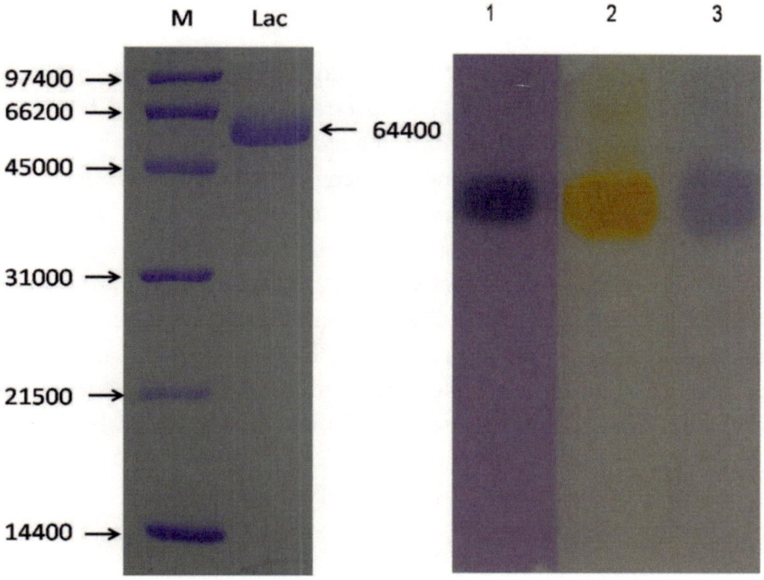

Figure 4. Electrophoresis analyses of purified *Pycnoporus sanguineus* CS 2 laccase by non-denaturing SDS- PAGE (left panel) and Native PAGE (right panel). Lanes M and Lac correspond to the Coomassie staining of molecular weight markers and purified laccase, respectively. The markers were phosphorylase b (97.4 kDa), serum albumin (66.2 kDa), ovalbumin (45 kDa), carbonic anhydrase (31 kDa), trypsin inhibitor (21.5 kDa), and lysozyme (14.4). On the right, lane 1 shows the Coomassie staining of purified laccase, and lanes 2 and 3, the activity staining with 2,6-DMP and the pair MBTH + DMAB, respectively.

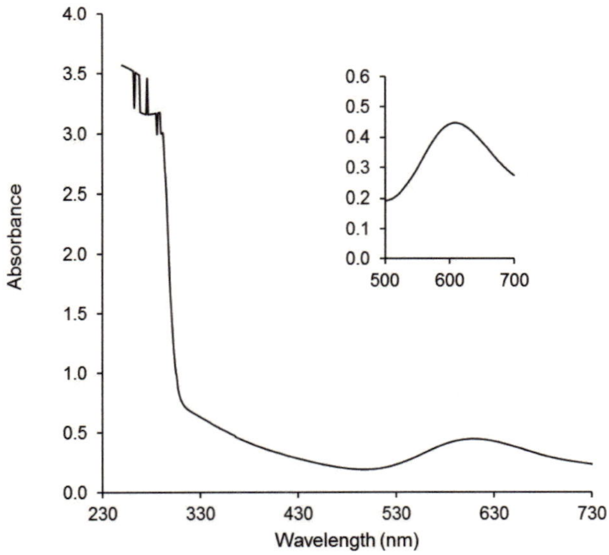

Figure 5. UV-Vis Spectrum of *Pycnoporus sanguineus* CS 2 Laccase. Assay was performed with a preparation of 25 μM laccase in bidistilled water. Insert shows the enlargement of the peak close to 610 nm.

In addition to blue laccases, other "atypical" forms of the enzyme named "yellow" laccases and "white" laccases have been reported. In the first case, it has been proposed that a variation in the redox state of Cu-T1 centers, by the presence of endogenous ligands, decreases the absorbance at 600 nm, without altering the spectral characteristics of the Cu-T2 and Cu-T3 centers, resulting in a yellow color [6]. In white laccases, like the one produced by *Pleurotus ostreatus*, it has been informed the presence of a single copper atom, which is accompanied by two of zinc and one of iron, instead of the regular four copper atoms [7]. In both, yellow and white laccases, the protein structure is similar to that found in blue laccases, but the changes in the redox state of the active site (whereas by the presence of the endogenous mediator or by the substitutions in the Cu centers), enables them to oxidize directly substrates of higher redox potential. According to all the above, the laccase produced by *P. sanguineus* CS 2 corresponds to a blue laccase, most likely containing the regular composition of Cu in its catalytic centers, but like atypical laccases is capable of acting on substrates of higher redox potential.

3.3. pH and temperature dependence

Enzyme was further characterized for its pH and temperature dependence. The effect of pH on laccase activity was studied using some of the most common laccase substrates, including the phenolic 2, 6-DMP, *o*-dianisidine and SGZ, as well as the non-phenolic ABTS. In general, laccase exhibited optima activity in the zone of pH between 3.0 and 4.5, depending on the particular substrate, then it declined in a gradual way towards the neutral zone of pH, and was completely lost at pH 6.5. Optimal pH values were 3.5, 3.5, 4.5 and 3.0 for 2, 6-DMP, *o*-dianisidine, SGZ and ABTS, respectively (Figure 6). These results were similar to those reported in literature for most of the fungal laccases [2]. It is known that biphasic pH-activity profiles with phenolic substrates (as the one showed by SGZ), are a consequence of two opposite effects: one generated by the difference in the redox potential between the reducer substrate and the Cu-T1 centers, when changing from acidic to neutral conditions. The other one is directly associated to the inhibitory action of OH^- ions over the activity of Cu-T2/T3 centers. For non-phenolic substrates as ABTS, the first effect should be minimal and the inhibition by OH^- reflects the monotonic decrease in the enzyme activity [12, 34].

The influence of temperature on *P. sanguineus* CS 2 laccase activity was investigated with 2, 6-DMP (2 mM), at pH 4.0, in the zone, from 20–80 °C. The profile temperature-activity showed a gradual increase from the lower limit at 20 °C to achieve an optimal value at 65 °C, and declined as temperature approached 80 °C. However, the enzymatic activity in these conditions remained relatively high compared to the value showed under optimal conditions (with a level close to 85%) (Figure 7). Indicating that *P. sanguineus* CS 2 laccase is a thermotolerant enzyme [35]. These data were evaluated according to the Arrhenius model in order to estimate the energy of activation (E_a) for the system. This parameter has been relatively little studied in thermotolerant laccases. The calculated E_a value (16.2 kJ/mol) for *P. sanguineus* CS 2 laccase is similar to the values reported for other thermotolerant laccases, as that for *Myceliophora thermophila* (19 kJ/mol) [36] and for the recombinant laccase from *Coprinus cinereus* (14 kJ/mol) [37], but smaller than those calculated in this report for other laccases, which apparently did

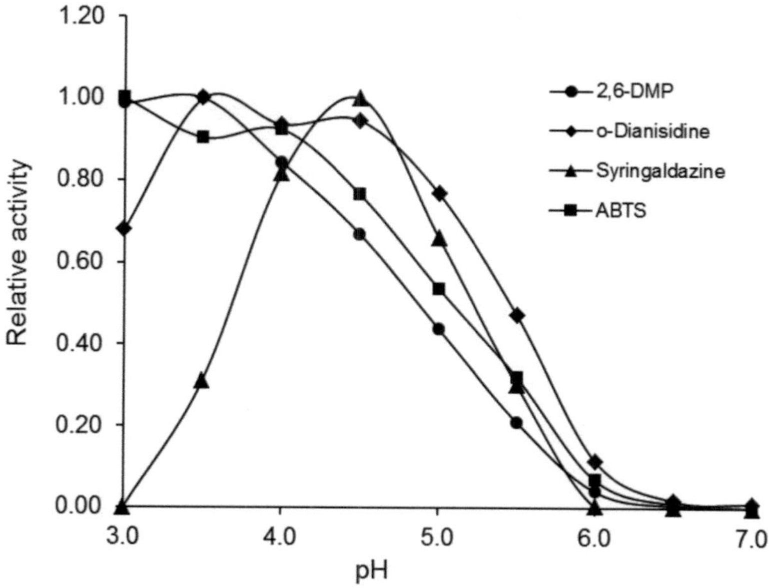

Figure 6. pH versus activity profiles of *Pycnoporus sanguineus* CS 2 laccase on various substrates. Assays were done by triplicate in 200 mM citrate-phosphate buffer at indicated pH, with 2, 6-DMP (2 mM), *o*-dianisidine (0.66 mM), SGZ (0.05 mM) and ABTS (2 mM).

not show a direct relationship between thermotolerance and the magnitude of E_a. On the other hand, the function showed a change in slope in the high temperature zone (50-70 °C) before the enzyme denaturing breaking zone. This effect could correspond to a decrease in the E_a of the system, caused by a thermotropic transition of the protein conformation, which should facilitate the limiting step of the reaction. Other possibility would be the coexistence of two enzyme populations, one of them showing an increased activity by temperature and the other being totally inactivated by thermal denaturing. These alternatives should be further explored.

Thermostability is a desirable property of industrial enzymes. Curves of temperature-stability of *P. sanguineus* CS 2 laccase showed that enzyme retained practically all of its activity after incubation for 8 h at 50 and 60 °C. Moreover, when incubations at 60 °C were extended to 24 h, the laccase retained 98% of its original activity (not shown). The enzyme also retained almost 50% of its activity after 4h at 70 °C. Inactivation curve showed a first-order decaying behavior (correlation > 0.96), with a calculated half-life ($t_{1/2}$) of 3.85 h [corresponding to a constant of thermal inactivation (*k*) of 0.187 h^{-1}]. To the best of our knowledge, this is one of the highest $t_{1/2}$ values found in laccases from mesophilic fungi. It is known that most of typical fungal laccases lose their activity in a few minutes at 60 °C [3, 15, 30, 38].

In comparison to laccases isolated from other *Pycnoporus* species, $t_{1/2}$ value at 70 °C here described is higher than those reported for laccase I (0.13 h) and laccase II (2.06 h) from *Pycnoporus* sp SYBC-L1 [13, 30], and for the laccase from the thermotolerant *P. sanguineus* CeIBMD001 (0.21 h) [16]. Native laccase also seems to be more resistant to thermal inactivation than *P. sanguineus* SCC 108 laccase ($t_{1/2}$ = 3.33 h at 65 °C) reported by [31] and the *P. sangui-*

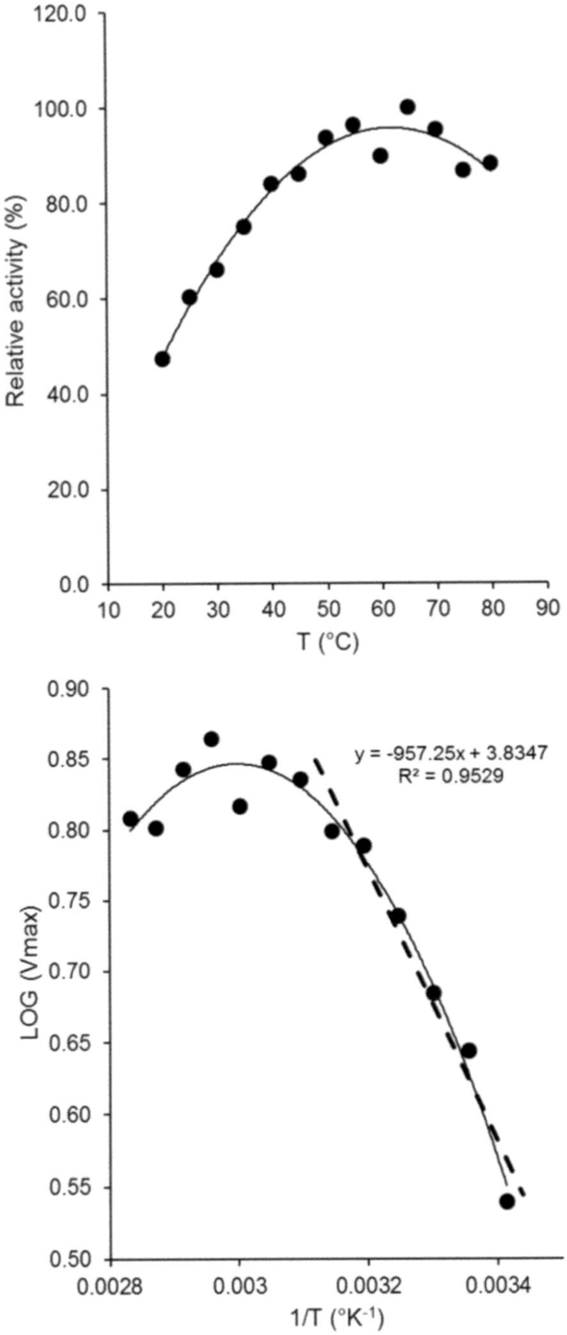

Figure 7. Effect of temperature on *Pycnoporus sanguineus* CS 2 laccase activity. Assays were performed by triplicate in 200 mM citrate-phosphate buffer at pH 4.0. Reaction rates were measured under saturating substrate concentrations (2 mM 2,6-DMP). The fitting line in lower panel shows the results of the Arrhenius analysis of data, corresponding to: LOG (Vmax) = [(-E_a /2.303 RT) + constant].

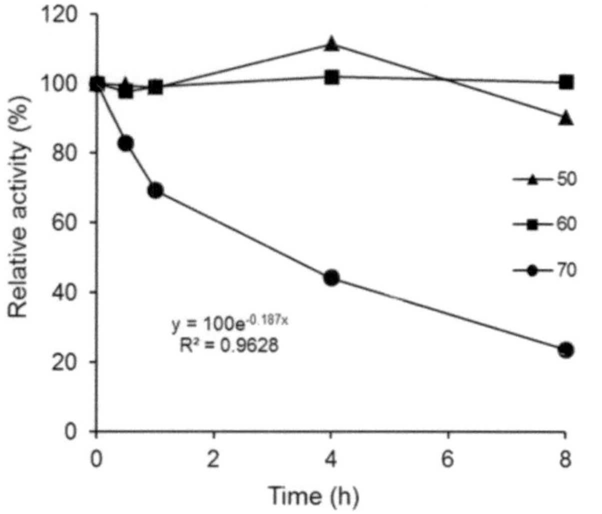

Figure 8. Effect of temperature on *Pycnoporus sanguineus* CS 2 laccase stability. Incubations were performed at various temperatures in distilled water. Aliquots were withdrawn at the indicated times and initial rates measured at 25 °C in 0.2 mM citrate/phosphate buffer at pH 4.0 with 2 mM 2, 6-DMP.

neus CCT- 4518 laccase studied in [39], which lost 60% of its initial activity after 2 h at 70 °C. Interestingly, three laccases from tropical or subtropical strains of *Pycnoporus* species (*P. sanguineus* BRFM 902, *P. sanguineus* BRFM 66, and *P. coccineus* BRFM 938) of different geographic regions (French Guinea, China and Australia, respectively) with remarkable thermal resistance have been recently reported by a research group in France [13]. A relationship between *P. sanguineus* CS 2 laccase with these and other *Pycnoporus* laccases already described was established by comparing the aminoacid sequences of an internal protein fragment (peptides 2+3+4, Table 2) from the native laccase with those sequences deposited at GenBank. Aminoacid sequence of *P. sanguineus* CS 2 laccase showed 99 % similarity to *P. sanguineus* BRFM 902 laccase, 93% to *P. coccineus* BRFM 938 [13], *P. cinnabarinus* PM laccases [5], and *P. sanguineus* BRFM 66 laccase [13], but only 84 % to *Trametes cinnabarina* [40] and 71% to *P sanguineus* CeIBMD001 laccase [16]. These results highlight the importance of *Pycnoporus* species biodiversity for the prospection for new thermostable laccases.

3.4. Kinetic properties

The kinetic properties of enzyme were studied with some typical substrates. The values of the Michaelis constant (K_m), catalytic constant (K_{cat}) and specificity constant (K_{cat}/K_m), were calculated by the Lineweaver-Burk method. Laccase showed the highest affinity and molecular activity, on ABTS (K_m = 23 mM, K_{cat} = 221 s^{-1}) compared to *o*-dianisidine (K_m = 44 mM, K_{cat} = 197 s^{-1}), and 2, 6-DMP (K_m =41 mM, K_{cat} = 88 s^{-1}). So, in terms of catalytic efficiency the best substrate resulted ABTS (K_{cat}/K_m = 9.4 x 10^6 s^{-1} M^{-1}) followed by *o*-dianisidine (K_{cat}/K_m = 4.5 x 10^6 s^{-1} M^{-1}) and 2, 6-DMP (K_{cat}/K_m = 2.2 x 10^6 s^{-1} M^{-1}). These results are summarized in Table 3, comparing the values of specificity constants (K_{cat}/K_m) for these substrates with those reported for other *Pycnoporus* laccases, the native enzyme showed higher values for all assayed substrates except for the reported laccase II from *Pycnoporus sp* SYBC-L1 [30]. Like typical laccases, the enzyme of *P. sanguineus* CS 2, showed activity on a variety of substrates, such as the phenolic 2, 6-DMP, *o*-dianisidine and SGZ, as well as the non-phenolic ABTS.

Peptide	Amino acid sequences
1	EAVVVNGITPAPLIAGKK
2*	GPFVVYDPNDPQASLYDIDNDDTVITLADWYHLAAKVGQR
3*	FPLGADATLINGLGR
4*	TPGTTSADLAVIKVTQGK
5	YSFVLDASQPVDNYWIRANPPFGNVGFAGGINSAILR
6	SAGSSEYNYDNPVFR

* Contiguous peptides of the internal laccase fragment used in alignments

Table 2. Amino acid sequences corresponding to internal peptides of *Pycnoporus sanguineus* CS2 laccase

Substrate	K_m (μM)	V_{max} (μmol/min/ml)	k_{cat} (s^{-1})	k_{cat}/K_m (s^{-1}/M^{-1})
2,6-DMP	41	500	88	2.16×10^6
o-dianisidine	44	1111	197	4.49×10^6
ABTS	23	1250	221	9.38×10^6

Table 3. Kinetics constants of *Pycnoporus sanguineus* CS 2 laccase

Among these substrates this laccase showed preference for ABTS and this characteristic was consistent with most of fungal laccases [32, 41, 42]. Unexpectedly the substrate saturation graphics with SYR showed a sigmoidal-like behavior instead of the common hyperbolic one (not shown). This result could be explained considering a kinetic mechanism of positive cooperativity as that described for monomeric mnemonical enzymes [43], where a conformational change of interacting enzyme at the end of the first catalytic cycle, reacts more readily with a second substrate molecule than other free-enzyme. Other factor contributing to this result could be the presence of ethanol in routinely SYR assay affecting the substrate solubility and/or enzyme activity. Whether mechanistic on phenomenological, this observation must be taken into account in future works, considering the relevance of this substrate in laccase characterization.

3.5. Dye decolorization

As revealed by the activity staining of native gels shown above, *P. sanguineus* CS 2 laccase was also able to promote the oxidative coupling between MBTH and DMAB. This reac-

http://dx.doi.org/10.5772/56374

tion has been considered as indicative of the ability of some laccases to catalyze reactions requiring a higher redox potential, as in the case of the enzymatic decolorization of many synthetic dyes. The non-phenolic azo MR [44, 45] and diazo RB 5 dyes [46] have been used as models for studying the ability of laccases to degrade recalcitrant compounds (Figure 9). Although general consensus is that laccases require meditators for acting over these dyes, *P. sanguineus* CS 2 laccase showed the capability to decolorize directly both compounds, but with different ability. Decolorization of MR and RB 5 reached a level of 70 %, and 15% respectively, after 4 h at 25 °C.

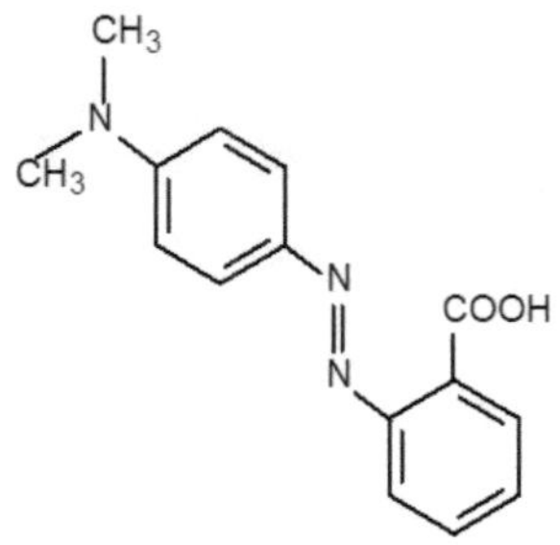

Methyl Red (Acid Red 2, CI: 13020)

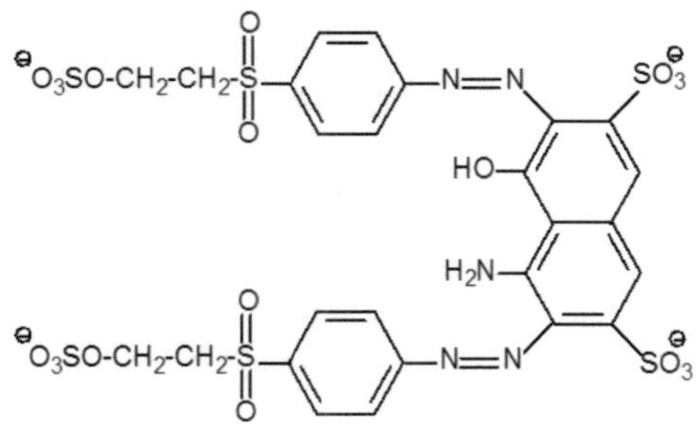

Reactive Black 5 (RB 5, CI: 20505)

Figure 9. Chemical structure of the recalcitrant methyl red and reactive black 5 dyes used in this study.

While *Ganoderma lucidum* [38], *Trametes trogii* [47] and *Lentinula edodes* [45] laccases were only able to decolorize RB 5 in the presence of mediators, a recent report state that three *Pycnoporus* laccases [13], were able to perform this decolorization in the absence of mediators, under similar conditions used in this work, with decolorization reaching from 29 to 45% after 52 h, at room temperature. The recalcitrance of RB 5 to laccase decolorization has been explained by its high redox potential or steric hindrances limiting accessibility of enzyme to –OH and $–NH_2$ groups in dye. As in this study native laccase attained around 70% decolorization after 20 h at room temperature, decolorization assays were performed at 60 °C taking advantage of its thermostability trying to overcome limiting factors. As expected, decolorization process was faster under the influence of tempera-

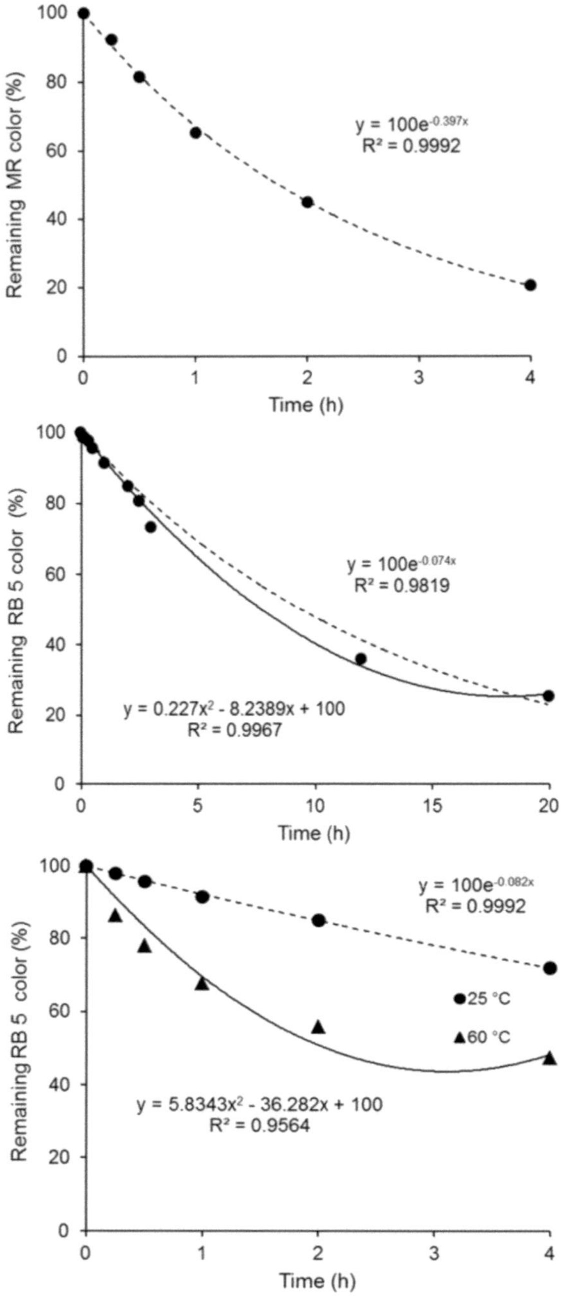

Figure 10. Decolorization of methyl red and reactive black 5 by *Pycnoporus sanguineus* CS2 laccase. Assays were performed by incubating 25 μM Methyl Red (upper panel) and 25 μM Reactive Black 5 (middle and lower panels) with laccase (5U/ml) in 0.2 mM citrate/phosphate buffer at pH 4.0 Aliquots were withdrawn from the assay mixture at the indicated times and remaining color was determined as described in text. Lines showed the best data fittings corresponding to the exponential first-order (dashed) or polynomial second-order decay functions (continuous).

ture, reaching around 50% after 4 h, although it also seems to be limited faster (Figure 10). While MR decolorization fitted an exponential first order decay model, RB 5 decolorization changes rapidly from this behavior to fit a polynomial second order model. This effect could be related to several factors as an increased enzyme inactivation by endogenous generated reaction intermediates and/or dead-end transformation products. This relationship must be investigated in future work. Nonetheless these results illustrate the potential of the thermostable *Pycnoporus sanguineus* CS2 laccase for practical applications.

4. Conclusion and future prospects

Its thermostability and ability for acting on high redox substrates and recalcitrant dyes, makes *Pycnoporus sanguineus* CS 2 laccase a good prospect for its application in industrial and environmental processes. This laccase could also be interesting as a model in studies associating structure-function of thermotolerant proteins from mesophilic microorganisms.

Acknowledgements

Authors thank the financial support provided by the Sistema de Fondos INNOVAPYME-CONACYT (Proyecto No. 139352). We also tank to Unidad de Proteómica, IBT-UNAM for assistance in peptide sequencing.

Author details

Sergio M. Salcedo Martínez[1,2], Guadalupe Gutiérrez-Soto[1,3], Carlos F. Rodríguez Garza[1], Tania J. Villarreal Galván[1], Juan F. Contreras Cordero[4] and Carlos E. Hernández Luna[1*]

*Address all correspondence to: carlosehlmx@yahoo.com

1 Autonomous University of Nuevo León, Laboratory of Enzymology, Faculty of Biological Sciences, San Nicolás de los Garza, N.L. México

2 Autonomous University of Nuevo León, Department of Botanic, Faculty of Biological Sciences, San Nicolás de los Garza, N.L. México

3 Autonomous University of Nuevo León, Department of Biotechnology, Faculty of Agronomy, San Nicolás de los Garza, N.L. México

4 Autonomous University of Nuevo León, Department of Microbiology and Immunology, Faculty of Biological Sciences, San Nicolás de los Garza, N.L. México

References

[1] Mayer AM, Staples RC. Laccase: New Functions for an Old Enzyme. Phytochemistry 2002;60 551-565.

[2] Baldrian P. Fungal Laccases-Occurrence and Properties. FEMS Microbiological Reviews 2005;20 1-28.

[3] Morozova OV, Shumakovich GP, Gorbacheva MA, Shleev SV, Yaropolov AI. ''Blue'' Laccases. Biochemistry (Moscow) 2007;72 1136–1412.

[4] Thurston CF. The Structure of Fungal Laccases. Microbiology 1994 ;140 19-26.

[5] Eggert C, Temp U, Eriksson K-EL. The Ligninolytic System of the White- rot Fungus *Pycnoporus cinnabarinus*: Purification and Characterization of the Laccase. Applied Environmental Microbiology 1996;62 1151–1158.

[6] Leontievsky AA, Vares T, Lankinen P, Shergill JK, Pozdnyakona NN, Myasoedova NM. Blue and Yellow Laccases of Ligninolytic Fungi. FEMS Microbiological Letters 1997;156 9-14.

[7] Palmieri G, Giardina, P, Bianco C, Sacloni A, Capasso A, Sannia G.A A Novel White Laccase from *Pleurotus ostreatus*. Journal of Biological Chemistry 1997;50 31301-31307.

[8] Pointing SB. Feasibility of Bioremediation by White–rot Fungi. Applied Microbiology and Biotechnology 2001;57 20-33.

[9] Reddy CA. The Potential of White-rot Fungi in the Treatment of Pollutants. Current Opinion in Biotechnology 1995;6 320-328.

[10] Rodríguez-Couto S, Toca-Herrera JL. Laccases in the Textile Industry. Biotechnology and Molecular Biolology Reviews 2006a;1 115-120.

[11] Rodríguez-Couto S, Toca-Herrera JL. Industrial and Biotechnological Applications of Laccases: A review. Biotechnology Advances 2006b;24 500-513.

[12] Xu F, Berka RM, Wahleithner JA, Nelson BA, Shuster JR, Brown SH, Palmer AE, Solomon EI. Site-directed Mutations in Fungal Laccase: Effect on Redox Potential, Activity and pH Profile. Biochemistry Journal 1998;334(1) 63-70.

[13] Uzan E, Nousiainen P, Balland V, Sipila J, Piumi F, Navarro D, Asther M, Record E, Lomascolo A. High Redox Potential Laccases from the Ligninolytic Fungi *Pycnoporus coccineus* and *Pycnoporus sanguineus* Suitable for White Biotechnology: from Gene Cloning to Enzyme Characterization and Applications. Journal of Applied Microbiology 2010;108 2199-2213.

[14] Rodgers CJ, Blanford CF, Giddens SR, Skamnioti P, Armstrong FA Gurr SJ. Designer Laccases: A Vogue for High-potential Fungal Enzymes?. Trends in Biotechnology 2009;28(2) 63-72.

[15] Jordaan J, Leukes WD. Isolation of a Thermostable Laccase with DMAB and MBTH Oxidative Coupling Activity from a Mesophilic White-rot Fungus. Enzyme and Microbial Technology 2003;33 212-219.

[16] Dantán-González E, Vite-Vallejo O, Martínez-Anaya C, Méndez-Sánchez M, González MC, Palomares LA, Folch-Mallol J. Production of Two Novel Laccase Isoforms by a Thermotolerant Strain of *Pycnoporus sanguineus* Isolated from an Oil-polluted Tropical Habitat. International Microbiology 2008;11 163–169.

[17] Pozdnyakova NN, Turkovskaya OV, Yudina EN Rodakiewicz-Nowak Y. Yellow Laccase from the Fungus *Pleurotus ostreatus* D1: Purification and Characterization. Applied Biochemistry and Microbiology 2006;42(1) 56-61.

[18] Hernández-Luna CE, Gutiérrez-Soto G, Salcedo-Martínez SM. Screening for Decolorizing Basidiomycetes in Mexico. World Journal of Microbiology and Biotecnology 2008;24 465-473.

[19] Ryvarden L. Genera of Polypores Nomenclature and Taxonomy. Synopsis fungorum Vol. 5. Fungiflora A/S Oslo; 1991.

[20] Guzmán G. Identificación de los Hongos: Comestibles, Venenosos, Alucinantes y Destructores de la Madera: Editorial Limusa S.A. México; 1980.

[21] Abadulla E, Tzanov T, Costa S, Robra K, Gübitz G. Decolorization and Detoxification of Textile Dyes with a Laccase from *Trametes hirsuta*. Applied and Environmental Microbiology 2000;66 3357-3362.

[22] Tien M, Kirk TK. Lignin Peroxidase of *Phanerochaete chrysosporium*. Methods in Enzymology 1988;161 238-248.

[23] Kuan C, Johnson J, Tien M. Kinetic Analysis of Manganese Peroxidase. The Reaction with Manganese Complexes. Journal of Biological Chemistry 1993;268 20064–20070

[24] Pickard MA, Vandertol H, Roman R, Vazquez-Duhalt R. High Production of Ligninolytic Enzymes from White-rot Fungi in Cereal Bran Flakes Liquid Medium. Canadian Journal of Microbioliology 1999;45 627-631.

[25] Dhouib A, Hamza M, Zouari H, Mechichi T, Hmidi R, Labat M, Martinez MJ, Sayadi S. Screening for Ligninolytic Enzyme Production by Diverse Fungi from Tunisia. World Journal of Microbiology and Biotechnology 2005 ;21 1415-1423.

[26] Zouari-Mechichi H, Mechichi T, Dhouib A, Sayadi S, Martínez AT, Martínez MJ. Laccase Purification and Characterization from *Trametes trogii* isolated in Tunisia: Decolorization of Textile Dyes by the Purified Enzyme. Enzyme and Microbial Technology 2006;39 141-148.

[27] Laemmli U. Cleavage of Structural Proteins During the Assembly of the Head of the Bacteriophage T4. Nature 1970;227 680-685.

[28] Garfin DE. Methods in Enzymology. Academic Press Inc. 1990.

[29] Guzmán G. Los Hongos del Edén Quintana Roo. (*Introducción a la Micología Tropical de México*). INECOL y CONABIO, Xalapa ; 2003.

[30] Wang ZX, Caia YJ, Liao XR, Taoc GJ, Lia YY, Zhanga F, Zhang DB. Purification and Characterization of Two Thermostable Laccases with High Cold Adapted Characteristics from *Pycnoporus* sp. SYBC-L1. Process Biochemistry 2010;45 1720-1729.

[31] Litthauer D, Jansen van Vuuren M, van Tonder A, Wolfaardt FW. Purification and Kinetics of a Thermostable Laccase from *Pycnoporus sanguineus* (SCC 108). Enzyme Microbial Technology 2007;40 563–568.

[32] Lu L, Zhao M, Zhang BB, Yu SY, Bian X-J, Wang W, Wang Y. Purification and Characterization of Laccase from *Pycnoporus sanguineus* and Decolorization of an Anthraquinone Dye by the Enzyme. Applied Microbiology and Biotechnology 2007;74 1232-1239.

[33] Hernández-Fernaud JR, Marina A, González K, Vázquez J, Falcón MA. Production, Partial Characterization and Spectroscopic Study of the Extracellular Laccase Activity from *Fusarium proliferatum*. Applied Microbiology and Biotechnology 2006 ;70 212-221.

[34] Madzak C, Mimmi MC, Caminade E, Brault A, Baumberger S, Briozzo P, Mougin C, Jolivalt C. Shifting the Optimal pH of Activity for a Laccase from the Fungus *Trametes versicolor* by Structure-based Mutagenesis. Protein Engineering 2006;334 63-70.

[35] Hildén K, Hakala TK, Lundell T. Thermotolerant and Thermostable Laccases. Biotechnological Letters 2009;31 1117–1128.

[36] Holm KA, Nielsen DM, Eriksen J. Automated Colorimetric Determination of Recombinant Fungal Laccase Activity in Fermentation Samples Using Syringaldazine as Chromogenic Substrate. Journal of Automated Chemistry 1998;20 199 –203.

[37] Schneider K, Caspersen MB, Mondorf K, Halkier T, Skov LK, Ostergaard PR, Brown KM, Brown SH, Xu F. Characterization of a *Coprinus cinereus* Laccase. Enzyme and Microbial Technology 1999;25 502–508.

[38] Murugesan K, Nam IH, Kim YM, Chang YS. Decolorization of Reactive Dyes by a Thermostable Laccase Produced by *Ganoderma lucidum* in Solid-state Culture. Enzyme and Microbial Technology 2007;40 1662-1672.

[39] García TA, Santiago MF, Ulhoa CJ. Studies on the *Pycnoporus sanguineus* CCT-4518 Laccase Purified by Hydrophobic Interaction Chromatography. Applied Microbiology and Biotechnology 2007;75 311–318.

[40] Antorini M, Herpoel-Gimbert I, Choinowski T, Sigoillot JC, Asther M, Winterhalter K, Piontek K. Purification, Crystallization and X-ray Diffraction Study of Fully Functional Laccases from Two Ligninolytic Fungi. Biochemistry and Biophysics Acta 2002;1594 (1) 109-114

[41] Ko EM, Leem YE, Choi HT. Purification and Characterization of Laccase Isoenzymes from the White-rot Basidiomycete *Ganoderma lucidum*. Applied Microbiology and Biotechnology 2001;57 98-102.

[42] Zang H, Zang Y, Huang F, Gao P, Chen J. Purification and Characterization of a Thermostable Laccase with Unique Oxidative Characteristics from *Tramentes hirsuta*. Biotechnology Letters 2009;31 837-843.

[43] Ricard J, Noat G. Kinetic Co-operativity of Monomeric Mnemonical Enzymes 'The Significance of the Kinetic Hill- coefficient. European Journal of Biochemistry 1985;152 557-564.

[44] Haibo Z, Yinglong Z, Feng H, Peiji G, Jiachuan C. Purification and Characterization of a Thermostable Laccase with Unique Oxidative Characteristics from *Trametes hirsuta*. Biotechnological Letters 2009;31 837-843.

[45] Nagai M, Sato T, Watanabe H, Saito K, Kawata M, Enei H. Purification and Characterization of an Extracellular Laccase from the Edible Mushroom *Lentinula edodes*, and Decolorization of Chemically Different Dyes. Applied Microbiology and Biotechnology 2002;60 327-335.

[46] Camarero S, Ibarra D, Martínez MJ, Martínez AT. Lignin-Derived compounds as Efficient Laccase Mediators for Decolorization of Different Types of Recalcitrant Dyes. Applied and Environmental Microbiology 2005;71(4) 1775-1784.

[47] Zeng X, Cai Y, Liao X, Zeng X, Li W, Zhang D. Decolorization of Synthetic Dyes by Crude Laccase from a Newly Isolated *Trametes trogii* Strain Cultivated on Solid Agroindustrial Residue. Journal of Hazardous Materials 2011;189 517-525.

Chapter 16

Scientific Swift in Bioremediation: An Overview

Ranjith N. Kumavath and Pratap Devarapalli

Additional information is available at the end of the chapter

http://dx.doi.org/10.5772/56409

1. Introduction

A pure environment gives a quality of life on earth. In ancient times, it was believed that people on earth had an unlimited abundance of land and resources; today, however, the resources in the world show a greater or lesser degree of our carelessness and negligence in using them. In many parts of globe, the problems associated with contaminated sites are now growing up. The actual cause of this scenario is result from past industrial activities when awareness of the health and environmental effects connected with the production, use, and disposal of hazardous substances were less well recognized than today. It became a global complication when the estimated number of contaminated sites became significant. There are several traditional methods which have been applied to overcome this inconvenience. From the list of ideas which have been applied the best ones are to completely demolish the pollutants if possible, or at least to transform them to innoxious substances. Bioremediation is an option that utilizes microbes to remove many contaminants from the environment by a diversity of enzymatic processes. However, it will not always be suitable as the range of contaminants on which it is effective is limited, the time scales involved are relatively long, and the residual contaminant levels achievable may not always be appropriate. we attempted to assist by providing information how the bioremediation is linked with cutting edge sciences like genomics, transcriptomics, proteomics, interactomics and bioinformatics.

Some new techniques in molecular biology particularly genetic engineering, transcriptomics, proteomics and interactomics offer remarkable promise as tools to study the mechanisms involved in regulation of mineralization pathways. The strategies need to be refined in which transcriptomics and proteomics data are combined together in order to understand the mineralization process in a meaningful way.These techniques show great promise in their ability to predict organisms' metabolism in contaminated environments and to predict the microbial assisted attenuation of contaminants to accelerate bioremediation. Bioinformatics technology has been developed to identify and analyse various components of cells such as gene and protein functions, interactions, metabolic and regulatory pathways. Bioinformatics analysis will facilitate and quicken the analysis of cellular process to understand the cellular mechanism to treat and control microbial cells as factories. The next decade will belong to understanding molecular mechanism and cellular manipulation using the integration of bioinformatics.

Bioremediation is an option that utilizes microbes to remove many contaminants from the environment by a diversity of enzymatic processes. The major positive shades of bioremediation are comparatively low-cost and techniques based on low-technology (Robb et al, 1995) which generally have a high public acceptance and can often be carried out on site. However, it will not always be suitable as the range of contaminants on which it is effective is limited, the time scales involved are relatively long, and the residual contaminant levels achievable may not always be appropriate. Varying degrees of success bioremediation has been used at a number of sites worldwide (Ajay et al, 2009). Here, we attempted to assist by providing information how the bioremediation is linked with cutting edge sciences like genomics, transcriptomics, proteomics, interactomics and bioinformatics (Fleming et al. 1993., Schena et al. 1998., Sikkema et al. 1995., Kuhner et al. 2005., Ellis et al. 2000).

2. Genetic analysis of genes involved

Examining the presence and expression of the key genes involved in bioremediation can yield more information on microbial processes than analysis of 16S rRNA sequences (Rogers and McClure. 2003). In general, there is a positive correlation between the relative abundance of the genes involved in bioremediation and the potential for contaminant degradation (Rogers and McClure. 2003., Schneegurt and Kulpa. 1998).

However, the genes for bioremediation can be present but not expressed. Therefore, there has been an increased emphasis on quantifying the levels of mRNA for key bioremediation genes. Often, increased mRNA concentrations can be, at least qualitatively, associated with higher rates of contaminant degradation (Schneegurt and Kulpa. 1998). For example, the concentrations of mRNA for *nahA* a gene involved in aerobic degradation of naphthalene were positively correlated with rates of naphthalene degradation in hydrocarbon-contaminated soil (Fleming et al. 1993).The reduction of soluble ionic mercury, Hg(II),to volatile Hg(0), is one mechanism for removing mercury from water; the concentration of mRNA for *merA* a gene involved in Hg(II) reduction was highest in mercury-contaminated waters with the highest rates of Hg (II) reduction (Nazaret et al. 1994). However, the concentration of *merA* was not always proportional to the rate of Hg (II) reduction (Nazaret et al. 1994., Jeffrey et al. 1996), illustrating that factors other than gene transcription can control the rates of bioremediation processes.

Highly sensitive methods that can detect mRNA for key bioremediation genes in single cells are now available (Bakermans and Madsen. 2002). This technique, coupled with 16S rRNA probing of the same environmental samples, could provide data on which phylogenetic groups of organisms are expressing the genes of interest. Analysis of the mRNA concentrations for genes other than those directly involved in bioremediation might yield additional insights into the factors that control the rate and extent of bioremediation. Sub-optimal nutrient levels, pH, salinity and other environmental factors can limit the growth and metabolism of organisms that are involved in bioremediation in contaminated environments. Ecological studies of phyto-plankton use molecular techniques to evaluate the stress response of photosynthetic microorganisms in the environment (Palenik and Wood. 1998). In a similar manner, evaluation of the metabolic state of bioremediating microorganisms through analysis of the mRNA concentrations for key genes that are involved in responding to stress could help to identify modifications to contaminated environments that might promote bioremediation.

3. Role of transcriptomics

The subset of genes transcribed in any given organism is called the transcriptome, which is a dynamic link between the genome, the proteome and the cellular phenotype. The regulation of gene expression is one of the key processes for adapting to changes in environmental conditions and thus for survival. Transcriptomics describes this process in a genome wide range. DNA microarrays are an extremely powerful platform in transcriptomics that enable determination of the mRNA expression level of practically every gene of an organism (Schena et al. 1998., Golyshin et al. 2003., Diaz. 2004) The most challenging issue in microarray experiments is elucidation of data (Dharmadi and Gonzalez. 2004). Often, hundreds of genes may be up- and/or down-regulated in a particular stress condition. In this context, several statistical issues become tremendously complex, including accounting for random and systematic errors and performing poor analysis.

4. Applications of DNA microarray

Even with the complete genome sequences of microorganisms with the potential for bioremediation (Golyshin et al. 2003., Tiedje. 2002., Heidelberg et al. 2002., Seshadri et al. 2005., Rabus et al. 2005), studies are not accelerating in a rapid manner. With the completed genome sequences, it is possible to analyse the expression of all genes in each genome under various environmental conditions using whole-genome DNA microarrays (Gao et al. 2004., Muffler et al. 2002., Schut et al. 2003). Such genome-wide expression analysis provides important data for identifying regulatory circuits in these organisms (Lovley. 2003., Rabus et al. 2005., Muffler et al. 2002). In the past, DNA microarrays have been used to evaluate the physiology of pure environmental cultures (Schut et al. 2003) and to monitor the catabolic gene expression profile in mixed microbial communities (Dennis et al. 2003). More than 100 genes were found to be affected by oxygen-limiting conditions when a DNA microarray was used to study changes in mRNA expression levels in Bacillus subtilis grown under anaerobic conditions (Ye et al. 2000). Sensitivity may often be a part of the problem in PCR-based cDNA microarrays, since only genes from populations contributing to more than 5% of the community DNA can be detected. Several parameters were evaluated to validate the sensitivity of spotted oligonucleotide DNA microarrays and their applicability for bacterial functional genomics (Denef et al. 2003). Optimal parameters were found to be 50-C6- amino-modified 70 mers printed on CMT-GAPS II substrates at a 40 *m*M concentration combined with the use of tyramide signal amplification labelling. Based on most of the known genes and pathways involved in biodegradation and metal resistance, a comprehensive 50-mer-based oligonucleotide microarray was developed for effective monitoring of biodegrading populations (Rhee et al. 2004). This type of DNA microarray was effectively used to analyze naphthalene-amended enrichment, and soil microcosms demonstrated that microflora changed differentially depending on the incubation conditions (Cho and Tiedje. 2002).A global gene expression analysis revealed the co-regulation of several thusfar- unknown genes during the degradation of alkylbenzenes (Kuhner et al. 2005). Besides this, DNA microarrays have been used to determine bacterial species, in quantitative applications of stress gene analysis of microbial genomes and in genome-wide transcriptional profiles (Muffler et al. 2002., Greene and Voordouw. 2003).

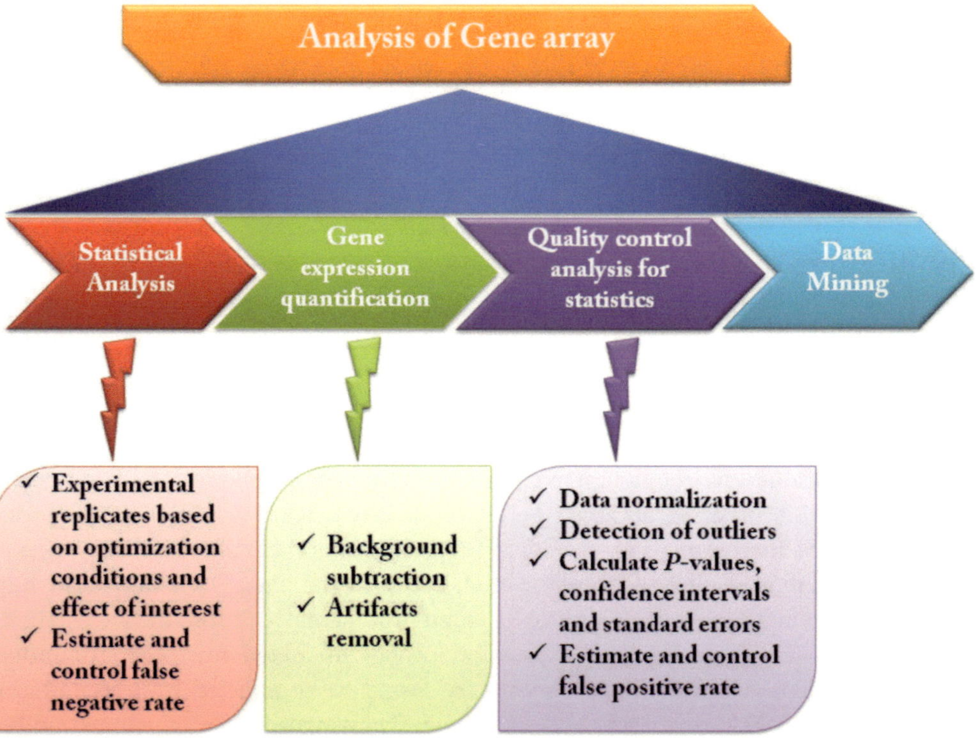

Figure 1. Work flow of gene array analysis. Diagrammatic representation of DNA microarray data analysis and relative limitations under each category of data analysis during data mining.

5. Foot prints of proteomics

The terms 'proteomics' and 'proteome' were introduced in 1995 (Wasinger et al. 1995), which is a key postgenomic feature that emerged from the growth of large and complex genome sequencing datasets. Proteomic analysis is particularly vital because the observed phenotype is a direct result of the action of the proteins rather than the genome sequence.Traditionally, this technology is based on highly efficient methods of separation using two-dimensional polyacrylamide gel electrophoresis (2-DE) and modern tools of bioinformatics in conjunction with mass spectrometry (MS) (Hochstrasser. 1995). However, 2-DE has been considered to be a limited approach for very basic and hydrophobic membrane proteins in compartmental proteomics. In bioremediation, the proteome of the membrane proteins is of high interest, specifically in PAH biodegradation, where many alterations in any site specific bacterium affects cell-surface proteins and receptors (Sikkema et al. 1995). The improvements in 2-DE for use in compartmental proteomics have been made by introducing an alternative approach for multidimensional protein identification technology (MudPIT) (Paoletti et al. 2004). MS has revolutionized the environmental proteomics towards the analysis of small molecules to peptides and proteins that has pushed up the sensitivity in protein identification by several orders of magnitude followed by minimizing the process from many hours to a few minutes (Aebersold and Mann. 2003). The advancement in MS techniques coupled with database searching have played a crucial role in proteomics for protein identification.

Matrixassociated laser desorption/ionization time-of-flight MS (MALDI-TOF-MS) is the most commonly used MS approach to identifying proteins of interest excised from 2-DE gels, by generation of peptidemass fingerprinting (Aebersold and Mann. 2003., Aitken and Learmonth. 2002., Landry et al. 2000). Surface-enhanced laser-desorption-ionization MS (SELDI-TOF-MS) is the combination of direct sample fractions on a chip integrated with MALDI-TOF-MS analysis (Merchant and Weinberger. 2000., Seibert et al. 2005). A variety of differentially expressed signature proteins were analysed using SELDITOF- MS in blue mussels (Mytilus edulis) exposed to PAHs and heavy metals (Knigge et al. 2004). The liquid chromatography MS (LC-MS) technique has begun to open a new analytical window for direct detection and identification of potential contaminants in water (Joo and Kim. 2005). In addition, the metabolites and degradation products have been taken into account to assess the fate of organic contaminants such as pesticides, surfactants, algal and cynobacterial toxins, disinfection by-products or pharmaceuticals in the environment and during water treatment processes (Joo and Kim. 2005).

6. Interaction of interactomics

Genome-wide mRNA profiling is unable to provide any information about the activity, arrangement, or final destination of the gene products, the proteins. Various proteomic approaches, on the other hand, can successfully provide the straight answers. It is very rare that any protein molecule acts as a unique pillar during the physiological response in bioremediation process of any contaminant when cellular proteins and various other related cellular expressions are on crest (Muffler et al. 2002., Kuhner et al. 2005., Eyers et al. 2004., Segura et al. 2005). In general, cellular life is organized through a complex protein interaction network, with many proteins taking part in multicomponent protein aggregation. The detection of these aggregated proteins, i.e. 'interactomics', is usually based upon affinity tag/pull down/MS/MS approaches at a proteome level (Lee and Lee. 2004., Coulombe et al. 2004., Gingras et al. 2005). Studies on protein–protein interaction and supermolecular complex formation represent one of the main directions of functional proteomics and/or second generation proteomics.

The growing demands of genomics and proteomics for the analysis of gene and protein function from a global bioremediation perspective are enhancing the need for microarray-based assays enormously. In the past, protein microarray technology has been successfully implicated for the identification, quantification and functional analysis of protein in basic and applied proteome research (Labaer and Ramachandran. 2005). Other than the DNA chip, a large variety of protein-microarraybased approaches have already been verified that this technology is capable of filling the gap between transcriptomics and proteomics (Liu and Zhu. 2005). However, in bioremediation, microarray-based protein–protein interaction studies still need to make progress to understand the chemotaxis phenomenon of any site specific bacterium towards the environmental contaminant.

7. Revolution of genomics

A drastic innovation in the study of pure cultures has been brought by the application of genomics to bioremediation. (Nierman & Nelson, 2002). Next generation genome sequencing techniques play a vital role in advancing the understanding of physiological and genomic features of microorganisms relevant to bioremediation. Complete, or nearly complete, genome sequences are now available for several organisms that are important in bioremediation (Table. 1). The notions of researches have been changed after the application of bioremediation to the advanced sciences like genomics which gave different answers. For example, molecular analyses have indicated that *Geobacter* species are important in the bioremediation of organic and metal contaminants in subsurface environments. The sequencing of several genomes of microorganisms of the genus *Geobacter,* as well as closely related organisms, has significantly altered the concept of how *Geobacter* species function in contaminated subsurface environments. For instance, before the sequencing of the *Geobacter* genomes, *Geobacter* species were thought to be non-motile, but genes encoding flagella were subsequently discovered in the *Geobacter* genomes (Childers et al. 2002) Further investigations revealed that *Geobacter metallireducens* specifically produces flagella only when the organism is growing on insoluble Fe(ra) or Mn(IV) oxides. Genes for chemotaxis were also evident in the *Geobacter* genomes, and experimental investigations have revealed that *G. metallireducens* has a novel chemotaxis to Fe(II), which could help guide it to Fe(III) oxides under anaerobic conditions. Pili genes are present and are also specifically expressed during growth on insoluble oxides (Childers et al.

Microorganism	Relevance to bioremediation	Web site for genome documentation
Dehalococcoides ethanogenes	Reductive dechlorination of chlorinated solvents to ethylene. The 16S rRNA gene sequence of D. ethanogenes is closely related to sequences that are enriched in subsurface environments in which chlorinated solvents are being degraded	http://www.tigr.org
Geobacter sulfurreducens Geobacter metallireducens	Anaerobic oxidation of aromatic hydrocarbons and reductive precipitation of uranium. 16S rRNA gene sequences closely related to known *Geobacter* species predominate during anaerobic in situ bioremediation of aromatic hydrocarbons and uranium.	http://www.jgi.doe.gov http://www.tigr.org
Rhodopseudomonas palustris	Main organism for elucidating pathways of anaerobic metabolism of aromatic compounds, and regulation of this metabolism.	http://www.jgi.doe.gov
Pseudomonas putida	Metabolically versatile microorganism capable of aerobically degrading a wide variety of organic contaminants. Excellent organism for genetic engineering of bioremediation capabilities.	http://www.tigr.org
Dechloromonas aromatica	Representative of ubiquitous genus of perchlorate-reducing microorganisms and capable of the anaerobic oxidation of benzene coupled to nitrate reduction.	http://www.jgi.doe.gov
Desulfitobacterium hafniense	Reductive dechlorination of chlorinated solvents and phenols. *Desulfitobacterium* species are widespread in a variety of environments.	http://www.jgi.doe.gov
Desulfovibrio vulgaris	Shown to reductively precipitate uranium and chromium. An actual role in contaminated environments is yet to be demonstrated.	http://www.tigr.org
Shewanella oneidensis	A closely related *Shewanella* species was found to reduce U(vi) to U(iv) in culture, but *Shewanella* species have not been shown to be important in metal reduction in any sedimentary environments.	http://www.tigr.org
Deinococcus radiodurans	Highly resistant to radiation and so might be genetically engineered for bioremediation of highly radioactive environments.	http://www.tigr.org

Table 1. Genomes of microorganisms pertinent to bioremediation.

2002). Genetic studies have indicated that the role of the pili is to aid in attachment to Fe(III) oxides, as well as facilitating movement along sediment particles in search of Fe(III).

This energy-efficient mechanism for locating and reducing Fe(ra) oxides in *Geobacter* species contrasts with the strategies for Fe(III) reduction in other well-studied organisms, such as *Shewanella* and *Geothrix* species. These other organisms release Fe(III) Chelators, which solubilize Fe(m) from Fe(m) oxides (Nevin and Lovley. 2002), and electron shuttling compounds, which accept electrons from the cell surface and then reduce Fe(m) oxides (Newman and Kolter. 2000., Nevin and Lovley. 2002).These strategies make it possible for *Shewanella* and *Geothrix* species to reduce Fe(III) without directly contacting the Fe(m) oxide. However, the synthesis of chelators and electron shuttles requires a significant amount of energy, and the lower metabolic energy requirements of the *Geobacter* approach is the probable explanation for the fact that *Geobacter* species consistently outcompete other Fe(III)-reducing microorganisms in several subsurface environments (Nevin and Lovley. 2002).Understanding this, and numerous other previously unsuspected physiological characteristics of *Geobacter* species, is important in guiding the manipulation of conditions in subsurface environments to optimize the ability of *Geobacter* species to remove organic and metal contaminants from polluted groundwater.

The study of the physiology of other microorganisms with bioremediation potential, the genomes of which have been sequenced, is now accelerating in a similar manner. With the completed genome sequences, it is possible using whole-genome DNA microarrays to analyse the expression of all the genes in each genome under various environmental conditions. Using pro-teomic techniques, it is possible to identify which proteins are expressed (Nierman & Nelson, 2002).Such genome-wide expression analysis provides important data for identifying regulatory circuits in these organisms (Baldi and Hatfield. 2002).This is significant as the mechanisms that control the regulation of the catabolic and respiratory genes that are the most important in bioremediation are largely unknown. As genetic systems for these environmentally significant organisms become available, it is possible to elucidate the function of the many genes of previously unknown function and to decipher bioremediation pathways. For example, the availability of the Geobacter genomes and a genetic system for these organisms is leading to the elucidation of which of the more than 100 c-type cytochromes that are apparent in the genome are important in electron transfer to metals (Lloyd et al. 2003., Leang et al. 2003).

Treatability study is a process, in which samples of the contaminated environment are incubated in the laboratory and the rates of contaminant degradation or immobilization are documented (Rogers and McClure. 2003). Giving little insight into the microorganisms that are responsible for the bioremediation, such studies provide an estimate of the potential metabolic activity of the microbial community. When bioremediation processes are researched in more detail, attempts are generally made to isolate the organisms responsible(Rogers. et al. 2003).The isolation and characterization of pure cultures has been, and will continue to be, crucial for the development and interpretation of molecular analyses in microbial ecology (Fig. 1).The recovery of isolates that are representative of the microorganisms responsible for the bioremediation process can be invaluable because, as outlined below, studying these isolates provides the opportunity to investigate not only their biodegradation reactions, but also other

http://dx.doi.org/10.5772/56409

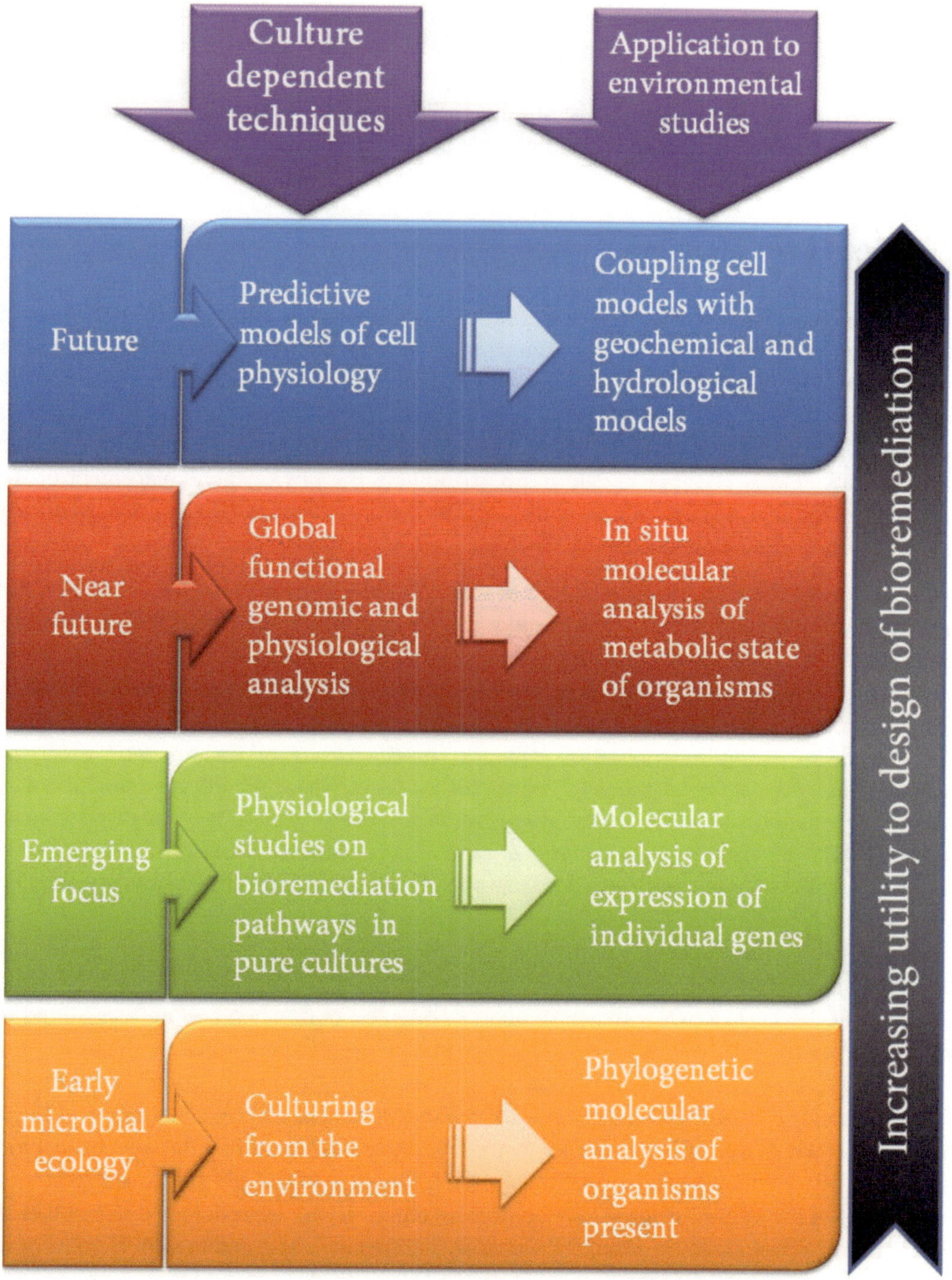

Figure 2. Evolution of increasingly sophisticated studies of pure cultures and their application to the study of microbial communities

aspects of their physiology that are likely to control their growth and activity in contaminated environments. However, before the application of molecular techniques to bioremediation, it was uncertain whether the isolated organisms were important in bioremediation *in situ*,or whether they were 'weeds' that grew rapidly in the laboratory but were not the primary organisms responsible for the reaction of interest in the environment.

8. The 16S rRNA approach

A significant advance in the field of microbial ecology was the finding that the sequences of highly conserved genes that are found in all microorganisms, most notably the 16S rRNA genes, could provide a phylogenetic characterization of the microorganisms that comprise microbial communities (Pace et al. 1986., Amann et al. 1995).This was a boon to the field of bioremediation because it meant that by analysing 16S rRNA sequences in contaminated environments, it was possible to determine definitively the phylogenetic placement of the microorganisms that are associated with bioremediation processes (Rogers and McClure. 2003., Watanabe and Baker. 2000)

One of the surprises from the application of the 16S rRNA approach to bioremediation has been the finding that, in some instances, microorganisms that predominate during bioremediation are closely related to organisms that can be cultured from subsurface environments (Lovley. 2001).This contrasts with the general dilemma in environmental microbiology that is, it can be difficult to recover the most environmentally relevant organisms in culture (Amann et al. 1995). For example, in polluted aquifers, in which microorganisms were oxidizing contaminants with the reduction of Fe (m) oxides, there was a significant enrichment in microorganisms with 16S rRNA sequences that were closely related to those of previously cultured Geobacter species (Rooney-varga et al. 1999., Snoeyenbos-West et al. 2000., Roling et al. 2001).Coupled with the fact that Geobacter species in pure culture are capable of oxidizing organic contaminants with the reduction of Fe(III) oxide (Lovley et al. 1989),this indicated that Geobacter species are important in contaminant degradation in situ. Geobacter species can also remove uranium from contaminated water by reducing soluble U(vi) to insoluble U(iv) (Lovley et al. 1991). 16S rRNA sequence analysis showed that, when acetate was added to uranium-contaminated groundwater to promote micro-bial reduction of U(vi), the number of Geobacter species increased by several orders of magnitude, accounting for as much as 85% of the microbial community in the groundwater (Anderson et al. In Press, Holmes et al. 2002) In aquifers in which the indigenous microbial community was degrading the solvent trichloroethene (TCE), 16S rRNA sequences that are ~99% identical to the 16S rRNA sequence of a pure culture of the TCE-degrader Dehalococcoides ethanogenes, were detected (Fennell et al. 2001., Richardson et al. 2002., Hendrickson et al. 2002).Marine sediments with high rates of anaerobic naphthalene degradation were found to be specifically enriched in microorganisms with 16S rRNA sequences closely related to NaphS2, an anaerobic naphthalene degrader that is available in pure culture (Hayes and Lovley. 2002). There was a close correspondence between the potential for aerobic degradation of the fuel oxygenate methyl tert-butyl ether (MTBE) in groundwater and the number of organisms with 16S rRNA sequences that had more than 99% similarity to the MTBE-degrading organism, strain PM-1,which is available in pure culture (Hristova et al. 2003).

The primary limitation of the 16S rRNA technique is that knowledge of the phylogeny of the organisms associated with bioremediation does not necessarily predict important aspects of their physiology (Pace. 1997., Achenbach and Coates. 2000). For example, microorganisms with 16S rRNA sequences closely related to the TCE-degrader D. ethanogenes can differ in the chlorinated compounds that they can degrade (He et al. 2003., Bunge et al. 2003), and predicting which of these compounds an uncultured organism will degrade might not be apparent from analysis of its 16S rRNA sequence alone (Hendrickson et al. 2002). Predicting physiology from phylogeny is even more difficult if there are no closely related organisms available in pure culture.

9. Comparative analysis of Omics in bioremediation

Based on an overall analysis of transcriptomics and proteomics, the comprehensive analysis of wholegenome sequencing is especially helpful to understand bioremediation-relevant microorganisms whose physiology has not yet been studied in detail. Global gene expression using DNA microarray technology, very much depends on the degree of coverage of the cellular mRNA and cellular proteins, whereas the coverage of the whole genome represents all the genes of an organism by definition. Cellular mRNA levels do not display as wide a dynamic range as the encoded proteins (Gygi et al. 1999). Thus, whole genome arrays are believed to provide a much more comprehensive overview of the actual gene expression pattern than proteomic studies.

According to global gene expression studies, both transcriptomics and proteomics support the view that the DNA array technologies record changes in gene expression more completely than the proteomics (Muffler et al. 2002., Kuhner et al. 2005., Eymann et al. 2002). Therefore, genomics data is deemed necessary to complement the proteomics approach (Hegde et al. 2003). However, proteomics would retain its central position in functional transcriptomics and/or genomics. The protein molecules, but not the mRNAs, are the key players in an on-site microbial mineralization reaction; the later are one of the highly unstable transmitters on the path from the genes to the ribosome, but each protein molecule represents the end product of gene expression (Kuhner et al. 2005). Complete protein profiling provides not only information on the individual organism, but also information on the fate and destination of protein molecules inside and outside the cell that can only be discovered via a joint transcriptomics, proteomics and interactomics approach (Figure 3).

10. Bioinformatics in bioremediation

MetaRouter is a system for maintaining heterogeneous information related to Biodegradation in a framework that allows its administration and mining (application of methods for extract-

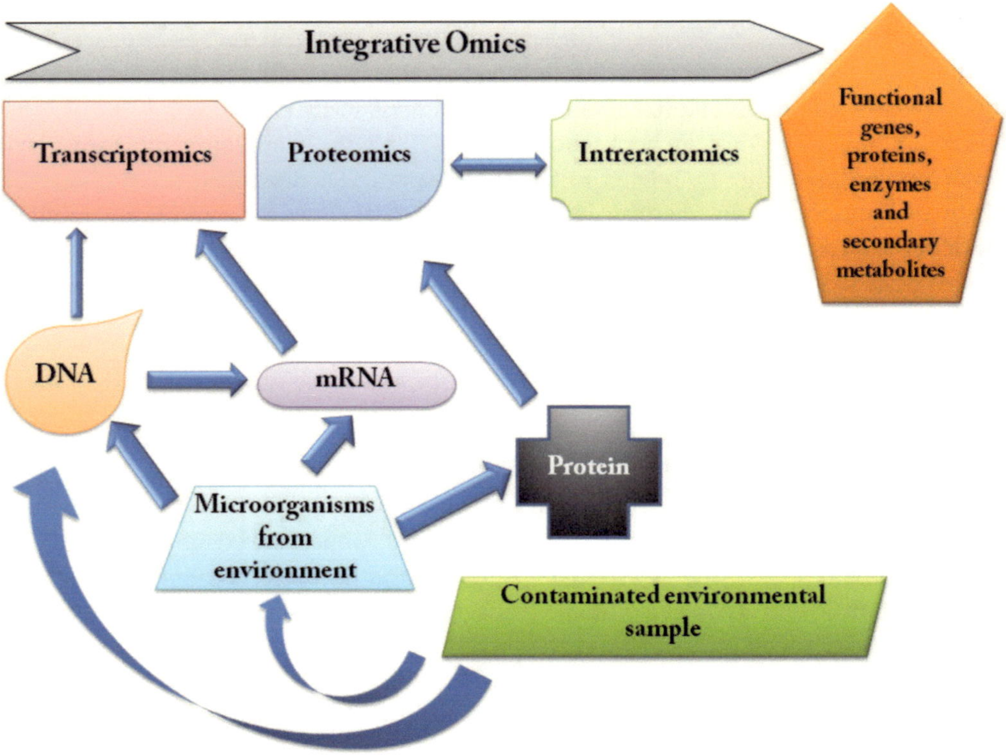

Figure 3. Omic technologies using a systematic biology approach to track the insights of bioremediation. DNA is directly extracted from contaminant environmental sites and from organisms will end up on transcriptomics (DNA microarrays). Transcriptomics will expend towards proteomics followed by interactomics. Extraction of protein from pure culture using 2-DE and protein microarray platforms will allow us to explore the new molecules of interest during mineralization process.

ing new data). It is an application intended for laboratories working in this area which need to maintain public and private data, linked internally and with external databases, and to extract new information from it. The system has an open and modular architecture adaptable to different customers. This multiplatform program, implemented in Postgre SQL (standard language for relational databases) and using SRS as an indexing system (used to connect and query Molecular Biology databases), works using a client/server architecture that allows the program to run on the user station or on the company server, so it can be accessed from any place in a secure way just by having a web browser.

The University of Minnesota Biocatalysts/Biodegradation Database (http://www.labmed.umn.edu/umbbd) begins its fifth year having met its initial goals. It contains approximately 100 pathways for microbial catabolic metabolism of primarily xenobiotic organic compounds, including information on approximately 650 reactions, 600 compounds and 400 enzymes, and containing approximately 250 microorganism entries. It includes information on most known microbial catabolic reaction types and the organic functional groups they transform. Having reached its first goals, it is ready to move beyond them. It is poised to grow in many different ways, including mirror sites; fold prediction for its sequenced enzymes; closer ties to genome and microbial strain databases; and the prediction of biodegradation pathways for compounds it does not contain (Ellis et al. 2000).

11. Approaches of systems biology

The rise of genomic technologies and systems biology provide fresh approaches to currently untactable biological processes that are at the root of serious environmental problems. One formidable challenge in this respect is the biological fate of the nearly 8 operons, etc. implicated in this process. The biodegradation database of the University of Minnesota documented new chemical compounds (~40 000 predominant) which are common in modern Organic and Industrial Chemistry. A large number of microbial strains are able to grow on environmental pollutants (about 800 today). Bioremediation was studied from a molecular biology point of view, characterizing the chemical reactions, genes; University of Minnesota has made a pioneering effort in putting together nearly every aspect of our current knowledge on biodegradation pathways and in developing systems for dealing with that data e.g. to learn rules for predicting biodegradative features. Yet, most information available in the literature of microbial biodegradation of xenobiotics and recalcitrant chemicals deals with duos consisting of one pollutant versus one strain and thus, lacks essential aspects of the natural scenarios, like the interchange of genes between bacteria or their metabolic cooperation. This study of genomes and 'functionomes' from a community point of view (in contrast to organism point of view) is leading, for example, to the sequencing of 'genomes' of communities and ecosystems, instead of single organisms. These circumstances expose the need to qualify and to represent the information available in biodegradation databases in a fashion in which the entire known biodegradative potential of the microbial world can be crossed with the whole collection of compounds known to be partially or totally degraded through (mostly) bacterial action (Kitano 2002).

12. Conclusion

The application of omic sciences to the study of bioremediation is clearly in its infancy. There are many technical issues that will need to be addressed before some of the more novel approaches, such as environmental genome sequencing and arrays. To elucidate the function of most genes recovered from the environment, it will be necessary to recover the relevant organisms and study gene function in pure culture. Microorganisms closely related to those that predominate in some contaminated environments are already available in culture, and the careful replication of environmental conditions during isolation will probably yield more. Microorganisms that typically comprise about one-fourth of the marine microbial community, but the presence of which had only previously been detected from 16S rRNA sequences. This search for previously uncultured organisms can be greatly accelerated with high-throughput culturing and screening strategies.

Some new techniques in molecular biology particularly genetic engineering, transcriptomics, proteomics and interactomics offer remarkable promise as tools to study the mechanisms involved in regulation of mineralization pathways. The applications of these techniques are still in their infancy, but the amount of data that is continuously being generated by today's genomics and proteomics technocrats needs to be organized in a stepwise manner within informative databases. The strategies need to be refined in which transcriptomics and proteomics data are combined together in order to understand the mineralization process in a meaningful way. These techniques show great promise in their ability to predict organisms' metabolism in contaminated environments and to predict the microbial assisted attenuation of contaminants to accelerate bioremediation. Bioinformatics technology has been developed to identify and analyse various components of cells such as gene and protein functions, interactions, metabolic and regulatory pathways. Bioinformatics analysis will facilitate and quicken the analysis of cellular process to understand the cellular mechanism to treat and control microbial cells as factories. The next decade will belong to understanding molecular mechanism and cellular manipulation using the integration of bioinformatics.

Author details

Ranjith N. Kumavath* and Pratap Devarapalli

*Address all correspondence to: RNKumavath@gmail.com; RNKumavath@cukerala.edu.in

Department of Genomic Sciences, School of Biological Sciences, Central University of Kerala, P.O. Central University, Kasaragod, India

References

[1] Achenbach, L. A, & Coates, J. D. Disparity between bacterial phylogeny and physiology comparing 16S rRNA sequences to assess relationships can be a powerful tool, but its limitations need to be considered. ASM News(2000). , 66, 714-715.

[2] Aebersold, R, & Mann, M. Mass spectrometry-based proteomics. Nature (2003). , 422, 198-207.

[3] Aitken, A, & Learmonth, M. Protein identification by in-gel digestion and mass spectrometric analysis. Mol Biotechnol (2002). , 20, 95-7.

[4] Ajay, S, Ramesh, C. K, & Owen, P. W. Advances in Applied Bioremediation, Soil Biology, Springer-Verlag Berlin Heidelberg (2009). DOI

[5] Amann, R. I, Ludwig, W, & Schleifer, K. H. Phylogenetic identification and in situdetection of individual microbial cells without cultivation. Microbiol. Rev. (1995). , 59, 143-169.

[6] Anderson, R. T, et al. Stimuating the in situ activity of Geobacter species to remove uranium from the groundwater of a uranium contaminated aquifer. Appl. Environ. Microbiol. (in the press).

[7] Bakermans, C, & Madsen, E. L. Detection in coal tar waste- contaminated groundwater of mRNA transcripts related to naphthalene dioxygenase by fluorescent in situ hybridization with tyramide signal amplification. J. Microbiol. Meth. (2002). , 50, 75-84.

[8] Baldi, P, & Hatfield, G. W. DNA Microarrays and Gene Expression, From Experiments to Data Analysis and Modeling. (Cambridge Univ. Press, Cambridge, UK,) (2002). , 135-176.

[9] Bunge, M, et al. Reductive dehalogenation of chlorinated dioxins by an anaerobic bacterium. Nature. (2003). , 421, 357-360.

[10] Cairney, T. Contaminated Land, Blackie, London (1993).

[11] Childers, S. E, Ciufo, S, & Lovley, D. R. Geobacter metallireducens accesses Fe (III) oxide by chemotaxis. Nature. (2002). , 416, 767-769.

[12] Cho, J. C, & Tiedje, J. M. Quantitative detection of microbial genes by using DNA microarrays. Appl Environ Microbiol (2002). , 68, 1425-30.

[13] Coulombe, B, Jeronimo, C, Langelier, M. F, et al. Interaction networks of the molecular machines that decode, replicate, and maintain the integrity of the human genome. Mol Cell Proteomics (2004). , 3, 851-6.

[14] Denef, V. J, Park, J, Rodrigues, J. L, et al. Validation of a more sensitive method for using spotted oligonucleotide DNA microarrays for functional genomics studies on bacterial communities. EnvironMicrobiol (2003). , 5, 933-43.

[15] Dennis, P, Edwards, E. A, Liss, S. N, et al. Monitoring gene expression in mixed microbial communities by using DNA microarrays. Appl EnvironMicrobiol (2003). , 69, 769-78.

[16] Dharmadi, Y, & Gonzalez, R. DNA microarrays: experimental issues, data analysis, and application to bacterial systems. Biotechnol Prog (2004). , 20, 1309-24.

[17] Diaz, E. Bacterial degradation of aromatic pollutants: a paradigm of metabolic versatility. IntMicrobiol (2004). , 7, 173-80.

[18] Ellis, L. B, Hershberger, C. D, & Wackett, L. P. The University of Minnesota Biocatalysis Biodegradation database: microorganisms, genomics and prediction. Nucleic Acids Res. (2000). , 28(1), 377-9.

[19] Eyers, L, George, I, Schuler, L, et al. Environmental genomics: exploring the unmined richness of microbes to degrade xenobiotics. Appl Microbiol Biotechnol (2004). , 66, 123-30.

[20] Eymann, C, Homuth, G, Scharf, C, et al. Bacillus subtilis functional genomics: global characterization of the stringent response by proteome and transcriptome analysis. J Bacteriol (2002). , 184, 2500-20.

[21] Fennell, D. E, Carrol, A. B, Gossett, J. M, & Zinder, S. H. Assessment of indigenous reductive dechlorinating potential at a TCE-contaminated site using microcosms, polymerase chain reaction analysis and site data. Environ. Sci. Technol. (2001). , 35, 1830-1839.

[22] Fleming, J. T, Sanseverino, J, & Sayler, G. S. Quantitative relationship between naphthalene catabolic gene frequency and expression in predicting PAH degradation in soils at town gas manufacturing sites. Environ. Sci. Technol. (1993). , 27, 1068-1074.

[23] Gao, H, Wang, Y, Liu, X, et al. Global transcriptome analysis of the heat shock response of Shewanella oneidensis. J Bacteriol (2004). , 186, 7796-803.

[24] Gingras, A. C, Aebersold, R, & Raught, B. Advances in protein complex analysis using mass spectrometry. J Physiol (2005). , 563, 11-21.

[25] Golyshin, P. N. Martins Dos Santos VA, Kaiser O, et al. Genome sequence completed of Alcanivorax borkumensis, a hydrocarbon-degrading bacterium that plays a global role in oil removal from marine systems. J Biotechnol (2003). , 106, 215-20.

[26] Greene, E. A, & Voordouw, G. Analysis of environmental microbial communities by reverse sample genome probing. JMicrobiolMethods (2003). , 53, 211-9.

[27] Gygi, S. P, Rochon, Y, Franza, B. R, et al. Correlation between protein and mRNA abundance in yeast. MolCell Biol (1999). , 19, 1720-30.

[28] Hayes, L. A, & Lovley, D. R. Specific 16S rDNA sequences associated with naphthalene degradation under sulfate- reducing conditions in harbor sediments. Microb. Ecol. (2002). , 43, 134-145.

[29] He, J, Ritalahti, K. M, Aiello, M. R, & Loffler, F. E. Enrichment culture and identification of the reductively dechlorinating population as a Dehalococcoides species. Appl. Environ. Microbiol. (2003). , 69, 996-1003.

[30] Hegde, P. S, White, I. R, & Debouck, C. Interplay of transcriptomics and proteomics. Curr Opin Biotechnol (2003). , 14, 647-51.

[31] Heidelberg, J. F, Paulsen, I. T, Nelson, K. E, et al. Genome sequence of the dissimilatory metal ion-reducing bacterium Shewanella oneidensis. Nat Biotechnol (2002). , 20, 1118-23.

[32] Hendrickson, E. R, et al. Molecular analysis of Dehalococcoides 16S ribosomal DNA from chloroethene- contaminated sites throughout North America and Europe. Appl. Environ. Microbiol. (2002). , 68, 485-495.

[33] Hochstrasser, D. F. Proteome in perspective. Clin Chem Lab Med (1998). Sikkema J, deBont JAM, Poolman B. Mechanisms of membrane toxicity of hydrocarbons. Microbiological Rev 1995;59:201-22., 36, 825-36.

[34] Holmes, D. E, Finneran, K. T, & Lovley, D. R. Enrichment of Geobacteraceae associated with stimulation of dissimilatory metal reduction in uranium-contaminated aquifer sediments. Appl. Environ. Microbiol. (2002). , 68, 2300-2306.

[35] Hristova, K, Gebreyesus, B, Mackay, D, & Scow, K. M. Naturally occurring bacteria similar to the methyl tert-butyl ether (MTBE)-degrading strain PM1 are present in MTBE- contaminated groundwater. Appl. Environ. Microbiol. (2003). , 69, 2616-2623.

[36] Jeffrey, W H, Nazaret, S, & Barkay, T. Detection of the merA gene and its expression in the environment. Microbial Ecol. (1996). , 32, 293-303.

[37] Joo, W. A, & Kim, C. W. Proteomics of Halophilic archaea.JChromatogr BAnalytTechnol Biomed Life Sci (2005). , 815, 237-50.

[38] Kitano, H. Systems Biology: a brief overview, Science. (2002). , 295, 1662-1664.

[39] Knigge, T, Monsinjon, T, & Andersen, O. K. Surface-enhanced laser desorption/ionization-time of flight-mass spectrometry approach to biomarker discovery in blue mussels (Mytilus edulis) exposed to polyaromatic hydrocarbons and heavy metals under field conditions. Proteomics (2004). , 4, 2722-7.

[40] Kuhner, S, Wohlbrand, L, Fritz, I, et al. Substrate-dependent regulation of anaerobic degradation pathways for toluene and ethylbenzene in a denitrifying bacterium, strain EbN1. J Bacteriol (2005). , 187, 1493-503.

[41] Labaer, J, & Ramachandran, N. Protein microarrays as tools for functional proteomics. Curr Opin Chem Biol (2005). , 9, 14-9.

[42] Landry, F, Lombardo, C. R, & Smith, J. W. A method for application of samples to matrix-assisted laser desorption ionization time-of-flight targets that enhances peptide detection. Anal Biochem (2000). , 279, 1-8.

[43] Lee, W. C, & Lee, K. H. Applications of affinity chromatography in proteomics. Anal Biochem (2004). , 324, 1-10.

[44] Liu, W. T, & Zhu, L. Environmental microbiology-on-a-chip and its future impacts. Trends Biotechnol (2005). , 23, 174-9.

[45] Lloyd, J. R, et al. Biochemical and genetic characterization of PpcA, a periplasmic c-type cytochrome in Geobacter sulfurreducens. Biochem. J.(2003). , 369, 153-161.

[46] Lovley, D. R. Anaerobes to the rescue. Science (2001). , 293, 1444-1446.

[47] Lovley, D. R. Phillips EJP, Gorby YA. & Landa ER. Microbial reduction of uranium. Nature(1991). , 350, 413-416.

[48] Lovley, D. R. Cleaning up with genomic: applying molecular biology to bioremediation. Nat RevMicrobiol (2003). , 1, 35-44.

[49] Lovley, D. R, et al. Oxidation of aromatic contaminants coupled to microbial iron reduction. Nature. (1989). , 339, 297-299.

[50] Merchant, M, & Weinberger, S. R. Recent advancements in surface-enhanced laser desorption/ionization-time of flight-mass spectrometry. Electrophoresis (2000). , 21, 1164-77.

[51] Muffler, A, Bettermann, S, Haushalter, M, et al. Genomewide transcription profiling of Corynebacterium glutamicum after heat shock and during growth on acetate and glucose. J Biotechnol (2002). , 98, 255-68.

[52] Nazaret, S, Jeffrey, W. H, Saouter, E, Von Haven, R, & Barkay, T. merA gene expression in aquatic environments measured by mRNA production and Hg(II)volatilization. Appl. Environ. Microbiol. (1994). , 60, 4059-4065.

[53] Nevin, K. P, & Lovley, D. R. Mechanisms for accessing insoluble Fe(III) oxide during dissimilatory Fe(III) reduction by Geothrix fermentans. Appl. Environ. Microbiol. (2002). , 68, 2294-2299.

[54] Nevin, K. P, & Lovley, D. R. Mechanisms for Fe(III) oxide reduction in sedimentary environments. Geomicrobiol. J. (2002). , 19, 141-159.

[55] Newman, D. K. Kolter. A role for excreted quinones in extracellular electron transfer. Nature.(2000). , 405, 94-97.

[56] Nierman, W. C, & Nelson, K. E. Genomics for applied microbiology. Adv. Appl. Microbiol. (2002). , 51, 201-245.

[57] Pace, N. R. A molecular view of microbial diversity and the biosphere. Science(1997). , 276, 734-740.

[58] Pace, N. R, Stahl, D. A, Lane, D. J, & Olsen, G. J. The analysis of natural populations by ribosomal RNA sequence. Adv. Gen. Microbiol. Ecol. (1986). , 9, 1-55.

[59] Palenik, B, & Wood, A. M. In Molecular Approaches to the Study of the Ocean (ed. Cooksey, K. E.) (Chapman & Hall, London,). (1998). , 187-205.

[60] Paoletti, A. C, Zybailov, B, & Washburn, M. P. Principles and applications of multidimensional protein identification technology. Expert Rev Proteomics (2004). , 1, 275-82.

[61] Rabus, R, Kube, M, Heider, J, et al. The genome sequence of an anaerobic aromatic-degrading denitrifying bacterium, strain EbN1. ArchMicrobiol (2005). , 183, 27-36.

[62] Rhee, S. K, Liu, X, Wu, L, et al. Detection of genes involved in biodegradation and biotransformation in microbial communities by using 50-mer oligonucleotide microarrays. Appl EnvironMicrobiol (2004)., 70, 4303-17.

[63] Richardson, R. E, Bhupathiraju, V. K, Song, D. L, Goulet, T. A, & Alvarez-cohen, L. Phylogenetic characterization of microbial communities that reductively dechlorinate TCE based upon a combination of molecular techniques. Environ. Sci. Technol. (2002)., 36, 2652-2662.

[64] Robb IIIAJHoggatt PR, Mobil EP. A Cost effective Bioremediation strategy using low technology resources for reclamation of dry land hydrocarbon contamination: A case study. Exploration and Production Environmental Conference, March (1995). Houston, Texas. 978-1-55563-449-0, 27-29.

[65] Rogers, S. L, & Mcclure, N. in Bioremediation: A criticial review (eds Head, I. M., Singleton, I. & Milner, M. G.) (2003). Horizon Scientific Press, Wymondham, UK,).

[66] Roling WFMvan Breukelen BM, Braster M, Lin B. & van verseveld, HW. Relationships between microbial community structure and hydrochemistry in a landfill leachate-polluted aquifer. Appl. Environ. Microbiol. (2001)., 67, 4619-4629.

[67] Rooney-varga, J. N, Anderson, R. T, Fraga, J. L, Ringelberg, D, & Lovley, D. R. Microbial communities associated with anaerobic benzene mineralization in a petroleum-contaminated aquifer. Appl. Environ. Microbiol. (1999)., 65, 3056-3063.

[68] Schena, M, Heller, R. A, Theriault, T. P, et al. Microarrays: biotechnology's discovery platform for functional genomics. Trends Biotechnol (1998)., 16, 301-6.

[69] Schneegurt, M. A, & Kulpa, C. F. Jr The application of molecular techniques in environmental biotechnology for monitoring microbial systems. Biotechnol. Appl. Biochem. (1998)., 27, 73-79.

[70] Schut, G. J, Brehm, S. D, Datta, S, et al. Whole-genome DNA microarray analysis of a hyperthermophile and an archaeon: Pyrococcus furiosus grown on carbohydrates or peptides. J Bacteriol (2003)., 185, 3935-47.

[71] Schut, G. J, Zhou, J, & Adams, M. W. DNA microarray analysis of the hyperthermophilic archaeon Pyrococcus furiosus: evidencefor a new type of sulfur-reducing enzyme complex. J Bacteriol (2001)., 183, 7027-36.

[72] Segura, A, Godoy, P, Van Dillewijn, P, et al. Proteomic analysis reveals the participation of energy- and stress-related proteins in the response of Pseudomonas putida DOT-T1E to toluene. J Bacteriol (2005)., 187, 5937-45.

[73] Seibert, V, Ebert, M. P, & Buschmann, T. Advances in clinical cancer proteomics: SELDI-ToF-mass spectrometry and biomarker discovery. Brief Funct Genomic Proteomic (2005)., 4, 16-26.

[74] Seshadri, R, Adrian, L, Fouts, D. E, et al. Genome sequence of the PCE-dechlorinating bacterium Dehalococcoides ethenogenes. Science (2005). , 307, 105-8.

[75] Sikkema, J. deBont JAM, Poolman B. Mechanisms of membrane toxicity of hydrocarbons. Microbiological Rev (1995). , 59, 201-22.

[76] Snoeyenbos-west, O. L, Nevin, K. P, & Lovley, D. R. Stimulation of dissimilatory Fe(III) reduction results in a predominance of Geobacterspecies in a variety of sandy aquifers. Microb. Ecol. (2000). , 39, 153-167.

[77] Tiedje, J. M. Shewanella-the environmentally versatile genome. Nat Biotechnol (2002). , 20, 1093-4.

[78] Wasinger, V. C, Cordwell, S. J, Cerpa-poljak, A, et al. Progress with gene-product mapping of the Mollicutes: Mycoplasma genitalium. Electrophoresis (1995). , 16, 1090-4.

[79] Watanabe, K, & Baker, P. W. Environmentally relevant microorganisms. J. Biosci. Bioeng. (2000). , 89, 1-11.

[80] Ye, R. W, Tao, W, Bedzyk, L, et al. Global gene expression profiles of Bacillus subtilis grown under anaerobic conditions. J Bacteriol (2000). , 182, 4458-65.